EEG Signal Processing and Feature Extraction

Li Hu • Zhiguo Zhang
Editors

EEG Signal Processing and Feature Extraction

Springer

Editors
Li Hu
CAS Key Laboratory of Mental Health
Institute of Psychology, Chinese
Academy of Sciences
Beijing, China

Department of Psychology
University of Chinese Academy
of Sciences
Beijing, China

Zhiguo Zhang
School of Biomedical Engineering, Health
Science Center
Shenzhen University
Shenzhen, Guangdong, China

ISBN 978-981-13-9112-5 ISBN 978-981-13-9113-2 (eBook)
https://doi.org/10.1007/978-981-13-9113-2

This Springer imprint is published by the registered company Springer Nature Singapore Pte Ltd.
The registered company address is: 152 Beach Road, #21-01/04 Gateway East, Singapore 189721, Singapore

Contents

Contributors

Yang Bai Department of Basic Medical Science, School of Medicine, Hangzhou Normal University, Hangzhou, Zhejiang, China

Qing Cai School of Electrical and Information Engineering, Tianjin University, Tianjin, China

Fengyu Cong School of Biomedical Engineering, Faculty of Electronic Information and Electrical Engineering, Dalian University of Technology, Dalian, China

Zhongke Gao School of Electrical and Information Engineering, Tianjin University, Tianjin, China

Ruolei Gu CAS Key Laboratory of Behavioral Science, Institute of Psychology, Chinese Academy of Sciences, Beijing, China
Department of Psychology, University of Chinese Academy of Sciences, Beijing, China

Li Hu CAS Key Laboratory of Mental Health, Institute of Psychology, Chinese Academy of Sciences, Beijing, China
Department of Psychology, University of Chinese Academy of Sciences, Beijing, China

Gan Huang School of Biomedical Engineering, Health Science Center, Shenzhen University, Shenzhen, China

Huibin Jia Key Laboratory of Child Development and Learning Science of Ministry of Education, School of Biological Sciences & Medical Engineering, Southeast University, Nanjing, Jiangsu, China

Xu Lei Sleep and Neuroimaging Centre, Faculty of Psychology, Southwest University, Chongqing, China

Fengnian Li Yanshan University Library, Yanshan University, Qinhuangdao, Hebei, China

Jiewei Li Department of Electrical and Electronic Engineering, The University of Hong Kong, Hong Kong, China

Xiaoli Li State Key Laboratory of Cognitive Neuroscience and Learning, Beijing Normal University, Beijing, China

Zhenhu Liang Institute of Electric Engineering, Yanshan University, Qinhuangdao, Hebei, China

Xuejing Lu CAS Key Laboratory of Mental Health, Institute of Psychology, Chinese Academy of Sciences, Beijing, China
Department of Psychology, University of Chinese Academy of Sciences, Beijing, China

Weiwei Peng School of Psychology, Shenzhen University, Shenzhen, Guangdong, China

Yanbo Sun School of Information Science and Engineering, Yanshan University, Qinhuangdao, Hebei, China

Yiheng Tu Department of Psychiatry, Massachusetts General Hospital and Harvard Medical School, Charlestown, MA, USA

Xinmin Wang School of Electrical and Information Engineering, Tianjin University, Tianjin, China

Zhenhao Wei School of Information Science and Engineering, Yanshan University, Qinhuangdao, Hebei, China

Dong Wen School of Information Science and Engineering, Yanshan University, Qinhuangdao, Hebei, China

Xiaolei Xia CAS Key Laboratory of Mental Health, Institute of Psychology, Chinese Academy of Sciences, Beijing, China
Department of Psychology, University of Chinese Academy of Sciences, Beijing, China

Yuxuan Yang School of Electrical and Information Engineering, Tianjin University, Tianjin, China

Zhiguo Zhang School of Biomedical Engineering, Health Science Center, Shenzhen University, Shenzhen, Guangdong, China

Yanhong Zhou School of Mathematics and Information Science and Technology, Hebei Normal University of Science and Technology, Qinhuangdao, Hebei, China

Chapter 1
Introduction

Li Hu and Zhiguo Zhang

Abstract As measurements of real-time electrodynamics in the human brain is evolving, a series of electroencephalogram (EEG) signal processing techniques have been developed rapidly. This book is aimed to provide the conceptual, mathematical, and implementational knowledge from EEG neural basis to almost all mainstream EEG signal processing and feature extraction methods in a comprehensive, simple, and easy-to-understand way.

Keywords Electroencephalogram (EEG) · EEG signal processing · Feature extraction · Methodology · Implementation

1.1 Historical Background

The last frontier of biological sciences—their ultimate challenge—is to understand the biological basis of consciousness and the mental processes by which we perceive, act, learn, and remember. Our brain is the key core within these processes. We begin with the fact that the brain weighs approximately 2.5 pounds but costs nearly 40% to 60% of blood glucose (Squire and Zola-Morgan 1988) and a considerable amount of oxygen equalling to what our muscles consumed during actions per day. How is such disproportionate amount of energy consumed? The answer is that it is used to produce electricity including synchronized and collective actions of small and large groups of neurons linked by axonal and dendritic connections. Each

L. Hu (✉)
CAS Key Laboratory of Mental Health, Institute of Psychology, Chinese Academy of Sciences, Beijing, China

Department of Psychology, University of Chinese Academy of Sciences, Beijing, China
e-mail: huli@psych.ac.cn

Z. Zhang
School of Biomedical Engineering, Health Science Center, Shenzhen University, Shenzhen, Guangdong, China
e-mail: zgzhang@szu.edu.cn

L. Hu, Z. Zhang (eds.), *EEG Signal Processing and Feature Extraction*,
https://doi.org/10.1007/978-981-13-9113-2_1

neuron acts like a dynamically oscillating battery that is recharged in a continuous manner (Steriade 1995). Those locally connected neurons recruit neurons in neighborhood with a sequential production of electrical potentials. More than 140 years ago, Richard Caton found the existence of such electrical activity of the brain (Caton 1875). His work is regarded as the first attempt at the electrophysiology of the brain and marking the birth of the electroencephalogram (EEG) (Brazier 1961). In the 1920s, Hans Berger demonstrated that the EEG could be measured from the human scalp without opening the skull (Berger 1929). The word *electroencephalogram* was proposed for depicting brain electrical activity in human beings. In 1934, Adrian and Mathews identified the "alpha rhythm" consisting of regular oscillations around 10 to 12 Hz. However, the neural sources of the alpha rhythm remained unclear until 43 years later, when Lopes da Silva et al. demonstrated that the alpha rhythm is generated at different depths within the visual cortex in dog (Lopes Da Silva and Storm Van Leeuwen 1977). Considering the intricacy of the underlying neural basis and volume conduction, there is no surprise that the functional significance and mechanisms of the EEG generation remained debatable for a long time. Based on the current understanding, EEG is defined as the electrical activity, which is normally recorded at the scalp of the human brain, generated by the firing of neurons within the brain (Niedermeyer and Lopes da Silva 2005). To date, over 150,000 studies using EEG techniques have been published, and the field still continues to grow every year. Why is EEG so popular? As we currently know, cognitive, perceptual, linguistic, emotional, and motor processes are really fast, which occur within tens to hundreds of milliseconds. The high temporal resolution property of EEG makes it well suited to capture these fast, dynamic, and temporally sequenced cognitive events. For decades, measurements of real-time electrodynamics of the human brain have evolved, thus contributing to the development of a series of EEG signal processing techniques. Our goal in this book is to provide the conceptual, mathematical, and implementational (e.g., via MATLAB programming) bases of both EEG and EEG signal processing in a comprehensive, simple, and easy-to-understand way.

1.2 Why to Write This Book?

Admittedly, there are good resources for generally learning EEG and relevant studies (Regan 1989; Handy 2004; Luck and Kappenman 2011; Sanei and Chambers 2013; Cohen 2014), as well as online tutorials (e.g., the EEGLAB website) that cover introductions of using software tools and interpreting results from analyses. However, on one side, these resources are less practical for understanding the mathematical principles or possible pitfalls of these analyses and provide insufficient instructions for users who want to implement the methods themselves or further adapt to current existing methods. On the other side, there are resources containing various background knowledge in mathematics or physics, which is difficult and obscure for beginners, especially those without a mathematical or engineering background, to understand. Most of these resources are lack of practical implementation matters, such as how to perform statistics or how to deal with limited data, and unavailable to learners without a formal mathematical training.

Therefore, we sought to offer breadth by covering a range of current mainstream EEG signal processing and feature extraction techniques and depth by devoting whole chapters to principles and implement strategies. Learners can go through this book chapter by chapter and implement the examples in the programming materials using the provided sample data, thus developing a better understanding of why and how EEG signal processing is performed, how to interpret relevant results, and what the methodological and practical issues of feature extraction analyses are.

1.3 Who Wrote It and Who to Use?

A terrific lineup of researchers who have rich experience in the related topics contributed to all these chapters. They have a background of either psychology or engineering, and all of them have worked in the interdisciplinary fields of EEG signal processing for many years and have published many papers in well-regarded and peer-viewed journals. These authors have a good understanding of EEG itself, and meanwhile they also have a deep understanding of the signal processing methods. To help learners understand and use the knowledge in the book, MATLAB code and example data are attached in most chapters to provide demonstrations and hands-on exercises for them.

We expect that this book can be well understood by and be beneficial for learners at different levels, including cognitive neuroscientists, psychologists, and the like, who are intelligent and motivated to understand and implement data analyses but lack backgrounds in mathematics and engineering science to utilize the mathematically dense analysis resources or to read the raw computer code behind software packages. Anyway, we hope that the book can make them better prepare for learning new knowledge and skills that are not included in this book.

1.4 The Organization of the Book

The book is organized into four main sections. The first section contains three chapters (i.e., Chaps. 2, 3 and 4) devoted to elaborate fundamental knowledge about EEG. Chapter 2 mainly introduces the electrophysiological properties and acquisition of EEG; Chap. 3 provides details of evoked potential (EP) / event-related potential (ERP), which is extensively used in psychology, psychiatry, and neural engineering; and Chap. 4 discusses how to design experiments to elicit EP/ERP for specific research or clinical purpose. These three chapters are not only suitable for beginning EEG researchers who are new to this field without any mathematics or engineering science background but also for more advanced researchers who are familiar with the general knowledge but would like to develop a more sophisticated perspective. Each chapter in this section will be most useful if read in its entirety.

The second section contains Chaps. 5 and 6 focused on the most common and classical signal processing methods and routines for EEG. Chapter 5 gives a complete description of preprocessing routines, while Chap. 6 introduces the classical and the most frequently used signal processing techniques for EEG: spectral analysis and time-frequency analysis. These two chapters are preliminary but essential contents for EEG processing, and they are useful for beginners with a background of either brain science or engineering.

The third section is composed of a set of chapters (i.e., Chaps. 7, 8, 9, 10, 11, 12, 13 and 14) providing a comprehensive introduction of almost all mainstream signal processing and feature extraction methods for EEG, including blind source separation (Chap. 7), microstate analysis (Chap. 8), source localization (Chap. 9), single-trial analysis (Chap. 10), nonlinear neural dynamics (Chap. 11), connectivity analysis (Chap. 12), spatial complex network analysis (Chap. 13), and temporal complex network analysis (Chap. 14). Researchers who are interested in one specific topic can get useful information from the related chapter.

The fourth section (i.e., Chaps. 15, 16, 17, 18, 19 and 20) focuses on machine learning (including deep learning) and statistical analysis for EEG. These contents are essential in almost all EEG-related work, and they are normally used on EEG features to examine differences of these features between conditions/cohorts or to classify conditions/cohorts. Chap. 15 introduces basic concepts and classical algorithms of machine learning, and Chap. 16 briefly introduces the applications of deep learning on EEG. The purpose of Chap. 17 is to provide the basic idea about statistics. Chapters 18 and 19 are, respectively, used to introduce simultaneous EEG-fMRI processing in a multimodality setting and a useful EEG toolbox (Letswave), both of which are currently the leading trends of EEG analyses. Lastly, Chap. 20 summarizes the issues of current gap, trends, and future directions about EEG.

References

Berger H. Ueber das Elektrenkephalogramm des Menschen. Archives fur Psychiatrie Nervenkrankheiten. 1929;87:527–70.

Brazier MAB. A history of the electrical activity of the brain; the first half-century. New York: Macmillan; 1961.

Caton R. The electric currents of the brain. Br Med J. 1875;2:278.

Cohen MX. Analyzing neural time series data: theory and practice. Cambridge, MA: MIT Press; 2014.

Handy TC. Event-related potentials: a methods handbook. Cambridge, MA: Bradford Books; 2004.

Lopes Da Silva FH, Storm Van Leeuwen W. The cortical source of the alpha rhythm. Neurosci Lett. 1977;6(2–3):237–41.

Luck SJ, Kappenman ES. The Oxford handbook of event-related potential components. Oxford: Oxford University Press; 2011.

Niedermeyer E, Lopes da Silva FH. Electroencephalography: basic principles, clinical applications, and related fields. 5th ed. Philadelphia: Lippincott Williams & Wilkins; 2005.

Regan D. Human brain electrophysiology: evoked potentials and evoked magnetic fields in science and medicine. New York: Elsevier; 1989.

Sanei S, Chambers JA. EEG signal processing. New York: Wiley; 2013.

Squire LR, Zola-Morgan S. Memory: brain systems and behavior. Trends Neurosci. 1988;11 (4):170–5.

Steriade M. Brain activation, then (1949) and now: coherent fast rhythms in corticothalamic networks. Arch Ital Biol. 1995;134(1):5–20.

Chapter 2
EEG: Neural Basis and Measurement

Xiaolei Xia and Li Hu

Abstract This chapter aims to provide background knowledge about electroencephalogram (EEG) origin and measurement. First, a brief introduction of neural basis of EEG is summarized. Then, issues about volume conduction and source estimation of EEG are discussed. Finally, the fundamentals of EEG measurement and the methods for improving performance of EEG measurement are provided.

Keywords Electroencephalogram · Neural basis · Electrocorticogram · Local field potential · Measurement

2.1 Neural Basis of EEG

Based on the current understanding, electroencephalogram (EEG) is defined as the electrical activity, which is generated by the firing of neurons within the human brain, and normally recorded at the brain scalp (Niedermeyer and Lopes da Silva 2005). It is commonly recognized that the EEG originates from summed synchronized synaptic activities in populations of cortical neurons, with a main contribution from pyramidal cells (Avitan et al. 2009; Holmes and Khazipov 2007; Kandel et al. 2013; Lopes da Silva 2009). Cortical neurons in the central nervous system (CNS) are electrically excitable cells, that is, the information transmitted and processed in these neurons is realized by electrochemical signaling, via specialized connections called synapses (Hildebrandt et al. 2008). Apart from the cell body, a typical neuron possesses several dendrites and an axon. The dendrites, extruding from the cell body, are primarily responsible for providing a large receptive area for synaptic input, thus receiving the electrical input to the cell body. The single axon, which arises from the cell body, mainly takes charge of the transmission of action potential, and provides

X. Xia (✉) · L. Hu
CAS Key Laboratory of Mental Health, Institute of Psychology, Chinese Academy of Sciences, Beijing, China

Department of Psychology, University of Chinese Academy of Sciences, Beijing, China
e-mail: xiaxl@psych.ac.cn; huli@psych.ac.cn

L. Hu, Z. Zhang (eds.), *EEG Signal Processing and Feature Extraction*,
https://doi.org/10.1007/978-981-13-9113-2_2

contacts with other nerve cells by the synapses, where signals are sent from the axon of one neuron to a dendrite of another (Squire 2008).

Most of the EEG signals are originated from cortical neurons, and there are two main types of electrical activities associated with cortical neurons, that is, *action potentials* and *postsynaptic potentials* (Rowan and Tolunsky 2003). The action potentials are discrete voltage spikes, which induce a brief local current (less than 10 ms) in the axon with a very limited potential field. In contrast, the postsynaptic potentials, which are voltages aroused by the binding of neurotransmitters to their receptors on the membrane of the postsynaptic neuron, are considerably longer (50–200 ms) in duration and greater in potential field, thus being considered as the primary contributors of EEG (Niedermeyer and Lopes da Silva 2005; Rowan and Tolunsky 2003).

The relationship between the cortical processing within the brain (microcosmic) and the electrical activity measured at the scalp (macroscopic) is not formatted in a straightforward manner. Cortical neurons that generate electrical fields measured at the scalp, must meet several conditions as following (Niedermeyer and Lopes da Silva 2005): (1) these cortical neurons are close to the scalp in distance; (2) the quantity of these cortical neurons should be large enough; (3) and these cortical neurons must be synchronously activated with a certain geometric configuration.

The pyramidal neurons, which are the main composition of cerebral cortex and in close proximity to the scalp, are highly polarized with the major orientation perpendicular to the cortical surface (Kandel et al. 2013). Therefore, the cortical pyramidal neurons are the primary generators of the scalp EEG (Niedermeyer and Lopes da Silva 2005).

2.1.1 Volume Conduction and Source Estimation

A current dipole causes currents to flow through the surrounding conductive medium (e.g., brain tissues) towards measurement surface (e.g., scalp). This is called volume conduction. The range of volume conduction depends on the current dipole and the conductive medium (Kajikawa and Schroeder 2011; Linden et al. 2011). Since these dipoles are not all the same and different biological tissues (e.g., brain, cerebrospinal fluid, skull, and scalp) have different conductivities, the extent of volume conduction could be remarkably diverse.

The voltage fluctuation measured from the scalp by electrodes reflects the total activities of all charged ions within the brain. Since there are many dipoles in the brain, and each dipole will influence the charge in almost all directions, the voltage fluctuation measured at any electrode on the scalp will be the sum of activities generated from many neural sources. This is known as spatial smearing of the signal and considered to be a by-product of the volume conduction (Freeman 1980; Jackson and Bolger 2014), which largely determines the spatial resolution of EEG. Specifically, no matter how many electrodes are used in an EEG recording system, the spatial resolution of EEG remains very poor, since the voltage fluctuations of any two adjacent points on the scalp could be quite similar.

It is of great importance to estimate the underlying neural sources through a certain voltage distribution recorded from the scalp, which is called the inverse problem of EEG. To note, since a given scalp voltage distribution could be generated from an unlimited number of possible neural sources, there is no single and unique solution for the source estimation, which is well-known as the "ill-posed" or "underdetermined" problem.

To date, two main approaches could provide a deep insight into the progress of signal transmission from the brain to the surface electrodes on the scalp, including the computer simulation and the experimental manipulation on epilepsy patients through surgery (Jackson and Bolger 2014). Compared to the surgery manipulation on patients, the computer simulations are much more flexible to estimate EEG neural sources. Particularly, to adopt several equivalent dipolar current sources representing the EEG, scalp voltage distribution has been shown to be efficient and useful for source estimation. However, the head models used in most simulation work are quite simplified, which cannot well represent the real human brain. In light of the poor spatial resolution of EEG, it is recommended to conduct the source estimation combining with other high spatial resolution techniques, e.g., the magnetic resonance imaging (MRI). In fact, the combination of high-density EEG recordings with volume conductor models derived from the MR images has been proved to enhance the accuracy of the source estimation, especially adopted in the non-spherical parts of the head (Fuchs et al. 2007). In comparison, the surgery manipulation on the real brain in epilepsy patients could lead to biased results with a much more limited and lower external validity, as the single-point current source injected into the epilepsy patients is quite different from the electrical activation within a real brain of healthy controls in most cases (Jackson and Bolger 2014).

2.1.2 EEG, ECoG, and LFP

Historically, electroencephalogram (EEG) has been referred to a recording of fluctuating electrical waveforms at the scalp. Electrocorticogram (ECoG) has been referred to a similar recording obtained directly from the cortex, and the local field potential (LFP; also known as intracranial EEG) has been referred to a recording from a small-sized electrode inserting in the brain (Fig. 2.1). All of them are main recordings of postsynaptic potentials from different numbers of neuronal populations (Buzsáki et al. 2012).

As one of the oldest techniques, EEG is widely used in current neuroscience studies (Niedermeyer and Lopes da Silva 2005; Nunez and Srinivasan 2006). Since EEG is recorded by several single electrodes on the scalp and integrated over a large area (10 cm^2 or more), it could be regarded as a spatiotemporally smoothed version of LFP (Buzsáki et al. 2012). Due to the effects of volume conduction, signal transmitted from the current source to the recording electrode is normally attenuated

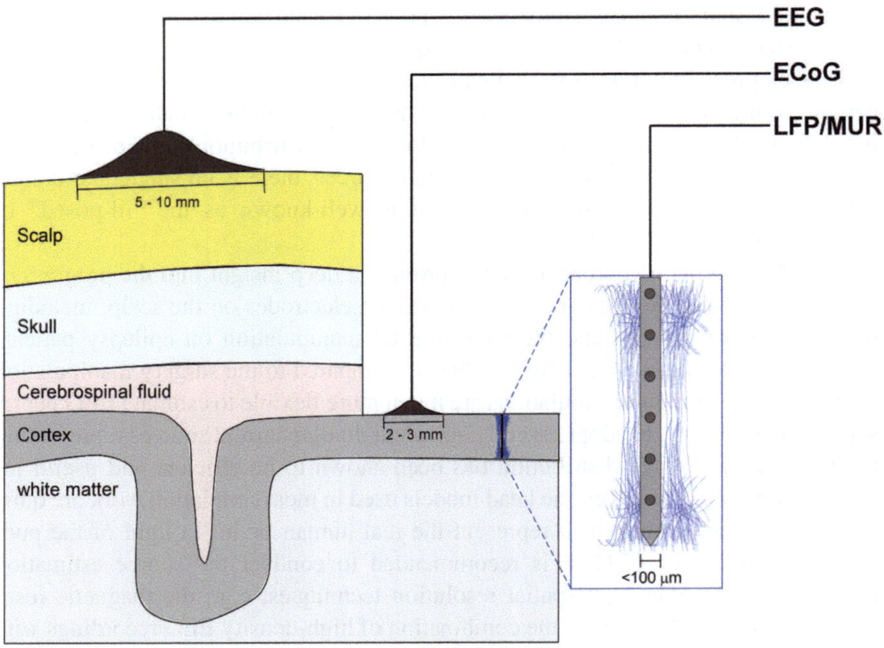

Fig. 2.1 The figure shows the differences of electrode placements among three types of electrical recording techniques, with the EEG electrodes placed on the scalp, the ECoG electrodes on the cortical surface, and the LFP/MUR electrodes inside the cortex, respectively

and distorted. As a result, the relationship between EEG and the firing patterns of the contributing neurons is unclear in most cases (Nunez and Srinivasan 2006).

Notably, ECoG is getting more and more popularly applied in clinical settings and animal studies (Engel et al. 2005). By recording potentials directly from the surface of the cortex, the distortion of signals through the skull and other intermediate tissue is avoided. In the meantime, the spatial resolution of the recorded electric field can be substantially improved (e.g., <5 mm^2) by using closely spaced subdural grid or strip electrodes (Bazhenov et al. 2011).

In contrast to EEG and ECoG, which mainly record electrical activity at the superficial layer of the cortex, LFP mainly samples electrical activity at the deeper layers within the cerebrum. By inserting delicate microelectrodes (made by metal, glass, or silicon probes) into the brain, LFP records a wideband signal (direct current to 40 kHz) from a small neuronal volume (Buzsáki et al. 2012). It should be noted that the wideband signal contains both postsynaptic potentials and action potentials. Therefore, the recordings of action potentials from neuronal populations are usually called multiunit recordings (MURs), and the recordings of postsynaptic potentials from neuronal populations are commonly referred to as LFP (Luck 2014). The spatial resolution of LFP is very high, since the distance between the recording sites and sources is really short and microelectrodes have minimal impact on brain tissues. Thus, it is possible that almost all electrical activities of neuronal populations in a small volume can be precisely recorded, when the density of recording sites is large enough (Buzsáki et al. 2012). Additional information about the intracellular dynamics can also be inferred from the extracellular potentials (Buzsáki et al. 2012; Gold et al. 2006; Henze et al. 2000).

2.2 EEG Measurement

To obtain a high quality of EEG signals, EEG measurement system should consist of several necessary elements, including the electrodes with conductive media, amplifiers with filters, analog-to-digital (A/D) converter, and recording device (Teplan 2002). Specifically, the microvolt signals recorded from electrodes at the scalp were transformed by amplifiers into signals within a proper range of voltage. Then, signals were transformed by converter from an analog format to a digital one and finally stored via the recording device.

2.2.1 Recording Electrodes

2.2.1.1 Electrode Type

In general, electrodes can be classified into three types according to their different roles in the recording (i.e., active, reference, and ground electrodes). The voltage of each single EEG electrode can be regarded as potential changes over time between the active electrode (A) and the reference electrode (R) (Teplan 2002). Theoretically, the reference electrode should be set at the remote position, with an absolute potential of zero. Thus, the potential variability between A and R (equals to A - R) can largely reflect the electrical activity near A. Actually, such a perfect reference electrode does not exit, and the reference site is not electrically neutral in most cases. Consequently, the potential difference between an active electrode and a reference electrode reflects the electrical activity at both sites. The ground electrode is mainly used for reducing the noise produced by the connection to the ground circuit. Most EEG recording systems consist of several active electrodes, one reference electrode, and one ground electrode.

The EEG recording electrodes play important roles in obtaining high-quality data. However, there are many types of electrodes, and each type has different characteristics (Teplan 2002). The most commonly used electrodes are Ag-AgCl electrodes, which are mainly made of silver covered with a thin coating of silver chloride. Generally, Ag-AgCl electrodes are applicable and suited in most situations, as this type of electrode can accurately record minimal changes in potentials due to its good electrical properties (Kutas 1997; Luck 2014; Picton et al. 1995; Picton et al. 2000; Rosler et al. 1995).

To obtain good electric conductivity and low contact impedance, the conductive medium (gel or saline) is necessary to fill the space between the electrodes and the scalp in most cases. The gel will build a more stable conductive connection between the electrode and the scalp than the saline, which is quite helpful to reduce artifacts generated from movements and skin surface. However, it takes a long time to apply the gel on the scalp and is difficult to clean up. On the contrary, the saline is more convenient for use but less stable in maintaining conductive connection.

The conventional wet electrodes (those electrodes with gel or saline as conductive medium) have several limitations in EEG data acquisition. First, skin preparation (e.g., skin exfoliation) is needed to reduce the impedance at the electrode-scalp junction. The procedure is uncomfortable or even painful for some subjects. Moreover, it may cause allergic reaction and infection (Picton et al. 2000). Second, it should be cautious about the consumption of gel, as overflow of gel may generate short circuit between two near electrodes to affect signal transduction (Lin et al. 2011). Third, the gel residuals are hard to clean up. To solve these problems, several kinds of dry electrodes have been developed in recent years (Fonseca et al. 2007; Griss et al. 2001; Griss et al. 2002; Chiou et al. 2006; Kim et al. 2009; Lin et al. 2011; Matthews et al. 2007; Matthews et al. 2008; Ruffini et al. 2006). Most of them were made by the microelectromechanical systems technique, and others were made by fabric-based or foam-based materials. These dry electrodes have shown satisfactory performance in EEG recording and may be more popularly applied in the future (Baek et al. 2008; Beckmann et al. 2010; Gruetzmann et al. 2007; Hoffmann and Ruff 2007; Lin et al. 2011; Zhang and Tao 2008).

2.2.1.2 Electrode Number

In some cases, except of reference and ground electrodes, a single active electrode is adequate in some clinical applications. However, for most EEG/ERP studies, it is necessary to record EEG signals from multiple active electrodes at different sites simultaneously. According to the topographies obtained from the multielectrode EEG recording, we can decompose a piece of data into different components, optimize the feature extraction of EEG, and identify some possible artifacts (Picton et al. 2000).

How many electrodes should be used in routine EEG recordings? Some researchers proposed that the number of recording sites is dependent on the spatial frequencies presented in the scalp recordings (Srinivasan et al. 1998); others suggested that recording from 32 active electrode sites is appropriate for most experiments (Luck 2014). Indeed, the use of high-density electrode arrays can improve the spatial resolution, such as the multichannel configurations comprised up to 256 active electrodes (Teplan 2002). However, recording from multiple electrodes is more expensive and more time-consuming for data acquisition and data analysis. It should be noted that the quality of EEG data is likely to be worse, as the number of electrodes increases considering that it is more difficult to detect and solve problems with more electrodes (Luck 2014).

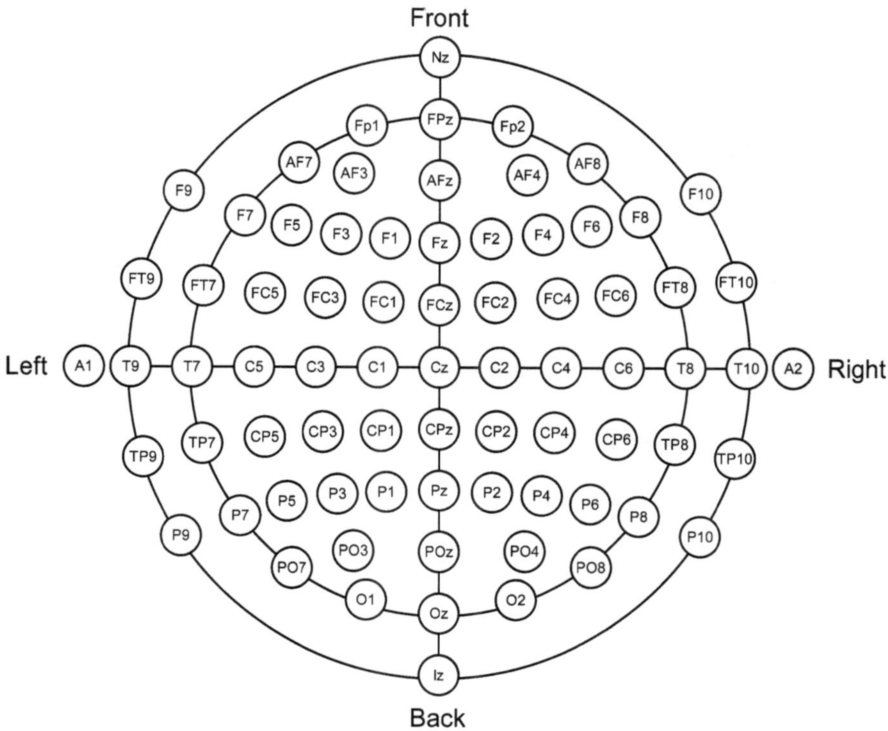

Fig. 2.2 International 10–20 System for electrode placement

2.2.1.3 Electrode Placement

In 1958, a standard protocol for EEG electrode placement was formulated by the International Federation in Electroencephalography and Clinical Neurophysiology (Jasper 1958) and was continually revised by the American Electroencephalographic Society later (Acharya et al. 2016; American Clinical Neurophysiology 2006; Knott 1993). It is called the International 10–20 System, which is widely used for defining electrode placement.

This system contains standardized locations of 75 electrodes on the scalp (Fig. 2.2). The benchmarks are first defined in the system (i.e., the nasion, inion, and two preauricular points). Locations of other electrodes can be determined according to those datum points. The electrodes are placed at 10% and 20% points along lines of latitude and longitude, and this is the reason why the system is called the 10–20 system. The name of each electrode contains two parts, that is, one or two letters and a number. The letter represents the general brain region of the electrode (Fp, frontal pole; F, frontal; C, central; P, parietal; O, occipital; and T, temporal). The number stands for the distance from the midline; the larger the number is, the greater distance it is from the midline (electrodes on the midline are labeled with a "z" for zero). Odd numbers are used at the left hemisphere, and even numbers are used in the right hemisphere (left and right side is from the point of view of a subject).

2.2.1.4 Impedances

High impedance can distort the EEG signals and introduce more noise. Most commercial EEG recording devices are equipped with impedance monitors. In order to decrease signal distortions, impedances at each electrode-scalp interface should be less than 10 K Ohms (Picton et al. 2000). The tolerance for high impedance becomes greater when amplifiers have high-input impedances and good common mode rejection (Taheri et al. 1994). Practically, impedance of the whole circuit comprised of two electrodes is measured, but built-in impedance checks usually display results already divided by two. It is desirable to control of all impedances after finishing each single measurement (Teplan 2002).

2.2.1.5 Reference Electrode

Several different locations, which can be used as reference for EEG recording, are reported in literature, such as the vertex (Cz), C7, linked-ears, linked-mastoids, ipsilateral-ear, contralateral-ear, and tip of the nose (Teplan 2002). In fact, any single site can be used as the reference during data acquisition, as researchers can always re-reference data offline. Notably, reference-free techniques and infinity reference are also commonly used in literature, including the common average reference, weighted average reference, and source derivation (Yao 2001; Yao et al. 2005).

2.2.2 Amplifiers and Converters

The amplifiers of the EEG recording system pick up electric voltage from electrodes and amplify the microvolt signals until they are compatible with A/D converters. Then the converters transform these signals from analog to digital form. To achieve this, amplifiers should satisfy several specific requirements. For example, they should provide selective amplification and rejection, that is, to amplify the physiological signal and to reject superimposed noise. Furthermore, the amplifiers should protect both subjects and equipment from damages. The basic requirements that a biopotential amplifier has to satisfy include (Nagel 2000):

1. The physiological process to be monitored should not be influenced in any way by the amplifier.
2. The measured signal should not be distorted.
3. The amplifier should provide the best possible separation of signal and interferences.

4. The amplifier has to offer protection for the subject away from any hazard of electrical shock.
5. The amplifier itself has to be protected against the damages possibly resulted from high-input voltages, which occur during the application of defibrillators or electrosurgical instrumentation.

The input signal to the amplifier consists of five components (Nagel 2000; Teplan 2002): the desired biopotentials, undesired biopotentials, a power line interference signal of 50/60 Hz and its harmonics, interference signals generated by the tissue/electrode interface, and noise. A good-quality amplifier should reject a large proportion of the signal interferences. The desired biopotential is called the *differential signal*, since the voltage is quite different between the two measuring electrodes of the differential amplifier. The line frequency interference signal is commonly referred to as the *common-mode signal*, as the voltage between the two input electrodes is almost the same. The extent to reject the common-mode signal is a vital performance characteristic of a biopotential amplifier. The ratio of the gain of differential mode over the gain of the common mode is referred to as the *common-mode rejection ratio*. In order to acquire high-quality data, the common-mode rejection used in an EEG recording system should be at least 100 dB.

The amplifier gain is a ratio calculated by dividing the output signal by the input signal. In order to provide best signal quality and proper voltage level for further data processing, the amplifier should provide a gain of 100–50,000 (considering the highest gain may not be the best option, combination of more parameters is demanded, e.g., the range of the A/D converter, sampling rate, noise) and maintain the best possible signal-to-noise ratio (Nagel 2000; Teplan 2002). In order to reduce the impact of electrically noisy environment, differential amplifiers should have high common-mode rejection ratios (at least 100 dB) and high input impedance (at least 100 M Ohms) (Nagel 2000; T. W. Picton et al. 2000; Teplan 2002).

In general, the amplifier contains an analog (hardware) filter unit. It is necessary to carry out a high-pass filtering to decrease potentials at low frequencies coming from bioelectric flow (e.g., breath) (Teplan 2002). Correspondingly, the half-amplitude cutoff frequency of the analog filter unit usually lies between 0.1 and 0.7 Hz. In addition, a low-pass filter is needed to ensure that the signal is band limited. The cutoff frequency for low-pass filtering depends on the highest frequency of interest. It should be noted that if the cutoff frequency of a low-pass filtering is greater than one half of sampling rate, the signal would be distorted, which is called aliasing. Thus, the sampling rate should be large enough, at least double of the highest frequency of interest. However, it doesn't mean that a large sampling rate is always the better option, considering that the sampling rate is proportional to the size of the obtained data and the time of further data processing. In general, a sampling rate of between 250 Hz and 1000 Hz is sufficient, and there is no need to record signal faster than 1000 Hz in EEG data acquisition.

After being processed by the amplifier, analog signals of each channel are repeatedly sampled at a fixed time interval and turned into a digital form by an A/D converter. The A/D converter is connected to a recording system, in which all data can be saved. The resolution of the A/D converter is the number of levels that are discriminated over a particular range, usually expressed as a power of 2 (bits) (Picton et al. 2000). Generally, the resolution of the A/D converter should be at least 12 bits (discerning 4096 value levels) (Teplan 2002). It is necessary to have a greater precision for an A/D converter if there is a large DC shift (Picton et al. 2000). In fact, voltage resolution over 20 bits is recommended. With this resolution, we can filter the low frequencies offline with a digital filter, instead of filtering the low frequencies using an analogy filter during data acquisition. If the resolution is less than 20 bits, a high-pass filter with a half-amplitude cutoff between approximately 0.01 and 0.1 Hz should be used to avoid going beyond the limit of the system (Luck 2014).

In the past, a typical EEG recording system contained an amplifier unit and a separate A/D converter unit. Nowadays, these two units are integrated together, and there is usually only one single analog filter in most cases (i.e., antialiasing filter). This is a good progress since digital filters are superior to analog filters. In other words, it is always better to filter offline rather than online if the recording system meets requirements.

After data being stored, digital filtering (offline) can be a useful tool to improve the signal-to-noise rate of the data. There are a lot of options related to the filter methods, from the traditional linear filtering to the novel nonlinear filtering methods. The choice of filter depends on the objectives set on the signal processing. Usually, the finite impulse response (FIR) filters can be a good choice, because they would not distort wave phases (Teplan 2002). A good filter should be designed to have little impact on the signal properties.

2.2.3 Artefacts

The data obtained from EEG recording system is composed of the EEG signal and a variety of noises. All of these non-EEG signals are referred to as artifacts. In contrast to EEG signals, artifacts usually have higher amplitudes and different morphologies. The artifacts in an EEG recording can be divided into two categories, either subject-related or technique-related (Teplan 2002). Subject-related artifacts are undesired physiological signals that may vastly distort the EEG data, including electromyogram (EMG), electrocardiogram (ECG), and artifacts caused by minor body movements, eye movements, and sweating. Technique-related artifacts are electrical signals caused by the surrounding environment, including power line interference (50/60 Hz), impedance fluctuation, cable movements, broken wire contacts, too much electrode paste/jelly or dried pieces, and low battery.

Some artifacts are tiny and steady (e.g., ECG), but others are large and transient (e.g., blinks). There are two main kinds of techniques for removing artifacts in the data processing (Luck 2014). For large and transient artifact, it can be easily detected and removed through discarding the contaminated EEG epochs (i.e., artifact rejection). For small and constant artifact, it is recommended to estimate the influence of these artifacts on the EEG/ERPs and use correction procedures (e.g., independent component analysis, ICA) to remove the artifacts from the EEG data (i.e., artifact correction). In addition, some useful solutions to reduce these artifacts could be done during data acquisition (see Sect. 2.2.4). For example, the technique-related noise may be reduced by EEG recording system through passive shielding and active electrode system (Jackson and Bolger 2014).

2.2.4 Suggestions on EEG Measurement

The EEG system records electrical activity generated from both the brain and other noise sources. To obtain data with high quality, some suggestions are provided below to reduce the disturbance of noise, thus improving performance of EEG measurement.

2.2.4.1 Recording Environment

Special shielded chambers are helpful to reduce the impact of electric background (especially alternating current line interference at 50/60 Hz), with other electronic devices far away from the shielded area (e.g., the amplifiers connected to alternating current power). However, these electrically isolated chambers are not always necessary, if you have an active electrode system. By placing the amplifier close to the electrodes, the active electrode systems can amplify the signal before transducing it along an unshielded wire, which greatly decreases the electrical noise introduced from the surrounding recording environment (Jackson and Bolger 2014; Luck 2014; MettingVanRijn et al. 1996). If you don't have neither a shielded chamber nor an active electrode system, you can use the notch filter to reduce the electrical noise at a narrow frequency band around 50/60 Hz.

If a video displayer is used in your experiments, the distance between the displayer and the subject should be far enough (100–200 cm is recommended) (Luck 2014). The chair is better to be angle-adjustable so that subjects can adjust themselves to a comfortable position. It is better to use DC-powered lights instead of AC-powered lights in the EEG recording room, as the battery-powered LED lights can decrease the electrical noise. The temperature of recording room should be suitable and controllable, so an air conditioner is needed. It is even better to use a sound-attenuating chamber to shield acoustic noise. Additionally, it is advised to keep the EEG recording environment the same for all subjects unless the environment itself is a designed independent variable in the experiment.

2.2.4.2 Parameters of EEG Recording System

The advised parameters of EEG recording system are provided as follows:

1. Electrode type. Ag-AgCl disc electrodes with conductive paste are recommended. Dry electrodes will be a promising substitute in the future, although they may not be as stable as the wet electrodes.
2. Electrode number. 32–64 active electrodes are recommended in most cases.
3. Electrode placement. The International 10–20 System is the only recommended choice.
4. Reference electrode. Considering that the EEG data can be re-referenced offline, any single site could be used as the reference electrode during EEG recording, as long as the electrical circuit is well-connected.
5. Amplifier and converter. An amplifier with high input impedance (at least 100 M Ohms) and high common-mode rejection ratio (at least 100 dB) is recommended, whereas the converter with a voltage resolution over 20 bits is optimal.
6. Analog filter. If the voltage resolution of the converter is high enough, we do not advise to apply any analog filter during EEG data acquisition, as it is superior to filter the data offline using digital filters instead. If you don't have a high-resolution converter, a high-pass filter with a half-amplitude cutoff between approximately 0.01 and 0.1 Hz is recommended during data acquisition.
7. Sampling rate. A sampling rate between 250 Hz and 1000 Hz is sufficient in most cases.

2.2.4.3 Subjects

To obtain data with satisfactory quality, it is important to choose the right representative samples with appropriate spectrum of demographic characteristics and other factors (Teplan 2002). What's more, most EEG experiments tend to be long and boring; it is very important to keep your subjects relaxed and enjoyed and focus his attention on the task during data collection (Luck 2014). If your subjects lack motivation or get bored, their performance could be affected and probably become worse, with possible muscle or movement artifacts introduced. Some suggestions are provided to avoid the problem as follows (Luck 2014):

1. Chatting with the subject during the preparation is highly recommended. Given that the preparation of EEG experiments is usually time-consuming, it is better for you to have a good conversation with the subject on topics about the work, hobbies, and sports of the subject's interest. By doing so, the subject will feel relaxed and trusted and be more likely to do a good performance during the experimental task. The conversation can also provide a good chance to note the subjects about some essential information for completing the experiment, such as how to fulfill the task in a right way, how to control the eye blinks, and other relevant matters.
2. The duration of blocks of trials should be reasonable. No one can maintain his attention for a long time; thus the whole task should be divided into several blocks, with a short rest after each block. In most experiments, a typical block lasting for 5–7 minutes and separated by a short rest for 1–2 minutes, is optimal and recommended.

References

Acharya JN, Hani A, Cheek J, Thirumala P, Tsuchida TN. American Clinical Neurophysiology society guideline 2: guidelines for standard electrode position nomenclature. J Clin Neurophysiol. 2016;33(4):308–11. https://doi.org/10.1097/WNP.0000000000000316.

American Clinical Neurophysiology, S. Guideline 5: Guidelines for standard electrode position nomenclature. J Clin Neurophysiol. 2006;23(2):107–10.

Avitan L, Teicher M, Abeles M. EEG generator–a model of potentials in a volume conductor. J Neurophysiol. 2009;102(5):3046–59. https://doi.org/10.1152/jn.91143.2008.

Baek JY, An JH, Choi JM, et al. Flexible polymeric dry electrodes for the long-term monitoring of ECG[J]. Sensors Actuators A Phys. 2008;143(2):423–9.

Bazhenov M, Lonjers P, Skorheim S, Bedard C, Dstexhe A. Non-homogeneous extracellular resistivity affects the current-source density profiles of up-down state oscillations. Philos Trans A Math Phys Eng Sci. 2011;369(1952):3802–19. https://doi.org/10.1098/rsta.2011.0119.

Beckmann L, Neuhaus C, Medrano G, Jungbecker N, Walter M, Gries T, Leonhardt S. Characterization of textile electrodes and conductors using standardized measurement setups. Physiol Meas. 2010;31(2):233–47. https://doi.org/10.1088/0967-3334/31/2/009.

Buzsáki G, Anastassiou CA, Koch C. The origin of extracellular fields and currents—EEG, ECoG, LFP and spikes. Nat Rev Neurosci. 2012;13:407.

Chiou JC, Ko LW, Lin CT, et al. Using novel MEMS EEG sensors in detecting drowsiness application[C]//2006 IEEE Biomedical Circuits and Systems Conference. IEEE; 2006. p. 33–36.

Engel AK, Moll CK, Fried I, Ojemann GA. Invasive recordings from the human brain: clinical insights and beyond. Nat Rev Neurosci. 2005;6(1):35–47. https://doi.org/10.1038/nrn1585.

Fonseca C, Silva Cunha JP, Martins RE, Ferreira VM, Marques de Sa JP, Barbosa MA, Martins da Silva A. A novel dry active electrode for EEG recording. IEEE Trans Biomed Eng. 2007;54 (1):162–5. https://doi.org/10.1109/TBME.2006.884649.

Freeman WJ. Use of spatial deconvolution ot compensate for distortion of EEG by volume conduction. IEEE Trans Biomed Eng. 1980;27(8):421–9. https://doi.org/10.1109/TBME.1980.326750.

Fuchs M, Wagner M, Kastner J. Development of volume conductor and source models to localize epileptic foci. J Clin Neurophysiol. 2007;24(2):101–19. https://doi.org/10.1097/WNP.0b013e318038fb3e.

Gold C, Henze DA, Koch C, Buzsaki G. On the origin of the extracellular action potential waveform: a modeling study. J Neurophysiol. 2006;95(5):3113–28. https://doi.org/10.1152/jn.00979.2005.

Griss P, Enoksson P, Tolvanen-Laakso HK, Merilainen P, Ollmar S, Stemme G. Micromachined electrodes for biopotential measurements. J Microelectromech Syst. 2001;10(1):10–6. https://doi.org/10.1109/84.911086.

Griss P, Tolvanen-Laakso HK, Merilainen P, Stemme G. Characterization of micromachined spiked biopotential electrodes. IEEE Trans Biomed Eng. 2002;49(6):597–604. https://doi.org/10.1109/TBME.2002.1001974.

Gruetzmann A, Hansen S, Muller J. Novel dry electrodes for ECG monitoring. Physiol Meas. 2007;28(11):1375–90. https://doi.org/10.1088/0967-3334/28/11/005.

Henze DA, Borhegyi Z, Csicsvari J, Mamiya A, Harris KD, Buzsaki G. Intracellular features predicted by extracellular recordings in the hippocampus in vivo. J Neurophysiol. 2000;84 (1):390–400. https://doi.org/10.1152/jn.2000.84.1.390.

Hildebrandt J, Smith D, Great Pacific Media (Firm). The nervous system: neurons, networks, and the human brain. Colorado Springs: Great Pacific Media; 2008.

Hoffmann KP, Ruff R. Flexible dry surface-electrodes for ECG long-term monitoring. Conf Proc IEEE Eng Med Biol Soc. 2007;2007:5740–3. https://doi.org/10.1109/IEMBS.2007.4353650.

Holmes GL, Khazipov R. Basic Neurophysiology and the cortical basis of EEG. In: Blum AS, Rutkove SB, editors. The clinical neurophysiology primer. Totowa: Humana Press; 2007. p. 19–33.

Jackson AF, Bolger DJ. The neurophysiological bases of EEG and EEG measurement: a review for the rest of us. Psychophysiology. 2014;51(11):1061–71. https://doi.org/10.1111/psyp.12283.

Jasper HH. The ten-twenty electrode system of the international federation. Electroencephalogr Clin Neurophysiol. 1958;10:370–5.

Kajikawa Y, Schroeder CE. How local is the local field potential? Neuron. 2011;72(5):847–58. https://doi.org/10.1016/j.neuron.2011.09.029.

Kandel ER, Schwartz JH, Jessell TM, Siegelbaum S, Hudspeth AJ. Principles of neural science. 5th ed. New York/London: McGraw-Hill; 2013.

Kim YS, Baek HJ, Kim JS, Lee HB, Choi JM, Park KS. Helmet-based physiological signal monitoring system. Eur J Appl Physiol. 2009;105(3):365–72. https://doi.org/10.1007/s00421-008-0912-6.

Knott JR. Regarding the American electroencephalographic society guidelines for standard electrode position nomenclature: a commentary on the proposal to change the 10–20 electrode designators. J Clin Neurophysiol. 1993;10(1):123–8.

Kutas M. Views on how the electrical activity that the brain generates reflects the functions of different language structures. Psychophysiology. 1997;34(4):383–98. https://doi.org/10.1111/j.1469-8986.1997.tb02382.x.

Lin CT, Liao LD, Liu YH, Wang IJ, Lin BS, Chang JY. Novel dry polymer foam electrodes for long-term EEG measurement. IEEE Trans Biomed Eng. 2011;58(5):1200–7. https://doi.org/10.1109/TBME.2010.2102353.

Linden H, Tetzlaff T, Potjans TC, Pettersen KH, Grun S, Diesmann M, Einevoll GT. Modeling the spatial reach of the LFP. Neuron. 2011;72(5):859–72. https://doi.org/10.1016/j.neuron.2011.11.006.

Lopes da Silva F. EEG: origin and measurement. In: Mulert C, Lemieux L, editors. EEG - fMRI. Berlin/Heidelberg: Springer; 2009.

Luck SJ. An introduction to the event-related potential technique. Cambridge, MA: MIT Press; 2014.

Matthews R, McDonald NJ, Anumula H, Woodward J, Turner PJ, Steindorf MA, Chang K, Pendleton JM. Novel hybrid bioelectrodes for ambulatory zero-prep EEG measurements using multi-channel wireless EEG system. Berlin/Heidelberg: Springer; 2007.

Matthews R, Turner PJ, McDonald NJ, Ermolaev K, Manus T, Shelby RA, Steindorf M. Real time workload classification from an ambulatory wireless EEG system using hybrid EEG electrodes. Conf Proc IEEE Eng Med Biol Soc. 2008;2008:5871–5. https://doi.org/10.1109/IEMBS.2008.4650550.

MettingVanRijn AC, Kuiper AP, Dankers TE, Grimbergen CA. Low-cost active electrode improves the resolution in biopotential recordings. Paper presented at the Proceedings of 18th Annual International Conference of the IEEE Engineering in Medicine and Biology Society; 1996

Nagel JH. Biopotential amplifiers. In: The biomedical engineering handbook, vol. 2. Boca Raton: CRC/Taylor & Francis; 2000. p. 1300.

Niedermeyer E, Lopes da Silva FH. Electroencephalography: basic principles, clinical applications, and related fields. 5th ed. Philadelphia: Lippincott Williams & Wilkins; 2005.

Nunez PL, Srinivasan R. Electric fields of the brain: the neurophysics of EEG. 2nd ed. Oxford/New York: Oxford University Press; 2006.

Picton TW, Lins OG, Scherg M. The recording and analysis of event-related potentials, vol. 10; 1995.

Picton TW, Bentin S, Berg P, Donchin E, Hillyard SA, Johnson R, Miller GA, Ritter W, Ruchkin DS, Rugg MD, Taylor MJ. Guidelines for using human event-related potentials to study cognition: recording standards and publication criteria. Psychophysiology. 2000;37(2):127–52.. doi:undefined

Rosler F, Heil M, Hennighausen E. Distinct cortical activation patterns during long-term memory retrieval of verbal, spatial, and color information. J Cogn Neurosci. 1995;7(1):51–65. https://doi.org/10.1162/jocn.1995.7.1.51.

Rowan AJ, Tolunsky E. Primer of EEG: with a mini-atlas. Philadelphia: Butterworth-Heinemann; 2003.

Ruffini G, Dunne S, Farrés E, Marco-Pallarés J, Ray C, Mendoza E, Silva R, Grau C. A dry electrophysiology electrode using CNT arrays. Sensors Actuators A Phys. 2006;132(1):34–41. https://doi.org/10.1016/j.sna.2006.06.013.

Squire LR. Fundamental neuroscience. 3rd ed. Amsterdam/Boston: Elsevier/Academic; 2008.

Srinivasan R, Tucker DM, Murias M. Estimating the spatial Nyquist of the human EEG. Behav Res Methods Instrum Comput. 1998;30(1):8–19. https://doi.org/10.3758/BF03209412.

Taheri BA, Knight RT, Smith RL. A dry electrode for EEG recording. Electroencephalogr Clin Neurophysiol. 1994;90(5):376–83.

Teplan M. Fundamentals of EEG measurement. Measurement Sci Rev. 2002;2(2):1–11.

Yao D. A method to standardize a reference of scalp EEG recordings to a point at infinity[J]. Physiol Meas. 2001;22(4):693.

Yao D, Wang L, Oostenveld R, et al. A comparative study of different references for EEG spectral mapping: the issue of the neutral reference and the use of the infinity reference[J]. Physiol Meas. 2005;26(3):173.

Zhang H, Tao XM. Textile-structured electrodes for electrocardiogram AU – Xu, P. J Textile Progress. 2008;40(4):183–213. https://doi.org/10.1080/00405160802597479.

Chapter 3
Electroencephalography, Evoked Potentials, and Event-Related Potentials

Xuejing Lu and Li Hu

Abstract This chapter aims to provide background knowledge necessary to understand electroencephalography (EEG) technique and its applications. First, the information about how EEG, evoked potentials (EPs), and event-related potentials (ERPs) are generated and obtained are summarized. Next, a brief overview of EPs and ERPs is provided, and classical EP and ERP components that are applied in clinical and neuroscience studies are also described, with an emphasis on different sensory modalities through which the stimuli are presented. Finally, the pitfalls and promise in EP and ERP studies are discussed.

Keywords Electroencephalography (EEG) · Evoked potentials (EPs) · Event-related potentials (ERPs)

3.1 Spontaneous EEG Activity

The recordings of the spontaneous electrical activity, called spontaneous EEG activity, exhibit certain characteristic waveforms that dominate in a wide range of frequencies, and are normally applied in various clinical diagnoses (e.g., epilepsy, coma, and brain death) (Hughes 1994). The spontaneous EEG activity is normally classified according to the frequency bands. Empirically, spontaneous EEG activity is divided into five frequency bands (Fig. 3.1): delta (δ, < 4 Hz), theta (θ, 4–8 Hz), alpha (α, 8–13 Hz), beta (β, 13–30 Hz), and gamma (γ, > 30 Hz). The rhythmic activity at each of these frequency bands characterizes with a certain distribution over the scalp and a certain biological significance. For instance, the alpha waves are normally measured from the occipital region of cortex in an awake person with the

X. Lu · L. Hu (✉)
CAS Key Laboratory of Mental Health, Institute of Psychology, Chinese Academy of Sciences, Beijing, China

Department of Psychology, University of Chinese Academy of Sciences, Beijing, China
e-mail: luxj@psych.ac.cn; huli@psych.ac.cn

L. Hu, Z. Zhang (eds.), *EEG Signal Processing and Feature Extraction*,
https://doi.org/10.1007/978-981-13-9113-2_3

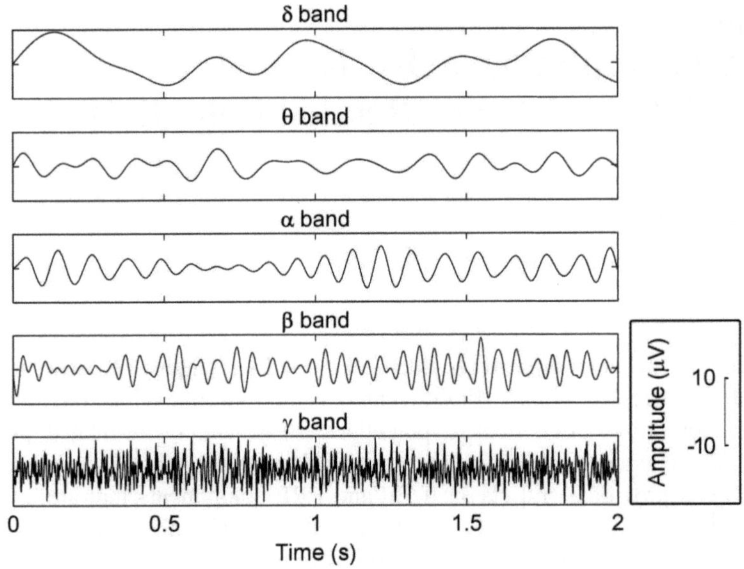

Fig. 3.1 Rhythmic EEG activity patterns. The rhythmic EEG activity is often divided into bands by frequency. Empirically, spontaneous EEG activity is divided into five frequency bands: delta (δ, < 4 Hz), theta (θ, 4–8 Hz), alpha (α, 8–13 Hz), beta (β, 13–30 Hz), and gamma (γ, > 30 Hz)

eyes closed, and attenuate with eyes opening or mental activity. In addition, with the increase of frequency, the EEG rhythms are prone to be characterized with a lower amplitude (Hughes 1994).

3.2 Evoked Potentials and Event-Related Responses

When a stimulus associated with specific sensory, cognitive, or motor event occurs, it disturbs the spontaneous EEG activity. Although such neural responses are embedded within the EEG, it is possible to extract these responses from the spontaneous EEG by means of a simple averaging technique or more sophisticated techniques, such as single-trial analysis and time-frequency analysis. Specifically, the averaged responses are called event-related potentials (ERPs) to denote the fact that they are electrical potentials and related to the particular events.

The early parts of brain responses are traditionally believed to be evoked by the presentation of a stimulus and to reflect brain activity that restrictedly relates to the basic sensory processing of the stimulus in the human brain (Callaway et al. 1978). From this perspective, researchers used the term "evoked potentials (EPs)" to refer to the measures of brain responses elicited by an external stimulus. However, accumulating evidence suggests that when time-locked to the psychological events, some of the brain responses, especially for those in late latencies, are more likely to reflect the cortical processing of psychological activities rather than simple sensory processes. In other words, the recorded brain responses, commonly referred as "event-related potentials", displayed a stable time relationship to a definable event, rather than just the evoked activity reflecting sensorimotor processes (Garnsey 1993; Niedermeyer and Lopes da Silva 2005; Rugg and Coles 1995). For these reasons, the ERP technique is popularly employed by physiologist, psychologist, and physicians for multiple applications (Begleiter et al. 1967; Deletis and Sala 2008; Gonzalez et al. 2009; Duncan et al. 2009).

3.3 An Overview of EPs and ERPs

As stated in the previous section, EEG is a very spatially coarse measure of brain activity with very low signal-to-noise ratio (SNR), and it cannot be used in its raw form to measure most of the neural processes that relate to a specific event or task. Fortunately, the event-related responses time-locked to the stimulus can be extracted from the continuous EEG data by means of a simple averaging technique (Pfurtscheller and Lopes da Silva 1999; Mouraux and Iannetti 2008). The time-locked brain responses represent as either phase-locked or non-phase-locked to the presentation of a stimulus. Since they are phase-locked to the stimulus onset, they can be easily observed in the time domain after averaging across several trials, known as ERPs. In contrast, if they are non-phase-locked to the stimulus onset, they would become invisible in the time domain by directly averaging across several trials. These non-phase-locked responses are called event-related oscillations (EROs), which represent the specific frequency changes of the ongoing EEG activities. EROs may appear either as a transient increase (event-related synchronization, ERS) or a transient decrease (event-related desynchronization, ERD) in synchrony of the underlying neuronal populations (Pfurtscheller and Lopes da Silva 1999; Mouraux and Iannetti 2008). In this chapter, we focus on the phase-locked EPs and ERPs.

How are ERPs generated? Some researchers showed that ERPs appear to be an additional activity and independent of the ongoing spontaneous EEG activity (Niedermeyer and Lopes da Silva 2005; Jervis et al. 1983; Schroeder et al. 1995), leading to the evoked model (Sauseng et al. 2007). Others, on the other hand, demonstrated that ERPs are generated (or partially generated) by the reorganization of phases of the ongoing EEG rhythmic activity (phase reset), leading to the phase reset model (Makeig et al. 2002; Min et al. 2007; Sauseng et al. 2007). Up to the present, researchers continue the debate about the neural fundamental of ERPs, either evoked model or the phase reset model. Min et al. (2007) observed that both partial phase reset and partial additive power contribute dynamically to the generation of ERPs and the pre-stimulus brain state exerts a prominent influence on event-related brain responses, suggesting that the combination of these two models may be a reasonable and comprehensive way to explain the generation of ERPs.

How to obtain an ERP and locate its neural generators? In most cases, the magnitude of ERPs (on the order of microvolts) is often several factors smaller than the magnitude of the background EEG activity (on the order of tens of microvolts) (Hu et al. 2010; Rugg and Coles 1995). Therefore, the identification of ERPs relies on signal processing methods for enhancing the SNR. One of the most widely used approaches to enhance SNR is the across-trial averaging in the time domain (Dawson 1951, 1954). Averaged ERPs often comprise monophasic deflections which are characterized by their polarity, latency, amplitude, and scalp distribution (Callaway et al. 1978; Mouraux and Iannetti 2008). The neural generators of each deflection of ERPs can be inferred from the scalp distribution using either the dipole source model (Valeriani et al. 2001) or the distributed source model (Michel et al. 2001). However, determining source locations from the scalp topography is a typical ill-posed inverse problem (Grech et al. 2008), where a large number of source configurations can be explained by a given scalp topography. Thus, additional assumptions (e.g., the number of dipole sources or spatial constraints from functional magnetic resonance imaging) are needed to obtain a unique solution. For this reason, the obtained source locations may be heavily biased by the assumptions, leading to limited applications (Mouraux and Iannetti 2008).

3.4 Common EP and ERP Components

3.4.1 Auditory Evoked Potentials

The presentation of an auditory stimulus evokes potentials generated in the cochlea. The first set of auditory responses occurs within a few milliseconds of a sudden sound onset, reflecting the flow of information from the cochlea through the brainstem and into the thalamus (see Fig. 3.2). Given the origin of these signals deep inside the brain, and the response typically shows small voltages (about 0.5 μV in amplitude), averaging thousands of trials is required to obtain clear brainstem auditory evoked potentials (BAEPs). These auditory brainstem responses are typically labeled with Roman numerals (waves I–VI) and probably represent the activation of acoustic nerve (wave I), cochlear nuclear nuclei (wave II), superior olives (wave III), lateral lemniscus tracts and nuclei (wave IV), and inferior colliculi (wave V). Wave VI presumably arises from the medial geniculate body but is not considered clinically useful (Guerreiro and Ehrenberg 1982). BAEPs are highly automatic and relatively unaffected by sleep, anesthesia, and even sufficient doses of barbiturates to induce coma (Goldie et al. 1981). Therefore, BAEPs can be used to assess the integrity of the auditory pathways, especially for newborns and infants. It is an extraordinary tool for early identification of hearing impairment or hearing loss, aiding in the early intervention.

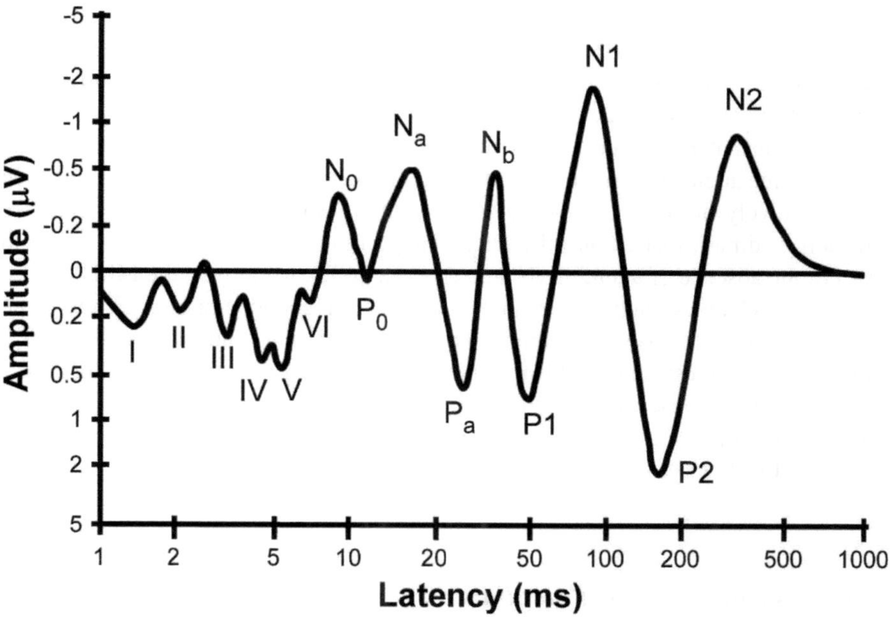

Fig. 3.2 Typical sequence of auditory evoked components. The waveform elicited by an auditory stimulus is shown over a period of time to demonstrate the auditory brainstem responses (waves I–VI), the mid-latency responses, and the long-latency responses

The BAEPs are followed by the mid-latency responses (MLRs; defined as responses between 10 and 50 ms), which probably arise at least in part from the medial geniculate nucleus and the primary auditory cortex, and the long-latency responses, which typically begin with the P50 (known as P1), N100 (known as N1), and P160 (known as P2) (Luck 2005). However, the term "long-latency responses" is confusing to some extent, because these responses are with long latencies in relation to mid-latency responses but with relatively short latencies when compared to high-level cognitive components such as P300 and N400. Therefore, some researchers also use the term "mid-latency" to describe evoked potential components occurring between 50 and 200 ms (Roth et al. 1980; Boutros et al. 2006).

The auditory N1 wave with a peak latency between 50 and 150 ms is one of the most prominent components during an auditory task. It has several distinct subcomponents that sum together to form the N1 peak, including a set of specific subcomponents that associated with the physical and temporal aspects of the stimulus and general state of the subject and a set of subcomponents which are not necessarily elicited by an auditory stimulus but depending on the conditions in which the stimulus occurs (Näätänen and Picton 1987). The auditory N1 is sensitive to attention (Woldorff et al. 1993), which can be manipulated by varying the task relevance of events or by inducing expectations that an event occurs at a particular time point (Lange 2013).

A repetitive nontarget stimulus can elicit an N2 deflection. When deviant stimuli are occasionally presented within the repetitive train, a larger N2 amplitude is evoked in response to the deviants. This effect can be divided into three subcomponents – N2a, N2b, and N2c. The N2a is an automatic effect that occurs mainly for auditory mismatches, even if they are task-irrelevant, which is more commonly known as the auditory mismatch negativity (MMN). Specifically, MMN is generated by a relatively automatic response to an auditory stimulus that differs from the preceding stimuli, that is, an infrequent change in homogeneous repetitive stimuli even in the absence of attention (Näätänen et al. 2007). Moreover, the MMN is also elicited by changes such as grammar violations or sound omission (Yabe et al. 1997). Although the MMN is considered an attention-independent process, emerging evidence suggests that the MMN can be modulated by attention (Sussman 2007). Nevertheless, because of its relatively high automatic nature, the MMN has been used to assess processing in individuals with difficulties in speaking or in making behavioral responses.

3.4.2 Visual Evoked Potentials

Typically, the first major visual ERP component is the P1 wave (see Fig. 3.3). However, sometimes the P1 is preceded by the C1 component, with an onset latency of 40–70 ms and peak latency of 60–100 ms. The C1 component is highly sensitive to basic visual stimulus parameters, such as contrast and spatial frequency, and it appears to be generated in the area V1 (primary visual cortex; Brodmann's area 17, striate cortex). Evidence showing that C1 is generated in the striate cortex within the calcarine fissure comes from studies demonstrating that the C1 reverses in polarity for upper vs. lower visual field stimulation (Jeffreys and Axford 1972).

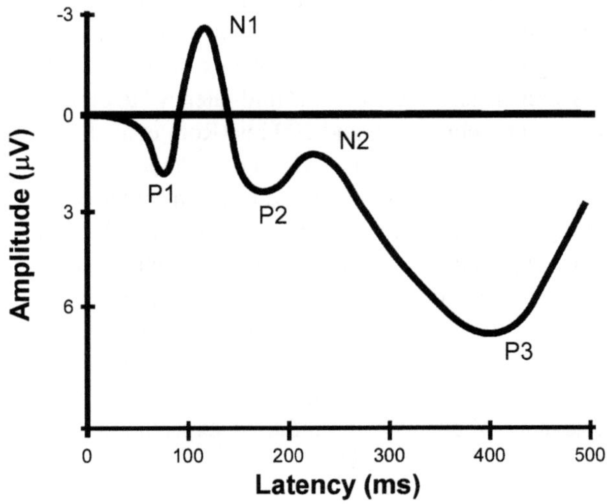

Fig. 3.3 Typical visual evoked components, including P1, N1, P2, N2, and P3 components

This reversal corresponds to the retinotopic organization of the striate cortex, in which the lower and upper visual hemifields are mapped in the upper and lower banks of the calcarine fissure. When the C1 wave is positive, it sums together with the P1 component, creating a single positive-going wave. Consequently, a distinct C1 wave is not usually observed unless upper-field stimuli are used to generate a negative C1 wave, which can easily be distinguished from the positive P1 wave (Luck 2005).

Visual P1 is maximal at lateral occipital electrode sites and onsets 60–90 ms poststimulus with a peak between 100 and 130 ms. Unlike the C1 component, the P1 component does not show a polarity reversal for upper vs. lower visual field stimulation. Although no clear consensus has yet been reached regarding the exact location of its sources, it is believed that the P1 component is mainly generated in extrastriate visual areas (Di Russo et al. 2002). On the other hand, like the C1 wave, the P1 wave is sensitive to variations in stimulus parameters and can be modulated by selective attention (Hillyard et al. 1998) as well as by the individual's state of arousal (Vogel and Luck 2000).

Following the P1 wave is the visual N1 wave, which is highly refractory. In other words, if a visual stimulus is presented at a certain location preceded by another stimulus at the same location at a short delay, the response to the second stimulus is greatly reduced. This effect can be accounted for by a model in which the neural populations that generate the N1 become refractory after their responses. According to this model, the ability of neurons to produce additional responses is diminished immediately following a response, from which point it gradually recovers. Like auditory N1, there are several visual N1 subcomponents. The earliest N1 subcomponent peaks at 100–150 ms poststimulus at anterior electrode sites, and there appear to be at least two posterior N1 components that peak at 150–200 ms poststimulus, arising from parietal cortex and lateral occipital cortex, respectively. It has shown that all three N1 subcomponents are influenced by spatial attention (Hillyard et al. 1998). In addition, the lateral occipital N1 subcomponent appears to be larger when subjects are performing discrimination tasks than when they are performing detection tasks, which has led to the proposal that this subcomponent reflects some sort of discriminative processing (Vogel and Luck 2000).

3.4.3 *Laser Evoked Potentials*

Brief radiant heat pulses, generated by infrared laser stimulators, excite selectively Aδ and C fiber free nerve endings located in the superficial layers of the skin (Bromm and Treede 1984; Carmon et al. 1976). Such stimuli elicit a number of brain responses that can be detected in the human EEG (Carmon et al. 1976). Laser evoked potentials (LEPs) are related to the activation of type-II A-mechano-heat nociceptors (Treede et al. 1998) and spinothalamic neurons located in the anterolateral quadrant of the spinal cord (Treede et al. 2003). They comprise a number of waves that are time-locked to the onset of the stimulus. As shown in

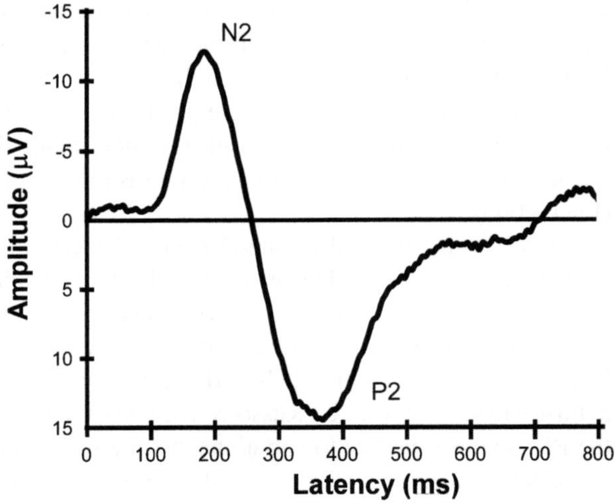

Fig. 3.4 An illustration of ERPs (N2 and P2 waves) evoked by nociceptive somatosensory inputs (i.e., laser stimuli) obtained at Cz electrode (reference to average)

Fig. 3.4, the largest response is a negative–positive vertex potential, that is, the N2 and P2 waves, peaking at approximately 200 and 350 ms when stimulating the hand dorsum (Bromm and Treede 1984). This complex is preceded by a smaller negative wave, that is, the N1 wave, which overlaps in time and space with the larger, subsequent N2 wave. The N1 is described as having a distribution maximal over the central or temporal region contralateral to the stimulated side (Kunde and Treede 1993; Treede et al. 2003; Cruccu et al. 2008). In order to isolate the N1 wave from the N2 wave, the N1 is usually detected at the contralateral central electrode (C3 or C4) referred to a frontal midline electrode (Fz or Fpz), an EEG montage that is also recommended for recording LEPs in clinical settings (Hu et al. 2010; Valentini et al. 2012). Several studies have shown that the N1, N2, and P2 waves reflect a combination of cortical activities originating from the primary and secondary somatosensory cortices, insula, and anterior cingulate cortex (ACC) (Garcia-Larrea et al. 2003; Cruccu et al. 2008).

While the N2 and P2 waves are characterized by a high SNR (with a peak-to-peak amplitude of several tens of microvolts when averaging 20–30 trials) (Carmon et al. 1980; Iannetti et al. 2005), the N1 wave has a smaller SNR and is thus more difficult to be detected. This difficulty is not only due to the fact that the N1 wave is generated by neural activities of smaller magnitude than those underlying the N2 and P2 waves (Treede et al. 2003; Cruccu et al. 2008), but also due to the fact that the N1 and N2 waves overlap in time and space with opposite polarities (Kunde and Treede 1993) and to the fact that temporal electrodes are often contaminated by artifacts related to the activity of the temporalis muscle. Therefore, the vast majority of physiological (Iannetti et al. 2003) and clinical (Treede et al. 2003) LEP studies conducted in the past decades have relied uniquely on measures of the N2 and P2 waves to investigate the nociceptive system. In recent years, a growing number of studies have started to explore experimental modulations of the latency and amplitude of the N1 wave and to characterize its functional significance (Legrain et al. 2002; Iannetti et al. 2008;

Mouraux and Iannetti 2009; Lee et al. 2009; Ellrich et al. 2007; Schmahl et al. 2004). Indeed, there is experimental evidence indicating that the N1 wave represents an early stage of sensory processing more directly related to the ascending nociceptive input (Lee et al. 2009), while the later N2 and P2 waves appear to reflect neural activities largely unspecific for the sensory modality of the eliciting stimulus (Mouraux and Iannetti 2009). For all these reasons, a more systematical examination of N1 has been recommended to enhance the sensitivity of LEPs in clinical applications (Cruccu et al. 2008; Treede et al. 2003).

When taking an extensive inspection to the LEP responses in the time domain, a small but clear contribution of somatosensory-specific activities to the very late part of the LEP is evident (Mouraux and Iannetti 2009; Liang et al. 2010). Although being anecdotally observed but overlooked for years by various EEG studies, such a distinct and prominent component has been successfully identified and extracted from the LEP waveforms, labeled as P4 (Hu et al. 2014). The P4 wave is characterized by lower SNR, smaller amplitude, and longer latency. Specifically, when elicited by hand stimuli, the P4 wave is maximal on the central–parietal electrodes contralateral to the stimulated side (Cc referenced to Fz), displayed as a positive amplitude with latency ranging from 390 to 410 ms. In contrast, when elicited by foot stimuli, the P4 wave is centrally distributed, with a maximum amplitude between Cz and Pz electrodes and a latency ranging from 430 to 450 ms (Hu et al. 2014). Notably, there is a tight physiological link between the neutral activities subserving the P4 and the N1 waves, as the peak latency and amplitude of P4 wave are heavily dependent on those of the N1 wave. Accordingly, compelling evidence has demonstrated that the P4 wave could share common neural sources with the N1 wave (Hu et al. 2010; Valentini et al. 2012; Hu et al. 2014), which has been regarded as the most crucial component of the LEPs related to the incoming somatosensory input (Lee et al. 2009; Mouraux and Iannetti 2009). Overall, aside from the former three main components in the LEP waveform, the P4 wave, which is generated (at least partly) from primary somatosensory cortex, has been recognized as the fourth separated component, reflecting a unique function of nociceptive information processing.

LEPs are considered as the best tool for assessing function of nociceptive pathways in physiological and clinical studies (Bromm and Treede 1991; Iannetti et al. 2001; Cruccu et al. 2008). LEPs can reflect the conduction abnormalities in the nociceptive pathways, from periphery to cortex, if lesions impinge any point in the spinothalamic system and the laser stimulus is applied to the corresponding cutaneous territory (Cruccu et al. 2008). At the peripheral level, LEPs can be used to assess the "small fiber disease" that is characteristic of loss in temperature and pain sensitivity and autonomous peripheral nerve function. The abnormality of LEPs often represents as the attenuation in amplitude or delay in latency or absence of responses in serious conditions (Agostino et al. 2000; Kakigi et al. 1991b; 1992). The latency and amplitude changes of LEPs were observed to correlate with the loss of Aδ fiber in sural nerve biopsy, and a complete loss of LEP responses peripheral neuropathies may indicate impairment of both Aδ and C fibers (Kakigi et al. 1991b; Treede et al. 2003). At the spinal cord level, LEPs can provide a functional assessment of spinal lesions. Previous studies showed that abnormality of LEPs was correlated with a variety of spinal lesions in patients with syringomyelia, arteriovenous malformations, and inflammatory myelopathies (Bromm et al. 1991; Treede et al. 2003; Kakigi et al. 1991a; Treede et al. 1991). At the brainstem level, LEPs may be abnormal (absence of responses, attenuation in amplitude, or delay in latency) because of lesions of the lower lateral brainstem (Treede et al. 2003). This is typically observed in Wallenberg's syndrome (also as known as lateral medullary syndrome), where pain and temperature sensation may be disturbed (Kanda et al. 1996). At the cortical level, lesions in the frontoparietal operculum (secondary somatosensory cortices or insula) were correlated with a reduction of pain sensitivity (Greenspan et al. 1999), thus contributing to abnormality of LEPs. For this reason, LEPs are also useful in patients with cortical lesions (Treede et al. 2003).

3.4.4 Somatosensory Evoked Potentials

Somatosensory evoked potentials (SEPs) are cortical and subcortical responses following the presentation of an electrical stimulation of large myelinated fibers (Aβ) (Cruccu et al. 2008) in the peripheral nerve (Devlin et al. 2006). The evoked tactile sensation is detected by these Aβ fibers, which ascend in the ipsilateral dorsal columns to the brainstem. This tactile information is then projected directly to the contralateral thalamus and terminated in the ventral posterior lateral nucleus (the primary somatosensory cortex, SI) (Kandel et al. 2000) and other somatosensory-related brain regions, which subserves mechanoreception (e.g., tactile recognition and vibration detection) (Treede 2007).

When imposing the electrical stimulation on the tibial nerve (normally using bipolar transcutaneous electrodes where cathode is placed midway between the medial border of the Achilles tendon and the posterior border of the medial malleolus and anode about 3 cm distal), the evoked tibial nerve SEPs can be recorded from the peripheral to cortical levels, including peripheral (N8), lumbar (N22), subcortical (P30), and cortical (P39) (Cruccu et al. 2008). At the cortical level, SEPs are usually recorded at the Cz electrode referred to a frontal midline electrode (Fz or Fpz). While the tibial nerve SEPs at the cortical level comprise a number of waves (P39–N50–P60) that are time-locked to the onset of the stimulus (see Fig. 3.5), only the P39 has definitely clinical value (Cruccu et al. 2008; Mauguiere et al. 1999). In contrast, when delivering the electrical stimuli to the median nerve (the recording anode should be located at the wrist crease and the cathode some 2 cm proximal), the evoked median nerve SEPs are recommended to be recorded from the peripheral (Erb's point) channel (N9), cervical channel (N13), parietal channel (P14 and N20) and frontal channel (P14, P20 and N30), respectively (Cruccu et al. 2008). Specifically, the electrodes at peripheral Erb's point must be placed within the angle formed by the posterior border of the clavicular head of the sternomastoid muscle and the clavicle, 2–3 cm above the clavicle (Erb's point). For cervical electrode, the anterior neck at the level of the glottis (Gl), which is commonly located over the 6th

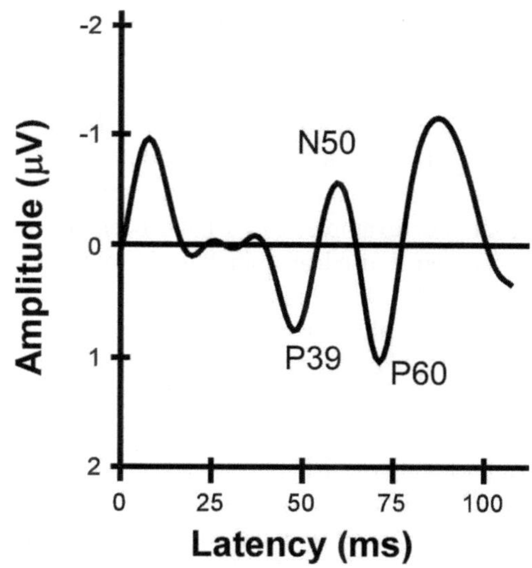

Fig. 3.5 A typical SEP waveform, which showed a cortical complex (P39–N50–P60, tibial nerve stimulus), reflected the brain responses evoked by the electrical somatosensory stimulus

(Cv6) or 7th (Cv7) cervical spinous process, is recommended as the recording reference. Additionally, the parietal scalp electrodes are placed at CP3 and CP4 and are designated as Pc (contralateral to stimulation) and Pi (ipsilateral to stimulation). Finally, the frontal scalp electrode should be located at the site Fz of the 10–20 system, or alternatively at F3–F4 (contralateral to stimulation). The above electrode locations are highly recommended for standard clinical recordings, thus minimizing stimulus artifact and ensuring high-quality data. One point should be emphasized that the later waves in SEPs are less reliable and probably affected by mental condition or cognitive status. Therefore, the obtained results should be interpreted with caution.

Because of the definable latency and amplitude of SEPs for a quantitative comparison throughout a procedure (Minahan 2002), SEPs are commonly used to examine the functional integrity of somatosensory pathways, from the peripheral sensory nerves to the sensory areas of the brain (Blum and Rutkove 2007). The abnormalities of SEP responses provide solid evidence for an impairment of the somatosensory system (Cruccu et al. 2008); thus SEPs have been widely used in both clinical diagnosis (Aminoff et al. 1988; Yiannikas and Vucic 2008; Zeman and Yiannikas 1989) and intraoperative neurophysiological monitoring (Minahan 2002; Luk et al. 2001; Hu et al. 2003; Nuwer 1998; Deletis and Shils 2002; Devlin et al. 2006).

Since electrical stimulation excites Aβ mechanoreceptors, SEPs are usually used to assess the function somatosensory pathways in both the peripheral and central nervous systems (Kraft et al. 1998; Cruccu et al. 2008). Similar with LEPs, the abnormality of SEPs often characterizes as the significant prolongation of inter-peak intervals or absence of obligate responses in serious conditions (Kraft et al. 1998). At the peripheral level, SEPs would be useful in diagnosing the peripheral nerves where severe disease occurs. The abnormality of SEPs has been previously reported in a variety of peripheral nerve disorders, including inflammatory polyradiculoneuropathies (Parry and Aminoff 1987), infectious disorders (Moglia et al. 1991), and toxic neuropathies (Krarup-Hansen et al. 1993). At the spinal cord level, when the disorders affected the ascending pathways of the spinal cord, which conduct tactile information, the abnormality can be recorded by SEPs over the spinal cord. For example, SEPs have been previously used to assess the impairments, which are caused by spinal cord tumors, in the somatosensory pathways (Livshits et al. 1992). In addition, SEPs can help delineating the boundaries of physiological unaffected neural tissue, such as syringomyelia (Anderson et al. 1986), thus assisting the therapeutist in surgery. At the brainstem level, SEPs are abnormal when lesions in the brainstem that involves the lemniscal pathways (Cruccu et al. 2008). At the cortex level, impairing the somatosensory pathways in the brain may cause the abnormality of SEPs. For example, tibial nerve SEPs could be abnormal in patients with multiple sclerosis (Davis et al. 1985; de Pablos and Agirre 2006; Loncarevic et al. 2008), represented as the prolongation of peak latencies and reduction or absence of peak amplitudes of SEPs (Kraft et al. 1998).

Importantly, SEPs are often used in the operating room for intraoperative monitoring in order to prevent neurological damage, to follow up physiological changes, and to locate the central sulcus (Deletis and Shils 2002; Cruccu et al. 2008). The real-time recording of SEPs during the intraoperative monitoring of spinal cord surgery would detect the temporary malfunctioning, which is caused by the prolonged insufficient blood supply to the spinal cord or mechanical compression, at an early stage, thus preventing irreversible spinal cord damage (Rossi et al. 2007; Wiedemayer et al. 2002).

3.4.5 Steady-State Evoked Potentials

Evoked potentials can be generated as a result of physical stimulation not only by an isolated, discrete stimulus that is presented at a relatively slow rate (i.e., a transient response) but also by a train of stimuli presented at a fast, fixed rate. Because the responses to such periodic stimuli can be very stable in amplitude and phase over time, those responses are referred as the steady-state evoked potentials (SSEPs), indicating that the neural system is in a steady state (Regan 1966). In addition, due to the fast stimulation rate, hundreds of trials can be collected in a very short period of time. As a result, steady-state ERPs are commonly used in the diagnosis of sensory disorders.

SSEPs are thought to result from an entrainment or resonance of a population of neurons responding to the stimulus at the frequency of stimulation or from the linear superposition of independent transient responses elicited by the fast repetition of the sensory stimulus (Herrmann 2001; Bohorquez and Ozdamar 2008). In other words, SSEPs are typically identified in the frequency domain as peaks appearing at the frequency of the repeated stimulus and/or at harmonics of that frequency. From this

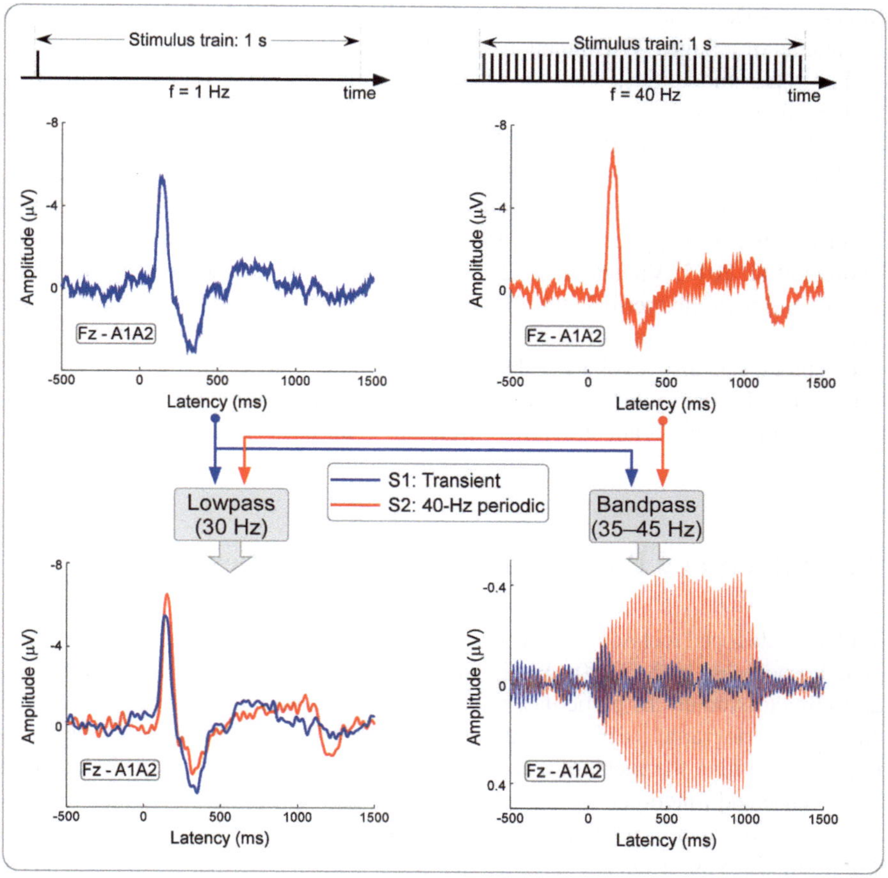

Fig. 3.6 A comparison between transient AEPs in response to a signal tone (top left) and SSAEPs in response to a train of tones presented at a rate of 40 Hz (top right). The waveforms (recorded at electrode Fz) obtained following transient and 40 Hz periodic stimulation are displayed. The SSAEPs, synchronized to 40-Hz auditory stimulation, were clearly presented using a 35–45 Hz bandpass filter (bottom panel). (Adapted with permission from Zhang et al. (2013))

perspective, different SSEP frequencies can be used to tag the different sensory inputs constituting a multimodal stimulus and, thereby, isolate the neural activity related specifically to each stream of sensory input (Nozaradan et al. 2012).

For auditory modality (see Fig. 3.6), a clear spectral peak appearing at 40 Hz was frequently reported in steady-state auditory evoked potentials (SSAEPs) (Bohorquez and Ozdamar 2008; Zhang et al. 2013). The SSAEPs evoked by stimulation of 40 Hz could be largely explained by the linear sum of transient auditory evoked potentials (AEPs), e.g., auditory brainstem response and MLRs. Compared with AEPs that lack frequency specificity, SSAEPs can be recorded in response to amplitude- and/or frequency-modulated tones. Such response follows the envelope of a complex stimulus, such as periodic modulation, or turning on and off, of a tone. Due to its frequency-specific nature, SSAEPs are clinically used to predict hearing sensitivity in patients of all ages (O'Donnell et al. 2013).

For other modalities, such as visual modality, periodic visual stimuli (e.g., flickers, reversing checkerboards, and moving gratings) can elicit steady-state visual evoked potentials (SSVEPs), which is typically generated in the V1 of visual cortex. SSVEPs are useful for both cognitive (e.g., visual attention and working memory) and clinical studies (e.g., aging, depression, anxiety, and schizophrenia). Recently, SSVEPs have been applied to SSVEP-driven brain–computer interface (BCI) system (Norcia et al. 2015; Vialatte et al. 2010). For nociceptive somatosensory modality, when rapid periodic thermal stimuli (i.e., trains of identical infrared laser pulses) are delivered to the dorsum of one's hand or foot, such selective activation of skin nociceptors can also evoke SSEPs, which are maximal at the scalp vertex and symmetrically distributed over both hemispheres, suggesting a radial source originating from midline brain structures (Mouraux et al. 2011; Colon et al. 2012).

3.5 Pitfalls and Promise in EP and ERP Studies

With the advanced technology development of electrical engineering, computer, and signal processing in the 1980s, EEG and ERPs were becoming popular in both basic and clinical studies. In particular, with the advent of techniques such as positron emission tomography (PET, a technique using a computerized radiographic technique to examine the metabolic activity in the brain) (Depresseux 1977; Leenders et al. 1984) and functional magnetic resonance imaging (fMRI, a technique to study brain function mainly related to changes in cerebral flow and cerebral blood oxygenation) (DeYoe et al. 1994; Villringer and Dirnagl 1995), EEG and ERP techniques are becoming more and more popular as an important complement to PET and fMRI techniques, as they can provide excellent temporal resolution. However, when using EEG and ERP techniques in examining the functional changes of the human brain, we need to pay attention to their advantages and limitations.

EEG is a direct measurement of neuronal activity with a sampling frequency as high as dozens of kHz (e.g., short-latency SEPs) (Ozaki et al. 1998; Inoue et al. 2001), thus resulting in extremely high temporal resolution (only a tenth of a millisecond). On the other hand, the spatial resolution of the EEG technique is poor and undefined. The limited spatial resolution is mainly caused by three factors: (1) the skin, skull, and meningeal layers interposed between the brain and the recording electrodes distort and exert a spatial low-pass filtering on neuronal electrical currents (volume conduction) (Nunez and Srinivasan 2006); (2) the recorded scalp signals have a spatial resolution in the order of centimeters (even using 128 channels EEG); and (3) the ill-posed property of the inverse problem when estimating the EEG/ERP sources from scalp recordings (Schroeder et al. 1995). For all these reasons, the discrimination between distinct but spatially neighboring neural sources is difficult, even not impossible.

Comparing to other neuroimaging techniques, which are used to investigate the brain function (e.g., PET and fMRI), EEG is much cheaper, since the equipment is relatively cheaper and the disposable supplies required for conducting an experiment are less. Moreover, in contrast to other electrophysiological techniques, EEG technique is noninvasive. For this reason, EEG recordings can be collected from subjects easily and repeatedly. Please note that both microelectrode recording (MER, a technique to record the electrophysiological activity, i.e., action potentials, from single neuron or several neurons) (Hua et al. 2000) and electrocorticography (ECoG, a technique to record the brain electrophysiological activity, i.e., postsynaptic potentials, by placing electrodes under the surface of the scalp and skull directly on the brain surface) (Kuruvilla and Flink 2003) are invasive.

As mentioned earlier, a large amount of neural activities cannot be measured at the scalp. If the generated electrical fields are "close field," thus forming tangential dipoles, EEG recorded at the scalp is not sensitive to these dipole fields. Even if those neurons generate an "open field" (dipole that are oriented perpendicular to the scalp electrodes), electrical activity may not be recorded by scalp EEG electrodes as neurons may be insufficiently synchronous both in time and in orientation. For example, the orientations of neurons in the thalamus are arranged in such a way that no electrical field can be detected outside itself (Nunez and Srinivasan 2006). Electrical activity in sulci, where opposite dipoles cancel out with each other because of the opposing cortical surfaces, cannot be detected using EEG technique in most cases. The resultant sensitivity contributes to the truth that some important neural activities cannot be detected using EEG technique. However, it also leads to the simplification of EEG recordings that can be easily analyzed and evaluated, which can be considered as an advantage of EEG technique per se.

References

Agostino R, Cruccu G, Romaniello A, Innocenti P, Inghilleri M, Manfredi M. Dysfunction of small myelinated afferents in diabetic polyneuropathy, as assessed by laser evoked potentials. Clin Neurophysiol. 2000;111:270–6.

Aminoff MJ, Olney RK, Parry GJ, Raskin NH. Relative utility of different electrophysiologic techniques in the evaluation of brachial plexopathies. Neurology. 1988;38:546–50.

Anderson NE, Frith RW, Synek VM. Somatosensory evoked potentials in syringomyelia. J Neurol Neurosurg Psychiatry. 1986;49:1407–10.

Begleiter H, Porjesz B, Gross MM. Cortical evoked potentials and psychopathology. A critical review. Arch Gen Psychiatry. 1967;17:755–8.

Blum AS, Rutkove SB. The clinical neurophysiology primer. Totowa: Humana Press; 2007.

Bohorquez J, Ozdamar O. Generation of the 40-Hz auditory steady-state response (ASSR) explained using convolution. Clin Neurophysiol. 2008;119:2598–607.

Boutros NN, Trautner P, Korzyukov O, Grunwald T, Burroughs S, Elger CE, Kurthen M, Rosburg T. Mid-latency auditory-evoked responses and sensory gating in focal epilepsy: a preliminary exploration. J Neuropsychiatry Clin Neurosci. 2006;18:409–16.

Bromm B, Treede RD. Nerve fibre discharges, cerebral potentials and sensations induced by CO2 laser stimulation. Hum Neurobiol. 1984;3:33–40.

Bromm B, Treede RD. Laser-evoked cerebral potentials in the assessment of cutaneous pain sensitivity in normal subjects and patients. Rev Neurol (Paris). 1991;147:625–43.

Bromm B, Frieling A, Lankers J. Laser-evoked brain potentials in patients with dissociated loss of pain and temperature sensibility. Electroencephalogr Clin Neurophysiol. 1991;80:284–91.

Callaway E, Tueting P, Koslow SH, National Institute of Mental Health (U.S.). Clinical Research Branch. Event-related brain potentials in man. Academic: New York; 1978.

Carmon A, Mor J, Goldberg J. Evoked cerebral responses to noxious thermal stimuli in humans. Exp Brain Res. 1976;25:103–7.

Carmon A, Friedman Y, Coger R, Kenton B. Single trial analysis of evoked potentials to noxious thermal stimulation in man. Pain. 1980;8:21–32.

Colon E, Legrain V, Mouraux A. Steady-state evoked potentials to study the processing of tactile and nociceptive somatosensory input in the human brain. Neurophysiol Clin. 2012;42:315–23.

Cruccu G, Aminoff MJ, Curio G, Guerit JM, Kakigi R, Mauguiere F, Rossini PM, Treede RD, Garcia-Larrea L. Recommendations for the clinical use of somatosensory-evoked potentials. Clin Neurophysiol. 2008;119:1705–19.

Davis SL, Aminoff MJ, Panitch HS. Clinical correlations of serial somatosensory evoked potentials in multiple sclerosis. Neurology. 1985;35:359–65.

Dawson GD. A summation technique for detecting small signals in a large irregular background. J Physiol. 1951;115:2p–3p.

Dawson GD. A summation technique for the detection of small evoked potentials. Electroencephalogr Clin Neurophysiol. 1954;6:65–84.

de Pablos C, Agirre Z. Trigeminal somatosensory evoked potentials in multiple sclerosis: a case report. Clin EEG Neurosci. 2006;37:243–6.

Deletis V, Sala F. Intraoperative neurophysiological monitoring of the spinal cord during spinal cord and spine surgery: a review focus on the corticospinal tracts. Clin Neurophysiol. 2008;119:248–64.

Deletis V, Shils JL. Neurophysiology in neurosurgery: a modern intraoperative approach. San Diego: Academic; 2002.

Depresseux JC. The positron emission tomography and its applications. J Belg Radiol. 1977;60:483–500.

Devlin VJ, Anderson PA, Schwartz DM, Vaughan R. Intraoperative neurophysiologic monitoring: focus on cervical myelopathy and related issues. Spine J. 2006;6:212S–24S.

DeYoe EA, Bandettini P, Neitz J, Miller D, Winans P. Functional magnetic resonance imaging (FMRI) of the human brain. J Neurosci Methods. 1994;54:171–87.

Di Russo F, Martinez A, Sereno MI, Pitzalis S, Hillyard SA. Cortical sources of the early components of the visual evoked potential. Hum Brain Mapp. 2002;15:95–111.

Duncan CC, Barry RJ, Connolly JF, Fischer C, Michie PT, Naatanen R, Polich J, Reinvang I, Van Petten C. Event-related potentials in clinical research: guidelines for eliciting, recording, and quantifying mismatch negativity, P300, and N400. Clin Neurophysiol. 2009;120:1883–908.

Ellrich J, Jung K, Ristic D, Yekta SS. Laser-evoked cortical potentials in cluster headache. Cephalalgia. 2007;27:510–8.

Garcia-Larrea L, Frot M, Valeriani M. Brain generators of laser-evoked potentials: from dipoles to functional significance. Neurophysiol Clin. 2003;33:279–92.

Garnsey SM. Event-related brain potentials in the study of language. Hove: L. Erlbaum; 1993.

Goldie WD, Chiappa KH, Young RR, Brooks EB. Brainstem auditory and short-latency somatosensory evoked responses in brain death. Neurology. 1981;31:248–8.

Gonzalez AA, Jeyanandarajan D, Hansen C, Zada G, Hsieh PC. Intraoperative neurophysiological monitoring during spine surgery: a review. Neurosurg Focus. 2009;27:E6.

Grech R, Cassar T, Muscat J, Camilleri KP, Fabri SG, Zervakis M, Xanthopoulos P, Sakkalis V, Vanrumste B. Review on solving the inverse problem in EEG source analysis. J Neuroeng Rehabil. 2008;5:25.

Greenspan JD, Lee RR, Lenz FA. Pain sensitivity alterations as a function of lesion location in the parasylvian cortex. Pain. 1999;81:273–82.

Guerreiro CA, Ehrenberg BL. Brainstem auditory evoked response: application in neurology. Arq Neuropsiquiatr. 1982;40:21–8.

Herrmann CS. Human EEG responses to 1-100Hz flicker: resonance phenomena in visual cortex and their potential correlation to cognitive phenomena. Exp Brain Res. 2001;137:346–53.

Hillyard SA, Vogel EK, Luck SJ. Sensory gain control (amplification) as a mechanism of selective attention: electrophysiological and neuroimaging evidence. Philos Trans R Soc Biol Sci. 1998;353:1257–70.

Hu Y, Luk KD, Lu WW, Leong JC. Application of time-frequency analysis to somatosensory evoked potential for intraoperative spinal cord monitoring. J Neurol Neurosurg Psychiatry. 2003;74:82–7.

Hu L, Mouraux A, Hu Y, Iannetti GD. A novel approach for enhancing the signal-to-noise ratio and detecting automatically event-related potentials (ERPs) in single trials. NeuroImage. 2010;50:99–111.

Hu L, Valentini E, Zhang ZG, Liang M, Iannetti GD. The primary somatosensory cortex contributes to the latest part of the cortical response elicited by nociceptive somatosensory stimuli in humans. NeuroImage. 2014;84:383–93.

Hua SE, Garonzik IM, Lee JI, Lenz FA. Microelectrode studies of normal organization and plasticity of human somatosensory thalamus. J Clin Neurophysiol. 2000;17:559–74.

Hughes JR. EEG in clinical practice. Boston: Butterworth-Heinemann; 1994.

Iannetti GD, Truini A, Galeotti F, Romaniello A, Manfredi M, Cruccu G. Usefulness of dorsal laser evoked potentials in patients with spinal cord damage: report of two cases. J Neurol Neurosurg Psychiatry. 2001;71:792–4.

Iannetti GD, Truini A, Romaniello A, Galeotti F, Rizzo C, Manfredi M, Cruccu G. Evidence of a specific spinal pathway for the sense of warmth in humans. J Neurophysiol. 2003;89:562–70.

Iannetti GD, Zambreanu L, Cruccu G, Tracey I. Operculoinsular cortex encodes pain intensity at the earliest stages of cortical processing as indicated by amplitude of laser-evoked potentials in humans. Neuroscience. 2005;131:199–208.

Iannetti GD, Hughes NP, Lee MC, Mouraux A. Determinants of laser-evoked EEG responses: pain perception or stimulus saliency? J Neurophysiol. 2008;100:815–28.

Inoue K, Hashimoto I, Nakamura S. High-frequency oscillations in human posterior tibial somatosensory evoked potentials are enhanced in patients with Parkinson's disease and multiple system atrophy. Neurosci Lett. 2001;297:89–92.

Jeffreys DA, Axford JG. Source locations of pattern-specific components of human visual evoked potentials. I. Component of striate cortical origin. Exp Brain Res. 1972;16:1–21.

Jervis BW, Nichols MJ, Johnson TE, Allen E, Hudson NR. A fundamental investigation of the composition of auditory evoked potentials. IEEE Trans Biomed Eng. 1983;30:43–50.

Kakigi R, Shibasaki H, Kuroda Y, Neshige R, Endo C, Tabuchi K, Kishikawa T. Pain-related somatosensory evoked potentials in syringomyelia. Brain. 1991a;114(Pt 4):1871–89.

Kakigi R, Shibasaki H, Tanaka K, Ikeda T, Oda K, Endo C, Ikeda A, Neshige R, Kuroda Y, Miyata K, et al. CO2 laser-induced pain-related somatosensory evoked potentials in peripheral neuropathies: correlation between electrophysiological and histopathological findings. Muscle Nerve. 1991b;14:441–50.

Kakigi R, Shibasaki H, Ikeda T, Neshige R, Endo C, Kuroda Y. Pain-related somatosensory evoked potentials following CO2 laser stimulation in peripheral neuropathies. Acta Neurol Scand. 1992;85:347–52.

Kanda M, Mima T, Xu X, Fujiwara N, Shindo K, Nagamine T, Ikeda A, Shibasaki H. Pain-related somatosensory evoked potentials can quantitatively evaluate hypalgesia in Wallenberg's syndrome. Acta Neurol Scand. 1996;94:131–6.

Kandel ER, Schwartz JH, Jessell TM. Principles of neural science. New York: McGraw-Hill Health Professions Division; 2000.

Kraft GH, Aminoff MJ, Baran EM, Litchy WJ, Stolov WC. Somatosensory evoked potentials: clinical uses. AAEM somatosensory evoked potentials subcommittee. American Association of Electrodiagnostic Medicine. Muscle Nerve. 1998;21:252–8.

Krarup-Hansen A, Fugleholm K, Helweg-Larsen S, Hauge EN, Schmalbruch H, Trojaborg W, Krarup C. Examination of distal involvement in cisplatin-induced neuropathy in man. An electrophysiological and histological study with particular reference to touch receptor function. Brain. 1993;116(Pt 5):1017–41.

Kunde V, Treede RD. Topography of middle-latency somatosensory evoked potentials following painful laser stimuli and non-painful electrical stimuli. Electroencephalogr Clin Neurophysiol. 1993;88:280–9.

Kuruvilla A, Flink R. Intraoperative electrocorticography in epilepsy surgery: useful or not? Seizure. 2003;12:577–84.

Lange K. The ups and downs of temporal orienting: a review of auditory temporal orienting studies and a model associating the heterogeneous findings on the auditory N1 with opposite effects of attention and prediction. Front Hum Neurosci. 2013;7:263.

Lee MC, Mouraux A, Iannetti GD. Characterizing the cortical activity through which pain emerges from nociception. J Neurosci. 2009;29:7909–16.

Leenders KL, Gibbs JM, Frackowiak RS, Lammertsma AA, Jones T. Positron emission tomography of the brain: new possibilities for the investigation of human cerebral pathophysiology. Prog Neurobiol. 1984;23:1–38.

Legrain V, Guérit JM, Bruyer R, Plaghki L. Attentional modulation of the nociceptive processing into the human brain: selective spatial attention, probability of stimulus occurrence, and target detection effects on laser evoked potentials. Pain. 2002;99:21–39.

Liang M, Mouraux A, Chan V, Blakemore C, Iannetti GD. Functional characterisation of sensory ERPs using probabilistic ICA: effect of stimulus modality and stimulus location. Clin Neurophysiol. 2010;121:577–87.

Livshits AV, Sokolova AA, Margishvili MG. The dynamics of somatosensory evoked potentials in patients with a spinal cord tumor. Zh Vopr Neirokhir Im N N Burdenko. 1992:19–21.

Loncarevic N, Tiric-Campara M, Mulabegovic N. Somatosensory evoked cerebral potentials (SSEP) in multiple sclerosis. Med Arh. 2008;62:80–1.

Luck S. An introduction to the event-related potential technique. Cambridge: MIT Press; 2005.

Luk KD, Hu Y, Lu WW, Wong YW. Effect of stimulus pulse duration on intraoperative somatosensory evoked potential (SEP) monitoring. J Spinal Disord. 2001;14:247–51.

Makeig S, Westerfield M, Jung TP, Enghoff S, Townsend J, Courchesne E, Sejnowski TJ. Dynamic brain sources of visual evoked responses. Science. 2002;295:690–4.

Mauguiere F, Allison T, Babiloni C, Buchner H, Eisen AA, Goodin DS, Jones SJ, Kakigi R, Matsuoka S, Nuwer M, Rossini PM, Shibasaki H. Somatosensory evoked potentials. The International Federation of Clinical Neurophysiology. Electroencephalogr Clin Neurophysiol Suppl. 1999;52:79–90.

Michel CM, Thut G, Morand S, Khateb A, Pegna AJ, Grave de Peralta R, Gonzalez S, Seeck M, Landis T. Electric source imaging of human brain functions. Brain Res Brain Res Rev. 2001;36:108–18.

Min BK, Busch NA, Debener S, Kranczioch C, Hanslmayr S, Engel AK, Herrmann CS. The best of both worlds: phase-reset of human EEG alpha activity and additive power contribute to ERP generation. Int J Psychophysiol. 2007;65:58–68.

Minahan RE. Intraoperative neuromonitoring. Neurologist. 2002;8:209–26.

Moglia A, Zandrini C, Alfonsi E, Rondanelli EG, Bono G, Nappi G. Neurophysiological markers of central and peripheral involvement of the nervous system in HIV-infection. Clin Electroencephalogr. 1991;22:193–8.

Mouraux A, Iannetti GD. Across-trial averaging of event-related EEG responses and beyond. Magn Reson Imaging. 2008;26:1041–54.

Mouraux A, Iannetti GD. Nociceptive laser-evoked brain potentials do not reflect nociceptive-specific neural activity. J Neurophysiol. 2009;101:3258–69.

Mouraux A, Iannetti GD, Colon E, Nozaradan S, Legrain V, Plaghki L. Nociceptive steady-state evoked potentials elicited by rapid periodic thermal stimulation of cutaneous nociceptors. J Neurosci. 2011;31:6079–87.

Näätänen R, Picton T. The N1 wave of the human electric and magnetic response to sound: a review and an analysis of the component structure. Psychophysiology. 1987;24:375–425.
Näätänen R, Paavilainen P, Rinne T, Alho K. The mismatch negativity (MMN) in basic research of central auditory processing: a review. Clin Neurophysiol. 2007;118:2544–90.
Niedermeyer E, Lopes da Silva FH. Electroencephalography : basic principles, clinical applications, and related fields. Philadelphia: Lippincott Williams & Wilkins; 2005.
Norcia AM, Appelbaum LG, Ales JM, Cottereau BR, Rossion B. The steady-state visual evoked potential in vision research: a review. J Vis. 2015;15:4.
Nozaradan S, Peretz I, Mouraux A. Steady-state evoked potentials as an index of multisensory temporal binding. NeuroImage. 2012;60:21–8.
Nunez PL, Srinivasan R. Electric fields of the brain : the neurophysics of EEG. Oxford: Oxford University Press; 2006.
Nuwer MR. Spinal cord monitoring with somatosensory techniques. J Clin Neurophysiol. 1998;15:183–93.
O'Donnell BF, Vohs JL, Krishnan GP, Rass O, Hetrick WP, Morzorati SL. The auditory steady-state response (ASSR): a translational biomarker for schizophrenia. Clin Neurophysiol. 2013;62:101–12.
Ozaki I, Suzuki C, Yaegashi Y, Baba M, Matsunaga M, Hashimoto I. High frequency oscillations in early cortical somatosensory evoked potentials. Electroencephalogr Clin Neurophysiol. 1998;108:536–42.
Parry GJ, Aminoff MJ. Somatosensory evoked potentials in chronic acquired demyelinating peripheral neuropathy. Neurology. 1987;37:313–6.
Pfurtscheller G, Lopes da Silva FH. Event-related EEG/MEG synchronization and desynchronization: basic principles. Clin Neurophysiol. 1999;110:1842–57.
Regan D. Some characteristics of average steady-state and transient responses evoked by modulated light. Electroencephalogr Clin Neurophysiol. 1966;20:238–48.
Rossi L, Bianchi AM, Merzagora A, Gaggiani A, Cerutti S, Bracchi F. Single trial somatosensory evoked potential extraction with ARX filtering for a combined spinal cord intraoperative neuromonitoring technique. Biomed Eng Online. 2007;6:2.
Roth WT, Horvath TB, Pfefferbaum A, Kopell BS. Event-related potentials in schizophrenics. Electroencephalogr Clin Neurophysiol. 1980;48:127–39.
Rugg MD, Coles MGH. Electrophysiology of mind : event-related brain potentials and cognition. Oxford: Oxford University Press; 1995.
Sauseng P, Klimesch W, Gruber WR, Hanslmayr S, Freunberger R, Doppelmayr M. Are event-related potential components generated by phase resetting of brain oscillations? A critical discussion. Neuroscience. 2007;146:1435–44.
Schmahl C, Greffrath W, Baumgartner U, Schlereth T, Magerl W, Philipsen A, Lieb K, Bohus M, Treede RD. Differential nociceptive deficits in patients with borderline personality disorder and self-injurious behavior: laser-evoked potentials, spatial discrimination of noxious stimuli, and pain ratings. Pain. 2004;110:470–9.
Schroeder CE, Steinschneider M, Javitt DC, Tenke CE, Givre SJ, Mehta AD, Simpson GV, Arezzo JC, Vaughan HG Jr. Localization of ERP generators and identification of underlying neural processes. Electroencephalogr Clin Neurophysiol Suppl. 1995;44:55–75.
Sussman ES. A new view on the MMN and attention debate. J Psychophysiol. 2007;21:164–75.
Treede RD. Das somatosensorische system. In: Robert F, Schmidt LF, editors. Physiologie des menschen. Heidelberg: Springer; 2007.
Treede RD, Lankers J, Frieling A, Zangemeister WH, Kunze K, Bromm B. Cerebral potentials evoked by painful, laser stimuli in patients with syringomyelia. Brain. 1991;114 (Pt 4):1595–607.
Treede RD, Meyer RA, Campbell JN. Myelinated mechanically insensitive afferents from monkey hairy skin: heat-response properties. J Neurophysiol. 1998;80:1082–93.
Treede RD, Lorenz J, Baumgartner U. Clinical usefulness of laser-evoked potentials. Neurophysiol Clin. 2003;33:303–14.

Valentini E, Li H, Bhisma Chakrabarti YH, Aglioti SM, Iannetti GD. The primary somatosensory cortex largely contributes to the early part of the cortical response elicited by nociceptive stimuli. NeuroImage. 2012;59:1571–81.

Valeriani M, Le Pera D, Tonali P. Characterizing somatosensory evoked potential sources with dipole models: advantages and limitations. Muscle Nerve. 2001;24:325–39.

Vialatte FB, Maurice M, Dauwels J, Cichocki A. Steady-state visually evoked potentials: focus on essential paradigms and future perspectives. Prog Neurobiol. 2010;90:418–38.

Villringer A, Dirnagl U. Coupling of brain activity and cerebral blood flow: basis of functional neuroimaging. Cerebrovasc Brain Metab Rev. 1995;7:240–76.

Vogel EK, Luck SJ. The visual N1 component as an index of a discrimination process. Psychophysiology. 2000;37:190–203.

Wiedemayer H, Fauser B, Sandalcioglu IE, Schafer H, Stolke D. The impact of neurophysiological intraoperative monitoring on surgical decisions: a critical analysis of 423 cases. J Neurosurg. 2002;96:255–62.

Woldorff MG, Gallen CC, Hampson SA, Hillyard SA, Pantev C, Sobel D, Bloom FE. Modulation of early sensory processing in human auditory cortex during auditory selective attention. Proc Natl Acad Sci. 1993;90:8722–6.

Yabe H, Tervaniemi M, Reinikainen K. Temporal window of integration revealed by MMN to sound omission. Neuroreport. 1997;8:1971–4.

Yiannikas C, Vucic S. Utility of somatosensory evoked potentials in chronic acquired demyelinating neuropathy. Muscle Nerve. 2008;38:1447–54.

Zeman BD, Yiannikas C. Functional prognosis in stroke: use of somatosensory evoked potentials. J Neurol Neurosurg Psychiatry. 1989;52:242–7.

Zhang L, Peng W, Zhang Z, Hu L. Distinct features of auditory steady-state responses as compared to transient event-related potentials. PLoS One. 2013;8:e69164.

Chapter 4
ERP Experimental Design

Ruolei Gu

Abstract In this chapter, I describe the basic principles of designing an event-related potential (ERP) experiment for psychology research and the rationales behind these principles. I also explain the challenges that researchers often encounter when trying to control potential confounding factors and keeping focus on the psychological processes of interest. I provide general suggestions to resolve these problems according to the literature together with my personal experience. Finally, I introduce some well-known experimental tasks that have been proven to reliably elicit specific ERP components. Researchers may consider developing their studies based on the classic paradigms.

Keywords Event-related potential · Experimental design · Experimental psychology · Cognitive process · Donders's subtractive method

This chapter mainly concerns about the experimental design issues when applying the event-related potential (ERP) method to investigate psychological processes. Consequently, the opinions proposed in this chapter are heavily influenced by experimental psychology, though I believe that they could also help ERP practices in other areas. Pioneer researchers such as Hallowell Davis and Pauline Davis mainly focused on using the ERPs as physiological indexes of sensory input (e.g., Davis 1939). To my knowledge, it was not until the mid-1960s that the ERP could be successfully employed to reveal neural correlates of human cognition (Sutton et al. 1965; Walter et al. 1964). Since then, ERP has been utilized to examine a wide range of topics (see the final part of this chapter). Given the popularity of ERP as a means

R. Gu (✉)
CAS Key Laboratory of Behavioral Science, Institute of Psychology, Chinese Academy of Sciences, Beijing, China

Department of Psychology, University of Chinese Academy of Sciences, Beijing, China
e-mail: gurl@psych.ac.cn

L. Hu, Z. Zhang (eds.), *EEG Signal Processing and Feature Extraction*,
https://doi.org/10.1007/978-981-13-9113-2_4

to probe neural correlates of cognition, it is necessary to clarify the fundamental principles of ERP experimental design. Otherwise, the theoretical significance of ERP data for psychology research would be in doubt. Several guidelines and committee reports provided by distinguished ERP experts have discussed these principles from different aspects (e.g., Duncan et al. 2009; Keil et al. 2014; Pernet et al. 2018; Picton et al. 2000), which have inspired the writing of this chapter.

The first question I suggest that a researcher should ask him/herself is whether ERP is suitable for his/her research interests. Answering this question requires the understanding of the pros and cons of the ERP method. As the readers have been aware of, the main advantage of ERP is the exquisite temporal resolution compared to most neuroimaging techniques. Consequently, the ERP is particularly useful when researchers try to untangle multiple cognitive processes and their corresponding neural signals in the time domain. For instance, researchers have debated on whether the single- or dual-process model is more suitable to describe the mechanism of recognition memory. Using ERP method, it has been found that parietal and mid-frontal ERP correlates (which indicate the processing of recollection and familiarity, respectively) emerge sequentially within hundreds of milliseconds, thus supporting the dual-process model (see Rugg and Curran 2007, for a review). In my opinion, an experienced ERP researcher would be particularly sensitive to research topics that involve temporally adjacent cognitive processes.

Meanwhile, the coarse spatial resolution means that the application of ERP could hardly be illuminating when researchers are interested in the activation pattern of a specific brain region. To overcome this disadvantage, source analysis has been rapidly developed, but its accuracy is inherently restricted considering that EEG source localization is an ill-posed problem (Grech et al. 2008). Furthermore, it is generally more difficult (though possible) to detect the activity of subcortical structures beneath the cerebral cortex, given that the source contribution of a dipole decreases as a function of its distance to the scalp surface. For an extreme example, brain stem potentials usually require thousands of trials for averaging (Cohen et al. 2011). One should be cautious about this problem when he/she uses the ERP to investigate human emotion and motivation, as they are closely associated with subcortical limbic areas such as the amygdala and basal ganglia. Indeed, a large body of ERP findings from emotion research indicates the cortical regulation of limbic activity, rather than the emotional response per se (Hajcak et al. 2010).

When a researcher is confident that the ERP is his/her optimal choice, there are three questions as follows: (1) Is the experimental design necessary and sufficient to elicit the psychological processes of interest? (2) Could the experimental design be reconciled with the recording and publication standards of the ERP? (3) Have potential confounding factors been adequately considered and controlled? Please see below for point-by-point interpretations of these questions.

4.1 Experimental Design and Cognitive Processes

A basic assumption of cognitive psychology is that information processing relies on internal (mental) representations (Münte et al. 2000). The construction, storage, retrieval, and manipulation of these representations are called cognitive processes. Based on the above understanding, a basic assumption of cognitive neuroscience is that cognitive processes could be captured at the neural level. From this perspective, ERP components are regarded as neural manifestations of specific cognitive processes, with their characteristics (particularly amplitude and latency) reflecting different aspects of those processes. For instance, P3 latency is widely considered as a biomarker of stimulus- or response-processing time (Verleger 1997).

As Picton et al. (2000) pointed out: "the general working hypothesis is that different cognitive processes are associated with different ERPs." However, it is often difficult to build one-on-one mappings between ERP components and cognitive processes. For example, the N2 component has been suggested to reflect numerous cognitive processes including error detection, conflict detection, response inhibition, and action monitoring. Although researchers argue that the N2 could be divided into subcomponents according to its scalp distribution, latency, and response pattern, each subcomponent has still been linked with various cognitive processes (for a review, see Folstein and Van Petten 2008). What makes it more complex is that the processing of a stimulus (even a very simple one) employs multiple cognitive processes in most cases, thus hindering the identification of the cognitive implication of ERP variations in response to the stimulus. Therefore, the emergence of a specific ERP component in grand-average waveforms should not be regarded as an evidence that the cognitive process of interest has been indeed elicited (i.e., the problem of reverse inference; see Poldrack 2006, 2008). Rather, the relationship between an ERP component and its underlying cognitive process could only be guaranteed by rigorous experimental designs.

In general, laboratory experiments manipulate psychological processes by modulating stimuli and task instructions. However, even researchers themselves may not understand accurately the cognitive mechanisms involved in the processing of specific experimental stimuli. Taking the relationship between facial stimuli and N170 as an example, the N170 is prominent in response to facial expression (especially human faces) but usually much smaller or even absent for non-face visual objects (Bentin et al. 1996). Accordingly, it seems reasonable to infer that the N170 reflects the process of face recognition. Nevertheless, whether the N170 is face-specific has actually been in hot debate (Rossion and Jacques 2008). Many researchers point out that faces could be different from other object categories in various perceptual features including symmetry, visual complexity, familiarity, and salience. It is thus possible that the processing of these features account for (at least in part) the N170 difference between faces and non-face stimuli. Relevantly, regarding the high visual similarity between members of the face category, it is also possible that a larger N170 is elicited by faces because of the smaller interstimulus perceptual variance within this category (Thierry et al. 2007). From this perspective, the N170 might not be selectively face-sensitive. Although these challenges seem to be annoying, researchers could strengthen the reliability of their interpretations of ERP results with appropriate experimental designs.

4.1.1 Stimulus Properties

Basically, experimental stimuli would better be in a perceptually simple and clear form. Stimulus features (including its content, intensity, duration, and location) associated with the research interest may be highlighted in some way; otherwise, they should be appropriately controlled in the experiment. Here, "control" generally means keep constant for each item within the same condition and across conditions. When it is practically impossible to "keep constant" certain parameter, researchers should try their best to reduce its variance within condition (unless it is purposely varied, e.g., a range of contrast levels to avoid predictability) and check whether its difference between conditions is statistically significant.

While visual stimuli lasting for tens of milliseconds are sufficient to induce detectable fluctuations in ERP waveforms, researchers who are interested in conscious and deliberate stages of stimulus processing usually use longer durations (e.g., 500–1500 ms). Visual stimuli (including pictures and words) are better placed around the center of computer screen to minimize eye movement artifacts (unless assessing the effects of peripheral visual stimuli). For picture stimuli, their obvious properties such as size, luminance, and contrast should be controlled if not manipulated; for word stimuli, these properties include the number of letters, phonemes and syllables, word frequency, and possibly semantic structure. Finally, for auditory stimuli, these properties include frequency, intensity (measured in decibels with the reference level), pitch, and discrimination acuity (see Pernet et al. 2018, for details). Basically, researchers are encouraged to select experimental stimuli from well-established databases (e.g., pictures from the International Affective Picture System [IAPS]) whenever possible, unless these databases are unsuitable for the research aim.

While the above paragraph focuses on the physical properties of a stimulus, its psychological properties might be modulated by task instruction. The instruction should be unambiguous, easy to follow, and fit the cognitive level and education level of participants. Stimulus meaning and response requirement should be explained in detail, particularly those associated with the research interest. Experimenters should give further instructions if participants are unfamiliar with the device for stimulus presentation or behavioral output (e.g., keyboard, button box, game pad). Multimedia presentation of task instruction might be helpful, especially for children. Participants could be provided with a small number of practice trials after task instruction, and a performance threshold could be set before they are allowed to begin the formal experiment. They could also be questioned to ensure that they understand the task as researchers expected. This point is particularly important when the experiment involves a cover story. For instance, participants may be told that they would finish the task together with other people, but the feedback "sent from another participant" is actually produced by a computer randomly (e.g., Luo et al. 2014). In this case, researchers should check (possibly after the experiment) if participants believe in the cover story.

The strict control of stimulus parameters and task instruction is to avoid confounds and misinterpretations of ERP components. For this purpose, researchers should make sure that "a given experimental effect has only a single possible cause" (Luck 2004). Here, the "cause" refers to the experimental manipulation that is expected to elicit the cognitive process of interest. When there is any uncontrolled physical or psychological stimulus difference between conditions aside from the experimental manipulation, it might induce additional cognitive processes that also manifest on ERP signals. It is inappropriate to assume that a small difference is too trivial to affect ERP results (Luck 2004). According to Steven Hillyard, the best way to avoid physical stimulus confounds is to use exactly the same physical stimuli across different conditions. This suggestion was named as "Hillyard principle" by Steven Luck, who hung it on his laboratory wall as a slogan (Luck 2014). An experiment that follows the Hillyard principle usually manipulates the meaning of stimuli via task instruction. For instance, participants might be provided with an asterisk ("*") as outcome feedback during a gambling task. In one condition, participants are told that the asterisk indicates that they win (or lose) in the current trial; in another condition, the asterisk indicates that the real outcome (win/loss) is unknowable to participants (i.e., ambiguous outcome). Using this task design, researchers could assume that the comparison of ERPs elicited by the asterisk between the two conditions reveals the difference in cognitive processing between unambiguous and ambiguous information. However, the Hillyard principle could only be applied in limited cases. In many studies which are theoretically important, it is necessary to use different stimuli across conditions. For instance, researchers may be interested in the processing of different pictures varied in emotional valence (positive vs. negative). In this case, physical stimulus confounds might be unavoidable (although one can still normalize some of the basic physical properties of pictures), but we may demonstrate that they are actually not responsible for the between-condition difference of ERP measures with additional data analyses or experiments (Luck 2004). The underlying logic of controlling stimulus properties is further elucidated below.

4.1.2 Condition Comparison

The processing of an individual stimulus inevitably involves multiple processes. Simply asking participants to watch a picture of a human face may elicit sensory, perceptual, attentional, emotional, social cognitive, and memory processes and so on. To make things more complicated, these processes are often overlapping in the time domain; thus an ERP component might reflect the influence of more than one process. Therefore, the ERP data in a single condition is hardly informative from the perspective of cognitive psychology.

To address this issue, cognitive psychologists have relied on Donders's subtractive method (named after the Dutch scientist Franciscus C. Donders) for a long time. As all mental operations take time, Donders proposed that differences in human reaction time could be used to infer differences in mental processing (Posner 2005). A modern version of Donderian method could be expressed in the following way: "one can create two conditions such that the conditions differ only in the process of interest and are equivalent with respect to all other processes" (Coles and Rugg 1995). Using Donders's subtractive method, researchers would be able to derive the neural correlates of a specific cognitive process from a deliberate experimental comparison. Suppose that the processing of a certain kind of stimuli involves the cognitive processes A, B, and C. When researchers focus on the process A, they may design a control condition which (presumably) contains only B and C. Therefore, the ERP differences between the experimental condition and control condition would be considered as the effect of A. This explains why stimulus properties need to be tightly controlled: if there are unplanned property variations between conditions, it is plausible that these conditions differ not only in the process A but also in additional cognitive components that are required for the processing of those properties. If that happens, it would be more difficult to determine the meaning of ERP results.

Another common practice is that all conditions involve the process A, but researchers are interested in comparing different variations of A (e.g., A1, A2). The key point of this method, as a modification to the Donderian approach, is to manipulate the functions of the same process rather than to add or delete different processes (see Stelmach 1982, for details). As mentioned above, an experiment may compare the ERPs elicited by positive (A1) and negative (A2) emotional pictures so as to see if electrophysiological activity is modulated by the processing of emotional valence. This experimental design would be beneficial from another type of control condition that contains A0 (e.g., neutral pictures), in which the level or amount of processing is assumed to be lowest and therefore could be treated as a baseline.

Here, please note that I introduce the subtractive method simply to help illustrate the rationale behind condition comparisons. The idea underlying Donderian logic (i.e., "pure insertion") assumes the relationship between cognitive components is linear (or additive); thus the effect of adding a new component is independent of other components. Nevertheless, the assumption may violate the observations that neural dynamics are nonlinear; when there is expected to be nonnegligible interactions between neural processes, researchers usually turn to factorial designs practically (i.e., taking on all possible combinations of levels across factors; see Friston et al. 1996, for details). This fact, however, does not harm the importance of Donderian method for the understanding of ERP experimental design. For examples of modern-day use of this method, please see Vidal et al. (2011).

Let's assume that we are interested in examining the face specificity of N170. According to Donderian logic, what kind of control condition should we use for our experiment? First, different kinds of stimuli (e.g., houses, vehicles, animals, flowers) have been selected to contrast with faces according to the literature. Apparently, these stimuli are hugely different from faces in many aspects, particularly sensory features. Among them, pictures of cars are frequently used in the control condition because of the high familiarity and complexity comparable to human faces (Rossion and Jacques 2008). However, car stimuli might still not be optimal regarding the issue of interstimulus perceptual variance. Some researchers compared upright and inverted faces, which contain exactly the same amount of sensory information and thus perfectly control all the aspects except the orientation (Itier et al. 2006). Nevertheless, an inverted face could still be easily perceived as a facial stimulus. Thus, any N170 difference between the upright and inverted conditions may not help determine if this component is face-specific. An elegant experiment design makes a comparison between upright and inverted Arcimboldo images: Arcimboldo was an Italian Renaissance artist best known for creating face-like paintings with non-facial objects such as fruits, vegetables, and flowers. Interestingly, the face perception would be much weaker or even disappear for inverted Arcimboldo images, as inversion disrupts holistic processing. Accordingly, comparing upright with inverted Arcimboldo portraits becomes an ideal test of the face specificity of N170 (e.g., Caharel et al. 2013).

Based on the aforementioned methodology, some researchers have taken a step further and suggested constructing difference waves between conditions so as to isolate ERP components that are selectively associated with experimental manipulation (Luck 2004). Indeed, the difference wave analysis has become a dominant approach for the measurement of certain ERP components, such as the feedback-related negativity (FRN; see Proudfit 2015, for a review). The FRN is widely considered to be a neural index of the processing of outcome valence (positive/negative). Based on the current understanding, many FRN researchers create a difference wave by subtracting the ERPs in the positive condition from those in the negative condition for each participant. The resulting difference wave in the same time window of the original FRN is supposed to be "purely" valence-related, since the common components that are insensitive to outcome valence have been removed. However, the difference wave analysis has its own limitations (see Luck 2004, for details). For instance, one might be interested in the modulating effect of a factor A. When this effect is significant on the FRN difference wave, it is possible that the A factor actually affects (1) positive outcome trials, (2) negative outcome trials, or (3) both. To resolve this problem, one would have to return to original waveforms to construct a factorial design.

4.1.3 Behavioral Measure

Another important way of investigating the cognitive implication of ERP components is to use behavioral measures. Here, potential measures include not only button press but also movements of the hands, fingers, eyes, and muscle contractions, as long as they are properly collected and analyzed (Pernet et al. 2018). While these behavioral responses might not be the focus of an ERP experiment, they could help clarify if the cognitive process of interest is elicited or the experimental manipulation is successful. For example, in each trial of the emotional Stroop task, participants are presented with an emotional (e.g., "death") or neutral word (e.g., "soil") in different colors. They are asked to name the color of the words ASAP while ignoring its meaning. However, previous studies have showed that there is an emotional delay in reaction time, that is, participants respond to emotional words more slowly than neutral words. This effect is generally interpreted as an attentional bias toward emotional stimuli (Williams et al. 1996). Therefore, in an ERP experiment that aims to investigate the electrophysiological mechanisms of the emotional Stroop effect, replicating the typical behavioral pattern (i.e., emotional delay) indicates that participants' attentional function has been indeed interfered by emotional processing during the task. Otherwise, even though there are ERP differences between conditions, researchers would meet difficulties to interpret these differences as the emotional influence on attentional function if they fail to get the typical behavioral result. This point is particularly notable when such a classic paradigm is applied. For these reasons, Picton et al. (2000) suggested assessing participants' behavior when possible. It should be noted, however, that the experimental environment of ERP studies is quite different from real-world naturalistic scenarios (Zhang 2018), which could affect behavioral performance in many ways. In addition, even though behavioral data shows that the cognitive process of interest is involved in the experiment, it is still undetermined whether the ERP differences between conditions really resulted from this process. In a word, behavioral evidence from an ERP experiment could be supportive, but not conclusive. Meanwhile, behavioral data involves a motor response component by default, which might interact with other cognitive components in the task.

4.2 Requirements of the ERP Method

As the readers may have learned from other chapters of this book, the ERP method has many specific requirements to ensure that the data are of good quality (see also Picton et al. 2000). The design of an ERP experiment for psychological research needs to follow not only the principles of experimental psychology but also these requirements. As a result, the ERP method is unsuitable for some research topics practically. Also, researchers might have to reluctantly modify some critical aspects of their original tasks derived from behavioral studies. Nonetheless, fulfilling the requirements of the ERP method is the premise of a successful and theoretically meaningful experiment. Below I summarize the most important requirements which should be taken into account for ERP experimental design.

4.2.1 Trial Number

The reliable delineation of an ERP component needs a sufficient number of repetitions of the stimuli (or responses) which elicit that component (the specific number varis across different ERP components; e.g., Cohen and Polich 1997; Marco-Pallares et al. 2011). For this reason, while one-shot game paradigms have been widely used in behavioral economics, only their iterated versions could be applied for ERP experiments. This point may fundamentally change the nature of the research question. For example, the classic Prisoner's Dilemma Game (PDG) asks two players to independently choose between "cooperate" and "defect," which jointly determined the decision outcome. Generally, people are more prone to defect their partners in the one-shot PDG; in contrast, they would be more prone to cooperate in the iterated PDG, since they have to deal with the same partner for many rounds (Raihani and Bshary 2011). In this case, although increasing the number of trials in each condition is necessary, there is a potential risk that researchers would deviate from their primary goal.

4.2.2 Stimulus Probability

Perhaps the most well-known and most frequently investigated ERP phenomenon is the "oddball" effect, that is, an improbable target stimulus elicits larger ERP amplitudes than standard stimuli (see also below). Traditionally, this oddball effect is often associated with the P3 component, but other components such as the N2 and FRN are also affected (Patel and Azzam 2005; San Martín 2012). Researchers suggest keeping the probabilities of stimulus/response categories constant within and across conditions, in order to prevent confounding the oddball effect with the experimental effects of interest (Picton et al. 2000). In some paradigms, however, stimulus probability might be inevitably modulated by other experimental variables. For example, participants may be asked to finish a reward learning task, in which they receive outcome feedback (positive/negative) of their decision in each trial. Although participants do not know the winning rule initially, they could make more optimal choices through trial-and-error learning. As the task moves on, the FRN elicited by negative feedback may become larger over time. Researchers may like to explain this result as a learning effect, but it can actually be interpreted more parsimoniously in terms of the oddball effect: as participants are more familiar with the winning rule, they would be less likely to receive negative feedback, which means the probability of negative feedback would decrease in the later stages of the task. Previous studies have tried various methods to control the influence of feedback probability during learning (e.g., Cohen and Ranganath 2007; Walsh and Anderson 2011).

4.2.3 Time-Locking

An ERP experiment requires that there is a time-locked relationship between the event and (1) its ERP effects and (2) its associated cognitive processes. First, the ERP waveforms prior to the time point zero of an event is supposed to reflect the default brain state, while those appearing after event presentation is supposed to reflect the impact of that event on electrocortical activity. Second, the cognitive process of interest is supposed to emerge within the same time window of event presentation. Fulfilling these requirements enables us to link the ERPs with the cognitive process being investigated. Here, the first requirement relies on triggering mechanism of the experimental procedure, while the second one relies on experimental procedure itself. In reality, the cognitive process may happen before its associated event. Imaging a simple gambling task in which participants choose between a high-risk and a low-risk option, both of which may result in wins or losses randomly. Using this task, we could investigate the ERPs elicited by outcome feedback (win/loss) and assume that they reflect the processing of outcome information, because participants are unable to predict the real outcome until its presentation. However, we may not assume that the ERPs elicited by gambling option (high−/low-risk) reflect the processing of option information, because the options are unchanged throughout the task. Hence, participants could actually assess options and make up their minds before option presentation (Gu et al. 2018). In this case, even though there are significant ERP results in the option stage, their cognitive meanings are unclear because there is probably no time-locked relationship between (objective) option presentation and (subjective) option assessment. Partly because of this reason, ERP studies on option assessment are scarce compared to those on outcome evaluation.

4.2.4 Time Interval

The intervals between events could be measured as either stimulus-onset asynchrony (SOV, the onset of the preceding stimulus to the onset of the next) or interstimulus interval (ISI, time distance from offset to onset). An ERP experiment requires sufficient intervals between adjacent events within each trial and across trials. Specifically, the ERPs elicited by an event usually need a few hundreds of milliseconds to return to the baseline level; before that time, if a second event appears immediately, then its associated ERP signals would likely be contaminated by the influence of its preceding event. Therefore, if a behavioral paradigm involves the presentations of several events within a very short time window, researchers often need to extend the intervals between these events when applying it for ERP research (for exceptions, see the steady-state visual evoked brain potential [SSVEP] technique; e.g., Keil et al. 2003). In addition, while fixed intervals generally do not cause problems in behavioral studies, ERP experiments often use varied intervals between the event being analyzed and its preceding event, which could be accomplished by inserting temporal jitters (i.e., time-varying intervals). This is because the expectation of a coming event generates anticipatory slow waves including the stimulus preceding negativity (SPN) and contingent negative variation (CNV) (Bocker et al. 1994). When the interval between two events A and B is fixed, participants would be able to anticipate the precise time of the onset of B after the presentation of A. In this case, a clear SPN/CNV would be observed in ERP waveforms and would significantly influence the baseline window of B.

4.2.5 Eye and Body Movement Control

Methods of controlling noncerebral artifacts have been systematically described in other chapters of this book. From the perspective of experimental design, researchers should consider whether their experimental setups and task requirements would generate unnecessary eye or body movements, especially in the time windows of the events being analyzed (Tal and Yuval-Greenberg 2018). For instance, in a gambling task which provides the chosen outcome and alternative outcome simultaneously, if the two outcomes are presented at the two vertical ends of the screen, participants would need to make saccades between these locations, which in turn generates large electroocular artifacts that could have been avoided.

4.3 Confounding Factors

I consider confounding factors as within-subject and between-subject variables that are not the interest of an experiment but have the potential to affect experimental results and therefore should be adequately controlled. The idea of eliminating the potential confounds of some within-subject variables has been introduced in the previous sections. Regarding other within-subject variables, I suggest that the most notable ones are *habituation* and *fatigue*, both of which manifest their effects over time (i.e., sequencing effects). Habituation happens because participants are exposed to the same stimuli for many times; fatigue happens usually when an experiment lasts for more than tens of minutes. Both factors may result in attenuated neural responses to stimuli in the latter half than the first half of the experiment. They might be inevitable for an ERP experiment which requires collecting a large number of trials for each condition. However, we may control their influence between conditions by using repeated measure designs (e.g., counterbalancing), which are the basis of experimental psychology and thus is not elaborated further.

Regarding between-subject variables, this section focuses on individual differences. It has been well-established that individual difference could manifest on the ERPs (Keil et al. 2014) and its influence may be stronger than within-subject factors in certain situations (Luck 2014). Thus, when individual differences are not the target of a study, they should be adequately regulated and reported. In the majority of ERP studies, the controlled individual differences include *age*, *sex*, and *handedness*. Indeed, these variables have been proven to play significant roles in individual ERP patterns. For instance, the amplitudes of the CNV, error-related negativity (ERN), and P3 all vary as a function of age among children between 7 and 17 years and young adults, possibly reflecting the effect of neural development (see Segalowitz and Davies 2004, for a review). Likewise, sex-related differences have been discovered in many ERP components including the N1, N2, P3, and late positive potential (LPP) (e.g., Gardener et al. 2013). In my opinion, however, the importance of the sex factor has not yet been fully appreciated. For one thing, the samples of many studies predominantly consist of male or female participants, but these studies usually assume that their conclusions are suitable for both genders. For another, many studies examined the potential modulating effect of age but not sex in statistical analyses (for the significance of the sex effect, see: Eliot 2011; Schiebinger 2014). Regarding handedness, limited studies have been devoted to explore the differences in ERPs between left- and right-handed persons (e.g., Beratis et al. 2009; Coulson and Lovett 2004; Nowicka et al. 2006). To my knowledge, most studies have exclusively employed right-handers as participants, which may help control individual difference in response bias but also have restricted researchers from understanding the potential role of handedness.

Aside from these variables, it is also necessary to control the *education level* of participants. Many studies have no problem with this issue, since they mostly recruit university undergraduates and graduates as participants. However, when a study contains elderly citizens or clinical samples, things might be more complicated. As pointed out by Picton et al. (2000), the importance of education level lies in the fact that it "is a reliable indicator of general cognitive abilities." Relevantly, the *level of specific cognitive abilities* should be assessed, controlled, and provided in the report when they are critical for the experimental task (Picton et al. 2000). In many tasks which are designed to investigate individual preference, when participants do not have sufficient cognitive abilities to meet task demands, their behavioral performance may not be explained in terms of preference. For instance, some studies using risk decision-making paradigms have found that elderly people are more risk-avoidant than young adults. However, one might suggest that elderly participants actually have deficits in reward learning abilities and working memory capacity, which have restricted them from understanding the potential benefits associated with choosing risky options. Finally, *socioeconomic status* also warrants careful attention, such as in social psychological research about self-esteem, fairness judgment, or social hierarchy.

Collectively, a "typical" sample in a "typical" ERP study usually consists of university students, age between 18 and 30, right-handed, having normal IQ, reading ability, and motor skills. Controlling participant characteristics in this way is expected to increase the level of homogeneity between participants within the same sample. In addition, they should have no history of medical (e.g., heart diseases), psychiatric (e.g., schizophrenia), or neurological (e.g., epilepsy) illness; also, they should be free of regular use of any substance affecting the central nervous system, especially on the experiment day. For the limitations of this participant screening strategy, please see the review from Henrich et al. (2010) for an example.

When the experimental design involves one or more between-subject factors (e.g., comparing between a clinical sample and a healthy control sample in neurological and psychiatric research), the aforementioned demographic variables should be matched between groups. The reason behind is similar with the Donderian approach of controlling irrelevant cognitive processes on the within-subject level. Keil et al. (2014) provide two helpful suggestions on this issue: first, researchers should be aware of the unexpected consequences of matching the selected characteristics of participants, since it might under the risk of systematically mismatch groups on other characteristics; second, a single control group might not be able to address all relevant characteristics, and multiple comparison groups might be needed in certain cases. Finally, when the between-subject factor is about a clinical issue, researchers should set up unambiguous, detailed inclusion and exclusion criteria of clinical and nonclinical groups. Readers are referred to the *Diagnostic and Statistical Manual of the American Psychiatric Association* (2013) for criteria of most psychiatric disorders, and a consultation or collaboration with clinical experts are highly recommended. While the criteria should be rigid, the consequence of applying them should also be fully evaluated. For instance, excluding participants with lifetime neurological or psychiatric illness may lead to rejecting up to 50% of elderly individuals, thus limiting the general implication of a study on aging (Pernet et al. 2018).

4.4 Classic Experimental Designs and Corresponding ERP Indexes

It turns out that designing a novel experimental task to examine the cognitive processes of interest and appropriately control confounding factors is a complicated issue. For this concern, Luck (2004) recommended using standard experimental manipulations that have been well studied in the literature. This strategy would be more likely to get reliable mappings between ERP components and cognitive processes than new paradigms (Picton et al. 2000). While researchers should not hesitate to develop new paradigms for their studies, they could also consider slightly modifying classic paradigms to fit specific research purposes, as long as the basic characteristics of those paradigms are unaffected. In this section, I introduce several classic experimental tasks that have been frequently employed in previous ERP studies. Mostly because of space limitation, I focus on those representative tasks that have been closely associated with a single ERP index, each of which is introduced following the description of its corresponding task (Fig. 4.1).

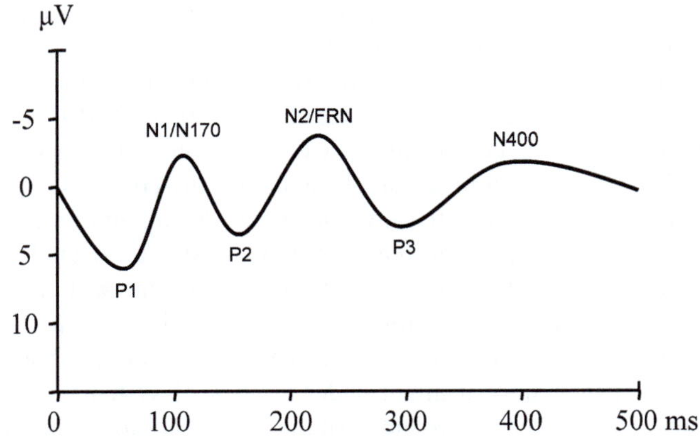

Fig. 4.1 A schematic of several important ERP components (negative voltages upward). Practically, different ERP components vary in their scalp distributions and might not be observable from the same waveform. Also note that the ERN temporally overlaps with the P1; thus they could not be jointly depicted in this waveform

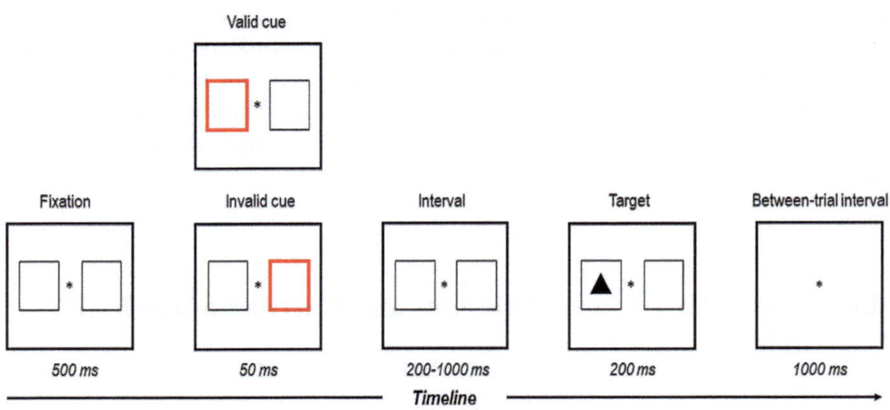

Fig. 4.2 An example of cue-target paradigm. (Inspired by Tian et al. 2011)

4.4.1 Cue-Target Paradigm

In the field of attention research, the cue-target paradigm (Fig. 4.2) requires participants to respond quickly to the position of a target. Prior to the target presentation, there is a preceding cue that carries information regarding the subsequent target stimuli (Neyedli and Welsh 2012). The cue could be valid (correctly indicates target location), invalid (provides erroneous information), or neutral (does not contain any information). Both the cue and the target could be research objects of various topics such as attention orientation, attention shift, and inhibition of return. When combining this paradigm with the ERP method, the P1 and N1 elicited by target presentation may indicate the attentional bias under the influence of cue information. Generally speaking, the core of this paradigm is that there is a one-on-one mapping between cue and target, which could be extended to other research fields. For instance, in the monetary incentive delay (MID) task, the cue indicates the amount of reward associated with successfully responding (Knutson et al. 2001; Schultz et al. 1998). In this case, cue presentation affects behavioral response by modulating motivation level rather than attention.

P1 & N1 The P1 and N1 are two important ERP indexes associated with sensory processing and attention orientation (Luck et al. 2000). In response to the presentation of visual stimuli, the P1 (100–130 ms poststimulus) and N1 (150–200) appear following the C1 (40–100 ms). However, since the C1 could only be observed in certain paradigms, the P1 is probably the first clear ERP peak that one is able to recognize in the average waveforms. Please be aware that their scalp distribution and latency vary as a function of stimulus modalities (visual vs. auditory; see Luck 2014). These two components, which are in the same frequency range (alpha band), are referred to as the "P1-N1 complex" by some researchers (Woodman 2010). From the perspective of cognitive psychology, both of them are mainly regarded as exogenous components modulated by physical stimulus attributes. Nevertheless, researchers believe that there is a functional dissociation between them: one possibility is that the P1 generally indicates bottom-up attentional processes elicited by incoming stimuli information, while the N1 is more likely to be associated with top-down selective attention toward task-relevant stimuli (Herrmann and Knight 2001). Although the P1 and N1 are supposed to reflect quick, coarse stages of stimulus processing, recent studies suggest that they also play an important role in more complex processes such as deliberate decision-making and social evaluation. For instance, one of my recent studies has revealed that the amplitude of P1 elicited by stage feedback could predict participants' decision (whether or not to take economic risk) in the next stage, indicating that the early attentional stage of stimulus processing participates in subjective risk preference (Gu et al. 2018).

4.4.2 *Face Presentation Paradigm*

Pictures of faces have been widely used for not only studies of face perception but also those of emotional processing, social decision-making, interpersonal relationship, ingroup/outgroup bias, and stereotypes. Many international databases (e.g., Matsumoto and Ekman's Japanese and Caucasian Facial Expressions of Emotion [JACFEE]) provide facial stimuli that have been edited to control irrelevant dimensions. Across different studies, faces could be vertically or horizontally oriented, inverted or contrast-reversal, with spatial frequency filtering or additional visual noises, or missing some inner features. In addition, they could be emotional or neutral, familiar (e.g., celebrity faces) or unfamiliar, in the same ethnic group with participants or not (Rossion and Jacques 2008). Faces could also be presented on an unconscious level (e.g., being backwardly masked: see Etkin et al. 2004). Finally, face recognition could be modulated by task demands, as researchers often assign some tasks (e.g., identifying the facial emotion) to attract participants' attention or for specific research goals. These manipulations are all theoretically interesting for psychology and social science and have been proven to influence ERP patterns (Hinojosa et al. 2015). For the design of appropriate control conditions, please see Sect. 4.1.2 for details.

N170 From the late 1980s, researchers have begun to realize that facial stimuli elicit specific electrophysiological signals compared to other visual stimuli. They first discovered a vertex positive potential (VPP) in the 140–180 ms time window, being more positive-going for faces than other kinds of objects. Later studies have found another negativity at bilateral occipitotemporal sites, which is temporally coincident with the VPP and probably reflects the negative counterpart of the same equivalent dipoles. This component, being more negative-going for faces than other stimuli, reaches its maximum at about 170 ms following stimulus onset and thus is called "N170" (Bentin et al. 1996). While the N170 and VPP are considered as flip sides of the same brain processes, the former one has received more attention from academia. This is because the neural origin of both components is supposed to be localized in the occipitotemporal cortex, which matches well with the scalp distribution of the N170. As mentioned above, many researchers believe that the N170 generally reflects a face-specific processing mechanism in the brain, but the debate still continues in many aspects (for details, see Hinojosa et al. 2015; Rossion and Jacques 2008).

4.4.2.1 Stimulus Response Compatibility (SRC) Paradigms

This concept refers to a large family of tasks including the Stroop, Simon, Eriksen flanker, and Navon, which are designed to investigate cognitive control (or "executive control")—the processes of storing, planning, and manipulating contextually relevant information to reach a specific goal in a top-down manner (Proctor and Vu 2006). These tasks are similar in the way that they all induce conflicts between different information (especially task-relevant and task-irrelevant ones): presenting color words in another incongruent color, displaying stimulus at an location that are incongruent with the correct location of button press, or showing incongruent visual features at the global and local levels (Simons 2010). Take the flanker task (Eriksen and Eriksen 1974) as an example: in the arrowhead version of this task (e.g., "> > < > >"), participants are required to respond to the direction of the center arrowhead. The surrounding flanker arrowheads, though they are irrelevant and should be ignored, usually interrupt the processing of the center target when their directions are incongruent with the center target and lead to prolonged reaction times and response errors. SRC paradigm are divided into two major categories according to the nature of conflict processing: stimulus-stimulus (S-S, compatibility between task-relevant and task-irrelevant stimuli) and stimulus-response (S-R; e.g., the Simon task). SRC paradigms could be employed to elicit the stimulus-locked N2 since they inherently involve conflict processing (Folstein and Van Petten 2008); meanwhile, response errors that occur in these tasks could elicit a response-locked ERN effect (Van Veen and Carter 2002).

4.4.2.2 Stop Signal Paradigm

Aside from conflict processing, inhibition processing is also a critical component of cognitive control (Miyake et al. 2000), which can be assessed by the stop signal and go/nogo paradigms (Verbruggen and Logan 2008). Here I focus on the stop signal paradigm because of space limitation. Participants perform a go task as they respond to the presentation of a target stimulus ASAP. Occasionally, a "stop" signal would appear shortly after the initial stimulus, which instructs participants to withhold their response. However, participants should not wait for the stop signal to occur. The need to stop a response elicits a fast control mechanism that inhibits action execution, and a frontocentral N2 would be generated accordingly by the stop signal (Verbruggen and Logan 2008). The target stimulus and the stop signal can be provided in different sensory channels (e.g., a visual target followed by an auditory stop signal). Researchers could manipulate behavioral performance by modulating task instruction (e.g., emphasizing reaction speed or accuracy) or the length of interval between the target and the stop signal (i.e., "stop signal delay [SSD]") (Folstein and Van Petten 2008).

N2 The N2 refers to the second negative peak (time window, 200–300 ms) in the averaged ERP waveforms and could be observed in many different paradigms (e.g., go/nogo, Stroop, stop signal, Eriksen flanker task; see Folstein and Van Petten 2008; Kok et al. 2004). This component is strongly connected with the P3 to a degree that they are sometimes referred to as the "N2-P3 complex," and researchers should be aware of their influence on one another. People believe that the N2 is a family consisting of multiple subcomponents, but the number of its members is under debate. Early studies suggest that there are at least an anterior N2a associated with attention orientation, a central N2b associated with conscious attention, and a frontocentral N2c associated with stimulus classification (Patel and Azzam 2005). Additionally, there might also be a N2pc that arises in the occipitotemporal region of the contralateral cortex as an index of attentional shift (for a review, see Patel and Azzam 2005). However, Folstein and Van Petten (2008) argued that the above categories need to be updated. Instead, they suggested that the visual N2 could be divided into a mismatch-related anterior N2 that encodes the detection of novelty or mismatch from mental template, another control-related anterior N2 associated with cognitive control (including response inhibition, response conflict, and error monitoring), and a third posterior N2 that reflect certain aspects of visual attention.

ERN The ERN is a response-locked rather than stimulus-locked ERP component, which peaks at about 50–100 ms after participants executed an erroneous response to a stimulus (Simons 2010). To avoid the influence of readiness potential, the ERN is baseline corrected with the average activity preceding the time point of stimulus rather than response (e.g., Eppinger et al. 2008). For correct trials, there is also a much smaller correct response negativity (CRN) in the same time window, which is topographically and morphologically similar with the ERN. The ERN is interpreted as the activity of an automatic response-monitoring system (which is modality nonspecific) rather than a conscious "oh-no!" effect, since subjective awareness is not necessary for ERN production. In contrast, the "error positivity (Pe)" that appears following the ERN is supposed to reflect awareness of the action error. According to the "reinforcement learning of the error-related negativity (RL-ERN)" theory, the ERN functions as a reward prediction error signal together with the FRN (see below). Holroyd and Coles (2002) conducted an experiment to show the relationship between the two components: in a probabilistic learning task, when there was a 100% mapping between an erroneous response and negative feedback, the ERN but not FRN emerged in the data; the reverse was true when the mapping was 50% (i.e., uncertain feedback). Nevertheless, some questions still remain. For instance, the ERN and FRN respond to certain variables in different patterns: while the ERN amplitude increases as a function of individual level of trait anxiety (Hajcak et al. 2003), the FRN decreases (Gu et al. 2010).

4.4.3 *Gambling Paradigm*

Monetary gambling task has a long history in the field of behavioral economics. To my knowledge, the study of Gehring and Willoughby (2002) was the first one to adopt this kind of task to investigate the ERPs associated with economic value processing. In its simplest form, each trial of Gehring and Willoughby (2002)'s gambling task (Fig. 4.3a) involves a forced choice between two options, which may or may not vary in risk level. After participants have made a decision, the corresponding outcome feedback would be presented, which is the focus of ERP analysis and usually generates an FRN and a P3 (San Martín 2012). There are at least

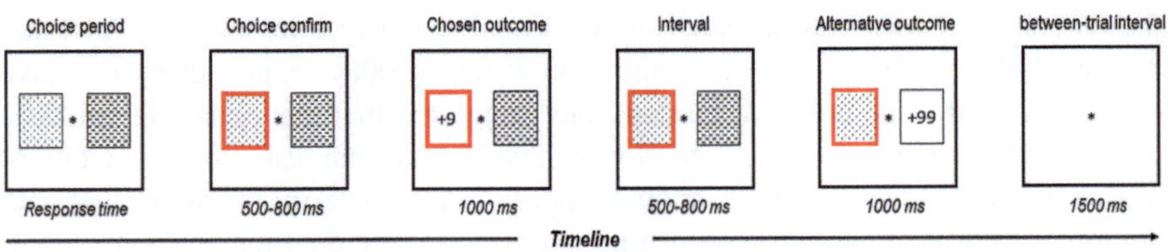

Fig. 4.3a An example of gambling paradigm. (Inspired by Yeung and Sanfey 2004)

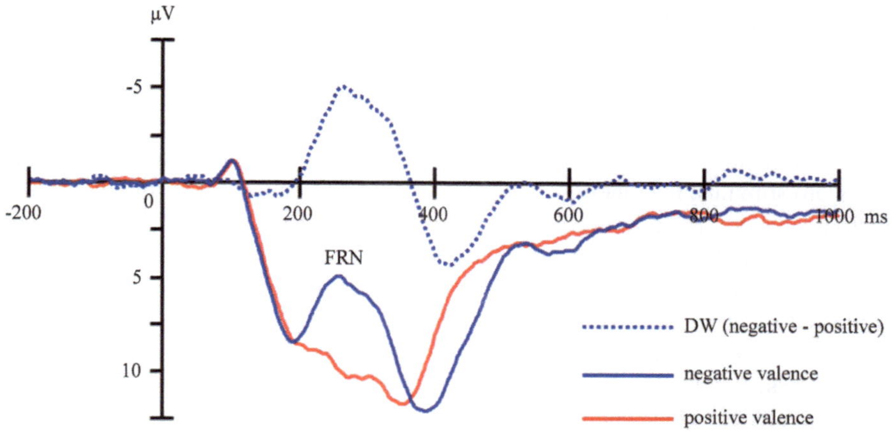

Fig. 4.3b Grand average waveforms elicited by gambling outcome (derived from the author's original data) and their difference wave (DW) that is often used to measure the FRN

two kinds of outcome feedback: positive (wins) and negative (losses), both of which appear on exactly 50% of trials, so as to control the effect of event probability (see Proudfit 2015, for a review). Numerous variations have been developed from the original task: there might be a cue presentation prior to the choice, which contains task-relevant information; more than two options may be provided, of which the appearance could be simple or complex; reward magnitude and winning probability may be manipulated as experimental variables; there might be restrictions on the gambling decision (e.g., time limitation); aside from winning and losing, participants may receive neutral (zero) or ambiguous feedback; and not only the chosen outcome but also the alternative outcome might be provided simultaneously or sequentially.

FRN The FRN (Fig. 4.3b) is a negative-going waveform that reaches its peak at about 200–300 ms following outcome presentation. Recent studies have found that cue presentation could also generate an FRN-like component (e.g. Liao et al. 2011). The FRN was first reported by Miltner et al. (1997) using a time estimation task; in their study, the FRN was larger for negative compared to positive performance feedback and was supposed to be a neural index of error detection. The importance of the FRN for neuroeconomics research had not been noticed until the study of Gehring and Willoughby (2002); they designed a simple gambling task (see above) and found out that the FRN (named as "medial-frontal negativity [MFN]" by Gehring & Willoughby) was larger for losses than wins. Combining these findings, Clay Holroyd and his colleagues proposed the RL-ERN theory in which the FRN is considered as a negative prediction error signal (whether the real outcome is worse than prior expectation) and is functionally analogous to the ERN. However, later studies have pointed out that the FRN amplitude is sensitive to both negative and positive prediction errors (i.e., a "salience prediction error" index; see Talmi et al. 2013). Also, it is possible that the FRN actually represents "reward positivity (RewP)" that is more positive-going for wins than losses (Proudfit 2015).

4.4.4 Oddball Paradigm

The most common version of the oddball paradigm involves two kinds of stimuli that are presented in a random series. These stimuli could be presented in either the visual or auditory modality. One of them occurs much more often and thus is called standard stimulus; the other one (i.e., the "oddball") is called target stimulus, which participants are asked to detect (may need an overt response; see Polich and Kok 1995). The probability of target varies across studies but is usually set between 0.1 and 0.2. Compared to standard stimuli, target stimuli elicit a larger N2, which is followed by a larger P3 (Folstein and Van Petten 2008). This paradigm is widely regarded as the most frequently used one in the history of ERP research (Picton et al. 2000). Based on the classic two-stimulus version, a three-stimulus oddball task has been developed, which contains an additional kind of distracter stimulus (Fig. 4.4a). The distracters also occur infrequently, but do not require participants to respond to. This three-stimulus version aims to distinguish between two cognitive processes: attention allocation and memory updating (see below).

P3 The P3 (Fig. 4.4b) is a large, broad, positive waveform and is one of the most noticeable late ERP components. Sutton et al. (1965) first discovered the P3 and reported that its peak emerged at about 300 ms following stimulus presentation. Therefore, this component is also widely called "P300," though its latency could actually be much longer than 300 ms in many studies (Luck 2014). As one of the most studied ERP components in the literature, the P3 has been associated with a large number of cognitive functions, including attention allocation, working memory updating, sensory discrimination, emotional and motivational processing, as well as social evaluation. It is also sensitive to various cognitive variables such as stimulus probability, stimulus complexity, and task difficulty (Duncan et al. 2009). For instance, the original study of Sutton et al. (1965) found that the P3 amplitude increased as a function of subjective degree of uncertainty. Finally, the P3 also plays an important role in clinical research and has been conceptualized as a biomarker for

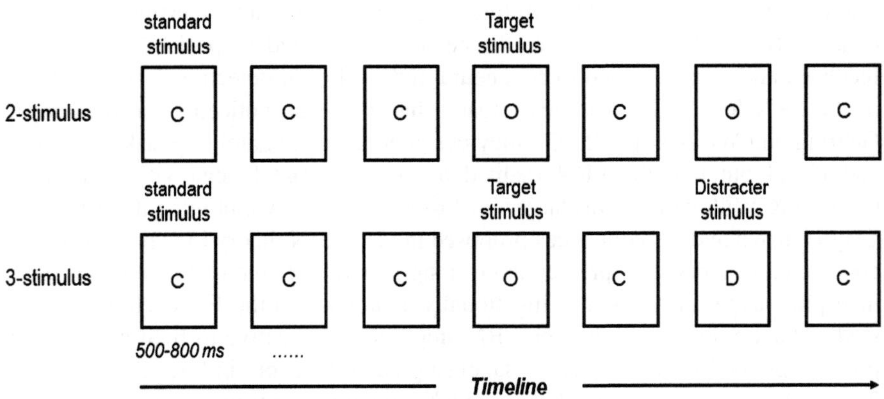

Fig. 4.4a An example of oddball paradigms. (Inspired by Polich and Criado 2006)

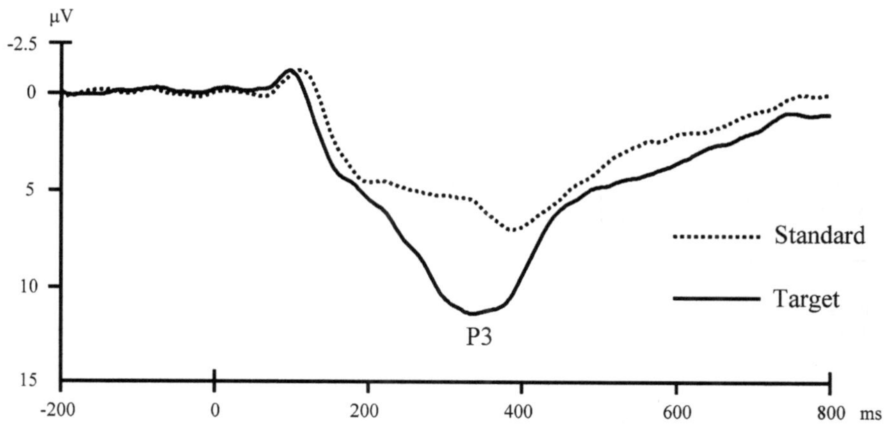

Fig. 4.4b Grand average waveforms elicited by oddball stimuli (derived from the author's data) in which a P3 component is observed

externalizing psychopathology (e.g., Iacono and Mcgue 2006). In the 1980s, the predominant interpretation of the P3 was that it indicates the updating of memory representation for the stimulus context (Donchin and Coles 1988). Nevertheless, this theory encountered difficulties to explain all follow-up findings as more and more cognitive functions have been associated with the P3. To reconcile this problem, John Polich and colleagues suggest that the P3 could be divided into two sub-components: the first one is a frontal P3a that reflects the activity of an attentional network responding to a novel stimulus; the second one is a parietal P3b that generates from the comparison between the current stimulus and mental representation stored in memory (Polich 2007; Polich and Criado 2006). Supporting this viewpoint, while the two-stimulus oddball task only elicit a typical P3, its three-stimulus version could produce a reliable P3a from the distracter and P3b from the target.

4.4.5 Sentence Comprehension Paradigm

There are also numerous variations within this category, aiming to explore different topics in the field of language research. Here I mainly focus on the standard semantic incongruence paradigm that has been designed to investigate the N400 component (Fig. 4.5). The development of this paradigm was based on Taylor's (1953) "cloze probability" paradigm for measuring readability (here, cloze probability means "the percentage of individuals who would continue a sentence fragment with that word"; cited from Kutas and Federmeier 2011). Stimuli used in the paradigm are visual sentences which may or may not contain unexpected or semantically incongruent endings (Kutas and Hillyard 1984). To reduce temporal overlaps of EEG signals associated with adjacent words, these sentences are typically presented one word at a

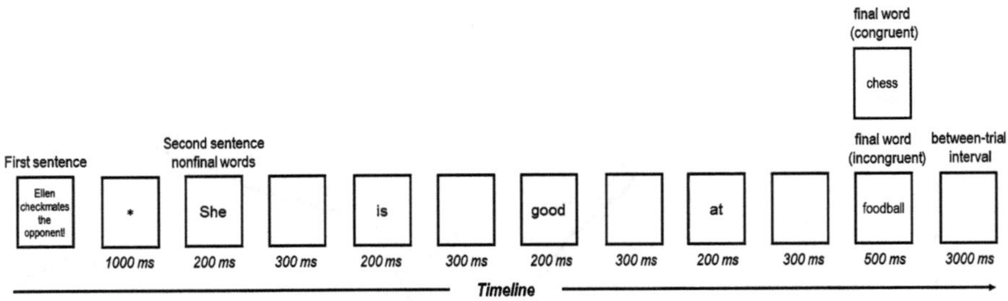

Fig. 4.5 An example of sentence comprehension paradigm. (Inspired by Federmeier and Kutas 1999)

time on the screen, with 200 ms for the duration of each word and another 300 ms ISI between them. The most important object for ERP analysis is the context-congruent or incongruent sentence final words, which accordingly may have a longer duration (e.g., 500 ms; see Federmeier and Kutas 1999). Researchers also recommend inserting a 3000–5000 ms interval between two sentences, allowing participants to blink without contaminating the recording epoch (Duncan et al. 2009). This paradigm has been successfully adopted to examine abnormalities of semantic processing in clinical research (Deldin et al. 2006).

N400 First reported by Kutas and Hillyard (1980), the N400 is a negative-going voltage deflection in the 300 and 600 ms poststimulus-onset window. This component has a very broad scalp distribution but is largest over midline centro-parietal sites (Kutas and Federmeier 2011). The most notable characteristic of the N400 is its sensitivity to semantic incongruence, although semantic anomalies are not essential for N400 elicitation. For example, participants first read an incomplete sentence, "Ellen checkmates the opponent! She is good at . . .," and then see its terminal word in different conditions, "chess" or "football." Consequently, "football" (which is incongruent with the preceding context) would elicit a larger N400 than "chess." This semantic context effect is robust regardless of language form (printed, spoken, or signed). Aside from the contextual factor, N400 amplitude is also affected by other lexical variables such as word frequency and concreteness (Duncan et al. 2009). Although N400-like potentials could be elicited by meaningful nonlinguistic stimuli such as line drawings, these potentials have slightly different scalp distributions compared to the verbal N400 (Duncan et al. 2009).

Acknowledgments The author sincerely thanks Jingwen Jin, Xuebing Li, Yongling Lin, Tingting Wu, and Jinfeng Ding for their helpful comments on the manuscript. The writing of this chapter was supported by the Major Program of the Chinese National Social Science Foundation (17ZDA324).

References

American Psychiatric Association. Diagnostic and statistical manual of mental disorders (DSM-5®). Arlington: American Psychiatric Publishing; 2013.

Bentin S, Allison T, Puce A, Perez E, McCarthy G. Electrophysiological studies of face perception in humans. J Cogn Neurosci. 1996;8(6):551–65. https://doi.org/10.1162/jocn.1996.8.6.551.

Beratis IN, Rabavilas A, Nanou ED, Hountala C, Maganioti AE, Capsalis CN, et al. Effect of initiation-inhibition and handedness on the patterns of the P50 event-related potential component: a low resolution electromagnetic tomography study. Behav Brain Funct. 2009;5:51. https://doi.org/10.1186/1744-9081-5-51.

Bocker KB, Brunia CH, van den Berg-Lenssen MM. A spatiotemporal dipole model of the stimulus preceding negativity (SPN) prior to feedback stimuli. Brain Topogr. 1994;7(1):71–88. https://doi.org/10.1007/BF01184839.

Caharel S, Leleu A, Bernard C, Viggiano MP, Lalonde R, Rebai M. Early holistic face-like processing of Arcimboldo paintings in the right occipito-temporal cortex: evidence from the N170 ERP component. Int J Psychophysiol. 2013;90(2):157–64. https://doi.org/10.1016/j.ijpsycho.2013.06.024.

Cohen J, Polich J. On the number of trials needed for P300. Int J Psychophysiol. 1997;25(3):249–55. https://doi.org/10.1016/S0167-8760(96)00743-X.

Cohen MX, Cavanagh JF, Slagter HA. Event-related potential activity in the basal ganglia differentiates rewards from nonrewards: temporospatial principal components analysis and source localization of the feedback negativity: commentary. Hum Brain Mapp. 2011;32(12):2270–1. https://doi.org/10.1002/hbm.21358.

Cohen MX, Ranganath C. Reinforcement learning signals predict future decisions. J Neurosci. 2007;27(2):371–8. https://doi.org/10.1523/JNEUROSCI.4421-06.2007.

Coles MGH, Rugg MD. Event-related brain potentials: an introduction. In: Rugg MD, Coles MGH, editors. Electrophysiology of mind: event-related brain potentials and cognition. New York: Oxford University Press; 1995. p. 1–26.

Coulson S, Lovett C. Handedness, hemispheric asymmetries, and joke comprehension. Cogn Brain Res. 2004;19(3):275–88. https://doi.org/10.1016/j.cogbrainres.2003.11.015.

Davis PA. Effects of acoustic stimuli on the waking human brain. J Neurophysiol. 1939;2(6):494–9. https://doi.org/10.1152/jn.1939.2.6.494.

Deldin P, Keller J, Casas BR, Best J, Gergen J, Miller GA. Normal N400 in mood disorders. Biol Psychol. 2006;71(1):74–9. https://doi.org/10.1016/j.biopsycho.2005.02.005.

Donchin E, Coles MGH. Is the P300 component a manifestation of context updating? Behav Brain Sci. 1988;11(3):355–72. https://doi.org/10.1017/S0140525X00058027.

Duncan CC, Barry RJ, Connolly JF, Fischer C, Michie PT, Naatanen R, et al. Event-related potentials in clinical research: guidelines for eliciting, recording, and quantifying mismatch negativity, P300, and N400. Clin Neurophysiol. 2009;120(11):1883–908. https://doi.org/10.1016/j.clinph.2009.07.045.

Eliot L. The trouble with sex differences. Neuron. 2011;72(6):895–8. https://doi.org/10.1016/j.neuron.2011.12.001.

Eppinger B, Kray J, Mock B, Mecklinger A. Better or worse than expected? Aging, learning, and the ERN. Neuropsychologia. 2008;46(2):521–39. https://doi.org/10.1016/j.neuropsychologia.2007.09.001.

Eriksen BA, Eriksen CW. Effects of noise letters upon the identification of a target letter in a nonsearch task. Percept Psychophys. 1974;16:143–9. https://doi.org/10.3758/BF03203267.

Etkin A, Klemenhagen KC, Dudman JT, Rogan MT, Hen R, Kandel ER, Hirsch J. Individual differences in trait anxiety predict the response of the basolateral amygdala to unconsciously processed fearful faces. Neuron. 2004;44(6):1043–55. https://doi.org/10.1016/j.neuron.2004.12.006.

Federmeier KD, Kutas M. A rose by any other name: long-term memory structure and sentence processing. J Mem Lang. 1999;41(4):469–95. https://doi.org/10.1006/jmla.1999.2660.
Folstein JR, Van Petten C. Influence of cognitive control and mismatch on the N2 component of the ERP: a review. Psychophysiology. 2008;45(1):152–70.
Friston KJ, Price CJ, Fletcher P, Moore C, Frackowiak RS, Dolan RJ. The trouble with cognitive subtraction. NeuroImage. 1996;4(2):97–104. https://doi.org/10.1006/nimg.1996.0033.
Gardener EK, Carr AR, Macgregor A, Felmingham KL. Sex differences and emotion regulation: an event-related potential study. PLoS One. 2013;8(10):e73475. https://doi.org/10.1371/journal.pone.0073475.
Gehring WJ, Willoughby AR. The medial frontal cortex and the rapid processing of monetary gains and losses. Science. 2002;295(5563):2279–82. https://doi.org/10.1126/science.1066893.
Grech R, Cassar T, Muscat J, Camilleri KP, Fabri SG, Zervakis M, et al. Review on solving the inverse problem in EEG source analysis. J Neuroeng Rehabil. 2008;5:25. https://doi.org/10.1186/1743-0003-5-25.
Gu R, Huang YX, Luo YJ. Anxiety and feedback negativity. Psychophysiology. 2010;47(5):961–7. https://doi.org/10.1111/j.1469-8986.2010.00997.x.
Gu R, Zhang D, Luo Y, Wang H, Broster LS. Predicting risk decisions in a modified Balloon Analogue Risk Task: conventional and single-trial ERP analyses. Cogn Affect Behav Neurosci. 2018;18(1):99–116. https://doi.org/10.3758/s13415-017-0555-3.
Hajcak G, MacNamara A, Olvet DM. Event-related potentials, emotion, and emotion regulation: an integrative review. Dev Neuropsychol. 2010;35(2):129–55. https://doi.org/10.1080/87565640903526504.
Hajcak G, McDonald N, Simons RF. Anxiety and error-related brain activity. Biol Psychol. 2003;64(1–2):77–90.
Henrich J, Heine SJ, Norenzayan A. The weirdest people in the world? Behav Brain Sci. 2010;33 (2–3):61–83;. discussion 83-135. https://doi.org/10.1017/S0140525X0999152X.
Herrmann CS, Knight RT. Mechanisms of human attention: event-related potentials and oscillations. Neurosci Biobehav Rev. 2001;25(6):465–76. https://doi.org/10.1016/S0149-7634(01)00027-6.
Hinojosa JA, Mercado F, Carretie L. N170 sensitivity to facial expression: a meta-analysis. Neurosci Biobehav Rev. 2015;55:498–509. https://doi.org/10.1016/j.neubiorev.2015.06.002.
Holroyd CB, Coles MGH. The neural basis of human error processing: reinforcement learning, dopamine, and the error-related negativity. Psychol Rev. 2002;109(4):679–709. https://doi.org/10.1037/0033-295X.109.4.679.
Iacono WG, Mcgue M. Association between P3 event-related brain potential amplitude and adolescent problem behavior. Psychophysiology. 2006;43(5):465–9.
Itier RJ, Latinus M, Taylor MJ. Face, eye and object early processing: what is the face specificity? NeuroImage. 2006;29(2):667–76. https://doi.org/10.1016/j.neuroimage.2005.07.041.
Keil A, Debener S, Gratton G, Junghofer M, Kappenman ES, Luck SJ, et al. Committee report: publication guidelines and recommendations for studies using electroencephalography and magnetoencephalography. Psychophysiology. 2014;51(1):1–21. https://doi.org/10.1111/psyp.12147.
Keil A, Gruber T, Muller MM, Moratti S, Stolarova M, Bradley MM, Lang PJ. Early modulation of visual perception by emotional arousal: evidence from steady-state visual evoked brain potentials. Cogn Affect Behav Neurosci. 2003;3(3):195–206. https://doi.org/10.3758/CABN.3.3.195.
Knutson B, Fong GW, Adams CM, Varner JL, Hommer D. Dissociation of reward anticipation and outcome with event-related fMRI. Neuroreport. 2001;12(17):3683–7. https://doi.org/10.1097/00001756-200112040-00016.
Kok A, Ramautar JR, De Ruiter MB, Band GP, Ridderinkhof KR. ERP components associated with successful and unsuccessful stopping in a stop-signal task. Psychophysiology. 2004;41(1):9–20. https://doi.org/10.1046/j.1469-8986.2003.00127.x.

Kutas M, Federmeier KD. Thirty years and counting: finding meaning in the N400 component of the event-related brain potential (ERP). Annu Rev Psychol. 2011;62:621–47. https://doi.org/10.1146/annurev.psych.093008.131123.

Kutas M, Hillyard SA. Reading senseless sentences: brain potentials reflect semantic incongruity. Science. 1980;207(4427):203–5. https://doi.org/10.1126/science.7350657.

Kutas M, Hillyard SA. Brain potentials during reading reflect word expectancy and semantic association. Nature. 1984;307(5947):161–3. https://doi.org/10.1038/307161a0.

Liao Y, Gramann K, Feng WF, Deak GO, Li H. This ought to be good: brain activity accompanying positive and negative expectations and outcomes. Psychophysiology. 2011;48(10):1412–9. https://doi.org/10.1111/j.1469-8986.2011.01205.x.

Luck SJ. Ten simple rules for designing and interpreting ERP experiments. In: Handy TC, editor. Event-related potentials: a methods handbook. Cambridge, MA: MIT Press; 2004. p. 17–32.

Luck SJ. An introduction to the event-related potential technique. 2nd ed. Cambridge, MA: MIT press; 2014.

Luck SJ, Woodman GF, Vogel EK. Event-related potential studies of attention. Trends Cogn Sci. 2000;4(11):432–40. https://doi.org/10.1016/S1364-6613(00)01545-X.

Luo Y, Wu T, Broster LS, Feng C, Zhang D, Gu R, Luo YJ. The temporal course of the influence of anxiety on fairness considerations. Psychophysiology. 2014;51(9):834–42. https://doi.org/10.1111/psyp.12235.

Marco-Pallares J, Cucurell D, Munte TF, Strien N, Rodriguez-Fornells A. On the number of trials needed for a stable feedback-related negativity. Psychophysiology. 2011;48(6):852–60. https://doi.org/10.1111/j.1469-8986.2010.01152.x.

Miltner WHR, Braun CH, Coles MGH. Event-related brain potentials following incorrect feedback in a time-estimation task: evidence for a "generic" neural system for error detection. J Cogn Neurosci. 1997;9(6):788–98. https://doi.org/10.1162/jocn.1997.9.6.788.

Miyake A, Friedman NP, Emerson MJ, Witzki AH, Howerter A, Wager TD. The unity and diversity of executive functions and their contributions to complex “frontal lobe” tasks: a latent variable analysis. Cogn Psychol. 2000;41(1):49–100. https://doi.org/10.1006/cogp.1999.0734.

Münte TF, Urbach TP, Düzel E, Kutas M, Boller F, Grafman J, Rizzolatti G. Event-related brain potentials in the study of human cognition and neuropsychology. In: Boller F, Grafman J, Rizzolatti G, editors. Handbook of neuropsychology, vol. 1. Amsterdam: Elsevier Science Publishers B.V; 2000. p. 139–235.

Neyedli HF, Welsh TN. The processes of facilitation and inhibition in a cue–target paradigm: insight from movement trajectory deviations. Acta Psychol. 2012;139(1):159–65. https://doi.org/10.1016/j.actpsy.2011.11.001.

Nowicka A, Marchewka A, Szatkowska I. Lateralization of repetition effects in event-related potentials to words in left- and right-handed women. Neurosci Lett. 2006;393(2–3):150–4. https://doi.org/10.1016/j.neulet.2005.09.055.

Patel SH, Azzam PN. Characterization of N200 and P300: selected studies of the event-related potential. Int J Med Sci. 2005;2(4):147–54. https://doi.org/10.7150/ijms.2.147.

Pernet C, Garrido M, Gramfort A, Maurits N, Michel C, Pang E, et al., Best practices in data analysis and sharing in neuroimaging using MEEG. Retrieved from OSF Preprints website: https://osf.io/a8dhx. https://doi.org/10.31219/osf.io/a8dhx; 2018

Picton TW, Bentin S, Berg P, Donchin E, Hillyard SA, Johnson R, et al. Guidelines for using human event-related potentials to study cognition: recording standards and publication criteria. Psychophysiology. 2000;37(2):127–52.

Poldrack RA. Can cognitive processes be inferred from neuroimaging data? Trends Cogn Sci. 2006;10(2):59–63. https://doi.org/10.1016/j.tics.2005.12.004.

Poldrack RA. The role of fMRI in cognitive neuroscience: where do we stand? Curr Opin Neurobiol. 2008;18(2):223–7. https://doi.org/10.1016/j.conb.2008.07.006.

Polich J. Updating P300: an integrative theory of P3a and P3b. Clin Neurophysiol. 2007;118(10):2128–48. https://doi.org/10.1016/j.clinph.2007.04.019.

Polich J, Criado JR. Neuropsychology and neuropharmacology of P3a and P3b. Int J Psychophysiol. 2006;60(2):172–85. https://doi.org/10.1016/j.ijpsycho.2005.12.012.

Polich J, Kok A. Cognitive and biological determinants of P300: an integrative review. Biol Psychol. 1995;41(2):103–46. https://doi.org/10.1016/0301-0511(95)05130-9.

Posner MI. Timing the brain: mental chronometry as a tool in neuroscience. PLoS Biol. 2005;3(2): e51. https://doi.org/10.1371/journal.pbio.0030051.

Proctor RW, Vu KPL. Stimulus-response compatibility principles: data, theory, and application. Boca Raton, FL: CRC Press; 2006.

Proudfit GH. The reward positivity: from basic research on reward to a biomarker for depression. Psychophysiology. 2015;52(4):449–59. https://doi.org/10.1111/psyp.12370.

Raihani NJ, Bshary R. Resolving the iterated prisoner's dilemma: theory and reality. J Evol Biol. 2011;24(8):1628–39. https://doi.org/10.1111/j.1420-9101.2011.02307.x.

Rossion B, Jacques C. Does physical interstimulus variance account for early electrophysiological face sensitive responses in the human brain? Ten lessons on the N170. NeuroImage. 2008;39 (4):1959–79. https://doi.org/10.1016/j.neuroimage.2007.10.011.

Rugg MD, Curran T. Event-related potentials and recognition memory. Trends Cogn Sci. 2007;11 (6):251–7. https://doi.org/10.1016/j.tics.2007.04.004.

San Martín R. Event-related potential studies of outcome processing and feedback-guided learning. Front Hum Neurosci. 2012;6:304. https://doi.org/10.3389/fnhum.2012.00304.

Schiebinger L. Scientific research must take gender into account. Nature. 2014;507(7490):9. https://doi.org/10.1038/507009a.

Schultz W, Tremblay L, Hollerman JR. Reward prediction in primate basal ganglia and frontal cortex. Neuropharmacology. 1998;37(4–5):421–9. https://doi.org/10.1016/S0028-3908(98)00071-9.

Segalowitz SJ, Davies PL. Charting the maturation of the frontal lobe: an electrophysiological strategy. Brain Cogn. 2004;55(1):116–33. https://doi.org/10.1016/S0278-2626(03)00283-5.

Simons RF. The way of our errors: theme and variations. Psychophysiology. 2010;47(1):1–14. https://doi.org/10.1111/j.1469-8986.2009.00929.x.

Stelmach GE. Information-processing framework for understanding human motor behavior. In: Kelso JAS, editor. Human motor behavior: an introduction. Hillsdale: Lawrence Erlbaum; 1982. p. 63–91.

Sutton S, Braren M, Zubin J, John ER. Evoked-potential correlates of stimulus uncertainty. Science. 1965;150(3700):1187–8. https://doi.org/10.1126/science.150.3700.1187.

Tal N, Yuval-Greenberg S. Reducing saccadic artifacts and confounds in brain imaging studies through experimental design. Psychophysiology. 2018;55(11):e13215. https://doi.org/10.1111/psyp.13215.

Talmi D, Atkinson R, El-Deredy W. The feedback-related negativity signals salience prediction errors, not reward prediction errors. J Neurosci. 2013;33(19):8264–9. https://doi.org/10.1523/JNEUROSCI.5695-12.2013.

Taylor WL. "Cloze procedure": a new tool for measuring readability. Journalism Bull. 1953;30 (4):415–33. https://doi.org/10.1177/107769905303000401.

Thierry G, Martin CD, Downing P, Pegna AJ. Controlling for interstimulus perceptual variance abolishes N170 face selectivity. Nat Neurosci. 2007;10(4):505–11. https://doi.org/10.1038/nn1864.

Tian Y, Klein RM, Satel J, Xu P, Yao D. Electrophysiological explorations of the cause and effect of inhibition of return in a cue-target paradigm. Brain Topogr. 2011;24(2):164–82. https://doi.org/10.1007/s10548-011-0172-3.

Van Veen V, Carter CS. The anterior cingulate as a conflict monitor: fMRI and ERP studies. Physiol Behav. 2002;77(4–5):477–82. https://doi.org/10.1016/S0031-9384(02)00930-7.

Verbruggen F, Logan GD. Response inhibition in the stop-signal paradigm. Trends Cogn Sci. 2008;12(11):418–24. https://doi.org/10.1016/j.tics.2008.07.005.

Verleger R. On the utility of P3 latency as an index of mental chronometry. Psychophysiology. 1997;34(2):131–56. https://doi.org/10.1111/j.1469-8986.1997.tb02125.x.

Vidal F, Burle B, Grapperon J, Hasbroucq T. An ERP study of cognitive architecture and the insertion of mental processes: donders revisited. Psychophysiology. 2011;48(9):1242–51. https://doi.org/10.1111/j.1469-8986.2011.01186.x.

Walsh MM, Anderson JR. Modulation of the feedback-related negativity by instruction and experience. Proc Natl Acad Sci U S A. 2011;108(47):19048–53. https://doi.org/10.1073/pnas.1117189108.

Walter WG, Cooper R, Aldridge VJ, McCallum WC, Winter AL. Contingent negative variation: an electric sign of sensorimotor association and expectancy in the human brain. Nature. 1964;203:380–4. https://doi.org/10.1038/203380a0.

Williams JM, Mathews A, MacLeod C. The emotional Stroop task and psychopathology. Psychol Bull. 1996;120(1):3–24. https://doi.org/10.1037/0033-2909.120.1.3.

Woodman GF. A brief introduction to the use of event-related potentials in studies of perception and attention. Atten Percept Psychophys. 2010;72(8):2031–46. https://doi.org/10.3758/BF03196680.

Yeung N, Sanfey AG. Independent coding of reward magnitude and valence in the human brain. J Neurosci. 2004;24(28):6258–64. https://doi.org/10.1523/JNEUROSCI.4537-03.2004.

Zhang D. Computational EEG analysis for hyperscanning and social neuroscience. In: Im CH, editor. Computational EEG analysis. Singapore: Springer; 2018. p. 215–28.

Chapter 5
EEG Preprocessing and Denoising

Weiwei Peng

Abstract In this chapter, we first introduce physiological and non-physiological artifacts embedded in the raw EEG signals, e.g., ocular related artifacts (physiological) and power line interference (non-physiological). Then, we introduce the montage to describe and apply the location of scalp electrodes in the context of EEG studies. Further, we describe several preprocessing steps that are commonly used in the EEG preprocessing, including filtering, re-referencing, segmenting, removal of bad channels and trials, as well as decomposition of EEG using independent component analysis. More specifically, appropriate band-pass filtering can effectively reduce superimposed artifacts from various sources which are embedded in the EEG recordings. Re-referencing is a linear transformation of the EEG data, through which noise in the reference electrodes could turn into noise in the scalp electrodes. Data epochs that are time-locked to the specific events of interest should be extracted to facilitate the investigation of task/stimulus-related changes in EEG. Trials contaminated by artifacts, as well as bad channels that are not functioning properly for various reasons, should be excluded from further analysis. Given that the EEG data recorded from scalp electrodes can be considered as summations of neural activities, and that artifacts are independent with each other, independent component analysis could be a powerful and efficient strategy to separate artifact from EEG signals.

Keywords Physiological artifacts · Non-physiological artifacts · Filtering · Re-referencing · EEG epochs · Artifact correction

While EEG recordings tend to contain noise and artifacts such as eye blinking or movement, EEG signals measured from the scalp are not necessarily to accurately represent signals originated from the brain. Therefore, it is very essential to apply preprocessing and denoising to the recorded EEG data. Generally, preprocessing steps include the transformations or reorganizations of the recorded EEG data by

W. Peng (✉)
School of Psychology, Shenzhen University, Shenzhen, Guangdong, China

L. Hu, Z. Zhang (eds.), *EEG Signal Processing and Feature Extraction*,
https://doi.org/10.1007/978-981-13-9113-2_5

removing bad or artifact-ridden data without changing clean data (transformation) and segmenting continuous raw signals without change of the data (reorganizations). Notably, the application of preprocessing steps largely depends on several aspects, including the goal of the study, the details of the experiment design, the equipment of recording EEG raw signals, as well as the advanced analysis you plan to perform. Here, we start the preprocessing procedures with an introduction of physiological and non-physiological artifacts embedded in the raw EEG signals.

5.1 Artifacts in EEG

As summarized in Table 5.1, artifacts embedded in the EEG recordings could be generally subdivided into two categories: physiological and non-physiological artifacts. Physiological artifacts include electrical potentials elicited by biological activities, which are largely generated from the physical part proximal to the head (e.g., eyes, muscles, and heart) or due to the subjects' movements. Non-physiological artifacts include the electrode-scalp interface, devices, and environment, which are generated anywhere near the EEG recording system. Whereas physiological artifacts that display characteristic electrical fields can be fairly easily identified, non-physiological artifacts that display various morphologies can either distort or obscure EEG activity.

5.1.1 Physiological Artifacts

Physiological artifacts generally originate from sources in the body. Some of the most common physiological artifacts are eye blinks, eye movements, head movements, heartbeats, and muscular noise (Islam et al. 2016). It is possible to detect and identify those artifacts if other biometric data are accessible for us, e.g., electrooculogram or eye-tracking data for the detection of eye blink and movement artifacts, electrocardiogram data for the detection of heartbeat artifacts, and accelerometer data for the detection of head movement artifacts.

Table 5.1 Physiological and non-physiological artifacts

Physiological artifacts	Non-physiological artifacts
Ocular-related artifacts due to eyes movements and blinking	Power line interference artifacts (50 Hz in Europe, and 60 Hz in the United States)
Electromyography artifacts due to frontalis and temporalis muscle activities	Electrode artifacts due to the poor placement of electrode on the scalp
Electrocardiographic artifacts due to heartbeats	Malfunction of any part within EEG recording system e.g., amplifiers
Scalp perspiration and movements	Digital artifacts such as loose wiring or loosening of circuit board connections

Physiological artifacts of ocular-related potential, for example, eye movements and eye blinks (as shown in Fig. 5.1a, b), are easily detected in conscious subject during the routine EEG recordings. When the eyes move, the electric dipole between the positively charged cornea and the negatively charged retina also moves. It thus yields a large electrical potential, typically characterized by anterior location, bilateral, and synchronized appearance (Jung et al. 2000). Ocular-related artifacts do not destroy the brain-generated EEG signal but linearly sum on top of the brain-generated EEG. Monitoring ocular activities, by placing electrodes above and below the eye, can help identify potentials that are in phase with cerebral and out of phase with extracerebral and ocular sources. There are several methods that have been proposed to successfully attenuate ocular-related artifacts but sparing brain activity, such as decomposition-based (Jung et al. 2000) and regression-based techniques (Gratton et al. 1983).

Electromyography artifacts (as shown in the Fig. 5.1c) are high-frequency activities and showed to be very spiky, but they are too fast to be an epileptic discharge. Principal sites of generating muscle artifacts are located at frontalis and temporalis muscles (Muthukumaraswamy 2013). Frontalis muscle is mainly involved in forced eye closure and photic stimulation. The contraction of frontalis muscle in periocular movement may elicit sustained or individual motor unit potentials, which can be appeared as "railroad tracks." Temporalis muscles become active in jaw clenching,

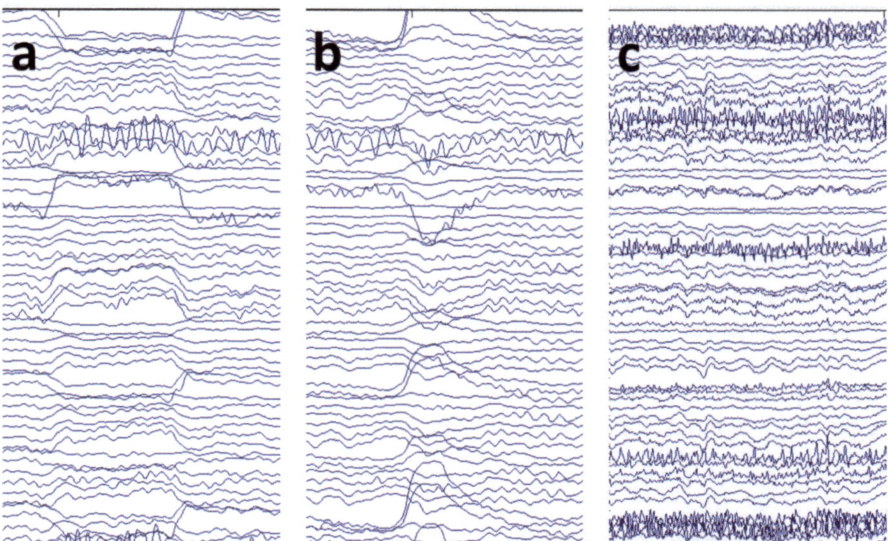

Fig. 5.1 Examples of ocular and electromyography artifacts. Ocular artifacts such as eye movements (**a**) and blinking (**b**) are represented as a large electrical potential, typically characterized by anterior location, bilateral, and synchronized appearance. Electromyography artifacts (**c**) such as muscle tension-related artifacts are high-frequency activities and showed to be very spiky

chewing, or bruxism. The contraction of temporalis muscles may elicit bursts of fast activities. Experimenters can instruct the subject to open the mouth, to relax the jaw and diminish this kind of muscle artifact.

Artifacts originated from the heart are variably present during routing EEG recordings (Nakamura and Shibasaki 1987). Referential montages would accentuate electrocardiographic artifact, especially when using ipsilateral ear reference with the larger intra-electrode distances. In contrast, referential linked ears montage can reduce the electrocardiographic artifact. Overweight subjects or patients with short stocky necks, as well as babies, may be predisposed to electrocardiogram artifacts, because the dipole is situated closer to the recording electrodes and is better able to transmit the current. While the electromyography artifacts "contaminate" the EEG, electrocardiographic artifacts are essential in interpreting physiologic functions that may occur during the recording session. Therefore, simultaneously recording the electrocardiogram activities during routine EEG recordings is essential to enable the recognition of the cardiac–cerebral relationship.

Scalp perspiration will also produce artifact by creating unwanted electric connections between electrodes. Perspiration artifact appears as very low-frequency (0.5 Hz) low-amplitude undulating waves (Barlow 1986). Changes in the DC electrode potential from perspiration may result in an unstable baseline (sweat sway) and crossing of tracings in adjacent channels. In addition, subjects' movements that cause the leads or electrodes to move can provide a large source of "physiologic" artifact on recordings. This is especially true for the awake and ambulatory patient or subject but is also quite notable in those who are agitated and confused.

These physiological artifacts can be handled by avoidance, rejection, or removal. To avoid the artifacts, experimenters often instruct subjects to suppress eye movement and blinking. Nevertheless, the occurrence of these artifacts is inevitable, as eye movements and blinks can be involuntary, especially in children and elderly people. To remove artifacts, several preprocessing techniques have been proposed to remove artifacts from EEG recording to improve the signal-to-noise ratio. For epoch-based method, the whole temporal segments of EEG contaminated by artifacts are rejected. For independent component analysis (ICA)-based method, artifact-related components are removed from the EEG signals. Nevertheless, in this way, a large amount of useful cerebral information in the EEG is also discarded, especially when the artifacts occur frequently.

5.1.2 Non-physiological Artifacts

Non-physiological artifacts generally originate from outside-world interference. One of the most common sources of non-physiological artifacts is the electric interference. It is the alternating mains power supply, at 60 Hz in the United States and 50 Hz in Europe. The electrical interference can be reduced by shielding cable, using a shielded recording rooms, or moving the subject far from the source of interference in the room. In addition, connecting the subject to ground, and keeping all electrode wires short, can also help to reduce the alternating current artifacts. Power line interference can also be removed by applying a notch filter at 50 or 60 Hz, and sometimes this kind of notch filter is pre-built in some headsets.

Another frequent cause of non-physiological artifact is electrode artifact due to the poor placement of electrode on the scalp. When an electrode moves and the electric double layer is disturbed, it would create a DC potential similar to discharging a capacitor (Barlow 1986), and the characteristic electrode "pop" occurs. "Pops" are sudden positive (or less commonly negative) discharges and usually show an initial high-voltage very steep deflection, followed by an exponential decay caused by the amplifier's low-frequency filter. The best way to reduce electrode artifact is a proper electrode application and maintenance. Electrodes should be cleaned well and inspected regularly to look for signs of corrosion between the electrode and conducting wire, damaged insulation, or broken lead wires. In addition, placing an additional electrode close to the electrode with suspected artifact may also help to determine whether it causes electrode artifact.

Malfunction of any part within EEG recording system can cause artifact. Modern amplifiers are small, low-power, and single-chip multichannel devices with solid-state integrated circuits. Amplifier "noise" is caused by thermal agitation of electrons in the amplifier circuits. Portable equipment is more susceptible to mechanical wear and tear, such as vibration or jolting of machines. Loose wiring or loosening of circuit board connections is also a common source of artifact, causing loss of signal or intermittent failures. Digital artifacts include aliasing (inadequate sampling rate), errors in analog to digital conversion, skew errors, multiplexing artifacts, and blocking (Blum 1998).

The easiest way to minimize the effect of these non-physiological artifacts is to adjust the environment (e.g., shielding the room, properly securing the electrodes). The influence of environmental artifacts can also be somewhat reduced by using active electrodes (electrodes that have an additional low-noise amplifier inside).

5.2 Montage

The International 10–20 System (left panel of Fig. 5.2) is an internationally recognized method to describe and apply the location of scalp electrodes in the context of EEG studies. This system is based on the relationship between the location of electrodes and the underlying areas of the brain, specifically the cerebral cortex. In the 10–20 system, the "10" and "20" refer to the fact that the actual distances between adjacent electrodes are either 10 or 20% of the total front–back or right–left distance of the skull (Herwig et al. 2003). This method was developed to maintain standardization across studies, to ensure that outcomes of clinical or research studies could be compiled, reproduced, and compared using this scientific method.

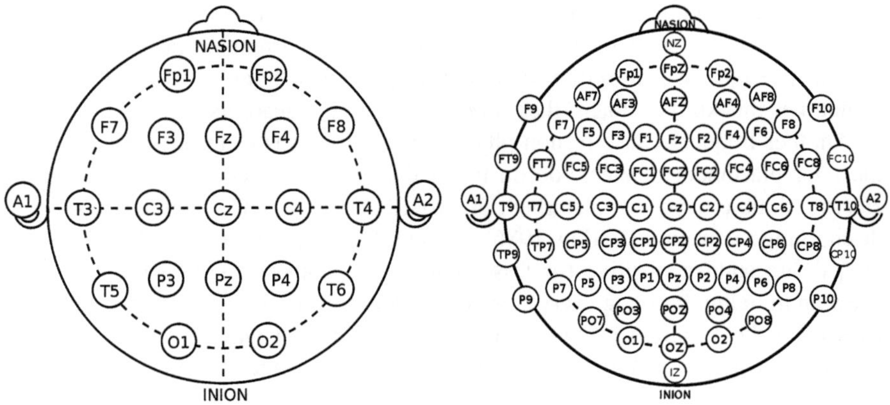

Fig. 5.2 The International 10–20 EEG Placement System. Left panel: The 10–20 system or International 10–20 System is an internationally recognized method to describe and apply the location of scalp electrodes in the context of EEG studies. The "10" and "20" refer to the fact that the actual distances between adjacent electrodes are either 10 or 20% of the total front–back or right–left distance of the skull. Each electrode placement site has a letter to identify the lobe or area of the brain it is reading from. Right panel: modified combinatorial nomenclature system. When recording a more detailed EEG using more electrodes, extra electrodes are interpolated using the 10% division, which fills in intermediate sites halfway between those of the existing 10–20 system

When recording a more detailed EEG using more electrodes, extra electrodes are interpolated using the 10% division (right panel of Fig. 5.2), which fills in intermediate sites halfway between those of the existing 10–20 system. This new electrode-naming system is more complicated, giving rise to the modified combinatorial nomenclature, short as MCN. This MCN system uses 1, 3, 5, 7, 9 for the left hemisphere and uses 2, 4, 6, 8, 10 for the right hemisphere, which represents 10%, 20%, 30%, 40%, 50% of the inion-to-nasion distance, respectively. Each electrode placement site on this MCN system has a letter to represent the specific lobe or area of the brain: frontal (F), temporal (T), parietal (P), occipital (O), and central (C). Even when there is no "central lobe" actually, the "C" electrodes can exhibit/represent EEG activity more typical of frontal, temporal, and some parietal–occipital activity. Suffixal (Z) sites referring electrodes placed on the midline sagittal plane of the skull (Fz, Cz, Pz, Oz), are present mostly for reference/measurement points. Nevertheless, these electrodes will not necessarily reflect or amplify lateral hemispheric cortical activities, as they are placed over the corpus callosum and do not represent hemispheric laterality.

Differences in electrical potentials between these electrodes constitute channels, and the combinations of different channels are called montages. There are two main types of montages: the bipolar and the referential montages (Hu et al. 2018). In the bipolar montage, channels are arranged in chains that follow an anterior-to-posterior or a transverse arrangement. The chains imply that the second lead in the first channel is the first lead in the second channel and so forth until the end of the chain. In such a configuration, external or environmental noise can easily be canceled out, as it measures the difference in electrode potential between contiguous electrodes, thus selectively amplifying the local potentials (Zaveri et al. 2006). In the referential montage, each channel represents the differential electrode potential of any given electrode to a single chosen electrode (reference electrode). In such a configuration, it has the advantage of detecting both local (near field) and distant (far field) potentials (Acharya et al. 2016) but is limited by the high susceptibility to the external noises. Compared with bipolar montage, the electrode potential amplitude of the deflection on a referential montage would take a closer representation of the absolute electrode potential recorded by the electrode. For a displaying purpose showing the EEG scalp maps in either 2-D or 3-D formats, or to estimate source locations for data components, information regarding the locations of the recording electrodes should be contained in the EEG dataset.

5.3 Filtering

Filtering is often necessary in the preprocessing pipeline, due to several reasons, e.g., presence of 50 Hz or 60 Hz line noise, high-frequency noises, as well as noises at very low frequencies. Although these noises and superimposed artifacts from various sources embedded in the EEG recordings, they can be effectively reduced by appropriate band-pass filtering. Therefore, digital filters can be applied to the raw EEG recordings and can significantly improve the interpretation of EEG contaminated by artifacts. Filtering EEG signals with certain frequencies was popular, such that either some frequencies are removed or possibly that some frequencies are remained.

According to which frequencies are remained or removed, there are four types of filters (as shown in Fig. 5.3): low-pass filter, high-pass filter, band-pass filter, and band-stop filter. For the low-pass filter, signals with low frequencies below a certain value are kept, while high frequencies greater than the certain value are removed/attenuated. For the high-pass filter, only signal with frequencies greater than a certain value are kept, while only those with low frequencies below the certain value are removed/attenuated. For the band-pass filter, signals with frequencies between a lower and upper bound are kept, while signals below the lower limit or greater than the upper limit are removed/attenuated. For the band-stop filter, signals with frequencies between a lower and upper bound are removed/attenuated, while signals below the lower limit or greater than the upper limit are kept.

When selecting the filters applied on the EEG signals, the frequency ranges of the artifacts embedded in the EEG recordings should be taken into consideration. For example, high-pass filters, with limit of 0.1 Hz, are applied to the EEG signals to remove low-frequency drifts; low-pass filters, e.g., with limit of 30 Hz, are applied to the EEG signals to remove high-frequency noise (e.g., interference due to muscular activities). To eliminate power line noise (50 Hz in Europe and Asia, 60 Hz in the United States), a band-stop filter is commonly used which removes signals within a

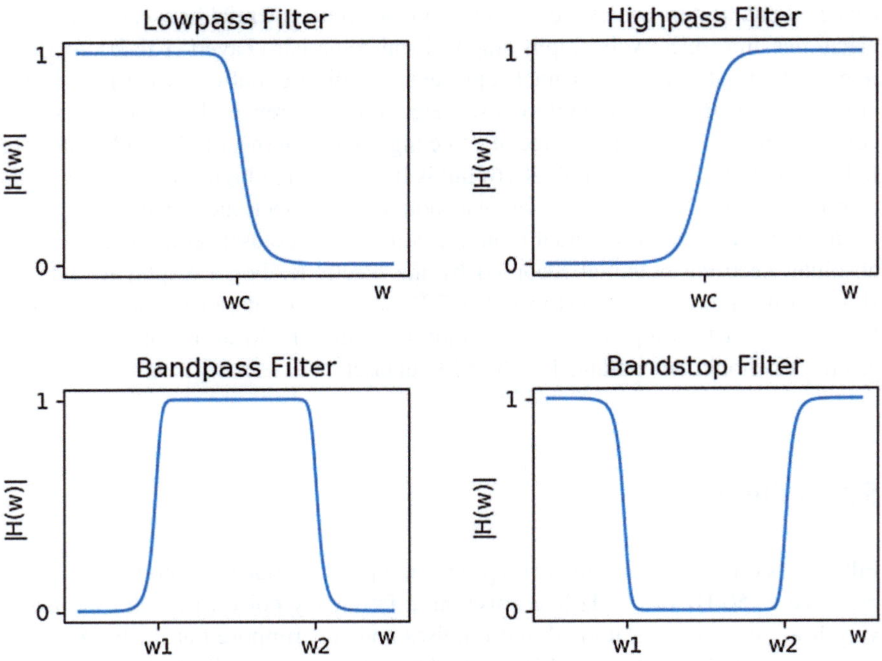

Fig. 5.3 Four types of filters: low-pass, high-pass, band-pass, and band-stop filters. A low-pass filter passes signals with a frequency lower than a selected cutoff frequency and attenuates signals with frequencies higher than the cutoff frequency, while a high-pass filter passes signals with a frequency higher than a certain cutoff frequency and attenuates signals with frequencies lower than the cutoff frequency. A band-pass filter passes frequencies within a certain range and attenuates frequencies outside that rånge, while a band-stop filter passes most frequencies unaltered and attenuates those in a specific range to very low levels

narrow frequency band and keeps signals within the rest of the spectrum almost undistorted. Beyond the consideration of frequency bands of artifacts, the selection of filters should also care about signals of which frequency range you are interested in. For example, if you are interested in the how the stimulation modulates the alpha signals, then a band-pass filter within alpha frequency range, such as 8–13 Hz range, is needed. In such a way, the signals within alpha frequency band are kept, and others outside this range are removed.

It is recommended to apply the filtering to the continuous EEG data as the first step of preprocessing pipeline, particularly before segmenting the continuous EEG data into epochs. Otherwise, it will introduce filtering artifacts at epoch boundaries if filtering the segmented EEG epochs. For example, we commonly apply the high-pass filter at 0.1 Hz to the continuous EEG data to minimize low-frequency slow drifts, but not the segmented EEG data. If applying the high-pass filter at 0.1 Hz to the segmented EEG data, it is done by filtering each segmented epoch separately, which will introduce filtering artifacts, since the edge artifact may last longer than the epochs (Cohen 2013).

5.4 Re-referencing

In the EEG recordings, there was a reference electrode, also termed as the common reference, e.g., one mastoid, linked mastoids, the vertex electrode, single or linked earlobes, or the nose tip. Since EEG data can also be re-referenced offline, the reference electrode during EEG recording is not very important. Particularly, in the recording systems with active electrodes (e.g., BIOSEMI Active Two) that record data without a recording reference, re-referencing offline is very necessary; otherwise there would be left 40 dB of unnecessary noise in the data (Delorme and Makeig 2004). Since referencing is a linear transformation, any activities present in the reference electrode will be reflected as activity in all other scalp electrodes. Therefore, when choosing a reference electrode, the reference electrodes should be properly placed and should have a good signal, because noise in the reference electrode will turn into noise in the scalp electrodes. Also note that when choosing a reference electrode, it is important that the reference electrode is placed far away from the locations of signals of interest; thus the reference electrode only has little influence on the signals of interest.

Mastoids (the bone behind the ear), the bilateral electrodes placed roughly behind the subject's ears, are frequently used as reference electrodes, since they are relatively far away from the locations of signals of interests. One of the ipsilateral mastoids, or the average of the bilateral mastoids, can be used. Nevertheless, referencing to one lateralized site is generally not recommended, because this will lead to a lateralization bias in the data. The average of the two earlobes is also commonly used. These are good choices because the reference electrodes are close to the other electrodes but record less brain activities. Nevertheless, these are not perfect references, because they can still measure neural activities at lateral temporal areas. The central electrode, Cz, is frequently used as the reference electrode, if the neural activities distant from that location are of interest and neural activities near that location are not of interest. In addition, the reference electrode could be theoretically placed anywhere such as the toe of the subject or a wall in the experiment room. Nevertheless, these are poor choices since the electrical potentials recorded from the reference electrode exhibit distinctive patterns compared with those from scalp electrodes. As a point at infinity is far from all the possible neural sources, Yao (2010) proposed a method to standardize a reference of scalp EEG recordings to a point at infinity, which could be considered as a reference electrode standardization technique (REST). It has been shown that REST is potentially effective for the most important superficial cortical region, and the standardization could be especially important in recovering the temporal information of EEG recordings.

For the EEG recording system with a large number of electrodes (e.g., >100), the electrode montage covers nearly the whole head, and an average reference is often recommended. The assumption of average reference is: the sum of the electric field values recorded at all scalp electrodes (sufficiently dense and evenly distributed) is always 0, and the current passing through the base of the skull to the neck and body is negligible. Nevertheless, this assumption is problematic, since the amount of electrode is not always dense and the distribution of electrodes over the head is not always even. Therefore, using average reference only makes sense when the EEG recording system has enough even channels so that the overall activity averages to 0. If you have less than 32 channels, consider using a different reference instead of average reference. Also note that the electrodes placed over nose tip or earlobe should be not included when estimating average reference.

5.5 Extracting Data Epochs and Removing Baseline Values

To particularly investigate sensory/cognitive event-related potentials, EEG data time-locked to the onset of specific events of interest were segmented to identify changes of EEG activities to the onset of sensory stimulation or cognitive tasks. In this segmentation procedure, the selection of the event onset (the time point "0") is important, which could be either stimulus onset or behavioral response made, represented as stimulus-related and response-related brain responses. The dimension of EEG data changed after the segmentation procedure (Fig. 5.4): 2-D of continuous EEG data (electrodes∗time) vs. 3-D of segmented EEG data (electrodes∗time∗trials). Nevertheless, the time duration in the continuous EEG data (recording time of the continuous EEG data) was much longer than that in the segmented EEG data (time window of extracting EEG segments). For the experiments in which several stimuli are presented with variable delays have multiple events that could be used as the time = 0 event, e.g., mismatch negative paradigm, it is recommended to extract the EEG data relative the earliest event so that all events are in the epoch or directly relative to the event of primary interest.

After extracting data epochs to specific experimental events, removing a mean baseline value from each epoch is necessary, since the electrical potential at baseline differs between data epochs. Baseline correction plays an important role in past and current methodological debates in ERP research, serving as a potential alternative to strong high-pass filtering. Nevertheless, baseline correction involves two inherent difficulties: (1) the choice of baseline interval and (2) the prior assumption of no systematic differences between conditions in the baseline interval.

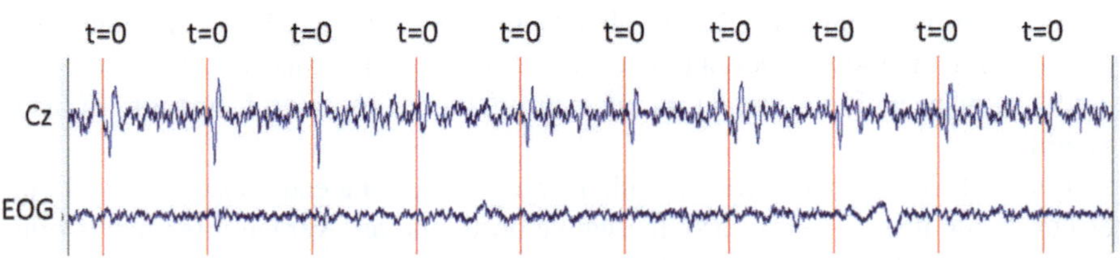

Fig. 5.4 Illustration of segmenting continuous EEG data into EEG epochs. EEG data time-locked to the onset of specific events of interest (t = 0) were segmented to identify changes of EEG activities to the onset of sensory stimulation or cognitive tasks

It is definitely necessary to extract data epochs and remove baseline values, thus for investigating stimulus/cognitive event-related potentials. Nevertheless, for the resting-state dataset, it is not necessary to perform data epoch extraction and baseline correction, even when sometimes the continuous EEG data were segmented into non-overlapping segments of a few seconds (e.g., 2 s) to facilitate the subsequent advanced analysis.

5.6 Removal and Interpolation of Bad Channels

Some of the EEG channels, especially for the high-density EEG channels, could not accurately provide the neurophysiological information of brain activities, when they are not properly placed on the scalp for various reasons. In this situation, these channels could be called as "bad channels," and it is important to exclude these channels from further analysis. When flagging the "bad channels," you could consider the following situations: (1) the channel is malfunctioning for some reasons; (2) the channel is improperly placed or does not have contact with the scalp; (3) two or more channels are bridged; and (4) the channel gets saturated. After flagging bad channels, you can directly remove these channels from further analysis.

Nevertheless, the dimension of EEG data matrix (channels∗time∗epochs) would change if you directly remove the bad channels. When you removed the "bad channels" for several subjects, the valid channels across subjects would be different (1 subject will have 63 electrodes, while another subject will have 64 electrodes), thus leading to inconveniencies to establish group-level EEG data. In addition, the removal of "bad channels" would also be problematic when you flag many "bad channels," and the valid channels are not enough to begin with the analysis, just because removing "bad channels" will result in loss of information. Due to these reasons, repairing these bad channels, by interpolating the "bad channels" based on the data from the "good channels," is an alternative method, instead of directly removing them. The most common interpolation of bad channels is by spherical splines (Greischar et al. 2004). It consists of the following steps: (1) projecting the good and bad electrodes onto a unit sphere; (2) computing a mapping matrix that maps N good channels to M bad channels; and (3) using this mapping matrix to compute interpolated data of the bad channels. The more channels you have, the more accurate the estimation of data on the bad channels will be. Nevertheless, the interpolated channel is no longer independent, and reduced the spatial resolution of the EEG, since it is a weighted sum of the activity of other channels and does not provide unique data. In the practical EEG preprocessing, the channels that are consistently flagged as bad channels could be removed, while the remaining bad channels could be repaired using interpolation method.

5.7 Removal of Bad Epochs

EEG epochs that are greatly contaminated by artifacts (e.g., eye blinks and movements), flagged as bad epochs, can be rejected away. For example, the EEG epochs can be rejected, if the subject closed his or her eyes for several hundred milliseconds or she or he may have been too tired on that trial to be focused on the task. Electromyography bursts are mainly located within 15–40 Hz in frequency, with relatively large amplitude and maximal electrodes around the face, neck, and ears. Epochs with excessive electromyography bursts in the EEG recordings should also be removed, when EEG signals at high frequencies (e.g., above 15 Hz) are of interest. Nevertheless, even you are not interested in the EEG data above 15 Hz, it is also recommended to remove the epochs with electromyography bursts, since electromyography bursts also indicate that the subject moved, sneezed, coughed, or giggled during that trial, but have not been engaged in activities of task during that trial.

The common approach to remove bad epochs is to visually inspect the data using an interactive viewer and mark the bad segments in the data manually. Nevertheless, there would be disagreement on whether the bad epochs should be rejected. Although well-trained experts are very likely to agree on the annotation of bad epochs, their judgement is subjected to fluctuations and not easy to be repeated. Another common and simple approach involving the rejection of bad epochs is the automatic rejection procedure, e.g., based on peak-to-peak signal amplitude values (Delorme and Makeig 2004; Oostenveld et al. 2011; Gramfort et al. 2014). When the peak-to-peak amplitude in the EEG data exceeds a pre-defined threshold, the respective EEG epoch could be considered as bad trial and removed in the following analysis. From the practitioner's standpoint, this approach seems quite easy to understand and simple to use. However, it can potentially lead to a large loss of data, consequently reducing the quality of the averaged ERP response. Another problem for automatic procedure to remove bad epochs is that the adopted criteria could not be appropriate for all subjects (may be appropriate for some subjects but not for others) and that the procedure could introduce both Type I (epochs you think should be retained are rejected) and Type II (epochs you think should be rejected are retained) errors.

Besides the removal of trials greatly contaminated by artifacts, sometimes we want to remove EEG epoch where subjects have made error behavioral response. It is concerned that epochs with error behaviors might influence both task performance and relevant brain activities. In addition, we may also want to remove epochs where subjects have made too slow (e.g., reactions times that are slower than three standard deviations from each subject's median reaction time) or too fast (e.g., reaction times less than 200 ms) response, concerning that the subject was not fully engaged in the experiment.

5.8 Removal of EEG Artifacts Using ICA

ICA was firstly created for dealing with the cocktail party problem, upon which you attempt to isolate a pertinent conversation from the noise of other conversations in a cocktail party (Hyvarinen and Oja 2000). Applying the ICA to the EEG data involves the decomposition of EEG time series data into a set of components. More specifically, EEG data are transformed to a collection of simultaneously recorded outputs of spatial filters applied to the whole multi-channel data, instead of a collection of simultaneously recorded single-channel data records. Thus, ICA is also a source separation technique that attempts to identify independent sources of variance in the EEG data (Anemuller et al. 2003).

In the original EEG data collected at single channels, each row of the recording data matrix represents the time course of summed in voltage differences between the respective channel and the references channels. After ICA decomposition, each row of the transformed data matrix represents the time course of the activity of one independent component that is spatially filtered from the channel data. The outputs of ICA procedure are statistically independent component (IC) waveforms, as well as matrix that transforms EEG data to IC data, and its inverse matrix to transform IC data back to EEG data. These outputs provide information about an IC's temporal and spatial properties. Concerning that ICA assumes an instantaneous relationship (e.g., common volume conduction) and that any relationship between EEG and EMG signals should involve propagation delays, it is recommended to only select EEG channels for ICA decomposition.

The EEG data recorded from scalp electrodes can be considered summations of real EEG signals and artifacts, which are independent of each other. ICA is therefore potentially a useful methodology to separate artifacts from EEG signals (Jung et al. 2000; Vorobyov and Cichocki 2002). To removal artifacts embedded in EEG recordings, the computed ICs are firstly classified as either artefactual or neural related components (Zou et al. 2016). If detected and flagged as artifact-related ICs, they can be subtracted from the recorded data, and the remaining data can be remixed. In the artifact correction, ICA is used to separate components in order to identify artifacts relevant with eye movements or heartbeats. These relevant ICs have characteristic shapes (topographies, time courses, and frequency spectra) and can often be identified automatically. That is, artifact-relevant components generally can be identified according to the topographies, across-trial temporal distributions, and frequency distributions of the components.

Abnormal topographies can be appeared as (1) power concentrated only in the frontal lobe in topography (ocular artifacts); (2) discontinued topography (noise artifacts); and (3) topography constrained within single electrode (electrode artifacts). Abnormal across-trial temporal distributions can be appeared as (1) inconsistent between epochs (without obvious peaks in average waveforms); (2) periodic waveform (power line interference); and (3) noisy pattern (similar to Gaussian noise). In addition, the frequency of artifact relevant components is in higher-frequency band (e.g., >30 Hz), while the frequency content of neural signals is in lower-frequency band (e.g., 5–20 Hz). Particularly, for components relevant with blink artifacts (as shown in Fig. 5.5), they have an anterior distribution, and their time courses are largely flat with occasional very high-amplitude spikes indicating artifacts of the eye muscles as they close and open.

Particularly, the application of ICA seems to be particularly useful in removing blinks and other oculomotor artifacts (Hoffmann and Falkenstein 2008; Plochl et al. 2012). Using ICA to correct artifacts is generally considered the best, since it does not assume orthogonal or gaussian behavior of the individual signals (Hyvarinen and Oja 2000). In contrast, PCA (principal component analysis) assumes that all signals are orthogonal and creates a succession of orthogonal base vectors, where each vector will account for as much variance as possible (Bugli and Lambert 2007). As a result, the first vector from PCA is significantly larger in magnitude than all the subsequent vectors. When the signal to noise ratio is low, important information in these subsequent vectors can get lost.

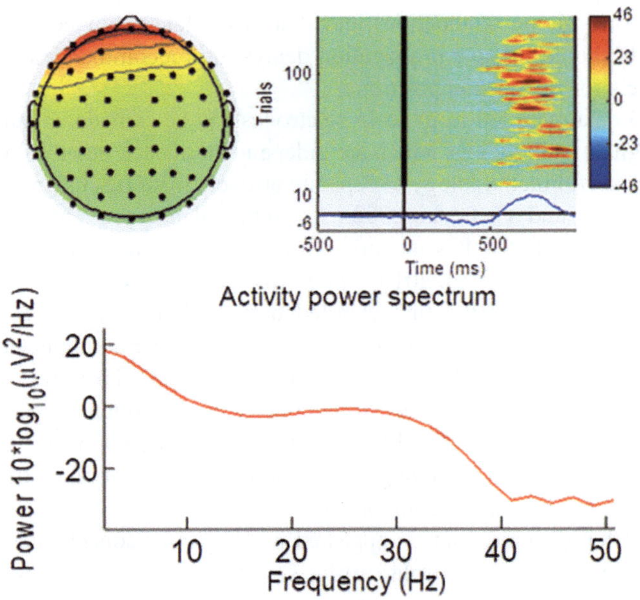

Fig. 5.5 Examples of components relevant with ocular artifacts. These components have typical characteristics, including anterior distributions and time course with occasional high-amplitude spikes indicating ocular artifacts

5.9 Summary

We have described several preprocessings that are commonly used in the EEG preprocessing, including filtering, re-referencing, segmenting, removal of bad channels and trials, as well as decomposition of EEG using ICA. As summarized in Table 5.2, the preprocessing pipeline below listed the commonly used preprocessing steps.

Admittedly, there is no universally adopted EEG preprocessing pipeline. It is suggested that investigators can make their own choice in choosing methods to transform or reorganize the raw data. Indeed, the application of preprocessing steps largely depend on the goal of the study, the details of the experiment design, the equipment of recording raw signals, as well as the advanced analysis you plan to

Table 5.2 Summary of preprocessing procedures

	Process	Remarks
1	Import data	From. CNT file
		Manual import of event and electrode information in the form of MATLAB arrays
2	Import channel locations	Load the channel location file describing channel positions
3	Remove empty channels	Remove HEOG, VEOG, M1, M2
4	Re-reference data	Compute common average
		Or reference to a specific channel(s)
5	Filter data	Low-pass filter at 30 Hz – >100 Hz
		High-pass filter at 1 Hz – >0.3 Hz
6–1	Extract epochs with correct responses	Extract epochs [−3.5 s, 0 s] around "correct responses"
6–2	Extract epochs around stimulus one	Extract epochs [−0.5 s, 1 s] around events
7	Baseline correction	Baseline latency range: [−0.5 s, −0.1 s] around events, that is, during fixation
8	Epoch rejection	Reject epochs with extreme values outside +/− 40 microvolts (higher threshold, e.g. 65/100uV)
		Reject epochs with abnormal trend >50 microV/epoch or R-squared >0.3
		Reject epochs with improbable data, outside 6 SD of single electrode or 5 SD of all electrodes
		Reject epochs with abnormal distribution, kurtosis outside 5 SD of mean kurtosis
		Aim to remove only small number of epochs (<10%)
9	Artifact removal	ICA on EEG epochs
		Manual component removal (see if it is inconsistent across trials and activation concentrate in frontal areas)
		2 components for eye and 1 component for sine wave
		Either remove as much epoch as possible or 2–4 components

perform. In other words, the details of applying certain preprocessing methods depend on what kind of the data that is being processed, how noisy the data is, and which technique would be used in the subsequent analysis.

Below are some suggestions that might help you choose appropriate preprocessing techniques. First, you can consider what kinds of artifacts might be present in your data and which ones you want to remove. For example, ocular-related artifacts such as eye movements and blinking could be considered a source of noise in many studies, but they could also reveal important patterns which are of interest in some studies. Second, you can consider which features you want to focus on. If you are interested in event-related potentials, you will need to have accurate temporal information. In contrast, you will need to have accurate spatial information if you are interested in motor imagery classification. Third, you can consider whether your analysis is performed online or offline. If you are interested in brain–computer interface and preprocessing data as soon as it arrives, you might not be able to use more computationally expensive methods.

References

Acharya JN, Hani AJ, Thirumala PD, Tsuchida TN. American clinical neurophysiology society guideline 3: a proposal for standard montages to be used in clinical EEG. J Clin Neurophysiol. 2016;33(4):312–6.

Anemuller J, Sejnowski TJ, Makeig S. Complex independent component analysis of frequency-domain electroencephalographic data. Neural Netw. 2003;16(9):1311–23.

Barlow J. Clinical applications of computer analysis of EEG and other neurophysiological signals. Handbook of EEG. Amsterdam: Elsevier; 1986.

Blum DE. Computer-based electroencephalography: technical basics, basis for new applications, and potential pitfalls. Electroencephalogr Clin Neurophysiol. 1998;106(2):118–26.

Bugli C, Lambert P. Comparison between principal component analysis and independent component analysis in electroencephalograms modelling. Biom J. 2007;49(2):312–27.

Cohen MK. Analyzing neural time series data: theory and practice. Cambridge, MA: MIT Press; 2013.

Delorme A, Makeig S. EEGLAB: an open source toolbox for analysis of single-trial EEG dynamics including independent component analysis. J Neurosci Methods. 2004;134(1):9–21.

Gramfort A, Luessi M, Larson E, Engemann DA, Strohmeier D, Brodbeck C, Parkkonen L, Hamalainen MS. MNE software for processing MEG and EEG data. NeuroImage. 2014;86:446–60.

Gratton G, Coles MG, Donchin E. A new method for off-line removal of ocular artifact. Electroencephalogr Clin Neurophysiol. 1983;55(4):468–84.

Greischar LL, Burghy CA, van Reekum CM, Jackson DC, Pizzagalli DA, Mueller C, Davidson RJ. Effects of electrode density and electrolyte spreading in dense array electroencephalographic recording. Clin Neurophysiol. 2004;115(3):710–20.

Herwig U, Satrapi P, Schonfeldt-Lecuona C. Using the international 10-20 EEG system for positioning of transcranial magnetic stimulation. Brain Topogr. 2003;16(2):95–9.

Hoffmann S, Falkenstein M. The correction of eye blink artefacts in the EEG: a comparison of two prominent methods. PLoS One. 2008;3(8):e3004.

Hu S, Lai YX, Valdes-Sosa PA, Bringas-Vega ML, Yao DZ. How do reference montage and electrodes setup affect the measured scalp EEG potentials? J Neural Eng. 2018;15(2):026013.

Hyvarinen A, Oja E. Independent component analysis: algorithms and applications. Neural Netw. 2000;13(4–5):411–30.

Islam MK, Rastegarnia A, Yang Z. Methods for artifact detection and removal from scalp EEG: a review. Neurophysiol Clin. 2016;46(4–5):287–305.

Jung TP, Makeig S, Humphries C, Lee TW, McKeown MJ, Iragui V, Sejnowski TJ. Removing electroencephalographic artifacts by blind source separation. Psychophysiology. 2000;37 (2):163–78.

Muthukumaraswamy SD. High-frequency brain activity and muscle artifacts in MEG/EEG: a review and recommendations. Front Hum Neurosci. 2013;7:138.

Nakamura M, Shibasaki H. Elimination of EKG artifacts from EEG records: a new method of non-cephalic referential EEG recording. Electroencephalogr Clin Neurophysiol. 1987;66 (1):89–92.

Oostenveld R, Fries P, Maris E, Schoffelen JM. FieldTrip: open source software for advanced analysis of MEG, EEG, and invasive electrophysiological data. Comput Intell Neurosci. 2011;2011:156869.

Plochl M, Ossandon JP, Konig P. Combining EEG and eye tracking: identification, characterization, and correction of eye movement artifacts in electroencephalographic data. Front Hum Neurosci. 2012;6:278.

Vorobyov S, Cichocki A. Blind noise reduction for multisensory signals using ICA and subspace filtering, with application to EEG analysis. Biol Cybern. 2002;86(4):293–303.

Yao DZ. A method to standardize a reference of scalp EEG recordings to a point at infinity. Physiol Meas. 2010;22(4):693–711.

Zaveri HP, Duckrow RB, Spencer SS. On the use of bipolar montages for time-series analysis of intracranial electroencephalograms. Clin Neurophysiol. 2006;117(9):2102–8.

Zou Y, Nathan V, Jafari R. Automatic identification of artifact-related independent components for artifact removal in EEG recordings. IEEE J Biomed Health Inform. 2016;20(1):73–81.

Chapter 6
Spectral and Time-Frequency Analysis

Zhiguo Zhang

Abstract EEG signals are typically characterized by oscillatory patterns at certain frequency bands. Normally, the EEG data, especially spontaneous EEG data, are analyzed in the frequency domain. The spectral analysis can transform EEG signals from time domain to the frequency domain, which can reveal how the power of EEG signals is distributed along frequencies. Furthermore, as EEG spectrum could substantially vary over time, joint time-frequency analysis is often used to reveal time-varying spectral activities of EEG. Particularly, time-frequency analysis is a powerful method to estimate the event-related EEG spectral patterns, i.e., event-related synchronization/desynchronization (ERS/ERD). In this chapter, I introduce some commonly used spectral estimation methods (e.g., the periodogram, the Welch's method, and the multitaper method) and time-frequency analysis methods (e.g., short-time Fourier transform and continuous wavelet transform). We also raise some practical issues and cautionary notes when using these methods on EEG data analysis, such as parameter tuning, visualization, and results reporting.

Keywords EEG · ERS/ERD · Fourier transform · Periodogram · Short-time Fourier transform · Wavelet transform

---

Electronic supplementary material The online version of this chapter (https://doi.org/10.1007/978-981-13-9113-2_6) contains supplementary material, which is available to authorized users.

Z. Zhang (✉)
School of Biomedical Engineering, Health Science Center, Shenzhen University, Shenzhen, Guangdong, China
e-mail: zgzhang@szu.edu.cn

L. Hu, Z. Zhang (eds.), *EEG Signal Processing and Feature Extraction*,
https://doi.org/10.1007/978-981-13-9113-2_6

6.1 Introduction

EEG originates from synchronous neural spikes, and it is in nature comprised of rhythmic activities in a wide spectrum of frequency (Niedermeyer and Lopes da Silva 2005; Buzsaki 2011; Cohen 2017). Therefore, EEG, especially spontaneous EEG, is commonly analyzed in the frequency domain, where one time-series signal is characterized by its periodicity. More precisely, it is necessary to describe how the signal power is distributed along frequency, which can be achieved by a wide variety of spectral estimation methods (Cohen 2014). Further, since EEG is modulated by a wide variety of internal states and external tasks, its spectrum is often unstable and substantially varies over time (Sanei and Chambers 2013; Schlogl 2000; Luck 2014). Particularly, in event-related experiments, sensory stimulation or cognitive tasks could increase or decrease EEG rhythmic activities at certain frequency bands, and these event-related spectral changes are known as event-related synchronization/(de)synchronization (ERS/ERD) (Pfurtscheller and Lopez da Silva 1999). The ERS/ERD are normally represented as time-variant spectral power in a joint time-frequency domain, which can be estimated and characterized by using a set of time-frequency analysis (TFA) methods (Roach and Mathalon 2008).

In the following, we will first present some basic concepts about frequency and spectrum. Then, we will briefly introduce the Fourier transform, which is the most fundamental tool for spectral estimation, and some other commonly used and well-established spectral estimation methods, such as the periodogram and the Welch's method. Next, the current mainstream TFA methods, including the short-time Fourier transform (STFT) and continuous wavelet transform (CWT), are introduced. Finally, we discuss how to estimate ERS/ERD of event-related EEG by using TFA and other relevant methods.

6.2 Spectral Estimation

6.2.1 Basic Concepts

Frequency is a fundamental and important parameter to describe the number of cycles of an oscillatory and vibratory waveform per unit time (Kay 1988; Stoica and Moses 2005). The unit of frequency is hertz (Hz), which means one cycle per second. A time-series signal, such as continuous EEG recording at one channel, can be represented and characterized either in the time domain as the change of signal amplitude (or other quantity) with respect to time or in the frequency domain as the change of signal power (or other quantity) with respect to frequency. Spectral estimation can transform a signal from time domain to frequency domain and can provide a description of the signal's power or energy distribution along frequency (Kay 1988; Stoica and Moses 2005). Spectral estimation is popularly used in a very wide range of signals that exhibit oscillatory and rhythmic patterns, and it is of particular importance in the analysis of EEG.

Mathematically, spectral estimation is defined as the estimation of the spectral density of a random process from a sequence of time samples of the process (Kay 1988). The purpose of spectrum estimation is to detect periodicities in the data, by observing peaks at the frequencies corresponding to these periodicities. Figure 6.1 shows a simulated signal consisting of three sinusoids, whose frequencies are, respectively, 2 Hz, 8 Hz, and 20 Hz and amplitudes are, respectively, 1, 0.8, and 0.5. The time-series signal in Fig. 6.1a clearly shows rhythms and periodicity, but the rhythms and periodicity can hardly be directly quantified from the time-domain representation. By using the Fourier transform, the signal is represented in the frequency domain (Fig. 6.1b), so the frequency and amplitude of these three sinusoids can be well determined.

In the context of EEG analysis, spectral estimation is normally applied on continuous EEG recordings in a time period to calculate the power of several certain rhythms: theta (1–4 Hz), delta (4–8 Hz), alpha (8–12 Hz), beta (12–20 Hz), gamma (>20 Hz).

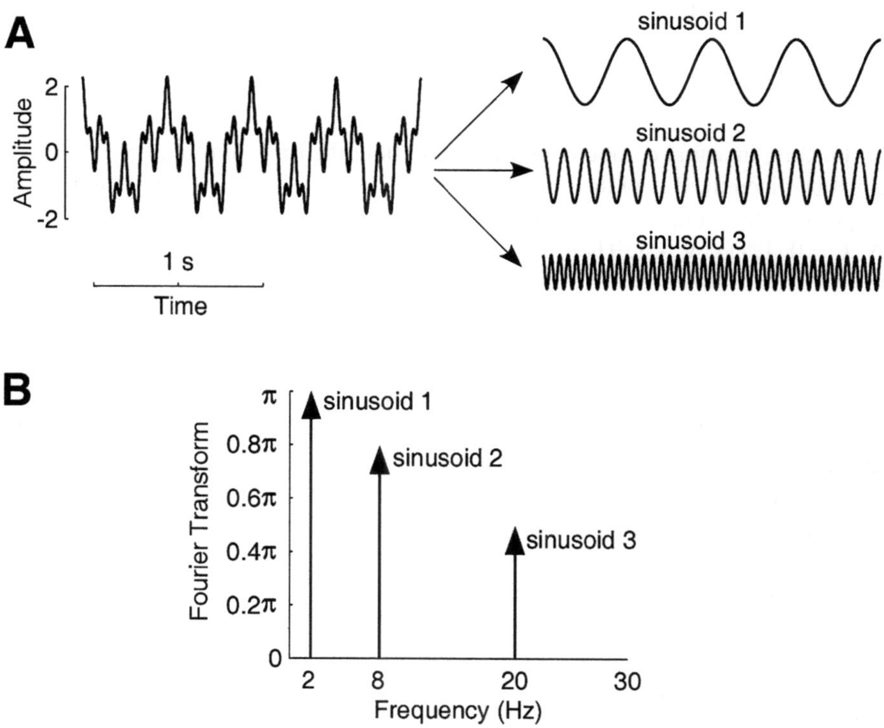

Fig. 6.1 Time-domain representation and frequency-domain representation of a simulated signal. (**a**) The simulated signal consists of three sinusoids, whose frequencies are, respectively, 2 Hz, 8 Hz, and 20 Hz and amplitudes are, respectively, 1, 0.8, and 0.5. (**b**) The signal's spectrum (calculated using the Fourier transform; see Sect. 6.2.2 for details) exhibits three peaks at 2 Hz, 8 Hz, and 10 Hz

The foundation of spectral estimation is the Fourier transform. The Fourier transform represents and characterizes a time-series signal as the summation of a series of sines and cosines. The details of Fourier transform are omitted here, and these details can be easily found in textbooks of signal processing, such as (Oppenheim et al. 1996; Mitra 2000; Proakis and Manolakis 2006). The Fourier transform for a continuous signal $x(t)$ is called continuous-time Fourier transform (CTFT), and it is calculated as

$$F(f) = \int_{-\infty}^{\infty} x(t)e^{-j2\pi f\tau}d\tau, \tag{6.1}$$

where f is the frequency with a unit of Hz. The frequency range of spectrum is from 0 to the half of the sampling rate (which is known as the Nyquist frequency). For example, the EEG signal in Fig. 6.2 has a sampling rate of 160 Hz; then the highest frequency that can be detected from the signal is 80 Hz.

Because real-world EEG obtained by any device is not continuous in time but sampled by analog-to-digital convertors at discrete time points, discrete-time Fourier transform (DTFT) is actually used instead of CTFT to perform the Fourier transform on EEG. DTFT of a discrete-time signal $x[n]$, $n = 1, 2, \cdots, N$ is calculated as

$$F(f) = \sum\nolimits_{n=1}^{N} x[n]e^{-j2\pi fn}. \tag{6.2}$$

Further, in actual computer computation, both the time and frequency domains must be discretized, and the number of sample points, N, must be a finite number. Supposing the sampling rate is F_s, the frequency domain is discretized as $f = kF_s/N$, $k = 0, 1, \cdots, N - 1$. Then, we have discrete Fourier transform (DFT) as

$$F[k] = \sum\nolimits_{n=0}^{N-1} x[n]e^{-j2\pi kn/N}. \tag{6.3}$$

For efficient computation, DFT is generally calculated using the celebrated fast Fourier transform (FFT) algorithm. In FFT, the number of sample points N is usually chosen as a power of 2, which can be achieved by zero padding (i.e., to add a number of zeros at the end of the original time-series data).

Based on the Fourier transform, we can calculate the power spectral density (PSD), which describes how the power of a random signal is distributed over different frequencies. PSD is sometimes simplified as "spectral density" or "power spectrum," and its unit is V^2/Hz or dB (i.e., $10\log_{10}(V^2/\text{Hz})$).

Spectral estimation is an extensively studied topic in the field of signal processing, and there are numerous spectral estimation methods developed for different types of signals and for different analysis purposes. In the following, we

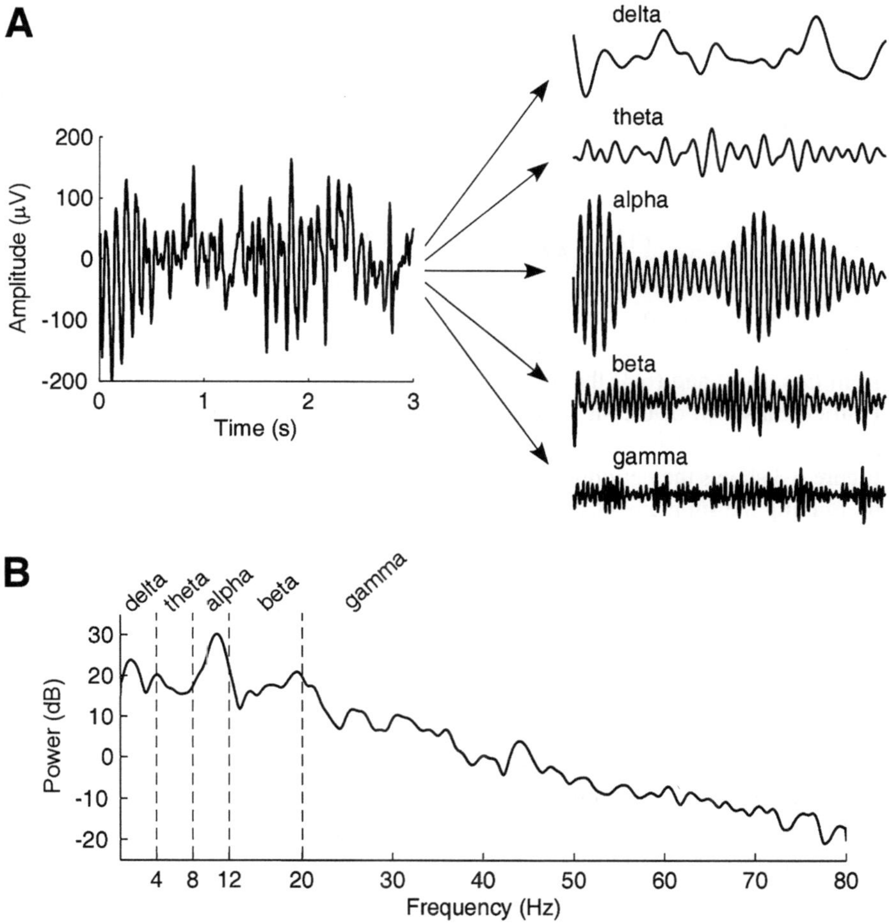

Fig. 6.2 Time-domain representation and frequency-domain representation (the spectrum) of an EEG signal with eyes closed. (**a**) The EEG signal is recorded at Oz and has a duration of 3 s and a sampling rate of 160 Hz. By using bandpass filtering with difference cutoff frequencies, the signal can be decomposed into five rhythms. (**b**) The spectrum of the EEG signal is calculated using the Welch's method (see Sect. 6.2.3 for details), and it shows how the signal power is distributed along frequency. It can be seen clearly the alpha rhythm has the highest power among five rhythms

will introduce some widely used spectral estimation methods of EEG, including (1) the periodogram, (2) the Welch's method, (3) the multitaper method, and (4) the autoregressive model-based spectral estimation.

6.2.2 Periodogram

Periodogram is a simple, yet popular, spectral estimation method. The periodogram of a discrete-time EEG signal $x[n]$, $n = 1, 2, \cdots, N$, with a sampling rate of F_s is calculated as

$$P(f) = \frac{1}{NF_s}\left|\sum_{n=1}^{N} x[n]w[n]e^{-j2\pi fn/F_s}\right|^2, \tag{6.4}$$

where $w[n]$ is a window function. This window function is used to assign different weights to samples, which are normally set to be "1" for all samples. It can be seen from Eqs. (6.3) and (6.4) that periodogram is simply the squared magnitude of the Fourier transform of the signal (multiplying by a constant). The calculation of the periodogram can be executed in MATLAB using the script "periodogram.m". Figure 6.3 shows the periodogram of an EEG signal, and the results can be obtained by running the supplementary MATLAB script – "demo_periodogram.m" attached in this book. Note that the periodogram in Fig. 6.3 is shown in two different scales: the linear scale and the logarithmic scale. The linear scale can highlight the predominant spectral peaks (such as the peak around 10 Hz in Fig. 6.3a) but will make other spectral components (especially in the high-frequency band) undistinguishable. On the other hand, by using a logarithmic scale, spectral components of different frequency bands are more visually comparable, but the spectral peaks cannot stand out. Because it is more often to examine EEG spectral power over a wide range of frequency, it is more common to show the periodogram (and spectrum estimated by other methods) in a logarithmic scale.

It should also be noted that, as EEG spectrum has a well-known "1/*f*" characteristic (i.e., the power decreases rapidly as a function of the frequency f), the

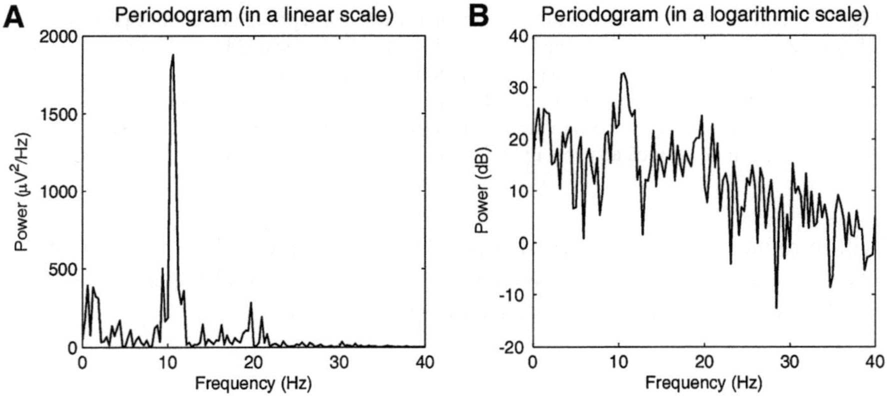

Fig. 6.3 Spectrum of the eyes-closed EEG signal (as shown in Fig. 6.2) estimated using the periodogram (with a rectangular window). (**a**) Periodogram shown in a linear scale. (**b**) Periodogram shown in a logarithmic scale

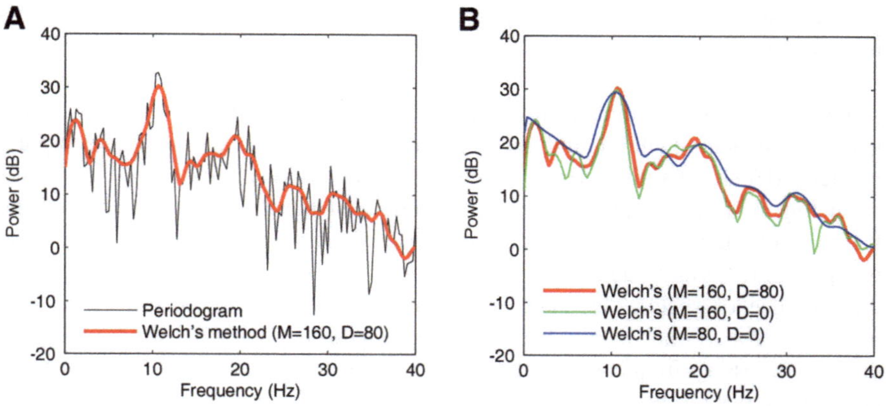

Fig. 6.4 Spectrum of the eyes-closed EEG signal (as shown in Fig. 6.2) estimated using the Welch's method with different parameters. (**a**) Comparison between the Welch's method and the periodogram. (**b**) Comparison between the Welch's method with different parameters of the window length and overlapping length

components at very low frequencies (e.g., <1 Hz) often dominate the spectrum. It is necessary to perform a detrending operation prior to spectral estimation. Detrending means to move a trend (usually a linear trend) from the signal, and it can be easily realized by the MATLAB function "detrend.m".

We can also see from Fig. 6.3 that the periodogram shows very high variance across frequency. Mathematically, the variance of the periodogram is the square of the periodogram, which is a constant independent of the number of samples N. It means that we cannot decrease the variance by using more data samples. As a result, the periodogram often exhibits very high variability, and spectral peaks may be difficult to be clearly observed and precisely located. For some EEG applications where a smooth representation of spectrum is needed, the periodogram may not be suitable.

6.2.3 *The Welch's Method*

To address the periodogram's problem of high variability, a number of smoothed versions of periodogram have been proposed, and these include smoothed periodogram (also known as the Daniell's method), averaged periodogram (also known as the Bartlett's method), and the Welch's method. Among these methods, the Welch's method is most popularly used in EEG analysis.

The Welch's method first divides the signal of N samples into K data segments of M samples, overlapping by D samples. If $D = M/2$, the overlap is 50%; if $D = 0$, the overlap is 0%. The overlapping segments are then windowed (i.e., multiplied by a symmetric bell-shaped window). Next, we can compute the DFT and get the periodogram of each windowed data segment. Then, it calculates the average of periodograms of all these data segments as the final spectral estimate. The Welch's method can lead to a reduction in the variance of spectral estimate as $1/K$.

The calculation of the Welch's method can be executed in MATLAB using the script "pwelch.m". Figure 6.4 shows the spectral density of an EEG signal estimated using the Welch's method, and the results can be obtained by running the supplementary MATLAB script "demo_welch.m" attached in this book.

In Fig. 6.4b, different parameters, the length of the data segments M and the length of overlapping D, lead to different spectral estimation results, because these two parameters jointly determine the number of data segments K and subsequently the variance of spectral estimate (note that the variance of spectral estimate is inversely proportional to K). First, the data segments should be long enough to provide necessary frequency resolution (the frequency interval is F_s/M); but a too large M implies too few segments, which cannot significantly decrease the variance of spectral estimation. Second, the length of overlapping D also determines K (a large D leads to a large L and vice versa) and subsequently the variance of the spectral estimate.

In addition, the window function used in the Welch's method can also influence the spectral estimates. A rectangular window has the best ability to resolve adjacent sinusoids (i.e., high resolution), but its wide spreading may mask important details at even lower levels (i.e., low dynamic range). Non-rectangular windows (such as the Hamming window, the Hann window, etc.) can redistribute leakage to places, where it does the least harm (high dynamic range) at the expense of increasing the leakage in the vicinity of the original component (low resolution). However, the window function will not largely influence the spectral estimation results in general. This issue is seldom discussed in the field of EEG analysis.

6.2.4 The Multitaper Method

The Welch's method decreases the variability of the periodogram by averaging the periodograms of windowed data segments. In the Welch's method, the window function is fixed, but the data segments are different. The multitaper method shares the similar idea of decreasing the variability by averaging periodograms of windowed data as the Welch's method, but in the multitaper method, the window functions are different, but the data segments are the same (all the data samples). Actually, the multitaper method averages modified periodograms obtained using a family of mutually orthogonal tapers, which are known as discrete prolate spheroidal sequences (Babadi and Brown 2014). In addition to mutual orthogonality, the tapers also have optimal time-frequency concentration properties. Each taper is multiplied by the data to form the tapered data, and then the periodogram of the tapered data is estimated. Subsequently, the periodograms of tapered data by using all tapers are averaged to produce spectral estimates with smaller variability.

The multitaper method can be executed in MATLAB using the script "pmtm.m". In "pmtm.m", we need to specify a time-half bandwidth product, nw, which determines

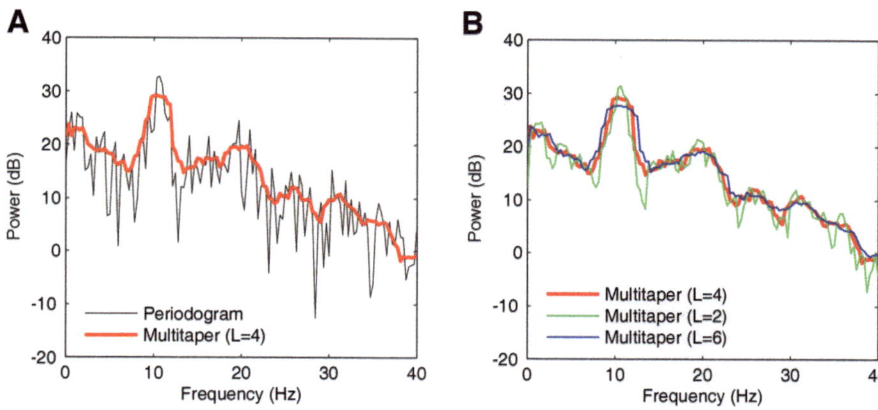

Fig. 6.5 Spectrum of the eyes-closed EEG signal (as shown in Fig. 6.2) estimated using the multitaper method with different parameters. (**a**) Comparison between the multitaper method and the periodogram. (**b**) Comparison between the multitaper method with different parameters of the number of tapers

the frequency resolution of the multitaper estimate. Because the number of tapers used in "pmtm.m" is (2∗nw-1) and the estimate variance is inversely proportional to the number of tapers, a large nw will lead to a smoother multitaper spectral estimate and vice versa. Figure 6.5 shows the spectral density of an EEG signal estimated using the multitaper method, and the results can be obtained by running the supplementary MATLAB script "demo_multitaper.m" attached in this book.

6.2.5 *Autoregressive Method*

The periodogram, the Welch's method, and the multitaper method all belong to a general class of non-parametric spectral estimation methods. Non-parametric methods estimate the spectrum purely from observed data, without assuming any particular model for the data. Unlike data-driven non-parametric methods, another class of methods is known as parametric spectral estimation, which first describes the signal under study using a specific model. Then the model parameters can be estimated from data, and finally the signal's spectrum can be calculated from the estimated model parameters. The most common model used for spectral estimation is the autoregressive (AR) model. An AR model of order P, which is denoted as AR (P), is written as

$$x[n] = \sum\nolimits_{p=1}^{P} a_p x[n-p] + e[n], \tag{6.5}$$

where a_p, $p = 1, 2, \cdots, P$, are AR coefficients and $e[n]$ is a random noise.

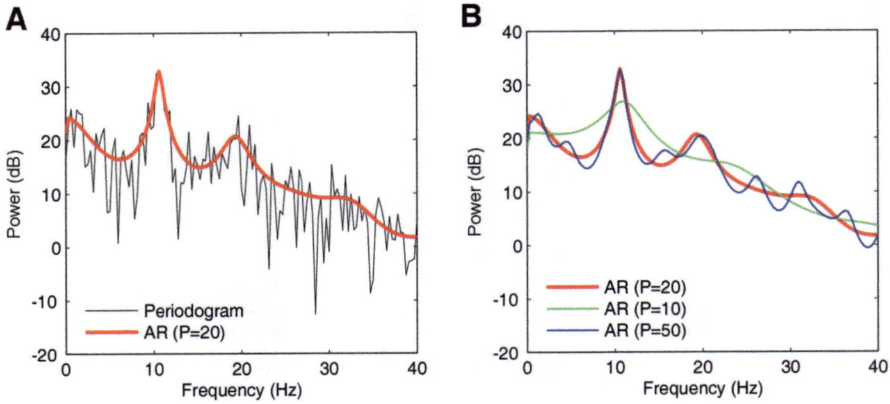

Fig. 6.6 Spectrum of the eyes-closed EEG signal (as shown in Fig. 6.2) estimated using the AR model (with different model orders) and the Yule-Walker method. (**a**) Comparison between the AR-based method and the periodogram. (**b**) Comparison between the AR model-based spectra with different model orders

The AR coefficients can be calculated from data samples using several methods, for example, the Yule-Walker method, the Burg's method, and the covariance method. For simplicity, we skip the detailed derivation of the estimation of the AR coefficients. Suppose the AR coefficients have been estimated, then the spectrum of the signal $x[n]$ can be calculated from these AR coefficients as

$$P(f) = \frac{\sigma^2}{\left|1 + \sum_{p=1}^{P} a_p e^{-j2\pi fp/Fs}\right|^2}, \tag{6.6}$$

where σ^2 is the variance of the white noise $e[n]$ and can also be estimated with abovementioned method, such as the Yule-Walker method.

AR spectral estimation using the Yule-Walker method can be executed in MATLAB using the script "pyulear.m". Figure 6.6 shows the spectral density of an EEG signal estimated based on the AR model and using the Yule-Walker method, and the results can be obtained by running the supplementary MATLAB script "demo_yulear.m" attached in this book.

Figure 6.6 also shows that different model orders greatly influence the estimation. A key problem in AR spectral estimation is to select the model order P. The higher the model order is, the less variance remains unaccounted. However, an excessive number of model parameters increases the statistical variability of their estimates. In another word, if P is chosen too large, spurious peaks in the spectral estimates will appear; if P is too small, the resolution is too low (not sufficient to resolve two adjacent sinusoids). In the field of signal processing, a rule of thumb is that the order is twice the number of sinusoids in the signal if the signal-to-noise ratio (SNR) is very high, and the optimal order decreases with the SNR. There are also some criteria for model order selection in AR spectrum. These include:

- Akaike information criterion (AIC):

$$AIC = N \ln\left(\sigma^2\right) + 2P \tag{6.7a}$$

- Bayesian information criterion (BIC):

$$AIC = N \ln\left(\sigma^2\right) + P \ln(N) \tag{6.7b}$$

- Final prediction error (FPE):

$$FPE = \sigma^2(N + P + 1)/(N - P - 1) \tag{6.7c}$$

In practice, we could also determine the model order by getting guidance from literature or trying out many model orders.

6.2.6 Comparison of Spectral Estimation Methods

Different spectral estimation methods have their unique properties and are suitable for different types of EEG signals and different research purposes. We first discuss the advantages and disadvantages of non-parametric methods and parametric methods, respectively. Non-parametric methods are based on data solely and do not require any priori definition of the model and its order. Hence, non-parametric methods are much simpler than parametric methods from a methodological and computational point of view. In contrast, as parametric methods are more statistically consistent, even on short segments, they can achieve reliable calculation of (physiologically interpretable) spectral parameters. Parametric methods do not need windowing and, thus, have no spectral leakage. Also, the frequency resolution of parametric methods is independent of number of data.

In EEG processing, non-parametric methods are more widely used, because the advantages of non-parametric methods match the characteristics of EEG signals. First, if the spectra are smooth, the non-parametric methods are adequate and more reasonable. Second, when additive noise is large, non-parametric methods are more accurate, because parametric methods are very sensitive to high noise levels. Third, if the length of data is long, non-parametric methods are sufficient and more reasonable. Because EEG signals typically have wide and smooth spectra, contain large amount of noise, and are sufficiently long (thousands and more samples), non-parametric methods are more suitable for EEG analyses. Of course, if model order is correctly specified, AR model-based spectrum can also achieve satisfactory results. Figure 6.7 shows the comparisons between some non-parametric and parametric methods on the eye-closed EEG. It can be seen clearly that, if an appropriate

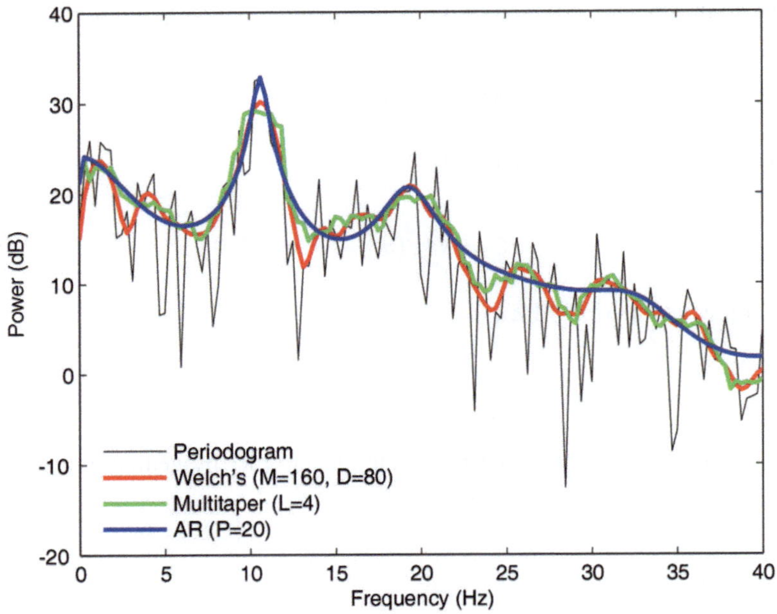

Fig. 6.7 Comparisons of spectral estimates of the eyes-closed EEG signal (as shown in Fig. 6.2)

model order is selected for the AR model (in this case, $P = 20$), AR model-based spectrum estimation is very similar to the estimate of the Welch's method.

Last but not the least, it is strongly advised to report parameters and settings (such as the number of FFT points, zero padding, windowing parameters, etc.) used in spectral estimation methods when reporting related results, because these parameters could largely influence the results (Pernet et al. 2018).

6.2.7 *Extraction of Spectral Features*

After an EEG signal's spectrum is estimated, it is still necessary to extract meaningful features from the spectral estimate. The most common EEG spectral features are band-limited power: spectral power within a certain frequency band. The frequency limits of a frequency band are determined either by conventional definition (i.e., the frequency bands of the delta, theta, alpha, beta, gamma rhythms) or by statistical analyses that can detect the most important frequency (e.g., at which the EEG power of two groups/conditions are different, as discussed in Chap. 17). Band-limited power can be calculated as the average or the summation of power values at frequency points within the specific frequency band of interest or concern. To eliminate the difference between individuals, relative power (the ratio between band-limited power and the total power in the whole frequency range) is often used. For example, alpha power (the band-limited power of the alpha rhythm from 8 Hz to 12 Hz) can be calculated as the spectral estimate averaged from all frequency points within the range 8–12 Hz, and relative alpha power is the ratio between the alpha power and the total power of the EEG signal.

Another type of important spectral features are the location, magnitude, and bandwidth of spectral peaks. For example, in the applications of brain-computer interfaces based on steady-state visual evoked potentials (SSVEP), it is necessary to locate the spectral peaks and extract the magnitudes of these peaks. Another example is to identify different alpha peaks of different individuals, for the inter-individual differences in alpha peaks could be indicative of an individual's traits or states. Lastly, sometimes we can also treat the spectrum as a random process and compute its statistics, such as the mean, variance, and entropy as features.

6.3 Time-Frequency Analysis

6.3.1 Basic Concepts

In spectral analysis, we have a basic assumption that the spectrum of an EEG signal is fixed and does not change over time. Apparently, this assumption oversimplifies the non-stationary characteristics and dynamic behaviors of EEG. EEG signals are actually highly non-stationary, which means that its statistical properties (e.g., the spectral density) substantially vary with time. Although time-varying characteristics of EEG spectrum could be attributed to noise or random disturbances, they could be modulated by experimental conditions or mental states and thus convey important information. For example, EEG spectrum will change considerably depending on various physiological and psychological states, e.g., eyes-open/closed (Barry et al. 2009), sleep (Fell et al. 1996), meditation (Cahn and Polich 2006). However, for non-stationary EEG, spectral estimation is unable to identify its time-variant spectral components and is unable to provide simultaneous time and frequency localization.

Figure 6.8 shows an EEG signal during two conditions (0–2 s, eyes-open; 2–4 s, eyes-closed) and its spectra in these two periods. Clearly, the signal in two periods has largely different spectra (black solid lines in Fig. 6.8b), particularly in the alpha band. But this difference cannot be observed from the spectrum of the whole period of signal (red dashed lines in Fig. 6.8b), because the spectrum does not contain any time information. To see how the spectrum changes with time, time-frequency analysis can provide a joint time-frequency distribution (TFD) of the signal power (i.e., the distribution pattern of power at each time point as well as at each frequency) (Boashash 2015). For example, in the TFD of Fig. 8C, the spectral power is strong around 10 Hz (frequency information) and in the range of 2–4 s (time information), and such joint time-frequency information indicates the experimental condition of eyes closed.

To reveal time-varying spectrum of non-stationary EEG signals, time-frequency analysis (TFA) techniques are needed. TFA comprises a batch of methods that

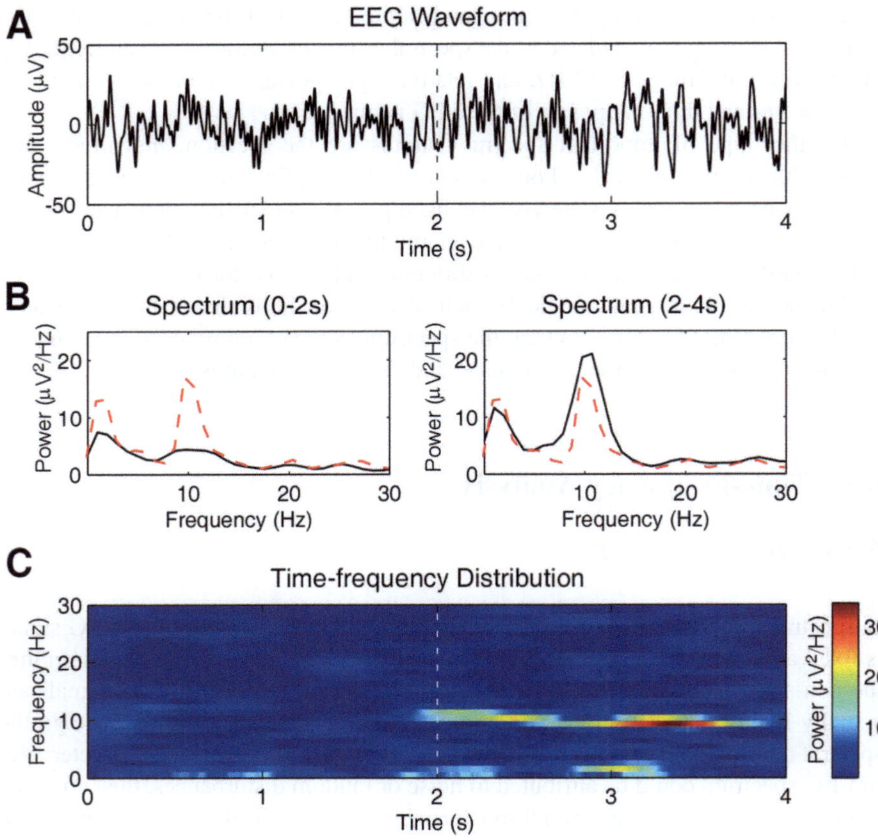

Fig. 6.8 An EEG signal recorded during two conditions: eyes open (0–2 s) and eyes closed (2–4 s). (**a**) The EEG waveform was recorded at Oz and with a sampling rate of 1000 Hz. (**b**) The spectrum of the whole period of EEG (0–4 s, shown as red dashed lines) and the spectra of the EEG signal during the eye-open (0–2 s) and eye-closed (2–4 s) conditions (shown as black solid lines), all of which are estimated using the Welch's method. It can be seen that the spectral estimates are largely different between two conditions, especially in the alpha band. (**c**) The TFD of the EEG signal, estimated using short-time Fourier transform (see Sect. 6.3.2 for details). The onset of alpha wave can be clearly observed around 2 s, indicating the start time of the eyes-closed condition

study a signal in both the time and frequency domains simultaneously. In general, there are two types of TFA techniques: time-frequency power distributions and time-frequency signal decomposition. Time-frequency power distributions can represent the amount of signal power assigned to a given time-frequency point, and some popular methods include the short-time Fourier transform (STFT) (Delorme and Makeig 2004) and the continuous wavelet transform (CWT) (Delorme and Makeig 2004; Mouraux and Iannetti 2008). Time-frequency signal decomposition methods decompose a signal into a set of additive components according to their distinct time-frequency properties, and some typical methods include the discrete wavelet transform (DWT) (Adeli et al. 2003), matching pursuit

(Durka 2007), and empirical mode decomposition (Sweeneyreed and Nasuto 2007). In the following, I mainly introduce two popular time-frequency power distributions methods for EEG: STFT and CWT.

6.3.2 *Short-Time Fourier Transform (STFT)*

STFT is a simple and common method for TFA of EEG, and it is based on a general sliding-window approach. The sliding-window approach assumes the non-stationary signal can be sliced into a number of short data segments, where the assumption of stationarity holds. In another word, although the signal has time-varying spectra, the spectrum of any one short data segment is fixed. Based on the idea of sliding window, conventional spectral estimation methods (such as the periodogram, the Welch's method, the AR method) can be performed on each of these segments. Spectral estimates of all windows can then be concatenated to form a spectral power distribution in the joint time-frequency domain. Figure 6.9 shows the steps of the sliding-window approach. These steps are also listed below:

1. Select a window function of finite length.
2. Place the window on top of the signal at $t = 0$.
3. Segment the signal using this window.
4. Compute the spectrum of the windowed data segment.
5. Incrementally slide the window along time.
6. Go to step 3, until the window reaches the end of the signal.

The most important properties of all TFA methods are their time resolution and frequency resolution. Time resolution and frequency resolution of a TFD are its capability to discriminate two signal components that are closely adjacent in the time domain or in the frequency domain. In STFT (and all other sliding-window approaches), there is a tradeoff between time-resolution and frequency resolution, which is known as the uncertainty principle of TFA. The uncertain principle tells us that one cannot know the exact time-frequency representation of a signal: time resolution and frequency resolution cannot be made arbitrarily small (either we resolve in frequency or in time, but not both). In another word, we cannot precisely know at what time instance a frequency component is located, and we can only know what interval of frequency are present in which time intervals.

Figure 6.10 shows the TFDs of an EEG signal by using STFT with different window sizes, and the results can be obtained by running the supplementary MATLAB script "demo_stft.m" attached in this book. We can see from Fig. 6.10 that different window sizes lead to different time-frequency resolution. A shorter window has better time resolution, so that the VEP component can be clearly observed and localized in time (around 150 ms) in Fig. 6.10b, where a window of 200 ms is used. But a shorter window also leads to poor frequency resolution, so the VEP component were dispersed in the frequency domain (from 0 to 15 Hz) and,

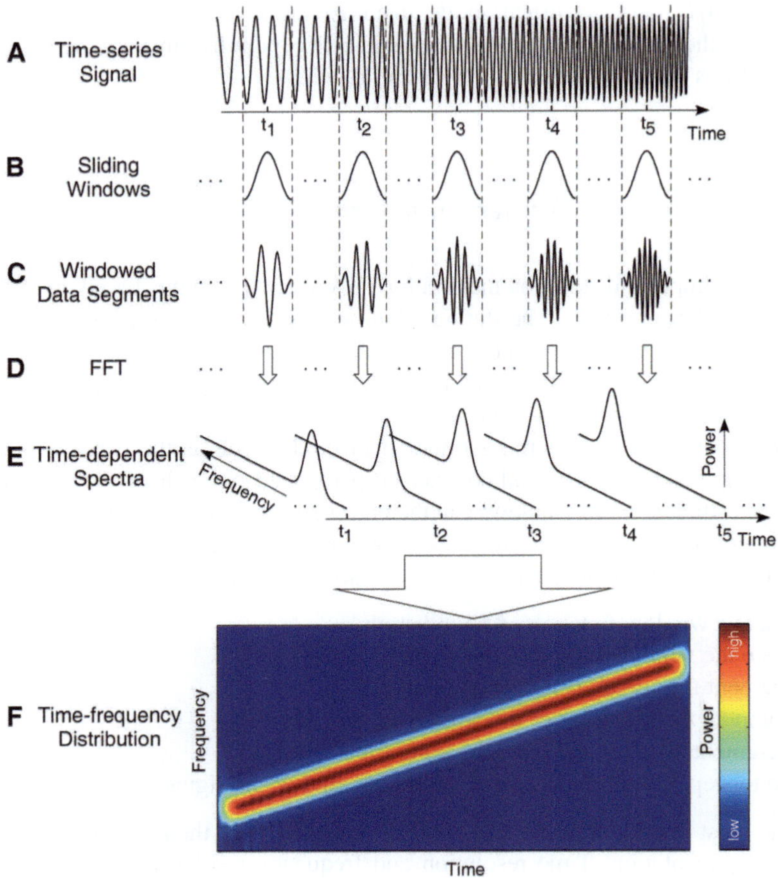

Fig. 6.9 Illustration of the general idea of the sliding-window approach (taking STFT for example). (**a**) The simulated signal has one linearly increased frequency. (**b**) Choose a symmetric, bell-shaped, and finite-length window, and slide it along the time axis. (**c**) At each time point (five points t1, . . ., t5 are shown here as examples), multiply the window function with the signal to generate a windowed data segment. (**d**) At each time point, use FFT (or other spectral estimation methods) to estimate the spectrum of the windowed data segment. E: The time-dependent spectra of all time points describe how the signal's spectral power changes with time. F: The time-dependent spectra are normally represented as a time-frequency data matrix, and the power of each point of the matrix is color coded. The data matrix is also known as time-frequency distribution (TFD), which characterizes how the signal power is distributed in the joint time-frequency domain

more seriously, the P300 (which has a low frequency around 2-3 Hz) cannot be observed in Fig. 6.10b. On the other hand, a longer window achieves good frequency resolution at the expense of poor time resolution. In Fig. 6.10d, where a window of 800 ms is used, the P300 component is very clear-cut in the time range of 300–600 ms and around 2–3 Hz. But the VEP is not clearly shown in Fig. 6.10d because the time resolution is so poor that some of the VEP power is leaked into the pre-stimulus range. In summary, window selection is the fundamental problem with

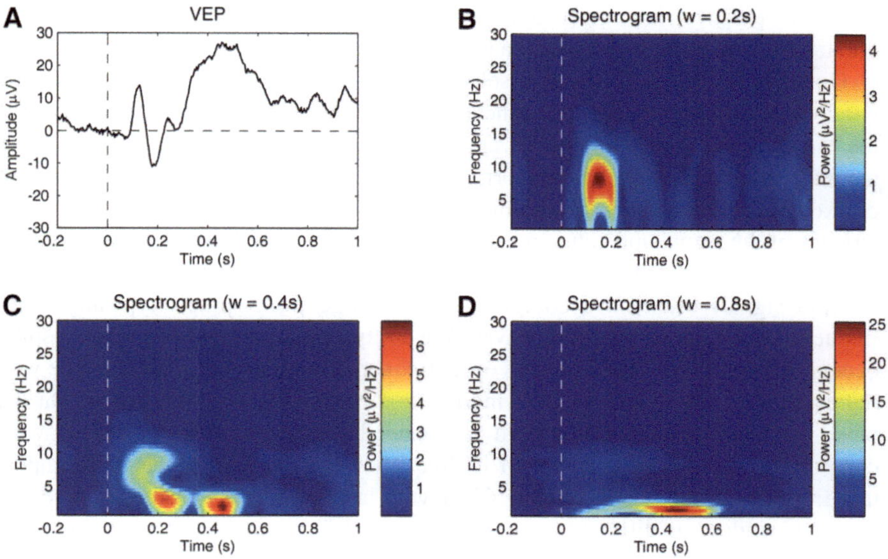

Fig. 6.10 STFTs with different window sizes of a visual evoked potential (VEP) signal. (**a**) The VEP signal was recorded with a sampling rate of 250 Hz at O1 in a visual oddball experiment. The VEP waveform shown here is the average of 30 trials. The original data is downloaded from https://vis.caltech.edu/~rodri/data/cg_o1t.asc. (**b**) STFT with a window size of 0.2 s. (**c**) STFT with a window size of 0.4 s. (**d**) STFT with a window size of 0.8 s

Table 6.1 The influence of window size on the resolution of STFT

Window size in time domain	Window size in frequency domain	Time resolution	Frequency resolution	Suitable for
Long	Short	Low	High	Slow-varying spectral components
Short	Long	High	Low	Transients, spikes, and discontinuities

STFT (and all sliding-window approaches). Table 6.1 summarizes the window sizes in different domains and their influence on the time resolution and frequency resolution of TFD.

6.3.3 Continuous Wavelet Transform (CWT)

The fundamental limitation of STFT is that the fixed-length window leads to fixed time-frequency resolution in the whole time-frequency domain. But, as discussed in Table 6.1, the window size should be varied in the time-frequency domain to optimize the time-frequency resolution for diversified spectral components in a signal. In practice, it is often the case that EEG signals are composed of short duration events of high-frequency and low-frequency events of long duration.

One useful strategy of window selection is to use an adaptive and variable time window that is short at high frequencies and long at low frequencies. Actually, it is the window selection strategy of another very popular time-frequency analysis method: continuous wavelet transform (CWT). CWT is able to address the problem of fixed time-frequency resolution in STFT according to the time-frequency characteristics of the signal and actual needs. More precisely, CWT uses long windows in the low-frequency range and short windows in the high-frequency range.

EEG signals often have components that range from spikes (typically high-frequency components) to slow waves. We are typically interested in when a spike occurs and don't care much what its exact frequency contents are, which can be well represented by CWT with small windows that have good time resolution. On the other hand, we usually don't need to locate the exact interval of a slow-wave component, and a good estimate of its frequency content is of much more importance, which can be obtained using CWT with large windows. Therefore, by using adaptive windows in different frequency ranges, CWT is suitable for analyzing time-frequency components of EEG signals.

The adaptive windows and flexible time-frequency resolution of CWT are achieved by using wavelets: basis functions that can both be localized in frequency (or "scale") and in time. A wavelet is a square-integrable function, which normally has an amplitude that begins at zero, increases, and then decreases back to zero. Some commonly used wavelets include Haar wavelets, Morlet wavelets, Daubechies wavelets, etc. For example, the Morlet wavelet is defined as (Cohen 2018).

$$\psi(t) = e^{-t^2/2} e^{2j\pi ft}. \tag{6.8}$$

where ω is the central frequency of the Morlet wavelet. From (6.8) we can see that the Morlet wavelet is actually a sinusoid $e^{2j\pi ft}$ weighted by a Gaussian kernel $e^{-t^2/2}$. Note that the definition of the Morlet wavelet may be slightly different in different books, which may take some scaling and shift parameters into consideration.

Given a "mother" wavelet as $\psi(t)$, we can make a basis to represent a signal by using scaled and shifted versions of $\psi(t)$, $\psi\left(\frac{1}{a}(t-\tau)\right)$, where α is a factor controlling the scaling or dilation and τ is a factor controlling the translation of position of the wavelet in the time domain. As the scale factor, α, expands the analyzed function in time, it thus allows for examination of the signal over a wide interval. Scale is inversely proportional to frequency: $\alpha = f_0/f$, where f_0 is the center frequency (mean of the lower cutoff frequency and the upper cutoff frequency) of the mother wavelet $\psi(t)$. A large scale will yield a smaller window in the frequency domain and better frequency resolution and vice versa. Scales can be arbitrarily set in CWT, but they normally are set to make frequency bins uniformly spaced. Figure 6.11 shows the scaling and shifting of a Morlet wavelet basis.

An EEG signal's CWT is calculated as the convolution of the signal with the wavelet basis function:

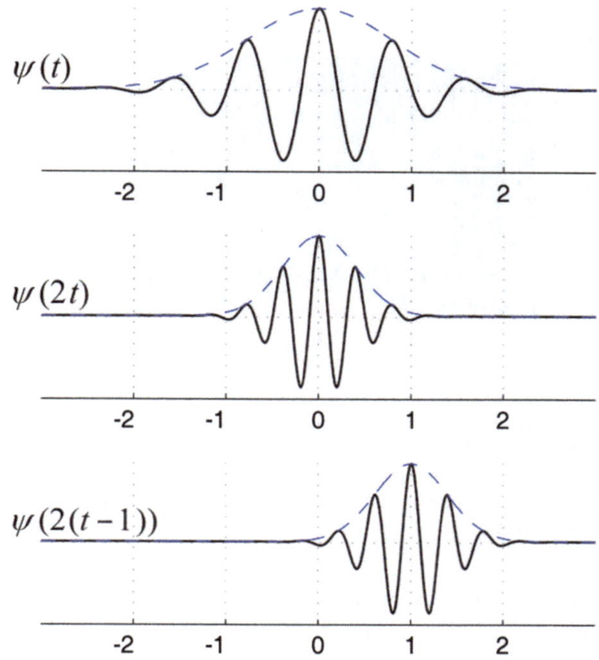

Fig. 6.11 Scaling and shifting of a Morlet wavelet basis

$$X(t, \alpha) = \mathrm{CWT}\{x(t)\} = \int_{-\infty}^{\infty} \frac{1}{\sqrt{\alpha}} x(\tau)\psi^{*}\left(\frac{1}{\alpha}(\tau - t)\right) d\tau. \tag{6.9}$$

The squared magnitude of $|X(t, \alpha)|^2$ is called the scalogram.

In CWT, the frequency resolution increases proportionally with the scale, while the time resolution decreases proportionally. Though we have not gained anything with respect to "total uncertainty" through CWT, we are still able to tune the separate time and frequency uncertainties according to our interests or the characteristics of the signal. Notably, CWT also has limitations in its time-frequency resolution. Considering that CWT has a poor time resolution for low-frequency components and a poor frequency resolution for high-frequency components, it is not suitable for identifying short-duration but low-frequency components, and long-duration but high-frequency components. In addition, the parameters of wavelets determine the time-frequency resolution of CWT and these parameters should also be carefully tuned for a better time-frequency representation.

Figure 6.12 shows the TFDs of the VEP signal (as shown in Fig. 6.10) estimated using Morlet CWT. The results can be obtained by running the supplementary MATLAB script "demo_mwt.m" attached in this book. In Fig. 6.11, we can see both the VEP and the P300 components. In the low-frequency range (<5 Hz), CWT uses a long window so that the P300 component can be revealed in the range of 300–600 ms and around 1–3 Hz. For VEP, which is in a relatively higher frequency band (5–10 Hz), a shorter window is used to better identify short-duration VEP

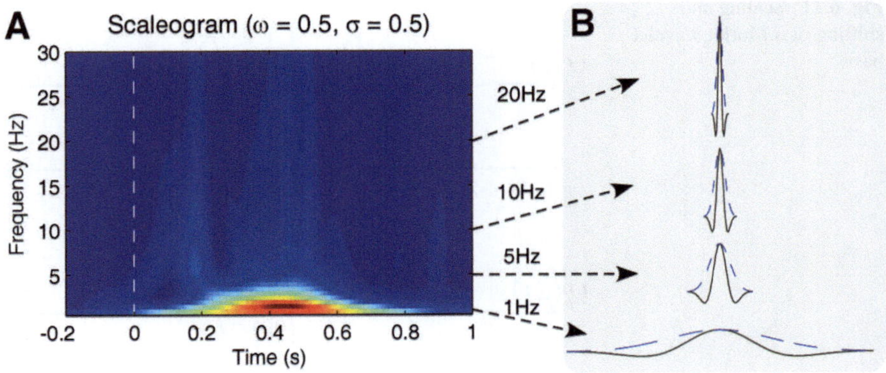

Fig. 6.12 CWT of visual evoked potentials (as shown in Fig. 6.10). (**a**) Scaleogram estimated using the Morlet wavelet with $\omega = 0.5$, $\sigma = 0.5$, which are, respectively, the central frequency and the spread of the mother wavelet. (**b**) The frequency-dependent wavelet bases (only the real parts of the wavelet functions are shown as black solid lines) and the corresponding window functions (blue dashed lines)

components. As mentioned earlier, different parameters of CWT (ω and σ in Morlet wavelet) will also influence the time-frequency resolution of TFD. Interested readers could set different parameters in "demo_mwt.m" to examine how these parameters change the time-frequency resolution of the TFDs.

6.3.4 Other Commonly Used TFA Methods

6.3.4.1 Time-Varying AR Model

The AR model is often used to provide a parametric spectral estimation of EEG. In the model of (5), the AR coefficients are fixed and do not change with time. As a consequence, the estimated spectrum is also unchanged. But, actually, because of the dynamic behavior of EEG, the AR coefficients should also vary with respect to time, resulting in a time-varying AR (TVAR) model:

$$x[n] = \sum\nolimits_{p=1}^{P} a_p[n]x[n-p] + e[n], \tag{6.10}$$

where $a_p[n]$, $p = 1, 2, \cdots, P$ are time-varying AR coefficients. To estimate $a_p[n]$, a number of adaptive filtering and Kalman filtering methods are available (Aboy et al. 2005; Arnold et al. 1998; Kaipio and Karjalainen 2002; Tarvainen et al. 2004; Khan and Dutt 2007). These methods can estimate the TVAR coefficients at each time instant in a recursive manner, which means $a_p[n]$ is estimated from $a_p[n-1]$. Once $a_p[n]$ is estimated, the time-varying AR spectrum can be calculated as

$$P(t,f) = \frac{\sigma^2(t)}{\left|1 + \sum_{p=1}^{P} a_p(t) e^{-j2\pi fp/Fs}\right|^2}, \tag{6.11}$$

where F_s is the sampling rate and $\sigma^2(t)$ is the instantaneous variance of $e(n)$.

The sliding-window approach can also be used to estimate the TVAR model. Compared with the sliding-window approach, adaptive filters have a lower computational complexity and are suitable for online processing. But adaptive filters have a lag in estimation because they only make use of past samples. In the TVAR-based TFA, the frequency resolution is mainly controlled by the model order P (which is the same as the case in the AR model-based spectrum). If P is too large, spurious peaks will appear; if P is too small, adjacent sinusoids cannot be resolved. The order P also has an influence on the time resolution, because P implies the number of past samples that contribute to the current sample. A larger P decreases the time resolution and vice versa.

6.3.4.2 Time-Frequency Decomposition Methods

Discrete wavelet transform (DWT) (Mallat 2008) is extensively used to decompose a signal into a series of additive components with different scales, and it will not produce a joint representation of signal power in the time-frequency domain. Unlike CWT, which can contain an infinite number of scales, DWT performs wavelet transform at finite discrete scales, and it leads to a more efficient decomposition of the signal. Normally, the scale is changed in powers of 2. Aside from DWT, matching pursuit (MP) and empirical mode decomposition (EMD) are also time-frequency decomposition methods that have been used for EEG. MP is an iterative algorithm that in each step finds an element from a set of functions (dictionary) that best matches the current residue of the signal decomposition (Durka 2007). EMD is a procedure for decomposition of a signal into a set of intrinsic mode functions, which are functions with the same number of extrema and zero crossings, with its envelopes being symmetric with respect to zero (Sweeneyreed and Nasuto 2007).

6.3.4.3 Latest Developments

Time-frequency analysis of EEG is still a field under development, because the fundamental limitation of time-frequency resolution and some practical requirements have not been perfectly solved. For example, in Zhang et al.'s work (Zhang et al. 2011), a local polynomial model-based method was proposed to achieve the optimal tradeoff between time resolution and frequency resolution in the whole time-frequency domain by using local polynomial basis with adaptive window size at each time-frequency point. In Kim et al.'s recent paper (Kim et al. 2018), the conventional multitaper method is extended to a TFA method, named state-space multitaper method, so that it can provide a statistical inference framework for TFA of EEG and other types of nonstationary time series.

6.4 Event-Related Synchronization/Desynchronization (ERS/ERD)

The most common application of TFA on EEG is to identify event-related synchronization/desynchronization (ERS/ERD) in event-related experiments. A huge body of experimental investigations have shown that various sensory stimuli or cognitive events do not only evoke EP/ERP but also induce transient power modulations of EEG spectrum. Such modulations are presented as an increase (ERS) or a decrease (ERD) of spectral power at a specific frequency band. The differences between EP/ERP and ERS/ERD are evident: EP/ERP are time-locked and phase-locked, and they are usually studied in the time domain, while ERS/ERD are time-locked but non-phase-locked, and they are usually studied in the time-frequency domain (Pfurtscheller and Lopez da Silva 1999; Mouraux and Iannetti 2008). As discussed in Chap. 3, time-domain averaging can effectively estimate EP/ERP, because these potentials are phase-locked and time-domain averaging can restrain the non-phase-locked noise and other components. However, ERS/ERD are not phase-locked, and they cannot survive from time-domain averaging. In the following, the estimation of ERS/ERD and related phase locking values will be introduced.

6.4.1 Estimation of ERS/ERD

ERS/ERD are also known as event-related spectral perturbation (ERSP), or can be identified from ERSP, which measures the event-related spectral changes relative to the spontaneous EEG baseline in a wide range of frequency and shows the spectral changes in the joint 2D time-frequency domain (Delorme and Makeig 2004). ERS and ERD can be identified from the ERSP as positive spectral changes and negative spectral changes, respectively. In this section, we will introduce how to estimate ERS/ERD from ERSP or baseline-corrected TFD of event-related EEG. It should be noted that ERS/ERD can also be estimated using other approaches (e.g., to estimate the spectral envelope at a certain frequency band), especially in the field of neural engineering, where computational time and single-trial estimation are key issues under consideration.

To estimate ERS/ERD or ERSP from multiple trials of EEG activities, we need first use TFA to transform single-trial EEG waveforms into TFDs and then average these single-trial TFDs to identify ERS/ERD. The procedure to estimate ERS/ERD is as follows:

1. Repeat the event of interest a given number of times.
2. Pre-processing to remove artifacts and noise.
3. Segment EEG into trials based on the onset time of events.
4. Time-frequency analysis.
5. Take average across trials and baseline correction.

By comparing above procedure to detect ERS/ERD with the procedure to estimate EP/ERP, we can see that except Step 4, which is only essential in the detection of ERS/ERD, other steps are the same. Similar to the estimation of EP/ERP, a reliable ERS/ERD component is normally detected from a number of trials. But ERS/ERD is not phase-locked to stimulation and will be smoothed out during across-trial averaging in the time domain. Therefore, we need to use TFA to transform ERS/ERD waveforms into the time-frequency domain and then use across-trial averaging in time-frequency domain to obtain the averaged TFDs, from which the ERS/ERD components could be identified. Figure 6.13 shows how to use TFA to estimate ERS/ERD of multiple trials of laser-evoked potentials (LEP) and compares different TFDs (of single LEP trials, of across-trial averaged LEP, averaged from TFDs of all LEP trials without or with baseline correction). A supplementary MATLAB script "demo_erserd.m" is attached in this book to demonstrate the procedure of ERS/ERD estimation.

TFA methods introduced in this chapter, such as STFT and CWT, can be used to estimate ERS/ERD. Window selection is crucial in the analysis of ERS/ERD, because the onset/offset time and the frequency range of ERS/ERD covey important information about the brain functioning. Frequency-dependent window (as in CWT) is recommended: short windows are used in high-frequency range (such as the gamma band) and long windows are used in low-frequency range (such as the theta and delta band). The window at one specific frequency should cover at least one cycle (usually 2–3 cycles) of ERS/ERD signals at this frequency. For example, to detect an ERS/ERD component at 10 Hz, the window size at 10 Hz should be at least 100 ms. But at very low-frequency bins, the window cannot be too long, for time resolution will be seriously degraded and, for example, some unwanted components could be seen even before event onset. In practice, it is often necessary to try out many possible parameters and to choose those parameters that can achieve the best visually or statistically reasonable TFDs.

6.4.2 Baseline Correction and Illustration of ERS/ERD

Another important issue in the detection of ERS/ERD is the baseline correction. Baseline correction is simple in EP/ERP estimation, but not trivial in the detection of ERS/ERD. This is mainly because ERS/ERD exists in a wide range of frequencies and the signal power at different frequencies is remarkably different. ERS/ERD is

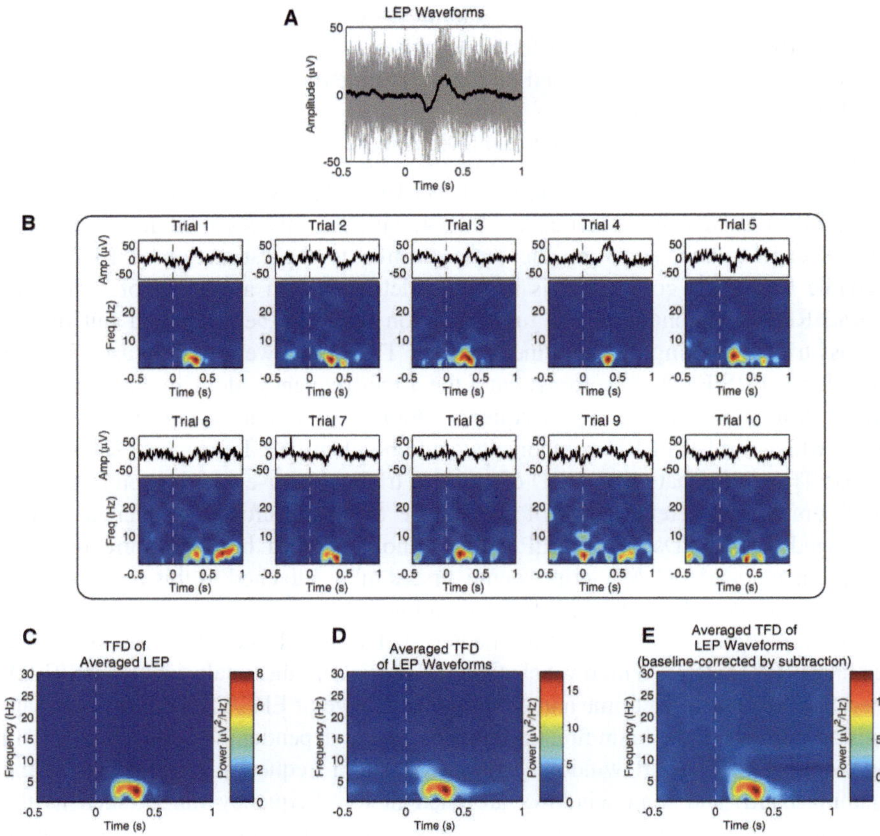

Fig. 6.13 TFA of multiple trials of laser-evoked potentials (LEP) for the detection of ERS/ERD. (**a**) 74 trials of LEP were recorded at Cz of one individual with a sampling rate of 256 Hz. More details about the experiment and the EEG data can be referred to (Hu et al. 2014). (**b**) LEP waveforms and TFDs (spectrograms) of randomly selected 10 trials. TFDs were obtained using STFT with a 400 ms Hamming window. (**c**) TFD of LEP averaged from 74 trials. (**d**) TFD averaged from 74 TFDs of single LEP trials. (**e**) TFD in D after baseline correction (subtraction). Subplots A, C, D, and E can be obtained by running the supplementary MATLAB script "demo_erserd.m" attached in this book

defined as the increase or decrease of signal power with respect to a baseline period. At each frequency, a baseline period is defined by the power values within a time window preceding the stimulus, and thus the baseline period is also called as the pre-stimulus period. In EP/ERP detection, the baseline values are subtracted from poststimulus values. Such a "subtraction" approach for baseline correction also works for ERS/ERD, but some other baseline correction approaches are also commonly used. Generally, the following four baseline correction approaches are often used for ERS/ERD estimation (Roach and Mathalon 2008):

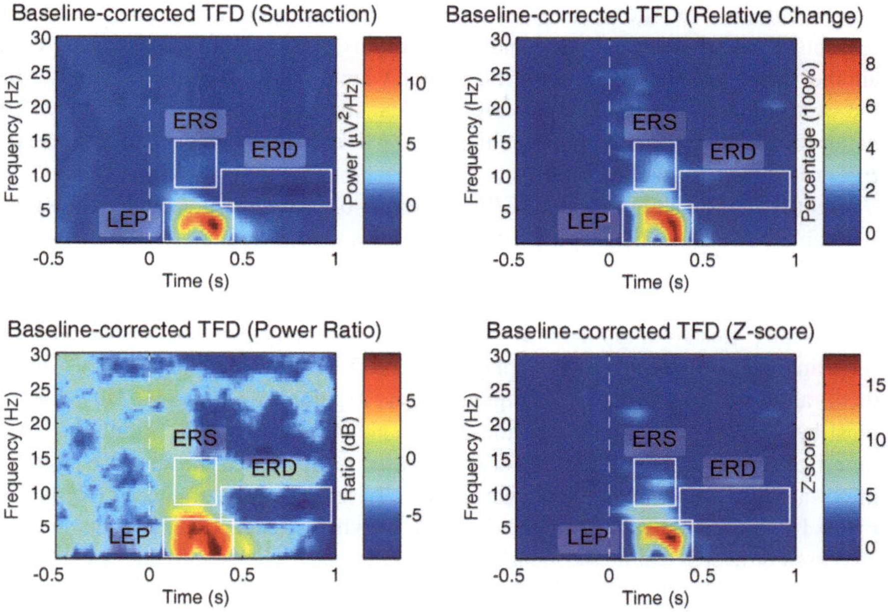

Fig. 6.14 Comparisons of TFDs obtained using four different baseline correction approaches for the detection of ERS/ERD. These results can be obtained by running the supplementary MATLAB script "demo_erserd.m" attached in this book

$$\text{Subtraction}: \quad P(t,f) - \bar{R}(f) \tag{6.12a}$$

$$\text{Relativec Change}: \quad [P(t,f) - \bar{R}(f)]/\bar{R}(f) \tag{6.12b}$$

$$\text{Power Ratio}: \quad \log_{10}(P(t,f)/\bar{R}(f)) \tag{6.12c}$$

$$\text{Z-score}: \quad [P(t,f) - \bar{R}(f)]/SD[R(f)] \tag{6.12d}$$

where $P(t,f)$ is the power value at a time-frequency point(t,f) and $R(t,f)$ is the baseline values with mean $\bar{R}(f)$ and standard deviation $SD[R(f)]$. A comparison among these four baseline correction approaches can be seen in Fig. 6.14.

Actually, expressing poststimulus EEG power relative to pre-stimulus EEG power entails two important and surprisingly neglected issues. First, it can introduce a significant bias in the estimation of ERD/ERS magnitude. In (Hu et al. 2014), it was shown that that expressing ERD/ERS as the average percentage of change calculated at single-trial level introduces a positive bias, resulting in an overestimation of ERS and an underestimation of ERD. This bias can be avoided using a single-trial baseline subtraction approach. Second, it confuses the contribution of pre- and poststimulus EEG power (Kim et al. 2018). Many studies have shown that pre-stimulus (baseline) EEG activities could modulate an individual's behavioral responses as well as neural responses to an event. Given that the variability in ERD/ERS is not only dependent on the variability in poststimulus power but also on the variability in pre-stimulus power, an estimation of the respective contribution of pre- and poststimulus EEG variability is needed. In another word, baseline correction (no matter which approach is used) may not be the best approach to quantify the event-related rhythmic activities, and it will be better to include baseline (pre-stimulus) EEG activities in the study of event-related EEG responses (Kim et al. 2018).

Another non-trivial issue in ERS/ERD estimation is how to present ERS/ERD in a suitable color scale. EEG spectral power follows the power law ($1/f$) and is highly unbalanced among frequency bands: low-frequency components have much higher power than high-frequency components. As a result, if we display EEG power at the same scale, weak components at high-frequency bands may be buried. Hence, sometimes it is necessary to display ERS/ERD at low-frequency bands (delta, theta, alpha, beta) and at high-frequency bands (gamma) separately. However, it must be mentioned that adjusting the color scale can only visually highlight those time-frequency regions where power is increased or decreased. From those visually important time-frequency regions, we can specify some regions of interest (ROIs). In Fig. 6.14, those rectangular boxes indicate ROIs where signal power is either largely increased or decreased. Subsequently, we can summarize the power values within all time-frequency points of a ROI by, for example, averaging across all time-frequency points, and the summarized power value can be used as the time-frequency feature of this ROI. It is definitely better to perform a rigorous statistical test at each time-frequency point to examine its significance and avoid possible bias caused by visual inspection. Please refer to Chap. 17 for more details about statistical analysis of time-frequency EEG data.

6.4.3 Phase Locking Value (PLV)

The estimation results of TFA (STFT or CWT) are actually complex values, but both spectrogram and scaleogram are the squared magnitude of TFA output, and they do not make use of the important phase information. An often-used phase feature is phase locking value (PLV) (Mouraux and Iannetti 2008), which is also known as inter-trial phase coherence (ITPC) (Delorme and Makeig 2004). PLV is an important complement to the time-frequency power, and it is calculated as

$$\mathrm{PLV}(t,f)=\frac{1}{M}\sum\nolimits_{i=1}^{M}\frac{F_i(t,f)}{|F_i(t,f)|}, \tag{6.14}$$

where $F_i(t,f)$is the complex time-frequency value from TFA. PLV is between 0 and 1: 0 reflects a completely random distribution of phase angles between trials, and 1 reflects perfectly synchronized phase angles across trials. Typically, ERP components are phase-locked, so the PLV values in ERP-related time-frequency regions are

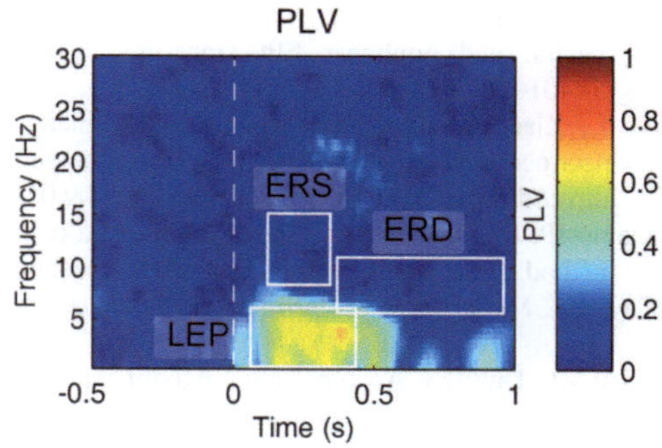

Fig. 6.15 PLV of the multiple trials of laser-evoked EEG signals (as shown in Fig. 6.13). The ROIs are defined based on TFDs of signal power in Fig. 6.14. This PLV result can be obtained by running the supplementary MATLAB script "demo_erserd.m" attached in this book

high (close to 1). But ERS/ERD components are non-phase-locked, so the PLV values in ERS/ERD-related time-frequency regions are close to zero. Figure 6.15 shows the PLV values of the LEP trials, which are used in Figs. 6.13 and 6.14.

References

Aboy M, Marquez OW, Mcnames J, Hornero R. Adaptive modeling and spectral estimation of nonstationary biomedical signals based on Kalman filtering. IEEE Trans Biomed Eng. 2005;52 (8):1485–9. https://doi.org/10.1109/TBME.2005.851465.

Adeli H, Zhou Z, Dadmehr N. Analysis of EEG records in an epileptic patient using wavelet transform. J Neurosci Methods. 2003;123(1):69–87. https://doi.org/10.1016/S0165-0270(02) 00340-0.

Arnold M, Miltner WH, Witte H, Bauer R, Braun C. Adaptive AR modeling of nonstationary time series by means of Kalman filtering. IEEE Trans Biomed Eng. 1998;45(5):553–62. https://doi.org/10.1109/10.668741.

Babadi B, Brown EN. A review of multitaper spectral analysis. IEEE Trans Biomed Eng. 2014;61 (5):1555–64. https://doi.org/10.1109/TBME.2014.2311996.

Barry RJ, Clarke AR, Johnstone SJ, Brown CR. EEG differences in children between eyes-closed and eyes-open resting conditions. Clin Neurophysiol. 2009;118(12):2765–73.

Boashash B. Time-frequency signal analysis and processing: a comprehensive reference. Boston: Academic; 2015.

Buzsaki G. Rhythms of the brain. Oxford: Oxford University Press; 2011.

Cahn BR, Polich J. Meditation states and traits: EEG, ERP, and neuroimaging studies. Psychol Bull. 2006;132(2):180–211. https://doi.org/10.1037/0033-2909.132.2.180.

Cohen MX. Analyzing neural time series data: theory and practice. Cambridge, MA: MIT Press; 2014.

Cohen MX. Where does EEG come from and what does it mean? Trends Neurosci. 2017;40 (4):208–18. https://doi.org/10.1016/j.tins.2017.02.004.

Cohen MX. A better way to define and describe Morlet wavelets for time-frequency analysis. bioRxiv. 2018; https://doi.org/10.1101/397182.

Delorme A, Makeig S. EEGLAB: an open source toolbox for analysis of single-trial EEG dynamics including independent component analysis. J Neurosci Methods. 2004;134(1):9–21. https://doi.org/10.1016/j.jneumeth.2003.10.009.

Durka P. Matching pursuit and unification in EEG analysis. Norwood: Artech House; 2007.

Fell J, Röschke J, Mann K, Schäffner C. Discrimination of sleep stages: a comparison between spectral and nonlinear EEG measures. Electroencephalogr Clin Neurophysiol. 1996;98 (5):401–10.

Hu L, Xiao P, Zhang ZG, Mouraux A, Iannetti GD. Single-trial time-frequency analysis of electrocortical signals: baseline correction and beyond. NeuroImage. 2014;84(1):876–87. https://doi.org/10.1016/j.neuroimage.2013.09.055.

Kaipio JP, Karjalainen PA. Estimation of event-related synchronization changes by a new TVAR method. IEEE Trans Biomed Eng. 2002;44(8):649–56. https://doi.org/10.1109/10.605421.

Kay SM. Modern spectral estimation: theory and application. Englewood Cliffs: Prentice Hall; 1988.

Khan ME, Dutt DN. An expectation-maximization algorithm based Kalman smoother approach for event-related desynchronization (ERD) estimation from EEG. IEEE Trans Biomed Eng. 2007;54(7):1191–8. https://doi.org/10.1109/TBME.2007.894827.

Kim SE, Behr MK, Ba D, Brown EN. State-space multitaper time-frequency analysis. Proc Natl Acad Sci U S A. 2018;115(1):E5–E14. https://doi.org/10.1073/pnas.1702877115.

Luck SJ. An introduction to the event-related potential technique. Cambridge, MA: MIT Press; 2014.

Mallat S. A wavelet tour of signal processing: the sparse way. Burlington: Academic; 2008.

Mitra SK. Digital signal processing: a computer-based approach. New York: McGraw-Hill Inc; 2000.

Mouraux A, Iannetti GD. Across-trial averaging of event-related EEG responses and beyond. Magn Reson Imaging. 2008;26(7):1041–54. https://doi.org/10.1016/j.mri.2008.01.011.

Niedermeyer E, Lopes da Silva FH. Electroencephalography: basic principles, clinical applications, and related fields. Philadelphia: Lippincott Williams & Wilkins; 2005.

Oppenheim AV, Willsky AS, Nawab SH. Signals and systems. London: Pearson; 1996.

Pernet C, Garrido M, Gramfort, MN, Michel C, Pang E, Salmelin R, Schoffelen JM, Valdes-Sosa PA, Puce A. Best practices in data analysis and sharing in neuroimaging using MEEG. 2018. https://osf.io/jh7mp/.

Pfurtscheller G, Lopez da Silva FH. Event-related EEG/MEG synchronization and desynchronization: basic principles. Clin Neurophysiol. 1999;110(11):1842–57.

Proakis JG, Manolakis DK. Digital signal processing: principles, algorithms and applications. Upper Saddle River: Pearson; 2006.

Roach BJ, Mathalon DH. Event-related EEG time-frequency analysis: an overview of measures and an analysis of early gamma band phase locking in schizophrenia. Schizophr Bull. 2008;34 (5):907–26. https://doi.org/10.1093/schbul/sbn093.

Sanei S, Chambers JA. EEG signal processing. New York: Wiley; 2013.

Schlogl A. The electroencephalogram and the adaptive autoregressive model: theory and applications. Aachen: Shaker Verlag GmbH; 2000.

Stoica P, Moses RL. Spectral analysis of signals. Upper Saddle River: Prentice Hall; 2005.

Sweeneyreed CM, Nasuto SJ. A novel approach to the detection of synchronisation in EEG based on empirical mode decomposition. J Comput Neurosci. 2007;23(1):79–111. https://doi.org/10.1007/s10827-007-0020-3.

Tarvainen MP, Hiltunen JK, Ranta-Aho PO, Karjalainen PA. Estimation of nonstationary EEG with Kalman smoother approach: an application to event-related synchronization (ERS). IEEE Trans Biomed Eng. 2004;51(3):516–24. https://doi.org/10.1109/TBME.2003.821029.

Zhang ZG, Hung YS, Chan SC. Local polynomial modeling of time-varying autoregressive models with application to time–frequency analysis of event-related EEG. IEEE Trans Biomed Eng. 2011;58(3):557–66. https://doi.org/10.1109/TBME.2010.2089686.

Chapter 7
Blind Source Separation

Fengyu Cong

Abstract Blind source separation algorithms have been widely used in the EEG signal processing. This chapter introduces the EEG model basis of blind source separation and details of three mainstream algorithms, i.e., principal component analysis (PCA), independent component analysis (ICA), and tensor decomposition, to provide a comprehensive review on this growing topic. The main focus will be on basic principles of applying ICA on continuous EEG data to remove artifacts, PCA, and tensor decomposition on ERP data to conduct group analysis. The introduction of current softwares specialized in PCA and ICA on EEG signal processing will also be covered.

Keywords Blind source separation · Independent component analysis · EEG · Event-related potentials · Back projection

7.1 Introduction of Blind Source Separation

Blind source separation (BSS) is to separate source signals from a set of mixed signals (observation), without the aid of relevant information (or with very little information) about the source signals or the mixing process (Cois 2009). Typically, the observations are obtained at the output from a set of sensors, where each sensor receives a different combination of source signals. There are different methods of BSS: principal component analysis (PCA), independent component analysis (ICA),

---

Electronic supplementary material The online version of this chapter (https://doi.org/10.1007/978-981-13-9113-2_7) contains supplementary material, which is available to authorized users.

F. Cong (✉)
School of Biomedical Engineering, Faculty of Electronic Information and Electrical Engineering, Dalian University of Technology, Dalian, China
e-mail: cong@dlut.edu.cn

L. Hu, Z. Zhang (eds.), *EEG Signal Processing and Feature Extraction*,
https://doi.org/10.1007/978-981-13-9113-2_7

nonnegative matrix factorization (NMF), and so on. Promising applications have already been found in the processing of communications signals (Chaumette et al. 1994; Mathew and Reddy 1995; Swindlehurst et al. 1997; van der Veen 1998), biomedical signals like ECG (De Lathauwer et al. 2000) and EEG (Makeig et al. 1996), and monitoring (Durso and Prieur 1997). However, BSS is now routinely performed on multidimensional data (Cichocki 2013; Kolda and Bader 2009), such as images and video signals (Zhao et al. 2013).

Figure 7.1 illustrates a fairly general BSS problem also referred to as blind signal decomposition or blind source extraction. We observe records of N sensor signals $\mathbf{X}(t) = [x_1(\mathrm{t}), x_2(\mathrm{t}), \cdots, x_N(\mathrm{t})]^T$ coming from a MIMO (multiple-input/multiple-output) and filtering system, where t is usually a discrete time sample and $(\cdot)^T$ denotes transpose of a vector. These signals are usually a superposition (mixture) of R unknown source signals $\mathbf{S}(t) = [s_1(\mathrm{t}), s_2(\mathrm{t}), \cdots, s_R(\mathrm{t})]^T$ and noises $\mathbf{E}(t) = [e_1(\mathrm{t}), e_2(\mathrm{t}), \cdots, e_N(\mathrm{t})]^T$. The primary objective is to estimate all the source signals $s_r(\mathrm{t})$ or only some of them with specific properties. This estimation is usually performed based on the output (sensor, observed) signals $x_n(\mathrm{t})$ in order to estimate sources. Usually, the inverse (unmixing) system should be adaptive in a way, provided with some tracking capability, to a nonstationary environment. Instead of estimating the source signals directly by projecting observed signals using the unmixing system, it is more convenient to identify an unknow mixing and filtering system (e.g., when the unmixing system does not exist, especially when the system is underdetermined, i.e., the number of observations is lower than the number of source signals, N < R) and simultaneously estimate the source signals by exploiting some priori information about the source signals and applying a suitable optimization procedure.

There appears to be something magic about blind source separation, since we are estimating the original source signals without knowing the parameters of the mixing

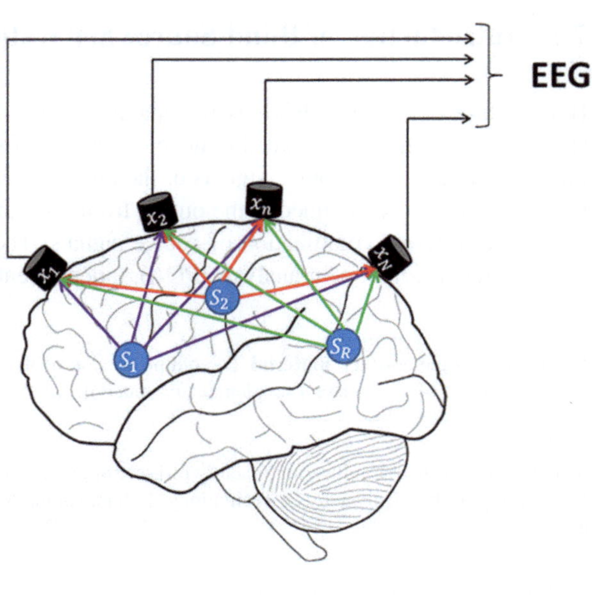

Fig. 7.1 BSS model for EEG signal

and filtering processes. It is difficult to imagine that one can estimate the source signals in this way. In fact, it is not possible to uniquely estimate the original source signals, without some priori knowledge. However, one can usually estimate these signals based on certain indeterminacies. In mathematical terms, the indeterminacies and ambiguities can be expressed as arbitrary scaling and permutation of the estimated source signals. These indeterminacies preserve, however, the waveforms of original sources. Although limitations of these indeterminacies seem to be rather severe, these limitations are not crucial in a great number of applications, since the most relevant information about the source signals is contained in the temporal waveforms or time-frequency patterns of the source signals, instead of embedding in their amplitudes or the arranged orders in the output from the system.

7.2 PCA and Rotation for ERP Analysis

7.2.1 The Theory of PCA and Rotation

Principal components analysis (PCA) is a multivariate technique that seeks to uncover latent variables responsible for pattern of covariation in numerical datasets (Harman 1976). It is a statistical procedure that uses an orthogonal transformation to convert a set of observations of possibly correlated variables into a set of values of linearly uncorrelated variable, referred as principal components.

The matrix $\mathbf{X} \in \mathbb{R}^{m \times n}$ (m $\geq$ n) can be written as follows:

$$\mathbf{X} = \mathbf{U}\boldsymbol{\Sigma}\mathbf{V}^T, \tag{7.1}$$

In specific,

1. $\mathbf{U} = [u_1 \quad u_2 \quad \cdots \quad u_n]$ is a unitary $(m \times n)$-matrix; each column of $\mathbf{U}$ is a principal component.
2. $\mathbf{V} = [\,v_1 \quad v_2 \quad \cdots \quad v_n\,]$ is a unitary$(n \times n)$-matrix, which means the loadings of the component.
3. $\boldsymbol{\Sigma}$ is a $(n \times n)$-matrix with the propertites of

   Pseudo-diagonality:

$$\boldsymbol{\Sigma} = \mathrm{diag}(\sigma_1, \sigma_2, \cdots, \sigma_n), \tag{7.2}$$

   Ordering:

$$\sigma_1 \geq \sigma_2 \geq \cdots \sigma_n \geq 0. \tag{7.3}$$

The σ_i are singular values of $\mathbf{X}$, referred to as the power of each principle component. Thus, the PCA could be used for dimension reduction.

For a long time, PCA has been used as a technique of data description and reduction to manage the copious quantities of measurements obtained in event-related potential (ERP) studies. Although PCA has limitations when applied to ERP data and is also sensitive to parameters such as component overlap and correlation (Dien 1998), it has been applied to numerous studies and gained considerable success (Dien 1999; Spencer et al. 2001; Squires et al. 1975).

Recognition of the limitations during PCA procedure has given rise to efforts to improve the process of technique itself. Note that the oblique rotation promax can generate more accurate results with correlated ERP components than the customary orthogonal rotation varimax (Dien 1998; Dien et al. 2005). The use of a covariance matrix for the relationship matrix and the inclusion of Kaiser normalization also yield refined results, in comparison to covariance loadings during rotation (Dien et al. 2005).

Varimax rotation is the most commonly used approach in PCAs (Kaiser 1958), although other rotation algorithms are available. In this approach, pairs of factors are rotated in the two-dimensional space formed by their two axes, so as to maximize the sum of the variance of the squared loadings. The factors are systematically rotated in pairs until changes are negligible. This procedure has considerable effects on factor loadings, which make the loading values as extreme as possible (either zero or extremely high). Such a principle is shared by other family members of the orthomax rotations. This is an appropriate criterion for ERP dataset (especially for a temporal PCA).

Promax is a method for quick rotation to oblique simple structure. Here, the promax algorithm is taken as an example. Starting with a matrix of factor loadings that has been rotated to orthogonal simple structure, a crucial problem was to formulate an ideal "pattern" matrix as a function for the orthogonal solution, which means to find a set of weights to a given factor loading. We define a matrix $\mathbf{P} = (p_{ij})$ as:

$$p_{ij} = \left| a_{ij}^{k+1} \right| / a_{ij}, \tag{7.4}$$

with $k > 1$. Each element of this matrix is, except of sign which remains unchanged, the kth power of the corresponding element in the row-column normalized orthogonal matrix. The least squares fit of the orthogonal matrix of factor loadings to the pattern matrix generated by Eq. (7.4) could be calculated as follows:

$$\boldsymbol{L} = \left(\boldsymbol{F}^T\boldsymbol{F}\right)^{-1}\boldsymbol{F}^T\boldsymbol{P}, \tag{7.5}$$

where $\boldsymbol{L}$ is the unnormalized transformation matrix of the reference vector structure, $\boldsymbol{F}$ is the orthogonal rotated matrix, and $\boldsymbol{P}$ is the matrix derived from the orthogonal matrix defined in Eq. (7.4). Equation (7.5) is the "Procrustes" equation described by Cattell (1976). The columns of $\boldsymbol{L}$ are normalized such that their sums of squares are equal to unity. This provides the transformation matrix from the orthogonal factors to the oblique reference vectors. From this point standard and well-known formulate (Harman 1976) can be used to compute the intercorrelations among the oblique primary factors and the matrix of primary factor loadings.

After rotating the loading matrix, the components could be calculated as follows:

$$\mathbf{Y} = \mathbf{L}\mathbf{X}, \tag{7.6}$$

L comes from any kind of rotated factors algorithm, including "orthomax," "equamax," "promax," and so on.

7.2.2 Introduction of ERP_PCA Toolbox

Aimed to improve the performance of group analysis on ERP data and help researchers master advanced signal processing methods (e.g., wavelet filter, FFT filter, and PCA), we developed the ERP_PCA toolbox.

7.2.3 Input

After preprocessing, the EEG data will be reshaped to a matrix with the size of M∗N for each stimulus, where M denotes the number of electrodes and N denoted the number of samples. As shown in Fig. 7.2a, there are three matrices for each stimulus, which can be reshaped to a third-order tensor with the size of M∗N∗P (Fig. 7.2b) for each subject, where P denotes the number of stimuli. Finally, the data for each subject can be linked together and form a fourth-order tensor with the size of M∗N∗P∗Q, where the Q denotes the number of subjects. The fourth-order tensor is the input of our toolbox to perform group analysis, which contains all the information about the ERP data among all subjects.

7.2.4 Features of ERP_PCA Toolbox

The conventional method of analyzing the ERP data at group level is calculating the grand averaged data across subjects, which can be conducted rapidly in the toolbox. Given that the averaged data across trials are still noisy, the wavelet filter and FFT filter were good for further filtering noisy ERP data (Cong et al. 2015b). As a consequence, the wavelet and FFT filter modules were added into the toolbox.

In general, a grand averaged waveform can be regarded as a sum of several ERP components. Thus, PCA and rotation were gradually used to extract individual ERP component, which were embedded in our toolbox. The rectangle box below shows the processing methods in ERP_PCA toolbox (Fig. 7.3).

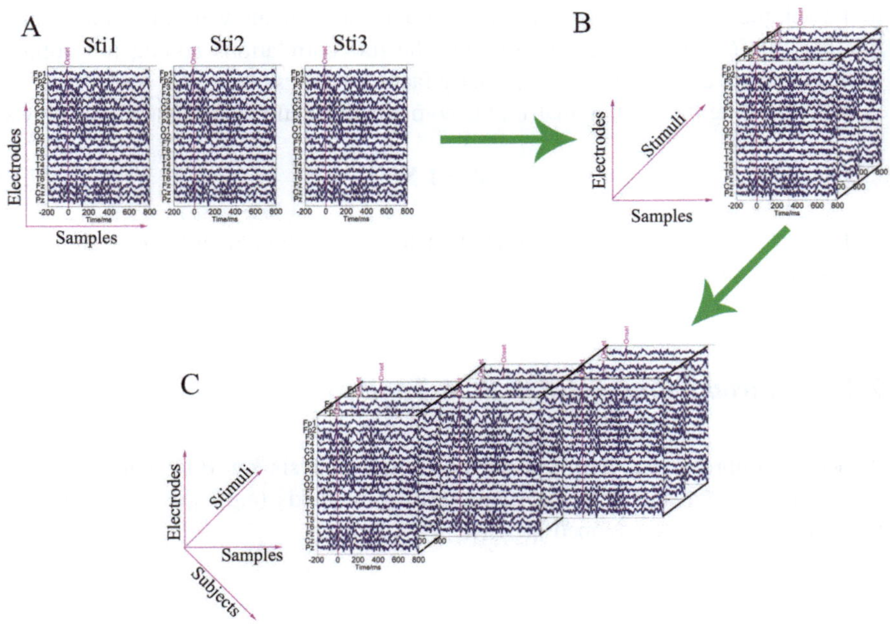

Fig. 7.2 Input of the ERP_PCA toolbox. (**a**), A matrix with the size of M∗N for each stimulus; (**b**), a third-order tensor with the size of M∗N∗P for each subject; (**c**)**:** a fourth-order tensor with the size of M∗N∗P∗Q for all the subjects

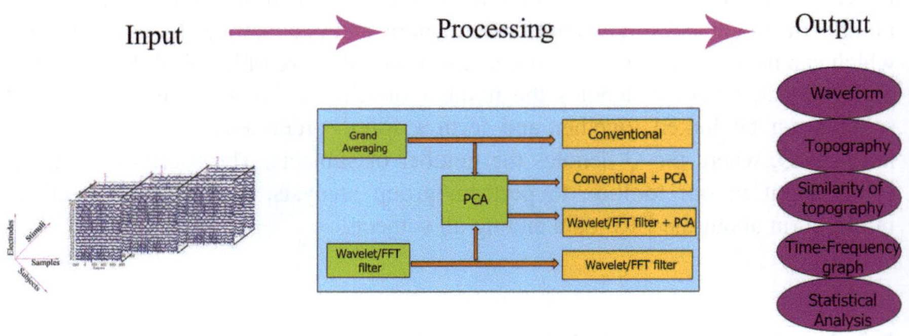

Fig. 7.3 The features of ERP_PCA toolbox

7.2.5 Output

The ERP_PCA toolbox provides a rapid output of processed data. The basic output involves (1) the waveform of grand averaged ERP data across subjects for each stimulus at certain electrode (Fig. 7.4a); (2) the topographies within a certain time window for each stimulus (Fig. 7.4b); (3) the similarity of topographies across subjects for each stimulus, which represents the homogeneity among the subjects

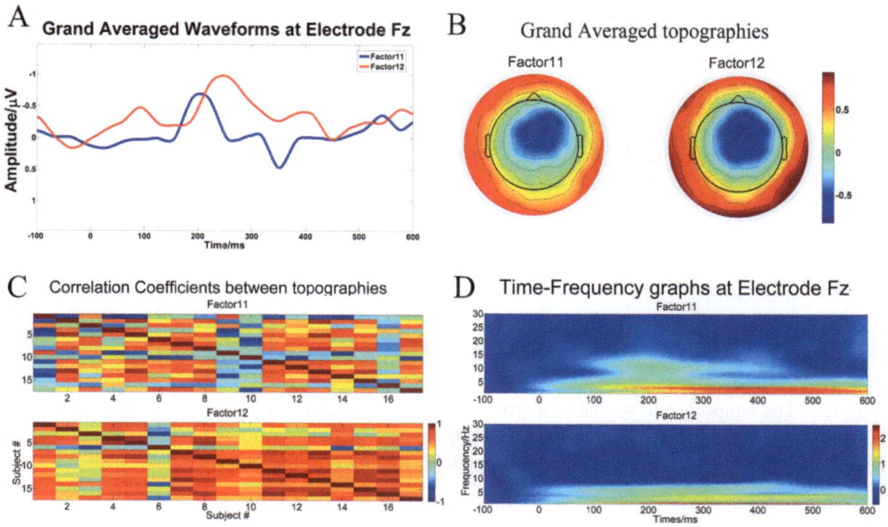

Fig. 7.4 Output of ERP_PCA toolbox. (**a**), The waveform of grand averaged ERP data across subjects for each stimulus at certain electrode; (**b**), the topographies within a certain time window for each stimulus; (**c**), the similarity of topographies across subjects for each stimulus, which represents the homogeneity among the subjects; (**d**), the time-frequency graphs for each stimulus based on wavelet decomposition

(Fig. 7.4c); and (4) the time-frequency graphs for each stimulus based on wavelet decomposition (Fig. 7.4d); in addition, (5) the toolbox enables users to output an excel of the amplitude of the interested component at single/several electrodes(s) for each stimulus, which can be conducted further statistical analysis on other platform, such as SPSS and R-studio.

7.2.6 A Detailed Example Using ERP_PCA Toolbox

PCA is a well-known method to extract individual component based on grand averaged ERP data across subjects. Here, we used the ERP data of a classic mismatch negativity (MMN) experiment to show how PCA is applied in extracting individual component. MMN is a pre-attention component mainly located at frontal electrodes and with the latency ranging from 100 to 250 ms. The experiment adopted a two-deviant oddball paradigm (small deviant stimuli and large deviant stimuli), and after subtracting the ERP responses to standard stimuli from mismatch responses, the difference waveforms for large and small deviant were drawn at electrode Fz (Fig. 7.5a). The interested time window was 150–300 ms, and the topographies (Fig. 7.5b) were obtained from the mean amplitude within this time window.

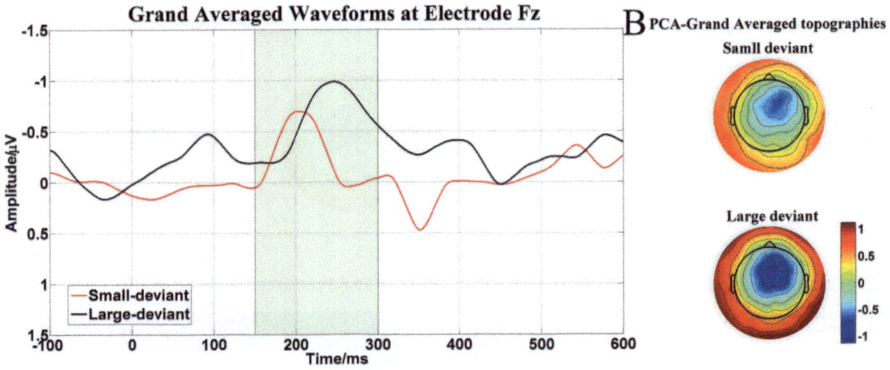

Fig. 7.5 The temporal and spatial properties of raw ERP data. (**a**), The waveforms of raw ERP data for small and large deviant stimuli; (**b**), the topographies calculated from the mean amplitude of the time window 150–300 ms for small and large deviant stimuli of the raw ERP data

The raw ERP waveform can be regarded as the sum of a few different ERP components; we want to extract single MMN component by PCA. After performing PCA, a few principle components were extracted. Here comes a question: How to select the interested components to extract single MMN component by back projection? The basic rules are selecting the principle components that meet both the temporal (latency and polarity) and spatial (topography) features of the interested ERP component (e.g., MMN). For extracting the MMN component, we selected the peaks of three principle components within the time window of 150–300 ms and the active region centered at frontal area in the topographies (Fig. 7.6).

Through back-projecting the three selected principle component, the single MMN component was extracted (Fig. 7.7a). The grand averaged topographies were obtained using the same time window of 150–300 ms (Fig. 7.7b).

7.3 Applying ICA on Continuous EEG Data

7.3.1 ICA Model

ICA is based on the linear model of the unknown source signals **S** and observed signal **Z**. The model could be described by the following equation:

$$\mathbf{Z} = \mathcal{A}\mathbf{S}, \tag{7.7}$$

where $\mathbf{Z} \in \mathcal{R}^{M \times N}$, $\mathbf{S} \in \mathcal{R}^{R \times N}$. $\mathcal{A} \in \mathcal{R}^{M \times R}$ with the full column rank is named as mixing matrix. **S** represents the source signals of ICA decomposition and R the number of components.

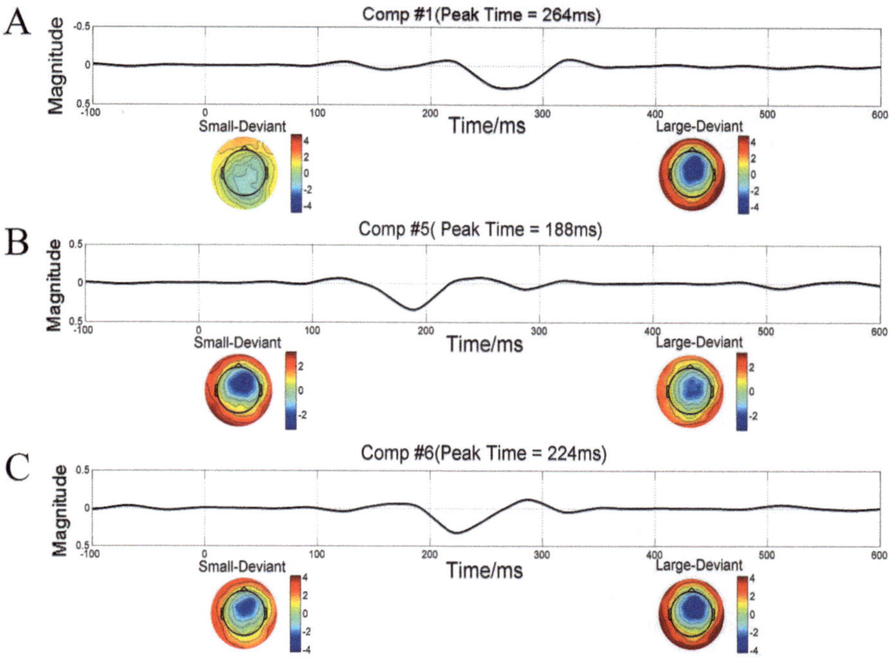

Fig. 7.6 The waveforms and topographies for three interested principle components
The components are selected based on the temporal (components peak within the time window of 150–300 ms) and spatial features (active region centered at frontal area)

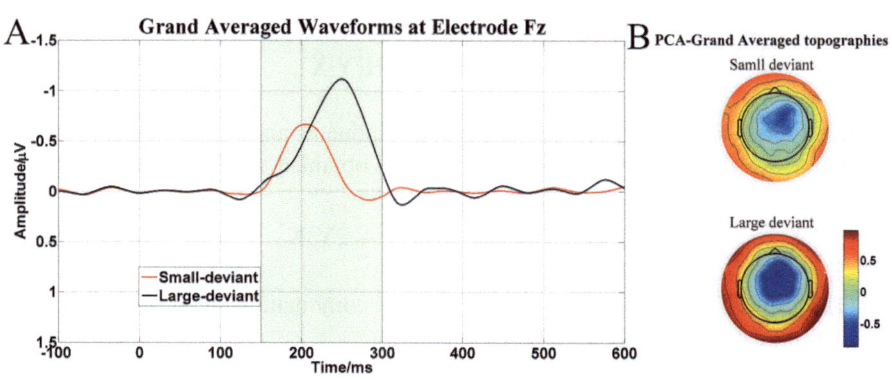

Fig. 7.7 The temporal and spatial properties of ERP data after using PCA to extract MMN. (**a**), the waveforms of ERP data after using PCA to extracted MMN for small and large deviant stimuli; (**b**), the topographies calculated from the mean amplitude of the time window 150–300 ms for small and large deviant stimuli of the ERP data after using PCA to extracted MMN

In this overdetermined model, the dimension of the observed signal M is assumed to be larger than that of the source signals R. Therefore, dimension reduction is necessary before data feed into ICA decomposition. Once model order (number of source components) was estimated, dimension reduction matrix was used to transform the model from overdetermined to determined, as follows:

$$\mathbf{X} = \mathbf{V}^{\mathbf{T}}\mathbf{Z} = \mathbf{V}^{\mathbf{T}}\mathcal{A}\mathbf{S} = \mathbf{A}\mathbf{S} \tag{7.8}$$

In the above formula, $\mathbf{A} \in \mathcal{R}^{R\times R}$, $\mathbf{A}=\mathbf{V}^{T}\mathcal{A}$, $\mathbf{X} \in \mathcal{R}^{R\times N}$, $\mathbf{V}^{\mathbf{T}} \in \mathcal{R}^{R\times M}$ is the dimension reduction matrix which derives from PCA. $\mathbf{V}^{\mathbf{T}}$ consists of the first R eigenvectors of the covariance matrix of the data matrix $\mathbf{Z}$.

The ICA decomposition model is as the following:

$$\mathbf{Y} = \mathbf{W}\mathbf{X}, \tag{7.9}$$

where $\mathbf{Y} = \mathcal{R}^{R\times N}$ is the component matrix, which is used to estimate source signals $\mathbf{S}$. ICA is an algorithm that finds unmixing matrix $\mathbf{W} \in \mathcal{R}^{\mathrm{R}\times\mathrm{R}}$ based on the independence of components.

According to difference between cost functions, Hyvärinen et al. proposed five different ICA algorithms, including algorithm based on non-Gaussian maximization, algorithm based on maximum likelihood estimation, algorithm based on minimum mutual information, algorithm based on tensor, and algorithm based on nonlinear decorrelation and nonlinear PCA (Hyvärinen et al. 2001). Beyond fixed-point-based FastICA (Hyvarinen 1999), max mutual information-based InfomaxICA (Sejnowski and Bell 1995) also have been widely used in all kinds of fields. The InfomaxICA is widely used due to its stability. The equation of mutual information is listed as follows:

$$\mathrm{I}(\mathbf{Y},\mathbf{X}) = \mathrm{H}(\mathbf{Y}) - \mathrm{H}(\mathbf{Y}|\mathbf{X}), \tag{7.10}$$

where I refers to the mutual information of components and H is the entropy. Based on the above formula, the iterative formula was obtained as follows:

$$\Delta\mathbf{W} \propto \left[\mathbf{W}^{T}\right]^{-1} + (1 - 2\mathbf{Y})\mathbf{X}^{T} \tag{7.11}$$

The estimation of mixing matrix needs the combination of $\mathbf{W}$ and dimension reduction matrix (Cong et al. 2014):

$$\mathbf{U} = \mathbf{V}\mathbf{B} = \mathbf{V}\mathbf{W}^{-1}, \tag{7.12}$$

where $\mathbf{B} = \mathbf{W}^{-\mathbf{1}}$ and $\mathbf{U} \in \mathcal{R}^{M\times R}$ is named as the coefficient matrix approximating the mixing matrix $\mathcal{A}$ in (7.7).

7.3.2 Stability of ICA Decomposition Under Global Optimization

Most of the ICA algorithms are self-adaptive, and an ICA algorithm may converge to local minima. Therefore, whether the results obtained by ICA decomposition are reproducible is crucial. If both Eqs. (7.8 and 7.9) are merged together, the global matrix **C** links the extracted components and the sources together as the following:

$$\mathbf{Y} = \mathbf{WX} = \mathbf{WAS} = \mathbf{CS}, \tag{7.13}$$

where $\mathbf{C} = \mathbf{WA}$.

As shown in Eq. (7.11), the ICA algorithm is adaptive. In theory, for global optimization, there is only one nonzero element in each column and each row of the global matrix **C**. Then, the global matrix equal to multiplication of a permutation matrix **P** and a diagonal matrix **D** as:

$$\mathbf{C} = \mathbf{PD} \tag{7.14}$$

$$\mathbf{Y} = \mathbf{CS} = \mathbf{PDS} \tag{7.15}$$

Subsequently, the coefficient matrix turns to be

$$\mathbf{U} = \mathbf{VB} = \mathbf{VW}^{-1} = \mathbf{VAD}^{-1}\mathbf{P}^{-1} = \mathbf{VV}^{T}\mathcal{A}\mathbf{D}^{-1}\mathbf{P}^{-1} = \mathcal{A}\mathbf{D}^{-1}\mathbf{P}^{-1}, \tag{7.16}$$

where, $\mathbf{WA} = \mathbf{PD}$, $\mathbf{A}=\mathbf{V}^{T}\mathcal{A}$, $\mathbf{VV}^{T}$ is an identity, $\mathbf{D}^{-1}$ is a diagonal matrix, and $\mathbf{P}^{-1}$ is a permutation matrix.

Under the global optimization, if one ICA algorithm is run multiple times with random initialization, the stability of coefficient matrix and component matrix is the same. Stability of extracted components represents the stability of ICA algorithm.

7.3.3 Stability of ICA Decomposition Under Local Optimization

Most ICA algorithms tend to converge to the local optimization as they are adaptive. The global matrix cannot be written as the production of a permutation matrix and a diagonal matrix. Subsequently:

$$\mathbf{Y} = \mathbf{CS} \neq \mathbf{PDS} \tag{7.17}$$

$$\mathbf{U} = \mathbf{VW}^{-1} \neq \mathcal{A}\mathbf{D}^{-1}\mathbf{P}^{-1} \tag{7.18}$$

the formula above means that some of the extracted components are still mixtures of some sources. In the estimation of coefficient matrix **U**, the matrix inverse of the unmixing matrix **W** cannot be avoided. It is widely acknowledged that the inverse operation can amplify the errors in **W**. Therefore, it is necessary to access whether both the coefficient matrix **U** and the component matrix **Y** are stably estimated or not.

For example, under the condition that ICA decomposition is run K times with random initialization, and the number of components is R, $R{*}K$ components can be fed into cluster algorithm. In ICASSO (Himberg et al. 2004), a parameter named Iq is calculated from the difference between intra-cluster similarities and inter-cluster similarities. In the application Iq is an index to evaluate the stability of component matrix, and we named as Comp_Iq.

In ICA decomposition, one component is always along with its associated coefficient vector. However, the stability of coefficient matrix and that of component are different. In order to assess the stability of the coefficient matrix, the memberships of component cluster result can be used to cluster coefficient vectors. The stability index of coefficient matrix could also be calculated. It is called Coef_Iq in this study.

The ICA algorithm is stable once both the component and the corresponding coefficient vector are stably extracted. The stability index of ICA algorithm is defined as

$$\text{Iq} = \text{Comp\_Iq} \times \text{Coef\_Iq} \tag{7.19}$$

The range of stability is from 0 to 1.0 means the component is extremely unstable. 1 means the component is stable. Comp_Iq work as the probability of the stability of the component matrix. Coef_Iq is the probability of the stability of the coefficient matrix. So the multiplication can be expressed as the stability of the whole (Zhang et al. 2018).

7.3.4 Subtraction to Reject One Component by Back Projection

For ICA mixed model:

$$\mathbf{X} = \mathbf{AS} \tag{7.20}$$

where $\mathbf{X} \in \mathbb{R}^{M \times N}$ is the signal to be decomposed. $\mathbf{A} \in \mathbb{R}^{M \times R}$ is mixing matrix. $\mathbf{S} \in \mathbb{R}^{R \times N}$ is the source signal. R represents the number of source signals.

ICA decomposition model:

$$\mathbf{Y} = \mathbf{WX} \tag{7.21}$$

where $\mathbf{W} \in \mathbb{R}^{R \times M}$ is unmixing matrix. The inverse matrix $\mathbf{B} = \boldsymbol{W}^{-1} \in \mathbb{R}^{M \times R}$ is used to estimate $\mathbf{A}$. $\mathbf{Y} \in \mathbb{R}^{R \times N}$ is used to estimate the source matrix $\mathbf{S}$. However, it is well-known that ICA has the magnitude and polarity indeterminacy. It is assumed that the kth component of the decomposition corresponds to the ith source signal. But $\boldsymbol{y}_{k,\,:} \neq \boldsymbol{s}_{i,\,:}, \boldsymbol{b}_{:,\,k} \neq \boldsymbol{a}_{:,\,i}$. But it can be proved that $\mathbf{E}_k = \boldsymbol{b}_{:,\,k} \cdot \boldsymbol{y}_{k,\,:} = \boldsymbol{a}_{:,\,i} \cdot \boldsymbol{s}_{i,\,:}$ under the condition of global optimal solution. That means rank-1 matrix $\mathbf{E}_k$ is ultimately decomposed by ICA. The proof process is as follows:

Global matrix:

$$\mathbf{C} = \mathbf{W}\mathbf{A} \tag{7.22}$$

Then:

$$\mathbf{Y} = \mathbf{C}\mathbf{S} \tag{7.23}$$

$$\mathbf{B}\mathbf{C} = \mathbf{A} \tag{7.24}$$

Under the condition of global optimal solution, each row and each column of $\mathbf{C}$ has only one nonzero element (Cong et al. 2011b). If the kth component of the decomposition corresponds to the ith source signal, then $c_{k,\,i}$ is nonzero.

$$\therefore \boldsymbol{y}_{k,:} = c_{k,i} \cdot \boldsymbol{s}_{i,:} \tag{7.25}$$

$$\boldsymbol{b}_{:,k} \cdot c_{k,i} = \boldsymbol{a}_{:,i} \tag{7.26}$$

Multiplying Eq. (7.25) with Eq. (7.26) on both sides,

$$\boldsymbol{b}_{:,k} \cdot c_{k,i} \cdot \boldsymbol{y}_{k,:} = \boldsymbol{a}_{:,i} \cdot c_{k,i} \cdot \boldsymbol{s}_{i,:} \tag{7.27}$$

The $c_{k,\,i}$ can be omitted on both sides of the equation because it is a real number:

$$\boldsymbol{b}_{:,k} \cdot \boldsymbol{y}_{k,:} = \boldsymbol{a}_{:,i} \cdot \boldsymbol{s}_{i,:} \tag{7.28}$$

Therefore, it is rank-1 matrix $\mathbf{E}_k = \boldsymbol{b}_{:,\,k} \cdot \boldsymbol{y}_{k,\,:}$ ($k = 1, 2, \cdots R$, R is the number of extracted components) that could be ultimately decomposed by ICA.

To reject the component, back projection result could be subtracted from the EEG recordings, as illustrated below (Cong et al. 2011a):

$$\mathbf{X}_{clear} = \mathbf{X} - \mathbf{E}_k \tag{7.29}$$

By this way electrooculogram (EOG) could be rejected from EEG.

7.3.5 ICA and Artifact Correction

ICA is an adaptive algorithm; consequently, it will work better if the artifacts being removed own large portion of the variance, such as blinks and large eye movement. After performing ICA on EEG data, the spatial and temporal features for each component were drawn. Here the basic features of independent components (ICs) for blinks and eye movement artifacts are shown (Luck 2014).

7.3.6 Blink

The eye seems like a battery, since there is constant electrical potential between cornea at the front of the eye and the retina at the back of the eye. Consequently, the source of the EOG is centered at frontal area and the potential gradually falling off toward the back of the head. When any eye movements occur, the constant potential will change, leading the voltage deflection at frontal area.

Figure 7.8 shows the basic features of an IC for blink artifact. As for the spatial feature, when the eyes blink, the constant potential will change greatly at the vertical direction. As Fig. 7.8a shows, the topography of the blink component will be either positive or negative at frontal area for the uncertain polarity of ICA algorithm and

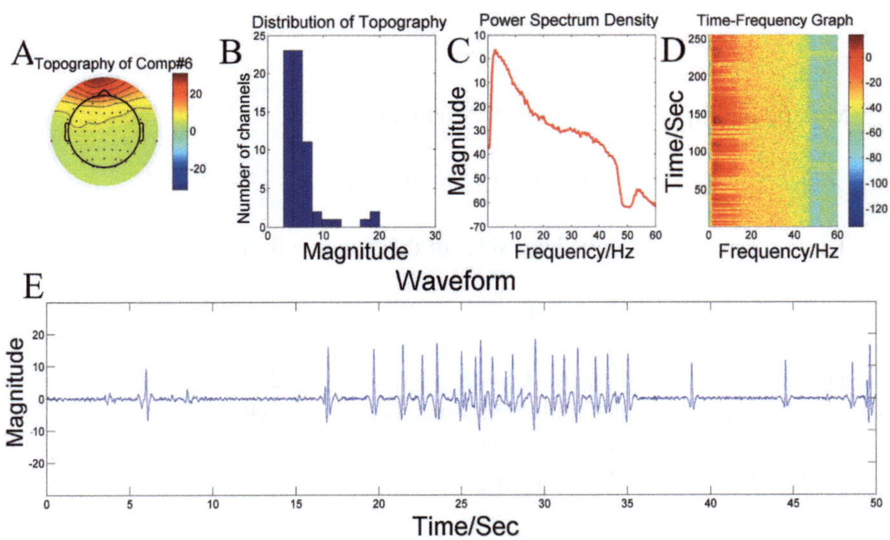

Fig. 7.8 Basic features of the independent component for blink artifact. (**a**) The topography of the blink component will be either positive or negative at frontal area and shows a tendency of decrease from frontal to occipital of the head; (**b**) the histogram of the above topography shows that frontal electrodes own pretty large coefficients; (**c**) the PSD of the IC for blink is mainly centered at low frequency band; (**d**) the time-frequency graph based on STFT method of the IC for blink; (**e**) the time series of the IC for blink shows the sudden monophasic deflection

shows a tendency of decrease from frontal to occipital of the head. Figure 7.8b is the histogram of the above topography, which implies that only few electrodes (frontal electrodes) own pretty large coefficients and the other electrodes own similar small coefficients. The eye blink response consists primarily of a monophasic deflection of 50–100 μ V with a typical duration of 200–400 ms. As a result, we can infer that compared with spontaneous EEG, the blink will be a sudden monophasic deflection lasting for 200–400 ms and the frequency mainly centered at 2.5–5 Hz. In terms of temporal feature, the time series were drawn in Fig. 7.8e, which shows the sudden monophasic deflection. Figure 7.8c shows the power spectrum density (PSD) of the IC of blink, which is mainly centered at low-frequency band. Figure 7.8d is the addition of time-frequency graph based on short time Fourier transform (STFT) method.

7.3.7 *Eye Movement*

Compared with blinks, the eye movement we referenced here will change the potential at the horizon direction. When the eyes move, the voltage becomes more positive over the side of the head that the eyes now point toward and will be reverse polarity at the opposite side, which seems like a scale bias to one side. As a result, the topography of the IC component for eye movement (Fig. 7.9a) shows reverse

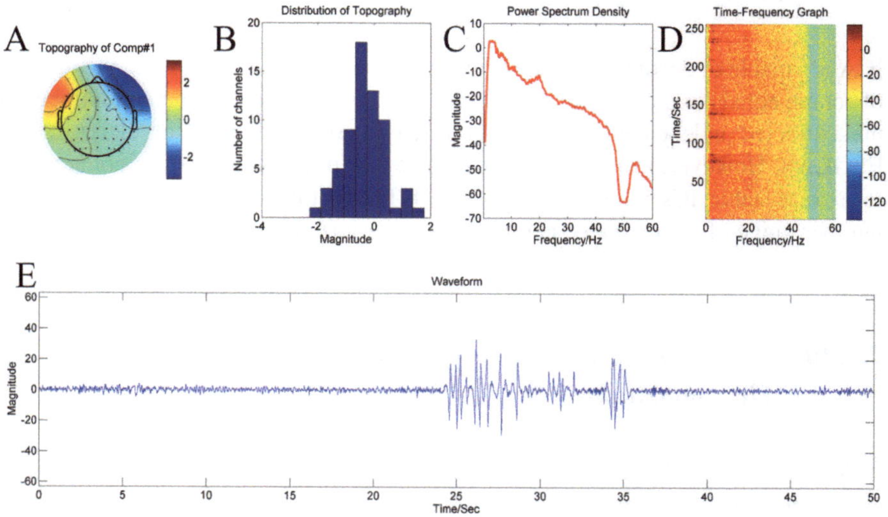

Fig. 7.9 Basic features of the independent component for eye movement artifact. (**a**), The topography of the eye movement component shows reverse polarity at the opposite sides centered at frontal electrodes; (**b**), the histogram of the above topography; (**c**), the PSD of the IC for eye movement; (**d**), the time-frequency graph based on STFT method of the IC for eye movement; (**e**), the time series of the IC for eye movement

polarity at the opposite sides centered at frontal electrodes, and the histogram of eye movement component (Fig. 7.9b) shows an almost uniform distribution. The vast majority of eye movements will be saccades (sudden shifts in eye position), which would consist of a sudden step from one voltage level to another voltage level, where it would remain until the eyes moved again. Consequently, the duration of the sudden deflection of IC for eye movement tends to be longer (Fig. 7.9e) than the blink component, and the PSD tends to center at a lower frequency band (Fig. 7.9c).

7.3.8 Introduction of ICA Toolbox

EEGLAB provides a manipulation to perform ICA on EEG data. In practice, there are several disadvantages of the ICA module in EEGLAB, which disregard the stability of ICs. In our ICA toolbox, we added the extraction of stable ICs. As mentioned in Sect. 7.3, the stability of ICs involved two parts: (1) determining the number of extracting ICs and (2) determining the times of running ICA decomposition. In addition, we improved the style of showing the results of ICs, namely, showing the features of one IC in one figure.

7.3.9 Determining the Number of Extracting Independent Components

ICA is capable of decomposing the determined model, which means that the number of source equals to the number of sensors. As a consequence, the default number of the extracted independent component is equal to the number of electrodes in EEGLAB. However, from our experience, the best practical number of components to extract stable ICs ranges from 15 to 30. In our toolbox, you can easily set a number/range of the extracted components to perform ICA (Fig. 7.11).

7.3.10 Determining the Times of Running ICA Decomposition

ICA is an adaptive algorithm, which means the results of ICs will be different among each run. To our knowledge, the ICs results are based on single running in EEGLAB. As ICASSO suggested, the ICA decomposition can be run many times to evaluate the stability of ICs. In our lab, the times of running ICA decomposition range from 15 to 50 to extract stable ICs on EEG data.

In our toolbox, you can easily set the times of running ICA decomposition (Fig. 7.11).

7.3.11 *Evaluating the Number of Stable Independent Components*

As ICASSO suggested, the Comp_Iq can be used to evaluate the number of stable ICs after performing ICA decomposition certain times at a certain number of components. In addition, the toolbox provides other indexes to evaluate the stability.

As mentioned in the theory part of ICA, the unmixing matrix was randomly initialized, which took several steps to convergence among each ICA decomposition. In consequence, the mean and standard deviant value of the steps to reached convergence were obtains across runs, as shown in Fig. 7.10a. With the increase of the number of components, the it took more steps to convergence, namely, it tended to be more difficult to convergence. In addition, the number of converged runs was obtained among all the runs. As Fig. 7.10b shows, among fifty runs, the unmixing matrix can't reach convergence-criteria within certain number of iteration (the threshold of iteration is 100 for FastICA and 512 for InfomaxICA) for each run after the number of extracted component was bigger than 25. The results suggested that the ICs might be unstable when the number of extracted component is above 25. As mentioned in 7.3, Fig. 7.10c illustrated the Comp_Iq among all the runs at certain number of components. In our Lab, the criterion was that the Comp_Iq for stable ICs should be more than 0.9. Figure 7.10d is the result of explaining variance of the first several components drawn from PCA.

Upon above results, the number of stable ICs was 25 in this study.

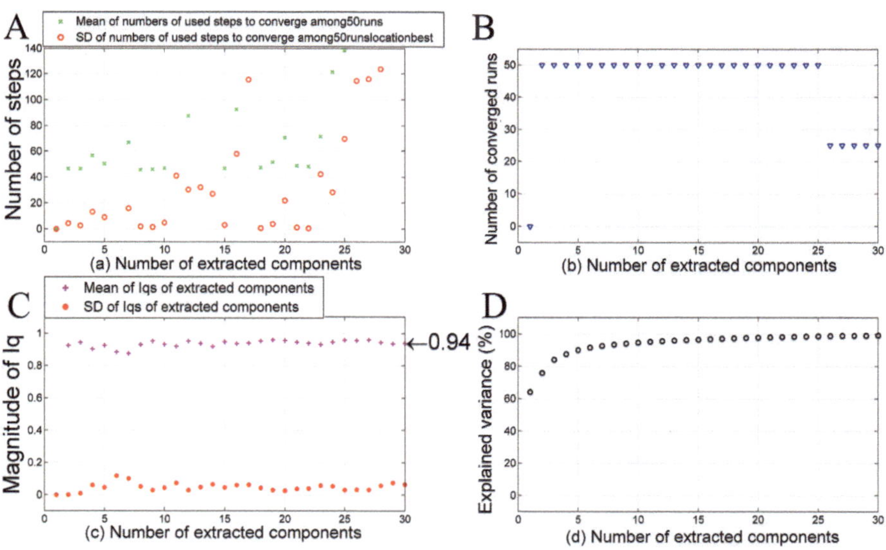

Fig. 7.10 Four indexes to evaluate the number of stable ICs. (**a**), the mean and standard deviant value of the steps to reached convergence were obtains across runs; (**b**), the number of converged runs is obtained among all the runs; (**c**), the Comp_Iq among all the runs at certain number of components; (**d**), the results of explaining variance of the first several components drawn from PCA

7.3.12 Showing the Results in a Detailed Way

The criterion of determining whether an IC is an artifact component is using the temporal (waveform) and spatial (topography) features. As we know, the waveforms and topographies are plotted in separated figures in EEGLAB, which means you have to check one IC in two figures. From the user's experience, it's pretty disturbing. In our toolbox, we plot the features of one IC in one figure and added three features of the IC: (1) the histogram of the topography; (2) the PSD of the IC; (3) and the time-frequency graph based on STFT method of the IC (Figs. 7.8 and 7.9).

7.3.13 Flowchart of ICA Toolbox

The ICA toolbox provides friendly dialogs to effectively perform ICA. The first step is choosing the EEG data, which you want to perform ICA (Fig. 7.11a). In addition, in our lab, we prefer to perform ICA on continuous EEG data (not the epoched EEG data), which retain the original temporal feature. The second step is setting several

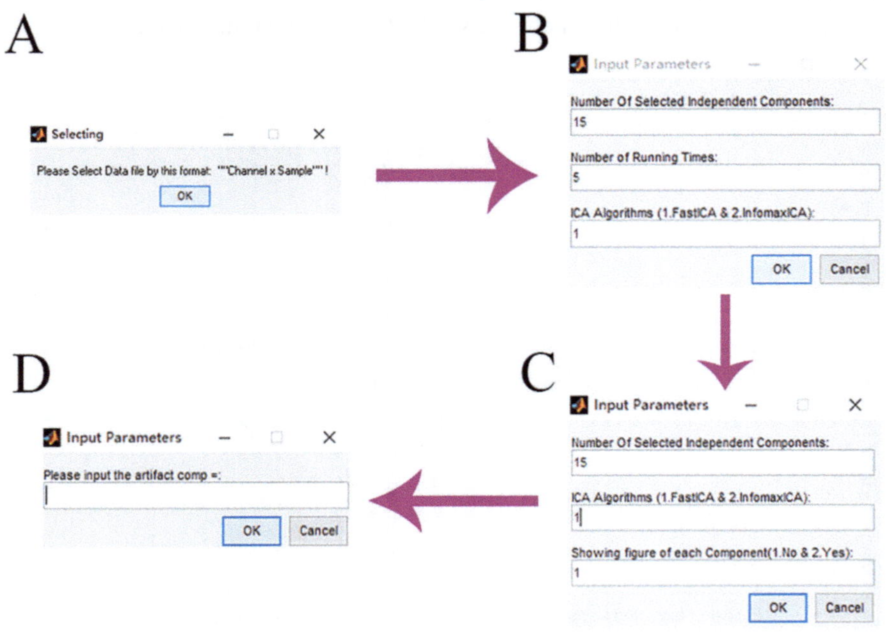

Fig. 7.11 The flowchart of ICA toolbox. (**a**), Choosing the EEG data; (**b**), setting several parameters for running ICA, which include the number of extracting ICs, the number of running times, and the algorithm to perform ICA; (**c**), showing the results of each IC; (**d**), inputting the index of the artifact component to remove

parameters for running ICA (Fig. 7.11b), which include the number of extracting ICs, the number of running times, and the algorithm to perform ICA. We suggested to use InfomaxICA to decompose continuous EEG data, if you have a high-performance workstation, because it's time-consuming.

The third step is showing the results of each IC (Fig. 7.11c), which will be saved as a PNG picture for each IC in the folder of the data you selected. The first two parameters should be the same as the numbers filled in the second step, and you can decide to plot the results of each IC in the current window or not. After checking the results of each IC, you can input the index of the artifact component to remove at the fourth step (Fig. 7.11d). The above steps are the basic procedures to conduct artifact removal on EEG data based on ICA in our ICA toolbox.

7.3.14 A Detailed Example Using the ICA Toolbox

Based on the features of time, space, frequency, and time-frequency domain, the artifacts of blinks and eye movement can be accurately detected and rejected. Figure 7.12 shows how ICA is applied in artifact correction. A 20-s period of EEG data with the sampling rate of 500 Hz is shown in Fig. 7.12a, and data epochs filled with blue are obvious blink artifacts. Then, the FastICA algorithm was performed on the Raw EEG data. After visual inspecting the ICs, the blink component (Fig. 7.12b) was selected out. Finally, the blink component was subtracted by back projection, and the clean data was obtained.

7.4 Tensor Decomposition for ERP Analysis

In EEG experiments, there are more modes than the two modes of time and space. For instance, analysis of EEG signal may compare responses recorded in different groups (e.g., comparison of responses between patients and healthy controls). Thus, at least one more mode appears, and it is the subject. Furthermore, in an experiment to elicit ERPs, there are modes of EEG trial (since several stimulus presentations are required) and stimulus presentation condition. This means the brain data collected by EEG techniques can be naturally fit into a multi-way array including multiple modes.

However, most applied computing tools for brain research are oriented for one-way or two-way data. Consequently, in order to facilitate the two-way signal processing methods, the extra methods besides the two modes of time and space are often concatenated (data are horizontally connected in plane) or stacked (data are vertically connected in a plane) with the time or the space mode for generating a matrix (Cong et al. 2013, 2014; Delorme and Makeig 2004; Eichele et al. 2011; Mode and Discovery 2012). This is often called unfolding a multi-way array into a matrix. For EEG data, such unfolding inevitably loses some potentially existing

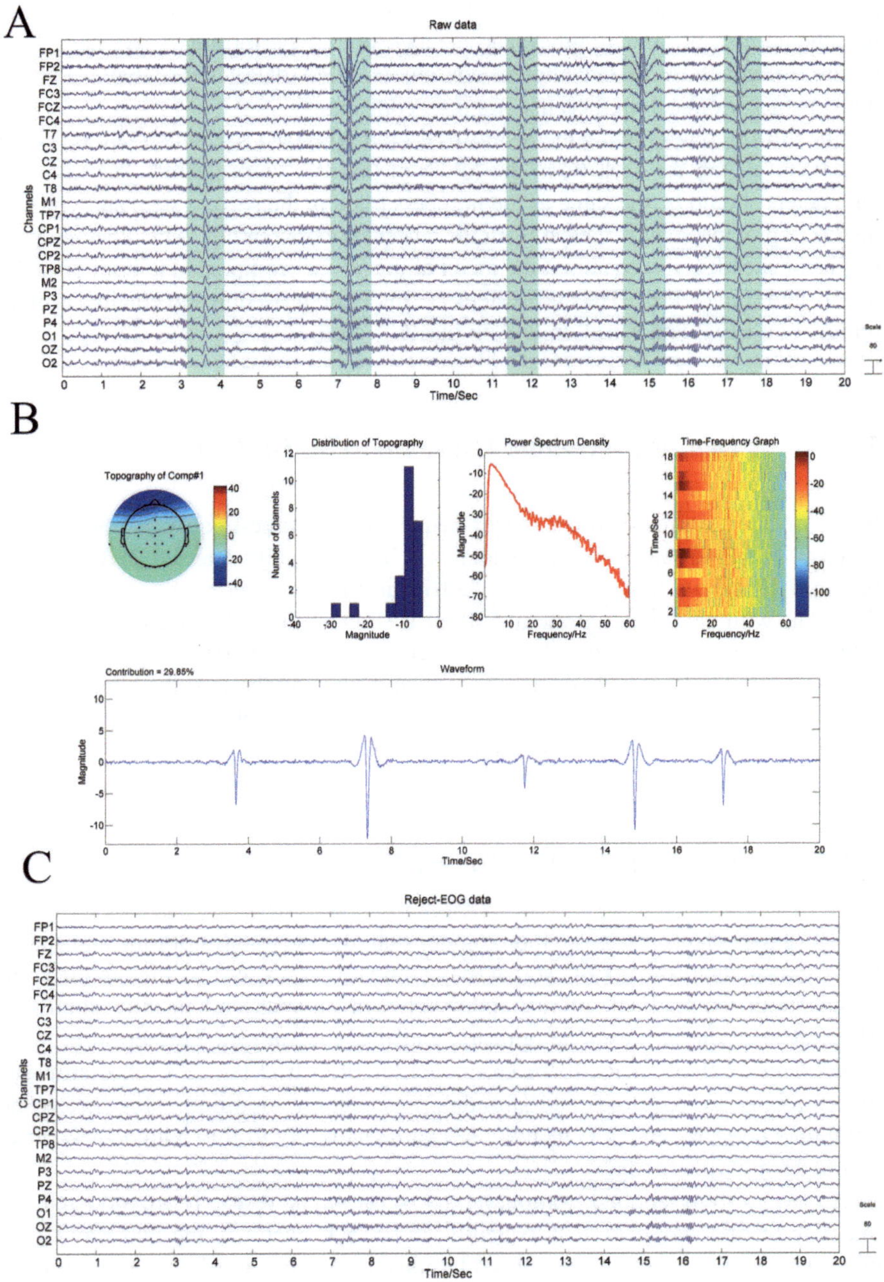

Fig. 7.12 Applying ICA in artifact correction. (**a**) The 20-s period of EEG data and data epochs filled with blue are obvious blink artifacts; (**b**) the selected blink component by visual inspecting each IC; (**c**) the clean EEG data after subtracting the blink component by back projection

interactions between/among the folded modes, such as time, frequency, and space modes. The interactions can be of research interest. Consequently, in order to appropriately reveal the interactions among multiple modes, the signal processing methods particularly for a multi-way array are naturally promising tools.

A multi-way array is named as a tensor (Cichocki et al. 2009; Kolda and Bader 2009). Recently, tensor decomposition has become surprisingly attractive for signal processing (Andrzej Cichocki 2013). Indeed, it has already been applied for analysis of ERPs in the 1980s (Mocks 1988).

Generally, for a given *N*th-order tensor $\underline{\mathbf{X}} \in \mathbb{R}^{I_1 \times I_2 \times \cdots \times I_N}$, the canonical polyadic decomposition (CPD) is defined as

$$\underline{\mathbf{X}} = \sum\nolimits_{r=1}^{R} \boldsymbol{u}_r^{(1)} \circ \boldsymbol{u}_r^{(2)} \circ \cdots \circ \boldsymbol{u}_r^{(N)} + \underline{\mathbf{E}} \tag{7.30}$$

$\mathbf{U}^{(n)} = \left[\boldsymbol{u}_1^{(n)}, \boldsymbol{u}_2^{(n)}, \cdots, \boldsymbol{u}_R^{(n)}\right] \in \mathbb{R}^{I_n \times R}$ denotes a component matrix for mode #n, and n = 1, 2, ⋯, N (Cong et al. 2015a).

The demo data was adopted from Cong et al. (2012), and the preprocessed data was a third-order tensor with the size of 60 (temporal samples) × 9 (channels) × 42 (the subjects are composed of 2 groups with 21 children per group). Then, the wavelet decomposition was performed at each channel for each subject to form a fourth-order tensor with the size of 60 (temporal samples) × 71 (frequency bins) × 9 (channels) × 42 (the subjects are composed of 2 groups with 21 children per group). According to the previous processing, 36 components were extracted for each mode.

Reckoned from the left to the right side, the first column shows three temporal components (#16–#18); the second column describes the three spectral components (#16–#18); the third column presents the three spatial components (i.e., topographies) (#16–#18); and the last column shows the three multi-domain feature components (#16–#18). The order of the components and the variance of a component in the NCPD are not determined. RD: reading disability. AD: attention deficit, adapted from Cong et al. (2012)

The criterion of selecting the components of interest was based on the temporal, spatial, and spectral properties of the interested ERP component. In this study, the component #20 was regarded as the desired multi-domain feature of the MMN shown in Fig. 7.13, because the corresponding temporal spatial and spectral components meet the MMN property requirements.

Obviously, the temporal, spectral, and spatial component matrices do not contain any information of the subjects, and they are common to all features. Variability among subjects exists in each feature component, which is a column of the feature component matrix. This is a characteristic of the multi-domain feature extraction of ERP, when NCPD is applied on the fourth-order ERP tensor of the TFR that includes the time, frequency, space, and feature modes, which can be performed further statistical analysis.

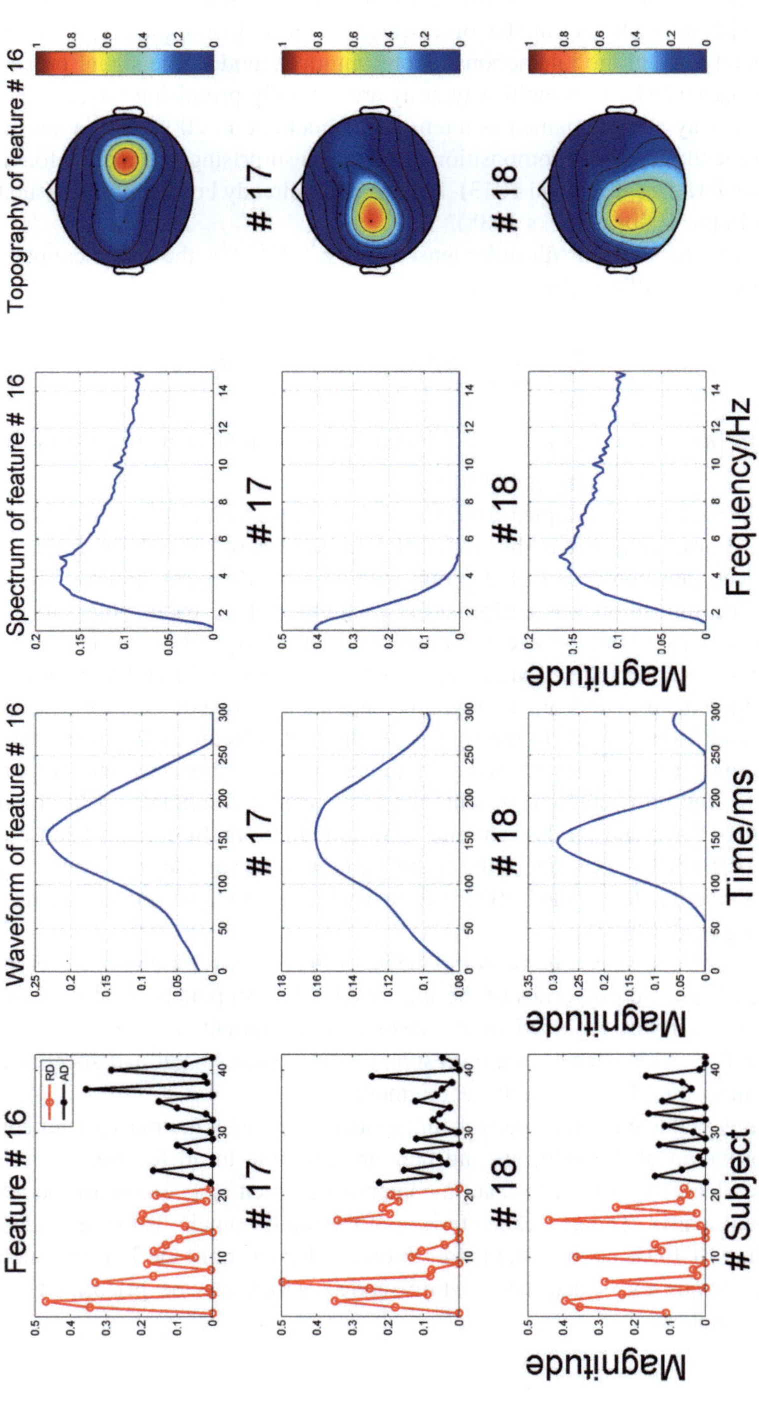

Fig. 7.13 The results of NCPD on a fourth-order ERP tensor of the TFRs

References

Cattell RB. The procrustes program: producing direct rotation to test a hypothesized factor structure. Comput Behav Sci. 1976;7:258–62.

Chaumette E, Comon P, Muller D. ICA-based technique for radiating sources estimation: application to airport surveillance. IEEE Proc Radar Sonar Navig. 1994;142(4):211.. 92363

Cichocki A. Tensor decompositions: a new concept in brain data analysis? arXiv Preprint arXiv: 1305.0395. 2013;50(2011):507–17.

Cichocki A, Zdunek R, Amari S. Nonnegative matrix and tensor factorization. IEEE Signal Process Mag. 2009;25(1):142–5.

Cois J. Blind signal separation: statistical principles. Proc IEEE. 2009;86(10):2009–25.

Cong F, Kalyakin I, Chang Z, Ristaniemi T. Analysis on subtracting projection of extracted independent components from EEG recordings. Biomed Tech (Berl). 2011a;56(4):223–34.

Cong F, Kalyakin I, Ristaniemi T. Can back-projection fully resolve polarity indeterminacy of independent component analysis in study of event-related potential? Biomed Signal Process Control. 2011b;6(4):422–6.

Cong F, Phan AH, Zhao Q, Huttunen-Scott T, Kaartinen J, Ristaniemi T, et al. Benefits of multi-domain feature of mismatch negativity extracted by non-negative tensor factorization from Eeg collected by low-density array. Int J Neural Syst. 2012;22(6):1250025.

Cong F, He Z, Hämäläinen J, Leppänen PHT, Lyytinen H, Cichocki A, Ristaniemi T. Validating rationale of group-level component analysis based on estimating number of sources in EEG through model order selection. J Neurosci Methods. 2013;212(1):165–72.

Cong F, Puoliväli T, Alluri V, Sipola T, Burunat I, Toiviainen P, et al. Key issues in decomposing fMRI during naturalistic and continuous music experience with independent component analysis. J Neurosci Methods. 2014;223:74–84.

Cong F, Lin Q-H, Kuang L-D, Gong X-F, Astikainen P, Ristaniemi T. Tensor decomposition of EEG signals: a brief review. J Neurosci Methods. 2015a;248:59–69.

Cong F, Ristaniemi T, Lyytinen H. Advanced signal processing on event related potentials (ERPs): World Scientific; 2015b. p. 131–87.

De Lathauwer L, De Moor B, Vandewalle J. Fetal electrocardiogram extraction by blind source. IEEE Trans Biomed Eng. 2000;47(5):567–72.

Delorme A, Makeig S. EEGLAB: an open source toolbox for analysis of single-trial EEG dynamics including independent component analysis. J Neurosci Methods. 2004;134(1):9–21.

Dien J. Addressing misallocation of variance in principal components analysis of event-related. Potentials. 1998;11(1):43–55.

Dien J. Differential lateralization of trait anxiety and trait fearfulness: evoked potential correlates. Personal Individ Differ. 1999;26:1998–2000.

Dien J, Beal DJ, Berg P. Optimizing principal components analysis of event-related potentials: matrix type, factor loading weighting, extraction, and rotations. Clin Neurophysiol. 2005;116:1808–25.

Durso G, Prieur P. Blind identification methods applied to Electricity C France's civil works and power plants monitoring. The 1997 IEEE Signal Processing Workshop on Higher-Order Statistics; 1997. p. 82–86.

Eichele T, Rachakonda S, Brakedal B, Eikeland R, Calhoun VD. EEGIFT: group independent component analysis for event-related EEG data. Comput Intell Neurosci. 2011;2011:1–9.

Harman HH. Modern factor analysis. Chicago: University of Chicago Press; 1976.

Himberg J, Hyvärinen A, Esposito F. Validating the independent components of neuroimaging time series via clustering and visualization. NeuroImage. 2004;22(3):1214–22.

Hyvarinen A. Fast and robust fixed-point algorithm for independent component analysis. IEEE Trans Neural Netw Learn Syst. 1999;10:626–34.

Hyvärinen A, Karhunenen J, Oja E. Independent component analysis. Neural Comput. 2001;13 (7):504.

Kaiser H. The Varimax criterion for analytic rotation in factor analysis. Psychometrika. 1958;23(3):187–200.

Kolda TG, Bader BW. Tensor decompositions and applications. SIAM Rev. 2009;51(3):455–500.

Luck SJ. An introduction to the event-related potential technique. Monogr Soc Res Child Dev. 2014;78(3):388.

Makeig S, Bell AJ, Jung T-P, Sejnowski TJ. Independent component analysis of electroencephalographic data. In: Advances in neural information processing systems; 1996. p. 145–51.

Mathew G, Reddy VU. Blind separation of multiple co-channel bpsk signals arriving at an antenna array. IEEE Signal Process Lett. 1995;2(9):176–8.

Mocks J. Topographic components model for event-related potentials and some biophysical considerations. IEEE Trans Biomed Eng. 1988;35:482.

Mode D, Discovery N. Multi-subject independent component analysis of fMRI: a decade of intrinsic networks, default mode, and neurodiagnostic discovery. IEEE Rev Biomed Eng. 2012;5:60–73.

Sejnowski TJ, Bell AJ. Information-maximization approach to blind separation and blind deconvolution. Technology. 1995;1159(1994):1129–59.

Spencer KM, Dien J, Donchin E. Spatiotemporal analysis of the late ERP responses to deviant stimuli. Psychophysiology. 2001;38(2):343–58.

Squires NK, Squires KC, Hillyard SA. Two varieties of long – latency positive waves evoke d by unpredictable auditory stimuli in man. Electroencephalogr Clin Neurophysiol. 1975;38 (4):387–401.

Swindlehurst AL, Goris MJ, Ottersten B. Some experiments with array data collected in actual urban and suburban environments. In: First IEEE signal processing workshop on signal processing advances in wireless communications: IEEE; 1997. p. 301–4.

van der Veen AJ. Algebraic methods for deterministic blind beamforming. Proc IEEE. 1998;86 (10):1987–2008.

Zhang Q, Hu G, Tian L, Ristaniemi T, Wang H, Chen H, Wu J, Cong F. Examining stability of independent component analysis based on coefficient and component matrices for voxel-based morphometry of structural magnetic resonance imaging. Cogn Neurodyn. 2018;12:461–70.

Zhao Q, Zhou G, Adali T, Zhang L, Cichocki A. Kernelization of tensor-based models for multiway data analysis: processing of multidimensional structured data. IEEE Signal Process Mag. 2013;30(4):137–48.

Chapter 8
Microstate Analysis

Huibin Jia

Abstract The microstate analysis, which is based on the clustering of electric field maps, has been proved to be an efficient technique that could fully use the spatial information of topographies of EEG/ERP signals. Typically, researchers found that only four kinds of distinct topographies (i.e., microstate classes) could explain most variances of the spontaneous EEG data. Moreover, each of the four commonly observed microstate classes was closely associated with the activity of distinct resting-state brain networks revealed by BOLD signals of resting-state fMRI. For ERP signals, the microstate segmentation can be used to identify the underlying ERP components and their latencies from the multichannel ERP waveforms. In this chapter, we illustrated the basic concepts of microstate analysis, the commonly used clustering algorithms, the metrics derived from microstate analysis, and how to perform microstate analysis using open-access tools.

Keywords Topographic maps · Spontaneous EEG · ERP · Spatial clustering

8.1 Introduction

Traditionally, the signal processing of EEG and ERPs is based on their features in time domain (e.g., the morphology of ERP waveforms) and frequency domain (e.g., spectral power and ERS/ERD) (Jia et al. 2017; Jeremy et al. 2014; Maxwell et al. 2015; Wu et al. 2016). In these approaches, a set of scalp electrodes were usually

---

Electronic supplementary material The online version of this chapter (https://doi.org/10.1007/978-981-13-9113-2_8) contains supplementary material, which is available to authorized users.

H. Jia (✉)
Key Laboratory of Child Development and Learning Science of Ministry of Education, School of Biological Sciences & Medical Engineering, Southeast University, Nanjing, Jiangsu, China

L. Hu, Z. Zhang (eds.), *EEG Signal Processing and Feature Extraction*,
https://doi.org/10.1007/978-981-13-9113-2_8

defined as regions of interest (ROIs) (Li et al. 2018). Thus, these traditional approaches failed to utilize rich spatial information inherent in EEG/ERP signals. In account of this aspect, the microstate analysis or the clustering of electric field maps has been recently developed and proved to be an efficient technique that could fully use the spatial information of topographic maps (Khanna et al. 2014). The microstate analysis is originated from the researches of Dietrich Lehmann and colleagues in 1987 (Lehmann et al. 1987). They found that the time series of scalp potential maps of spontaneous EEG signals are not a disorganized succession with irregularly varied topographical configurations, but rather an organized continuum with regularly changing configurations (i.e., it remains stable for a period of time typically ranging from 80 to 120 ms, followed by an abrupt alteration into a new configuration, and returns to its stability again) (Michel and Koenig 2017; Koenig et al. 2002). Note that, within the period of time for stable configuration, the strength of the scalp potential field may increase or decrease, but its topography remains stable. Since the scalp potential field reflects the momentary state of global activity of underlying brain networks and the topographical configuration indicates changes in the global coordination of cerebral activities, Dietrich Lehmann and his colleagues proposed that these stable periods should reflect the basic steps of information processing in human brain (i.e., "atoms of thoughts") and named them as functional microstates (Lehmann et al. 1998, 2005; Khanna et al. 2014). Although the initial studies were conducted on spontaneous EEG activities, similar findings (i.e., periods with quasi-stable potential landscape) have also been obtained in topographical ERP analysis (Murray et al. 2008). In general, each of these functional microstates in ERP waveforms may reflect certain ERP component (Hu et al. 2013).

For decades, researchers have developed several techniques in order to identify the underlying microstates (Pascual-Marqui et al. 1995; Gärtner et al. 2015; Murray et al. 2008). Although alternative approaches exist, most of the previous studies were based on spatial cluster analysis of scalp potential maps (Michel and Koenig 2017). In these techniques, the maps of all time points are assigned to a few classes (clusters), which could explain most variances in EEG/ERP activities. Then, functional microstates are determined post hoc by fitting the template map of identified classes back to the map of each time point. Finally, several temporal parameters of these microstate classes (e.g., the mean duration, occurrence rate per second, percentage of time coverage over the entirely recorded EEG signals) are computed (Khanna et al. 2014).

Researchers found that only 4~8 classes of distinct topographies were needed to explain most variances (~80%) of the spontaneous EEG data (Rieger et al. 2016; Pascual-Marqui et al. 1995; Van de Ville et al. 2010). In microstate segmentation of spontaneous EEG in eye-open or eye-closed resting state, four microstate classes (i.e., microstate class A, B, C, and D) have been consistently identified in different studies with notable similarities across studies (Gao et al. 2017). Simultaneous EEG-fMRI has been used to investigate correlations between EEG microstates and resting-state fMRI (Britz et al. 2010, 2014; Van de Ville et al. 2010). Britz et al. (2010) found that each of the four commonly observed microstate classes was closely associated with the activity of distinct resting-state brain networks revealed by BOLD signals of resting-state fMRI. Notably, the temporal parameters and topographical configurations of these four microstate classes could be influenced by neuropsychiatric disorders, age, personality, and cognitive states (e.g., eye-open and eye-closed) (Andreou et al. 2014; Britz et al. 2014; Gao et al. 2017; Hatz et al. 2015; Schlegel et al. 2012; Koenig et al. 2002; Seitzman et al. 2017).

For ERP signals, a traditional approach to identify underlying ERP components and their latencies from multichannel ERP waveforms is to visually inspect the ERP waveform of each channel and the topographic map of each time point. However, this approach has a series of limitations, such as being time-consuming, subjective, and reference-dependent. In contrast, the microstate segmentation (i.e., segmentation based on topographic ERP) provides a robust and objective approach to identify the underlying ERP components (Hu et al. 2013), since different scalp fields must have been generated by different configurations of generators in the brain and different ERP components must reflect distinct configurations of electric generators (Murray et al. 2008).

In the following, we will illustrate the basic concepts of microstate analysis, the commonly used clustering algorithms, the metrics derived from microstate analysis, and how to perform microstate analysis using open-access tools. The pipeline of microstate analysis is shown in Fig. 8.1.

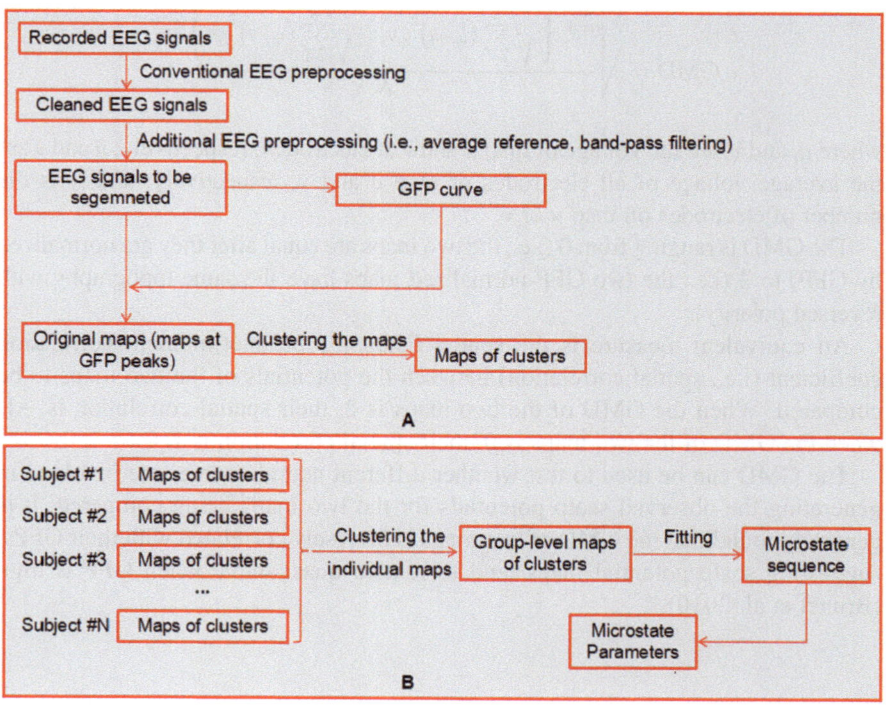

Fig. 8.1 The pipeline of classical microstate analysis for resting-state EEG signals, which is done through individual-level microstate identification (panel A) and following group-level microstate analysis (panel B)

8.2 Basic Concepts

Two global measures together with the waveforms and the scalp potential map are applied frequently in a set of studies using EEG/ERP microstate analysis. They are the global field power (GFP) and the global map dissimilarity (GMD) (Brunet et al. 2010).

The GFP is defined as the standard deviation of the electric potentials across all scalp electrodes of a given map, thus constituting a single, reference-independent measure of response strength. Note that the GFP is high in "hilly" maps (i.e., maps with pronounced peaks/troughs and steep gradients), but is low in "flat" maps (i.e., maps with shallow gradients) (Brunet et al. 2010). Moreover, high GFP is typically associated with relatively stable landscape configuration and high signal-to-noise ratio (SNR), while low GFP is commonly found in maps with low SNR, implying that the scalp configuration is changing into another new one.

The GMD is an index of configuration differences between two electric fields and is independent of the field strength (Brunet et al. 2010). It is defined as

$$\mathrm{GMD} = \sqrt{\frac{\sum_{i=1}^{N}\left\{\frac{u_i-\bar{u}}{\sqrt{\sum_{i=1}^{N}\left(u_i-\bar{u}\right)^2/N}} - \frac{v_i-\bar{v}}{\sqrt{\sum_{i=1}^{N}\left(v_i-\bar{v}\right)^2/N}}\right\}^2}{N}},$$

where u_i and v_i are the voltage of map u and v at electrode i, respectively; $\bar{u}$ and $\bar{v}$ are the average voltage of all electrodes of map u and v, respectively; and N is the number of electrodes on map u or v.

The GMD is ranging from 0 (i.e., the two maps are equal after they are normalized by GFP) to 2 (i.e., the two GFP-normalized maps have the same topography with reversed polarity).

An equivalent measure is the spatial Pearson's product-moment correlation coefficient (i.e., spatial correlation) between the potentials of the two maps to be compared. When the GMD of the two maps is 2, their spatial correlation is -1; when the GMD of the two maps is 0, their spatial correlation is 1.

The GMD can be used to test whether different neural sources are involved in generating the observed scalp potentials for the two maps being compared. It is generally found that the GMD of two maps is inversely correlated with their GFPs, suggesting scalp potential maps tend to remain quasi-stable when GFP is high (Brunet et al. 2010).

8.3 Spatial Clustering Algorithms in Microstate Analysis

During the past decades, several methods have been proposed in order to analyze/decompose the map configurations into a certain number of EEG/ERP microstates. In the vast majority of initial studies in this field, the spatial locations of negative and positive centroids in 2D or 3D scalp space (i.e., the locations of electrodes with negative extreme and positive extreme) were used as descriptors of each scalp map in EEG/ERP signals (Lehmann et al. 1987). Through observing the time series of these map descriptors, the moments where such descriptors significantly altered can be identified and used to define the borders between two successive microstates. Although this method has been proved to be useful in the initial studies, it has some limitations (Khanna et al. 2014). Firstly, complex electric maps often have more than one negative or positive centroid. Secondly, only limited spatial information (i.e., the locations of electrodes with negative and positive extremes) is used. Thus, for most related studies in these years, this method has been abandoned. Instead, two clustering algorithms were used. One was based on the k-means clustering (Pascual-Marqui et al. 1995). The other method was based on hierarchical clustering and was named as atomize and agglomerate hierarchical clustering (AAHC) (Murray et al. 2008). The basic ideas underlying these two kinds of methods are illustrated below.

Suppose there are n microstate classes of the EEG/ERP signals to be analyzed using k-means clustering-based microstate segmentation (Fig. 8.2). Although the

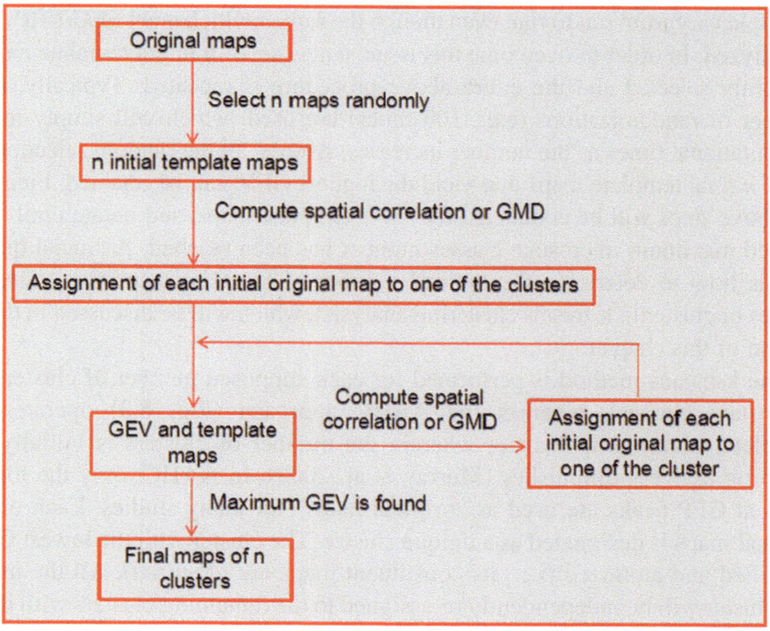

Fig. 8.2 The pipeline of k-means-based microstate identification

number of microstate classes can range from 1 to the total number of all data points in theory, the actual number of microstate classes usually ranges from 2 to 20 in typical multichannel EEG/ERP data. Steps of the k-means clustering-based microstate segmentation are as follows. Firstly, the GFP of each electric map is computed. As explained above, scalp potential maps tend to remain quasi-stable and have high SNR when GFP is high; thus, only the topographies at the GFP peaks are used as "original maps" in k-means clustering. Secondly, *n* scalp maps are randomly selected from these original maps and are later used as the "template maps" of the *n* microstate classes. Thirdly, the spatial correlation or GMD between each template map and each original map is computed, which provides spatial correlation series for each template map as a function of time. For each original map, one of the template maps yields highest spatial correlation or lowest GMD. According to these metrics, the global explained variance (GEV) of these template maps, which could quantify how well these template maps describe the total variance of the whole data, is calculated. Each of the *n* template maps is updated via averaging all the original maps that yielded highest spatial correlation or lowest GMD with a certain template map compared to all the other template maps. Then, the time series of spatial correlation or GMD for each updated template map and the resultant GEV are recomputed as above. These steps (i.e., updating template maps, recalculating spatial correlation or GMD series for each template map, and recalculating the GEV) are repeated until the GEV becomes stable (i.e., a maximum GEV is found). Note that, since the k-means clustering analysis is based on the random selection of initial template maps from all the original maps as seed clusters, the final results can in principle vary from run to run, even though the same multichannel EEG/ERP dataset is analyzed. In order to overcome this issue, a new set of *n* initial template maps are randomly selected and the entire above procedure is repeated. Typically, a high number of randomizations (e.g., 100 times) are used, which will simply increase computational times as the number increases. Among all the random selections, the set of *n* final template maps that yield the highest GEV will be retained. Finally, all the above steps will be conducted for $n + 1$ template maps, and iterate until a user-defined maximum microstate classes number has been reached. A crucial question here is how to determine the optimal number of template maps (i.e., microstate classes or cluster in k-means clustering analysis), which will be discussed in the next section of this chapter.

The k-means method is performed for each supposed number of clusters (i.e., microstate classes), whereas the AAHC approach (Fig. 8.3) operates in a completely bottom-up manner wherein the number of clusters is initially large and progressively diminishes (Murray et al. 2008). In AAHC, only the topographies at GFP peaks are used as "original maps" for many studies. Each of these original maps is designated as a unique cluster. The cluster with the lowest GEV is identified and atomized (i.e., its constituent maps are *atomized*). All the maps in this cluster will be independently re-assigned to the remaining clusters with highest spatial correlation. The AAHC then proceeds recursively by removing one cluster

Fig. 8.3 The pipeline of (T-)AAHC-based microstate identification

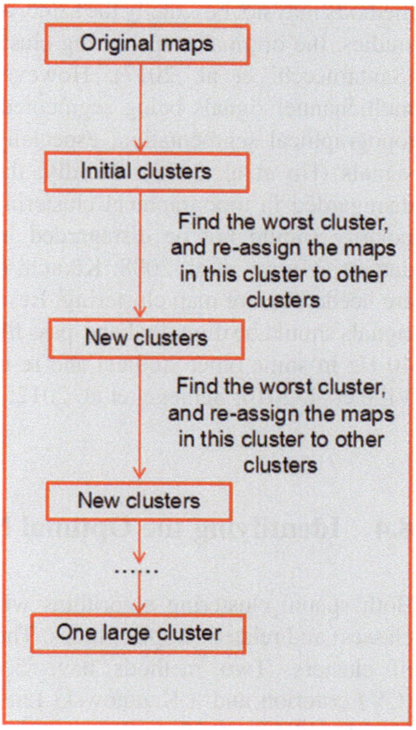

at a step and stops when all the original maps are merged into one single cluster. Supposing n clusters (i.e., microstate classes) exist in the EEG/ERP data being analyzed, the step when n clusters exist will be used in the next analysis. The template map of each cluster is obtained via averaging all the original maps belonging to that cluster. As is the case for k-means approach, we need to determine the optimal number of clusters in the next step.

A method closely related with the AAHC is the topographic atomize and agglomerate hierarchical clustering (T-AAHC) method. The only one difference between T-AAHC and AAHC is the assessment of the clustering quality and the definition of a "bad" cluster. As illustrated above, the "bad" cluster is defined as the cluster with the lowest GEV. However, in T-AAHC, the "worst" cluster is defined as the cluster that has the lowest sum of correlations between its members and prototypical map.

During the implementation of k-means clustering or (T-)AAHC, the following issues should be noted. Firstly, since a high number of randomizations are needed in k-means clustering, the computational time of k-means clustering is usually very long compared to the (T-)AAHC approach (Murray et al. 2008). Although these two methods often yield generally comparable results, especially for multichannel EEG/ERP activities with high SNR, the final template maps identified by the two methods may not be exactly the same, as have been shown below. Secondly, in many studies, the original maps during clustering are defined as the maps at GFP peaks (Santarnecchi et al. 2017). However, in some studies, all the maps in the multichannel signals being segmented are determined to be the original maps in topographical segmentation, especially when the dataset being analyzed is ERP signals (Hu et al. 2013). Thirdly, the polarity of each original map is usually disregarded in topographical clustering of resting-state EEG signals, whereas the polarity should not be disregarded in microstate analysis of multichannel ERP dataset (Murray et al. 2008; Kikuchi et al. 2011). Fourthly, some additional steps are needed before map clustering. Researchers suggested that the resting-state EEG signals should be digitally band-pass filtered between 2 and 20 Hz (or between 1 and 40 Hz in some other studies) and re-montaged against average reference (Van de Ville et al. 2010; Schlegel et al. 2012; Koenig et al. 2002).

8.4 Identifying the Optimal Number of Clusters

Both spatial clustering algorithms will identify a set of clusters (i.e., microstate classes) and related template maps. Then, we need to determine the optimal number of clusters. Two methods have been used in literatures: a cross-validation (CV) criterion and a Krzanowski-Lai (KL) criterion (Pascual-Marqui et al. 1995; Van de Ville et al. 2010; Santarnecchi et al. 2017; Michel and Koenig 2017). The first method seeks to optimize the ratio between the GEV and the degrees of freedom for a given cluster number, since increasing the number of clusters (i.e., microstate classes) would increase the GEV but decrease the degrees of freedom and vice versa. This method provides a value of CV for each cluster, and the minimum CV gives the optimal number of clusters. The second method works by first computing a quality measure of the clustering, and the highest value of this quality measure indicates the optimal map clustering. However, these two methods have some limitations. For example, the CV criterion is sensitive to the number of scalp electrodes. When high-density EEG is analyzed, a minimum CV usually could not be found. For the KL criterion, several prominent peaks may be observed, especially for resting-state EEG signals. In data analysis, both methods should be used in identifying the optimal number of clusters. Besides, several local maximum values may be observed in the results of KL criterion. In practice, the second maximum value of the KL curve may be more important than the first maximum value.

As mentioned above, the four clusters (i.e., microstate classes) were commonly identified in resting-state EEG data. Moreover, the four clusters exhibited highly similar configurations in various studies. Thus, many experimental and clinical studies fixed four clusters as a standard of resting-state EEG signal processing, in order to ensure consistency with previous studies (Gao et al. 2017; Michel and Koenig 2017).

8.5 Fitting the Template Maps to the Potential Maps

In this step, each scalp map of EEG/ERP dataset is assigned to one of the template maps. Two different approaches have been used in previous studies (Murray et al. 2008). In the first approach, the spatial correlation between each template map and the map of each time point is computed, and each map is assigned to the cluster with highest spatial correlation. In the second approach, we only compute the spatial correlation between *original maps* (i.e., the ones over GFP peaks) and template maps. If two adjacent *original maps* have highest spatial correlation for the same microstate class, then all the electric maps between these two adjacent *original maps* are assigned to the same microstate class (i.e., cluster). If the two adjacent *original maps* are assigned to different clusters, then the first half of maps between the two adjacent *original maps* are assigned to the preceding microstate class, whereas the last half of maps are assigned to the following microstate class. Through these fitting procedures of the first approach or the second approach, the scalp maps of all the time points will be re-expressed as an alternating sequence of functional microstates.

After this map fitting, temporal post-processing can be performed. For example, small segments could be rejected via temporal smoothing. Moreover, if the spatial correlation between template maps of two clusters is relatively high (e.g., higher than 0.9), these two clusters can be merged into one cluster (Brunet et al. 2010).

8.6 The Commonly Used Microstate Parameters

Many parameters can be derived from the microstate sequence. There are four commonly used microstate parameters: (1) the mean duration of a microstate class, (2) occurrence rate per second of a microstate class, (3) time coverage of a microstate class, and (4) the transition probability between adjacent microstate classes (Michel and Koenig 2017; Schlegel et al. 2012). Note that, for ERP maps, the second and the fourth metrics are not available since a given microstate class appears only once.

8.7 Available Tools of Microstate Analysis

There are some free and open-access tools that can be used in microstate analysis: (1) the LORETA software developed by Roberto D. Pascual-Marqui from the KEY Institute for Brain-Mind Research, University Hospital of Psychiatry, Zurich, Switzerland (Pascual-Marqui 2002) (http://www.uzh.ch/keyinst/loreta.htm); (2) the EEGLAB plugin for microstates developed by Thomas Koenig from the University of Bern, Switzerland (https://sccn.ucsd.edu/wiki/EEGLAB_Extensions_and_plug-ins); and (3) the CARTOOL software developed by Denis Brunet and colleagues from the Functional Brain Mapping Laboratory, Departments of Fundamental and Clinical Neurosciences, University Medical School, University of Geneva, Switzerland (Brunet et al. 2010) (https://sites.google.com/site/fbmlab/cartool).

The LORETA software has a module called "microstate segmentation," which can parse the topographic maps of EEG/ERP into microstates. Only k-means clustering-based microstate segmentation is available in the current version of LORETA. The EEGLAB plugin for microstates was developed as a plugin of the famous EEGLAB software, providing an easy-implementation way to conduct microstate analysis for EEGLAB users. Both k-means clustering and AAHC method can be found in this plugin. However, it could only be applied to the microstate analysis of resting-state EEG activities. We cannot perform ERP microstate segmentation using this EEGLAB plugin. Compared to the above two tools, the CARTOOL is a much versatile software. However, it could not recognize the files in EEGLAB format. Below, we will show how to identify the four commonly detected EEG microstates in spontaneous EEG activities using the EEGLAB plugin for microstates and how to identify ERP microstate in typical ERP waveforms using the CARTOOL software.

8.8 Example: Identifying Microstates in Spontaneous EEG

Basic Information of the Demo Datasets In the experiment, both eyes-closed and eyes-open resting-state EEG data were recorded about 5 min from the 64 scalp tin electrodes mounted in an elastic cap (Brain Products, Munich, Germany), with the sampling frequency of 500 Hz around 9:00 to 12:00 in the morning. Subject was introduced to eyes-closed first and then eyes-open. The EEG datasets consist of 20 4-min EEG recordings (eye-closed and eye-open condition), obtained from 10 volunteers in the Sleep and NeuroImaging Center of Southwest University, China. In preprocessing, the continuous EEG data were down-sampled to 250 Hz and digitally filtered within the 0.1–45 Hz frequency band using a Chebyshev type II filter. The filtered EEG recordings were re-referenced to average reference and then segmented to 2 sec. The segmentations with ocular, muscular, and other types of artifact were identified and excluded. We only retained the first 120 segmentations, constituting a 4-min EEG recording for each subject.

The EEGLAB plugin for microstates can be downloaded from https://sccn.ucsd.edu/wiki/EEGLAB_Extensions_and_plug-ins. A brief video on how to use this tool can be downloaded from http://www.thomaskoenig.ch/index.php/software/microstates-in-eeglab. In this plugin, there is a MATLAB script called "TestMSAnalyses.m," which is a demo script on how to use this tool. For users unfamiliar to MATLAB programming language, we have modified the above script into another script, i.e., the "restEEG_microstate_analysis.m."

In order to use this modified script, please follow the steps below:

1. This EEGLAB plugin for microstates does not have any external dependency, except EEGLAB. Please unzip the directory to the "plugins" folder under the *EEGLAB* folder, and add the paths via *addpath (genpath ('my_microstate0.3_directory))* or via the *set path* menu in MATLAB.
2. Open this script ("restEEG_microstate_analysis.m"), and run it.
3. A modal dialog box with prompt "Your version of DIPFIT" will open. Here, you should enter your version of DIPFIT, which is a plugin for EEGLAB. It can be found in the "plugins" folder under the "EEGLAB" folder. For EEGLAB 13.1, the version of DIPFIT is 2.2; thus, you should enter "2.2" in this case.
4. A folder selection dialog box with prompt "Path to the data of one condition" will open. This dialog box enables users to navigate to the folder (e.g., the "EC" folder in demo datasets), which stores the EEG data of all subjects in one condition. Before using this plugin for microstate analysis, you should create a folder for each condition/group and put all the preprocessed resting-state EEG datasets of a condition/group, which should be in EEGLAB format (∗.set and ∗.fdt), to the corresponding folder.
5. A modal dialog box with prompt "The name of this condition" will open. Here, you should enter the name of the condition. For example, if the "EC" folder has been selected in the above step, you should enter "EC" here.
6. A modal dialog box with prompt "The name of this group" will open. Here, you should enter the name of this group. Since there is only one group in demo datasets, you can enter any arbitrary input, such as "group 1."
7. A folder selection dialog box with prompt "Path to store the results" will open. This dialog box enables users to navigate to the folder where they are able to store the results of microstate analysis.
8. A modal dialog box with prompt "Any electrode to be deleted? True = 1 False = 0" will open. If there are any electrodes that should be deleted before microstate analysis (e.g., electrodes over bilateral mastoids and electrodes that can record non-brain signals), you should enter "1" here. Otherwise, enter "0" in this step.
9. If you entered "1" in the previous step, a modal dialog box will open. Here, you should enter the name of electrodes you want to exclude from microstate analysis.
10. After the above parameters have been entered, the EEGLAB will open, and the EEG datasets in the folder you specified will be loaded into EEGLAB. Some additional preprocessing will be conducted on these datasets, including loading the channel locations, average-referencing, and band-pass filtering between 2 and 20 Hz. Four modal dialog boxes will open sequentially, which are used to define the parameters for clustering. In the first dialog box with prompt "Using the maps of GFPPeaks as original maps? True = 1 False = 0," enter "1." In the second dialog box with prompt "Ignore the polarity of maps? True = 1 False = 0," enter "1." In the third dialog box with prompt "Number of restarts," enter "100." In the fourth dialog box with prompt "Use AAHC?

True = 1 False = 0," in case you want to use AAHC in map clustering, please enter "1"; otherwise enter "0" (in this case, k-means clustering will be used). Note that, since randomization of initial template map is not needed in AHHC, thus changing the input parameters of the third dialog box will not change the computational time if AAHC is selected in the fourth dialog box. After these parameters have been specified, the script will loop across all subjects to identify the individual clusters of all subjects.

11. After the individual clusters (i.e., individual microstate classes) of all subjects are identified, the script will combine the microstate maps across subjects. The group-level four microstate classes will pop up, and then we need to sort the four microstate classes in the canonical order: microstate class A, B, C, and D exhibit left-right orientation, right-left orientation, anterior-posterior orientation, and fronto-central maximum, respectively. The updated EEG datasets with microstate information for all the subjects will be stored in the path you specified in step 7.
12. A modal dialog box with prompt "Fitting based on GFP peaks? True = 1 False = 0" will open. For most cases, please enter "1."
13. Two files will also be stored in the path you specified in step 7. The names of these two files are "ResultsFromIndividualTemplates.csv" and "ResultsFromGrandGrandMeanTemplate.csv," respectively. The difference between these two files is during fitting each original map to the template maps of the four microstate classes (i.e., clusters), the individual template maps and group-level template maps will be used, respectively. These two files will store the microstate parameters (i.e., GEV and the duration, occurrence rate, time coverage, and transition probability between any two microstate classes).

8.9 Example: Identifying Microstates in Multichannel ERP

Here, the datasets in folder "Data_AVSP_20/L/datasets" will be used as demo datasets in ERP microstate segmentation. First of all, we need the group-level grand-average ERP activities in Brain Vision Recorder format. The "Grandaverage" plugin of EEGLAB, which can compute the ERP activities of specified EEGLAB datasets and can be downloaded from https://sccn.ucsd.edu/wiki/EEGLAB_Extensions_and_plug-ins, is used to obtain the grand-average ERP activities of the EEGLAB datasets in condition "L." The installation of this EEGLAB plugin is exactly the same as the microstate plugin shown above. In order to obtain the grand-average ERP activities in Brain Vision Recorder format, please follow the below steps:

1. Load all the 20 datasets in condition "L" at once via "File >> Load existing dataset."
2. Merge these datasets through "Edit >> Append datasets," and compute the group-level grand-average ERP activities through the "Grandaverage" plugin.

3. Re-extract data epochs via "Tools >> Extract epochs." Here, only the ERP activities between 0 and 1000 ms after the stimuli onset should be retained.
4. Reference to common average via "Tools >> Re-reference."
5. Export the ERP activities in Brain Vision Recorder format through "File >> Export >> Write Brain Vis. exchange format file." You should name the files storing grand-average ERP, such as "L_granderp_avgref." In this case, three files will be exported: "L_granderp_avgref.dat," "L_granderp_avgref. vhdr," and "L_granderp_avgref. Vmrk," which store the ERP data, the header information, and the markers, respectively.

After we have exported these ERP activities, we can use the CARTOOL software to conduct microstate segmentation. In this software, select "Tools >> EEG and Tracks >> Segmentation of EEG files," then a dialog with title "Segmentation" will open. This dialog box has two parts: "Files" and "Parameters."

The first part "Files" has a header "Files Presets" and three panels. For "Files Presets," select "ERPs/On Grand Mean(s)." In the first panel "Segmenting Groups of Files," hit "Add New Group of Files," and load the grand-average ERP generated in the previous step. In the second panel "Epochs," select "No epochs." In the third panel "Files Options," specify the directory where you want to store the results of microstate segmentation and the kinds of files you want to export. Then, hit "Next," which will open the second part "Parameters."

The second part "Parameters" has a header "Computation Presets" and four panels. For "Computation Presets," we can choose "EEG/Surface/ERPs/T-AAHC (recommended)" or "EEG/Surface/ERPs/K-Means (classical)" depending on which method you want to use. In this case, we choose the first method AAHC. For the first panel, we should do nothing. In the second panel "Data Preprocessing," select" Using Whole Data." In the third panel, we need to choose "Clustering method" (topographical AAHC [T-AAHC]) and "Range of Clusters" (from 1 to 20 clusters in this case) and decide what we want to do with time points with low correlations (Select "No Labeling if Below," "50"% here). In the fourth panel "Temporal Postprocessing," select "Sequentializing Segments," "Merging Correlated Segments" with "If Correlated above" 95%, "Segments Temporal Smoothing" with "Window Half Size" 3 and "Strength (Besag factor)" 10, and "Rejecting Small Segments" with "Shorter than or equal to" 20 TFs. Then, hit "Process."

After microstate segmentation is finished, a dozens of files will be generated in the folder you specified in the first part "Files" of dialog "Segmentation." Suppose the name of folder you specified is "Seg LGranderp." In this folder, you will find a file named as "Seg LGranderp.vrb." Open this file, and at the end of this file you will find the "Optimal Clustering" number. For the current ERP dataset, the optimal clustering number is 10. Then, we can open the files saving the results corresponding to cluster number that is 10. For the current data, the corresponding files are named as "Seg LGranderp.10.(12)." In the file name, "10" means we want to segment the ERP signals into *10* clusters (i.e., ERP microstates), and "12" means the actual number of identified microstates, since in many cases two nonadjacent segments are assigned to the same cluster during clustering. In the temporal postprocessing, the nonadjacent segments assumed to be in the same cluster are re-assigned to two different clusters. In order to see the results of microstate segmentation, you can drag the "Seg LGranderp.10.(12).seg" into the CARTOOL window. The second, third, fourth, and fifth cluster correspond to N1, N2, P2, and P4 component, respectively, which is very similar to the results in Hu et al. (2014).

The results of the above two examples have been shown in Figs. 8.4 and 8.5, respectively.

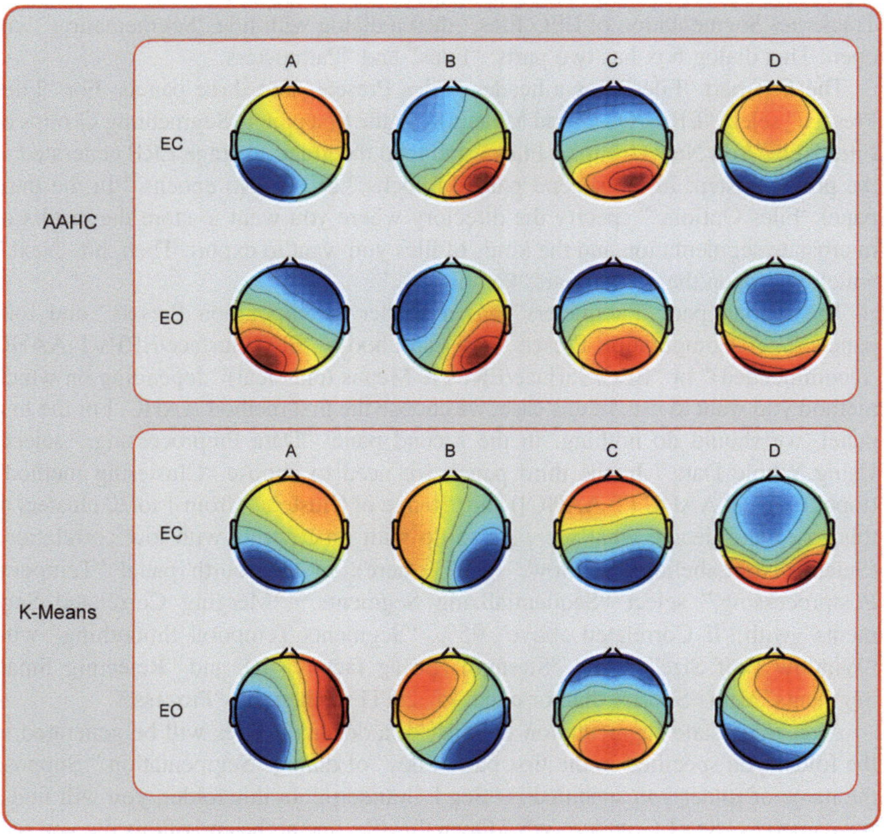

Fig. 8.4 The group-level maps of the four microstate classes (*A–D*) in the eye-closed (EC) and eye-open (EO) condition when the AAHC method or k-means clustering is used. Color should be used for this figure in print

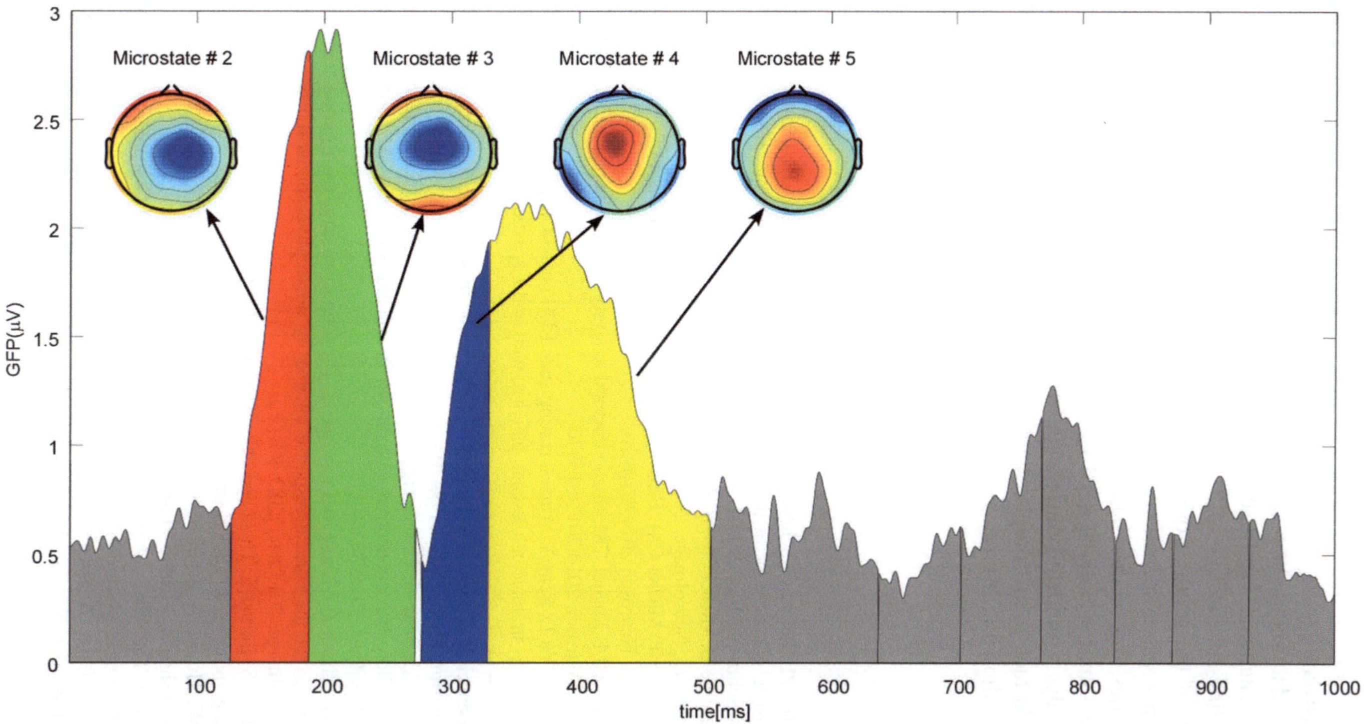

Fig. 8.5 The results of microstate segmentation conducted on group-level ERP waveforms of condition "L." The AAHC method identified 12 microstate classes, which are color-coded in the curve of GFP. The topographical configurations of cluster # 2, # 3, # 4, and # 5 are shown. These four clusters correspond to N1, N2, P2, and P4 component, respectively, as have been revealed in Hu et al. (2014)

8.10 Remarks and Conclusion

To sum up, several points of EEG/ERP microstate analysis should be mentioned. First, the EEG data should be preprocessed properly in order to obtain robust results. Second, the functional significance of the four commonly observed microstate classes in resting-state EEG is still unclear. Thus, you should be cautious about the interpretation of the results from microstate analysis in resting-state EEG. Third, although the microstate segmentation is mainly applied to resting-state EEG signals or ERP activities, this method can also be applied to time-frequency domain. More details can be found in Jia et al. (2015). Finally, we only provide the application of microstate analysis on identifying the underlying ERP components and latencies as an example. In fact, more complex data analysis based on ERP topography can be performed, as shown in Murray et al. (2008) for extended reading.

References

Andreou C, Faber PL, Leicht G, Schoettle D, Polomac N, Hanganu-Opatz IL, Lehmann D, Mulert C. Resting-state connectivity in the prodromal phase of schizophrenia: insights from EEG microstates. Schizophr Res. 2014;152(2–3):513–20.

Britz J, Van De Ville D, Michel CM. BOLD correlates of EEG topography reveal rapid resting-state network dynamics. NeuroImage. 2010;52:1162–70.

Britz J, Díaz HL, Ro T, Michel CM. EEG-microstate dependent emergence of perceptual awareness. Front Behav Neurosci. 2014;8(20):163.

Brunet D, Murray MM, Michel CM. Spatiotemporal analysis of multichannel EEG: CARTOOL. Comput Intell Neurosci. 2010;2011:813870.

Gao F, Jia H, Wu X, Yu D, Feng Y. Altered resting-state EEG microstate parameters and enhanced spatial complexity in male adolescent patients with mild spastic Diplegia. Brain Topogr. 2017;30(2):233–44.

Gärtner M, Brodbeck V, Laufs H, Schneider G. A stochastic model for EEG microstate sequence analysis. NeuroImage. 2015;104:199–208.

Hatz F, Hardmeier M, Benz N, Ehrensperger M, Gschwandtner U, Rüegg S, Schindler C, Monsch AU, Fuhr P. Microstate connectivity alterations in patients with early Alzheimer's disease. Alzheimers Res Ther. 2015;7(1):1–11.

Hu L, Valentini E, Zhang ZG, Liang M, Iannetti GD. The primary somatosensory cortex contributes to the latest part of the cortical response elicited by nociceptive somatosensory stimuli in humans. NeuroImage. 2013;84(1):383–93.

Jeremy H, Malone S, Bernat E. Theta and delta band activity explain N2 and P3 ERP component activity in a go/no-go task. Clin Neurophysiol. 2014;125(1):124–32.

Jia H, Peng W, Hu L. A novel approach to identify time-frequency oscillatory features in electrocortical signals. J Neurosci Methods. 2015;253:18–27.

Jia H, Li H, Yu D. The relationship between ERP components and EEG spatial complexity in a visual Go/Nogo task. J Neurophysiol. 2017;117(1):275–83.

Khanna A, Pascual-Leone A, Michel CM, Farzan F. Microstates in resting-state EEG: current status and future directions. Neurosci Biobehav Rev. 2014;49:105–13.

Kikuchi M, Koenig T, Munesue T, Hanaoka A, Strik W, Dierks T, Koshino Y, Minabe Y. EEG microstate analysis in drug-naive patients with panic disorder. PLoS One. 2011;6(7):65.

Koenig T, Prichep L, Lehmann D, Sosa PV, Braeker E, Kleinlogel H, Isenhart R, John ER. Millisecond by millisecond, year by year: normative EEG microstates and developmental stages. NeuroImage. 2002;16(1):41–8.

Lehmann D, Ozaki H, Pal I. EEG alpha map series: brain micro-states by space-oriented adaptive segmentation. Electroencephalogr Clin Neurophysiol. 1987;67(3):271–88.

Lehmann D, Strik WK, Henggeler B, Koenig T, Koukkou M. Brain electric microstates and momentary conscious mind states as building blocks of spontaneous thinking: I. Visual imagery and abstract thoughts. Int J Psychophysiol. 1998;29(1):1–11.

Lehmann D, Faber PL, Galderisi S, Herrmann WM, Kinoshita T, Koukkou M, Mucci A, Pascualmarqui RD, Saito N, Wackermann J. EEG microstate duration and syntax in acute, medication-naive, first-episode schizophrenia: a multi-center study. Psychiatry Res. 2005;138 (2):141–56.

Li H, Jia H, Yu D. The influence of vertical disparity gradient and cue conflict on EEG omega complexity in Panum's limiting case. J Neurophysiol. 2018;119(3):1201–8.

Maxwell CR, Villalobos ME, Schultz RT, Herpertz-Dahlmann B, Konrad K, Kohls G. Atypical laterality of resting gamma oscillations in autism spectrum disorders. J Autism Dev Disord. 2015;45(2):292–7.

Michel CM, Koenig T. EEG microstates as a tool for studying the temporal dynamics of whole-brain neuronal networks: a review. NeuroImage. 2017;180:577–93.

Murray MM, Brunet D, Michel CM. Topographic ERP analyses: a step-by-step tutorial review. Brain Topogr. 2008;20(4):249–64.

Pascual-Marqui RD. Standardized low-resolution brain electromagnetic tomography (sLORETA): technical details. Methods Find Exp Clin Pharmacol. 2002;24(Suppl D):5–12.

Pascual-Marqui RD, Michel CM, Lehmann D. Segmentation of brain electrical activity into microstates: model estimation and validation. IEEE Trans Biomed Eng. 1995;42(7):658–65.

Rieger K, Diaz Hernandez L, Baenninger A, Koenig T. 15 years of microstate research in schizophrenia – where are we? A meta-analysis. Front Psych. 2016;7:22.

Santarnecchi E, Khanna AR, Musaeus CS, Benwell CSY, Davila P, Farzan F, Matham S, Pascual-Leone A, Shafi MM. EEG microstate correlates of fluid intelligence and response to cognitive training. Brain Topogr. 2017;30:502–20.

Schlegel F, Lehmann D, Faber PL, Milz P, Gianotti LRR. EEG microstates during resting represent personality differences. Brain Topogr. 2012;25(1):20–6.

Seitzman BA, Abell M, Bartley SC, Erickson MA, Bolbecker AR, Hetrick WP. Cognitive manipulation of brain electric microstates. NeuroImage. 2017;146:533–43.

Van de Ville D, Britz J, Michel CM. EEG microstate sequences in healthy humans at rest reveal scale-free dynamics. Proc Natl Acad Sci U S A. 2010;107(42):18179–84.

Wu X, Jia H, Wang E, Du C, Wu X, Dang C. Vertical position of Chinese power words influences power judgments: evidence from spatial compatibility task and event-related potentials. Int J Psychophysiol. 2016;102:55–61.

Chapter 9
Source Analysis

Xu Lei

Abstract The scalp electroencephalogram (EEG) records electrical activity generated by the ensemble of a great number of pyramidal neurons within the brain. Sampling of electromagnetic brain signals in milliseconds has already been achieved. Unfortunately, the spatial resolution of EEG is very poor, which is limited by the relatively small number of spatial measurements (only a few hundred in EEG) and the inherent ambiguity of the underlying static electromagnetic inverse problem. In fact, localizing these potentials from the scalp EEG within the brain is an ill-posed inverse problem. This chapter firstly provides a brief introduction to the topic with an overview of the inverse problem from a signal processing perspective. In the next two sections, we describe the source models and head models of EEG source analysis, followed by various approaches to the inverse problem with which the properties of the neural current generators are estimated from the data. Finally, we discuss about the recent developments and the emerging signal processing issues of EEG data analysis.

Keywords Inverse problem · Forward problem · EEG · Source imaging

The scalp electroencephalogram (EEG) represents electrical activity generated by the ensemble of a great number of pyramidal neurons within the brain. When a large number of pyramidal neurons synchronously produce their post-synaptic potentials, significant electric currents start to flow in the surrounding tissue, which may be observable by the electrodes on the scalp. With advances in technology, sampling of electric brain signals at millisecond intervals has already been achieved. Unfortunately, the spatial resolution of EEG is very poor, which is limited by the relatively small number of spatial measurements (only a few hundred in EEG) and the inherent ambiguity of the underlying static electromagnetic inverse problem. In fact,

X. Lei (✉)
Sleep and Neuroimaging Centre, Faculty of Psychology, Southwest University, Chongqing, China
e-mail: xlei@swu.edu.cn

L. Hu, Z. Zhang (eds.), *EEG Signal Processing and Feature Extraction*,
https://doi.org/10.1007/978-981-13-9113-2_9

localizing these potentials from the scalp EEG within the brain is an ill-posed inverse problem (Helmholtz 1853). An introduction to the topic with an overview of the associated inverse problem from a signal processing perspective is provided below.

9.1 The Forward Problem

Two important concepts for EEG source analysis are the forward problem and the inverse problem (Fig. 9.1). The forward problem is an estimate of the topographical map that would result from activity of a dipole in a specific region of the brain with a specific orientation. The inverse problem is the estimate of what dipoles with different orientations and magnitudes could produce in an observed topographical map.

As the localization and orientation of dipole has been known, the output signal from the electrodes on the scalp can be solved analytically without any EEG data. Main factors that influence the accuracy of our calculation are the head model and the source model. For a particular source model, computation of the scalp potentials and induced electric fields requires solution of the quasi-static approximation of Maxwell's equations. The detail of the mathematical deductions can be found in Baillet's study (Baillet et al. 2001a). Analytic solutions exist for simplified geometries, such as when the head is assumed to consist of a set of nested concentric homogeneous spherical shells representing the brain, skull, and scalp. However, when the surface integrals are computed over realistic head shapes, these equations must be solved numerically. These models are routinely used in most clinical and research applications to EEG source analysis.

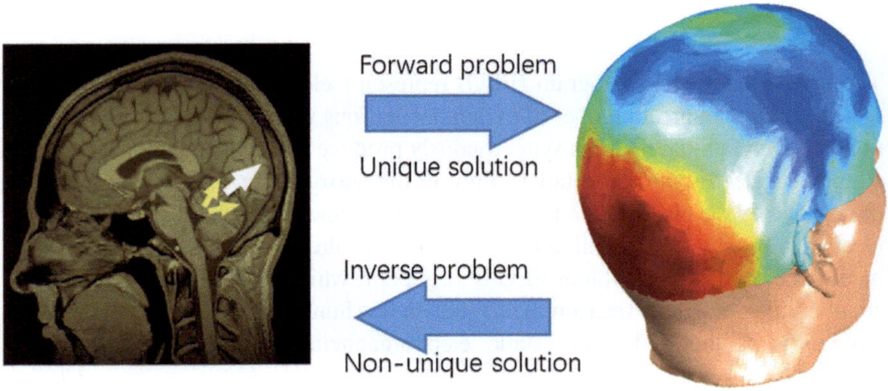

Fig. 9.1 Forward problem and inverse problem in EEG source analysis. The forward problem has a unique solution and can be obtained analytically without EEG data. The inverse problem has no unique solution because many different configurations of dipoles could produce the same pattern of topographical activity. Here, both the one white dipole and the two yellow dipoles could produce the same scalp recorded activity

9.1.1 Source Model

Source modeling includes how to model the source and determine the source space. There are at least three models to model the source, such as the charge model (Yao 1996), the dipole model (El Badia and Ha-Duong 2000), and multipolar models (Nolte and Curio 2000). The current dipole is a straightforward extension of the better-known model of the paired-charges dipole in electrostatics. In contrast, an equivalent charge model has similar effectiveness for imaging a pair of dipoles, while for imaging sources with special construction, the latter is superior to the equivalent current dipole (Yao 1996). Note that brain activity does not actually consist of discrete sets of current dipoles, but rather that the dipole is an approximate representation for synchronous activation of a large number of pyramidal neurons, extending over a few square centimeters of gray matter. However, an identifiability problem may arise when too many dipoles are utilized to represent a single large region of synchronous activation. These sources can be simplified to a multipolar model, which is derived by performing a Taylor series expansion of the Green function. An example of multipolar models was given in modeling the lateral extent of neuronal activity (Nolte and Curio 2000).

Another important topic is the distribution of source. If only a few restricted areas are assumed active synchronously and each area evolves with its specific temporal process, then the EEG potentials can be modeled by a small set of isolated sources. This localized single time-course hypothesis is both convenient and fairly realistic since neurons in a population are expected to be strongly coherent in time (Scherg and von Cramon 1986). However, the disadvantages are obvious for the isolated source model. For example, dipole position may be well approximated by the center of a cortical patch, but it is hard to represent the extension of source. In fact, dipole localization may be meaningless for a highly extended cortical region (Baillet et al. 2001b). Worse still, in the case of multiple sources, estimating the number of dipoles is often a nontrivial task (Benar et al. 2005).

The above-mentioned traps have led some scientists to prefer distributed sources where a large number of dipoles are placed evenly in the head, either in the whole brain volume (Hämäläinen and Ilmoniemi 1994) or along the surface of the cortex (Babiloni et al. 2003), the distributed source model. Because a dipole has three fixed orientations (along the cardinal x -, y -, and z-axes), forward problem can have different orientations in space. In some models, all three orientations are calculated, but for other models only dipoles that are perpendicular (normal) to the cortex are modeled. One optimal selection is based on the segmentation of the individual magnetic resonance imaging (MRI) into the gray and white matters. Because the interface between the gray and white matters forms a surface solution space of source imaging, the source space is constrained to the cortical surface, and the orientation of the source can be considered to be radial to the surface. The use of the individual MRI is necessary for clinical cases when deformations or lesions are presented. Determining the distribution of source is valuable to determine the Talairach coordinates of the dipoles, which is informed to draw conclusions about the activated neurophysiologic structures (Fig. 9.2).

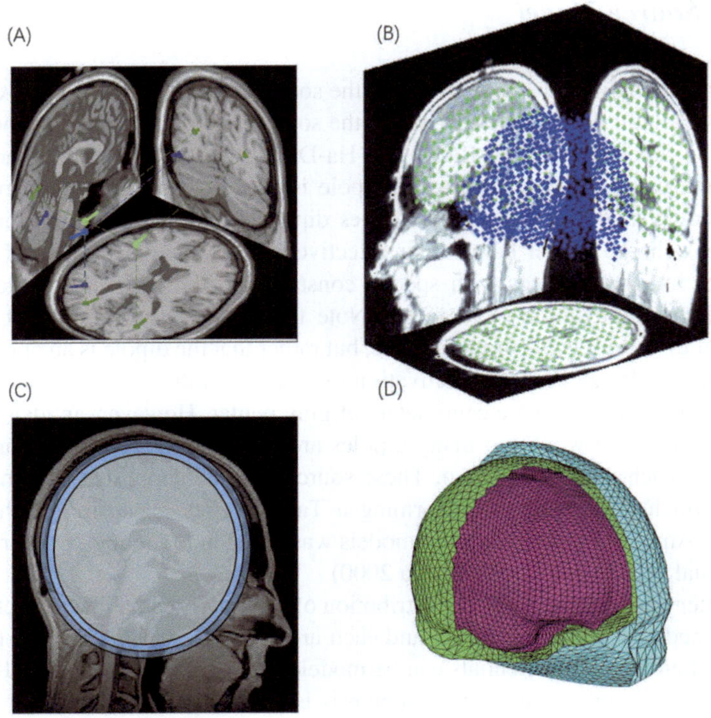

Fig. 9.2 The source space and the head model. (**a**) The dipole model assumes limited dipoles; their location, direction, and strength should be estimated. (**b**) The distributed source model. More than 1000 brain sources were constructed from an MRI (each blue dot is a location). (**c**) Three-shell spherical head model and (**d**) three-shell realistic head model. The three meshes correspond to the scalp, skull, and cortex

9.1.2 Head Model

Head model determines the measurements on the scalp after locating the sources at the given positions. Both the electromagnetic (permeability and conductivity) and geometrical (shape) features of the volume are considered in head model. The simplest and still widely utilized head model is the spherical model. Because of the uniform conductivity, the spherical model allows for an analytical solution of the forward problem (Yao 2000). However, the accuracy of source localization is limited in this model because the head is hard to be simplified to spherical and the conductivity is hard to be simplified to spatially uniform (Michel et al. 2004). An advanced model is the multi-shell spherical head model, which incorporated different conductivity parameters and considered local anisotropies, and it ameliorates the accuracy to a certain degree.

Several simulations have shown that accurate lead field computation can only be achieved by using realistic volume conductor models, such as the boundary element method (BEM) and the finite element method (FEM) (Valdes-Hernandez et al.

2009). For the BEM model, the interfaces are modeled as triangulations between compartments of equal isotropic conductivities (Hämäläinen and Sarvas 1989). Since individual MRI is now routinely performed as a part of most psychological protocols, this data is readily available. In this condition, surface boundaries for the brain, skull, and scalp were extracted from these images. Many automated and semi-automated methods exist for surface extraction from MR images (Fischl et al. 1999).

The FEM mode, in contrast, tessellates the whole volume and considers the anisotropic electrical conductivity of each element. Skull breaches and other structure anisotropies can be taken into account, especially for the deep sulci sources (Güllmar et al. 2010; van den Broek et al. 1998). However, the generation of realistic geometric models is a nontrivial task, requiring the accurate segmentation of MRI and stable tessellations of head compartments. Worse still, the information about tissue conductivities is rarely available with current technology. Because of its complexity of the individual FEM model, efforts have been made over recent years to combine the computational efficiency of template model with more accurate descriptions of the individual head shape. For example, in the condition that individual's MRI is not available, an approximate model (AM) was proposed (Valdes-Hernandez et al. 2009). Statistical comparisons revealed that the performance of AM was better than the usual average model in the MNI space, and it provided a convenient tool in large clinical studies.

9.2 The Inverse Problem

Given the observed pattern of topographical activity, the inverse problem reconstructed the most likely location, orientation, and magnitude of cortex source that generated this activity. This problem can be difficult to answer for both theoretical and practical reasons. At a theoretical level, there is no unique solution. This means that there are many possible – and possibly very different – configurations of brain activity that can produce the same topographical map (see Fig. 9.1). At a practical level, all methods for estimating the inverse solution require several assumptions and involve parameter selections, and those parameter selections may impact the results. The rest of this section introduces some of the commonly used methods for estimating solutions to the inverse problem, as well as to some of the assumptions and parameters involved in each method.

9.2.1 Equivalent Current Dipole

Equivalent current dipole (ECD) assumed the EEG signals are generated by a few isolated dipoles located at distinct positions. This may be a valid assumption for experiments that involve stimulation of primary sensory or motor cortices without concurrent cognitive or decision-making processes. Under the assumption of a few

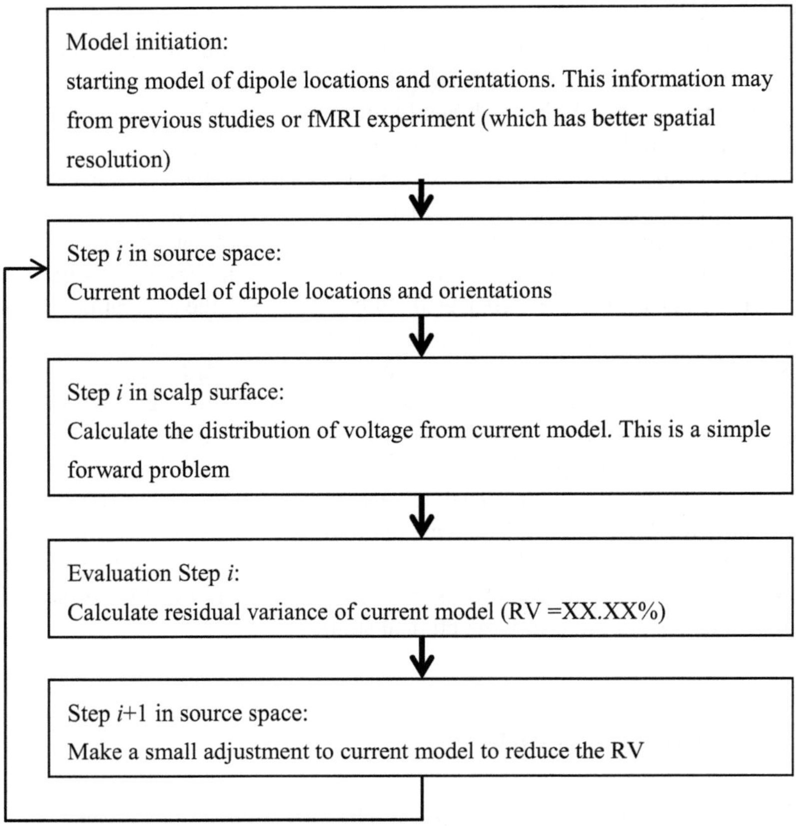

Fig. 9.3 The procedure of dipole fitting in equivalent current dipole model. The procedure includes five main steps. Model initiation: starting model of dipole locations and orientations. Iterated step in source space estimates current model of dipole locations and orientations. Iterated step in scalp surface calculates the distribution of voltage from current model. Evaluation step calculates residual variance of current model. The updated step in source space makes a small adjustment to current model to reduce the residual variance

isolated point sources, the sources can be uniquely identified mathematically and the parameters may be estimated by various nonlinear optimization algorithms (Lei et al. 2009). The target function usually includes minimizing the residual variance between simulated model and the recorded signal. Once the dipole location, orientation, and magnitude are estimated, a weight for each electrode can be determined such that the weighted sum of activity at all electrodes is the estimate of the dipole activity (Fig. 9.3).

Multiple approaches to dipole source localization exist. One is to fit dipoles to scalp recordings separately for each moment; this approach is known as moving dipole localization because both the dipole location and orientation can vary overtime. This is typically done on time-domain data (event-related potentials) of one time point, or the average of a time window, or the spatial pattern after ICA decomposition. Another approach is to fix the dipole location but allow the orientation to vary freely, or to fix both the dipole location and orientation but allow only the dipole magnitude to vary. This is implemented in BESA and SPM. Once the weights are constructed, a dipole time series can be created, and analyses can be performed on this time series, including time-frequency decomposition. Realistic head and brain models for the forward problem are possibly but not commonly used.

The advantage of EDM is its simplicity and the main drawback is that one needs to predefine the number of sources, which is unfeasible for many problems. In addition, the number of unknown parameters is smaller than the number of sensors. Furthermore, the optimization is often trapped in the local minima and the sources to be localized need to be of small number. Another disadvantage is the interpreting of one or a small number of dipoles. This rests on the rationale of the assumption that that dipole is the only active source in the brain. In addition, as some researches revealed, for complex cognitive processes or widespread epileptic discharges, the EDM assumption is inappropriate (Benar et al. 2005).

9.2.2 Beamformer and Scanning Methods

The beamforming approaches, which are originated from the radar and sonar signal processing, have recently been the most common algorithms in EEG source analysis (Gross et al. 2001; Sekihara et al. 2001; Van Veen et al. 1997). These approaches are designed to estimate the activity at one brain region by simultaneously minimizing the interference from other active regions. For this target, the selected operator optimizes a goal function that maximizes the ratio between activity and noise at each candidate region. The goal function contains the a priori information about the sources and a covariance matrix about the recorded signal. The method usually contains two steps. First, the sources are linearly estimated using an optimal direction that is a priori defined. Therefore, the beamforming approaches have the same basic limitations as the other linear inverse solutions. Second, the result is normalized by the noise power. The idea of normalization as a second step in the inverse solution has also been used in the sLORETA method. Simulation studies suggest that beamforming provides results with higher accuracy compared to other source analysis methods and that beamforming has the least amount of overestimation of the spatial extent of the activation (Dalal et al. 2008).

The main disadvantages of beamforming approaches are the number of parameters that must be set and the influences of those parameters on the results. These options include which algorithm to use; whether and which frequencies to analyze; whether to use time-domain covariances or frequency-domain cross-spectral densities; the length of the time window for computing the covariances and whether that window should change as a function of frequency; whether and how to regularize the covariance matrix; whether to compute weights based on all conditions or separately for each condition; what to use as a normalization baseline; how many voxels to estimate and where they are located; what type of forward model to use (for EEG, what values to use for skull and scalp conductance); whether to fix dipole orientations with respect to the cortex or use three cardinal orientations; if using three orientations, how to estimate a voxel's activity from projections onto the three orientations; and so on. The number of parameters and choices should not dissuade you from applying adaptive spatial filters, but it is important to consider the parameters of the analyses and their possible influences on the results.

A related but distinct technique is to use a scanning strategy. That is, all source locations are scanned one by one to quantify how well the scalp map that is forwardly computed from assuming a dipole at each location can be classified as either being in the signal subspace or being in the noise subspace, while the signal and noise subspaces are orthogonal to each other and estimated from EEG or MEG. An implementation of this approach is multiple signal classification (MUSIC), in which the nonlinear optimization process of finding the dipole location is avoided (Mosher et al. 1992). However, MUSIC's performance suffers if the sources to be localized are correlated, especially if they are close to one another. Variations of MUSIC, such as recursively applied and projected MUSIC and the first-principles vector localization method, mitigate this problem to a certain degree.

9.2.3 Distributed Source Model

The disadvantage of EDM is that the exact number of dipole sources generally cannot be determined a priori. In view of this problem, distributed source models have received increased attention, because they do not need any a priori assumption about source number. In these models, each point of a 3-D grid or 2-D surface of solution points is the source, and this implies that the number of sources is much larger than the number of electrodes. The following task is to infer a unique solution that explains the scalp measurements. Unfortunately, an infinite number of solutions can lead to the exactly same scalp measurements. This means that the inverse problem is highly underdetermined.

Based on DSM, EEG inverse problem can be stated as

$$Y = L\theta + \varepsilon_1, \tag{9.1}$$

where $Y \in R^{n \times s}$ is the scalp measurement with n sensors and s samples. $L \in R^{n \times d}$ is the lead field matrix and $\theta \in R^{d \times s}$ is the source dynamics in s samples for d dipoles. If sources are assumed to have free directions, L is a matrix of $n \times 3d$ and θ is the source solution matrix of $3d \times s$. 3 indicates the three orthogonal components of each dipole. The term ε_1 represents random fluctuations in sensor spaces. Equation (9.1) is a linear equation, and the EEG inverse problem is to solve this equation. Here n is usually much smaller than d; some additional constraints on the solution are needed. Here we introduced some typical algorithms.

The minimum norm (MN) solution is a general estimator for a 3-D brain source distribution in the absence of any a priori information (Hämäläinen and Ilmoniemi

1994). The only assumption in MN is that the solution should have minimum overall intensity, i.e., the smallest L2-norm. The solution has the simplest expression:

$$\theta = L^{\mathrm{T}}\left(LL^{\mathrm{T}}\right)^{+}Y, \tag{9.2}$$

where $^{+}$ denotes the Moore-Penrose pseudo-inverse and T denotes the transposition of a matrix. This is the minimum square multiply method, which is widely utilized in general linear model estimation. The MN solution provides a unique solution as only one combination of brain sources can have both the best fitting of the measurements and the lowest overall intensity. However, there is no physiological necessity to restrict the overall intensity with the lowest values. In addition, as superficial points can produce large voltage with very small intensity, the MN algorithm favors weak and localized superficial patterns. In contrast, deep sources are hard to be reconstructed and they may be incorrectly projected to some superficial sources.

Because of the superficial tendency of the MN, various weightings have been proposed to ameliorate the algorithm. One possible weighting is based on the norm of the columns of the lead field matrix (Lawson and Hanson 1974), and the related solution of weighted MN (WMN) is

$$X = W^{-1}A^{T}\left(AW^{-1}A^{T}\right)^{+}Y, \tag{9.3}$$

And the weighted matrix W is defined as

$$W = \mathrm{diag}(\|L_1\|, \|L_2\|, \ldots, \|L_d\|), \tag{9.4}$$

As the norm of the columns of the lead field matrix for deep source is small, they will obtain more weight during source imaging. There are additional constraints that can be added in this weighted matrix. A well-known example is the Laplacian weighted minimum norm algorithm (implemented in the LORETA software, Pascual-Marqui et al. 1994):

$$W = B \times \mathrm{diag}(\|L_1\|, \|L_2\|, \ldots, \|L_d\|), \tag{9.5}$$

where B denotes the discrete spatial Laplacian operator. Standard LORETA (sLORETA) further considered the post-process of the standardized current density power (Pascual-Marqui 2002):

$$\theta_l' = \theta_l^{T}\left\{[S_\theta]_{ll}\right\}^{-1}\theta_l, \tag{9.6}$$

where S_θ is the variance of the estimated current density:

$$S_\theta = W^{-1}A^{T}\left(AW^{-1}A^{T}\right)^{+}A, \tag{9.7}$$

The particular constraint in these LORETA family methods is that they select the weight with a smooth distribution by minimizing the spatial Laplacian, which measures the spatial roughness of the solution. For vector fields, because the smoothness is not uniquely defined, different definitions of smoothness will produce different optimally smoothed solutions (Mitiche et al. 1988). The smoothness constraint may be partially reasonable, as the activity in neurons in neighboring patches is always correlated. However, this assumption may be unsuitable as the spatial scale is mismatched between the distance of solution space and the distance of electrodes.

9.2.4 Sparse Solution

The disadvantage of LORETA family is that the result is very blurring, and the current neurophysiological functional localization idea indicates that the main neural electric activities should be sparsely localized (Gorodnitsky et al. 1995). A reasonable solution may be sparsely localized. Currently, there are at least two strategies to make the solution sparsely. One is to solve the $l_p (p \leq 1)$ norm solution directly, and this includes thel_0 and l_1 norm solution (Xu et al. 2007). Another method iteratively shrinks the solution space in order to get a sparse solution. Both strategies can change the initially underdetermined problem to a finally overdetermined problem. For example, some algorithms first obtain an initial solution based on LORETA and then shrink the solution gradually with special design of the iteration steps. Here we introduced the focal underdetermined system solver (FOCUSS) (Gorodnitsky et al. 1995) and the iterative Gaussian source imaging algorithm (GIA) (Lei et al. 2009) as examples of these algorithms.

FOCUSS is an energy localized iterative procedure (Gorodnitsky et al. 1995). By a linear transform $\theta = W\phi$, Eq. (9.1) is changed to

$$\begin{cases} min\|\phi\| \\ s.t: LW\phi = Y \end{cases} \tag{9.8}$$

The ith iteration of FOCUSS constructed the transform W_i as a diagonal matrix of the prior iteration solution θ_{i-1}. This iteration can be simplified as the following two equations:

$$W_i = 1/\text{diag}(\|L_1\|, \|L_2\|, \ldots, \|L_d\|) \times \text{diag}(\theta_{i-1}), \tag{9.9}$$

$$\theta_i = W_i{}^{-1} A^T \left(A W_i{}^{-1} A^T\right)^+ Y, \tag{9.10}$$

where W_i is constructed with the amplitudes of the elements of θ_{i-1}, and this weight matrix enhances some of the elements in θ_{i-1} having large amplitude and simultaneously decreases the remaining elements. $\|L_l\|$ is the norm of the lth column of the matrix L. During iteration procedure, Eq. (9.9) tends to decrease a large number of elements until they become zero. And this is the reason that the solution of FOCUSS is sparse.

Other constraints were introduced in the iteration procedure. For example, we assumed that the solution sources in a local region are approximately of the same orientation and with amplitudes with spatial Gaussian distribution (Lei et al. 2009). With the similar iteration step of FOCUSS, a new algorithm was proposed: Gaussian source imaging algorithm (GIA). The Gaussian source model (GSM) is the key step in GIA, and it iteratively determined the following two parameters:

First, determine the GSM center. Each center is defined as the grid node which meets two conditions: (a) its amplitude v is larger than 1% of the maximum value in the entire solution space; (b) v is the largest one among the nearest 6 neighbors. This step also determines the radius of GSM. In each direction (total of six directions), the distance from the center to the first node with increased amplitude is the radius of that direction. And the radius of GSM is the mean of all radii.

Second, adjust the amplitude at each GSM. The source strength is determined by the Gaussian function. For the ith GSM, δ_i is defined by the amplitude of center v_{max} and the mean value of its nearest neighbors $v_{neighbor}$:

$$\delta_i^2 = d^2 / \ln\left(v_{max}/v_{neighbor}\right), \tag{9.11}$$

where d is the space resolution, i.e., the distance of nearest neighbors. The amplitude of a neighbor node noted by position vector s_j is determined by.

$$Q_j = Q_i \times \exp\left(-\frac{\left\|s_i - s_j\right\|^2}{\sigma_i^2}\right) \tag{9.12}$$

The Q_j denotes the dipole moment of the center node i; its amplitude is the local maximum v_{max}. The s_i and s_j are the spatial position vectors of the grid nodes i and j, respectively. If v_j, norm of Q_j, is larger than that of the current solution at node j, norm of $\theta_i(j)$, Q_j, is set to node j. Otherwise, moment of node j will not be changed.

Note that above two steps are the smoothing and shrinking operations that modify the solution space. The iteration procedure is based on FOCUSS and it is guaranteed to converge as shown in reference (Gorodnitsky et al. 1995). For GIA, the initialization is important for correct source localization. We chose the smooth source distribution provided by LORETA as its initialization solution. Detailed discussion about convergence is given in Lei et al. (2009).

9.3 Bayesian Source Model

Bayesian inference is a general procedure for incorporating a priori information into the estimation of the sources. The method is conceptually simple, using basic laws of probability, making its application even to complicated problems relatively straightforward. For source analysis, Bayesian allows for a more detailed description of the anatomical and functional a priori information. The types of information include the neural current (Schmidt et al. 1999), the sparse focal nature of the sources (Phillips et al. 1997), combined spatial and temporal constraints (Baillet and Garnero 1997), and recently the resting-state spatial patterns (Lei et al. 2011). The parametric empirical Bayesian framework has been proved to be a promising tool for reliable estimation of EEG sources, because various priors can be used for source reconstruction (Lei et al. 2015).

9.3.1 Parametric Empirical Bayesian

Parametric empirical Bayesian (PEB) model (Lei et al. 2011; Mattout et al. 2006; Phillips et al. 2002) is used for EEG source imaging:

$$\begin{array}{ll} Y = L\theta + \varepsilon_1 & \varepsilon_1 \sim N(0, T, C_1) \\ \theta = 0 + \varepsilon_2 & \varepsilon_2 \sim N(0, T, C_2) \end{array} \tag{9.13}$$

where Y, L, θ, and ε_1 have the same definition as in Eq. (9.1), though they are random variables here.$N(\mu, T, C)$ represents a multivariate Gaussian distribution on a matrix, namely, $\varepsilon \sim N(\mu, T, C) \Leftrightarrow vec(\varepsilon) \sim N(\mu, T \otimes C)$, with mean μ and covariance $T \otimes C$. *vec* denotes the column-stacking operator, and $\otimes$ denotes the Kronecker tensor product. The term ε_2 represents random fluctuations in the source space. The temporal correlations are denoted by T. The spatial covariances of ε_1 and ε_2 are the mixtures of covariance component at sensor and source level, respectively. In the sensor level, $C_1 = \alpha^{-1} I_n$ was assumed to encode the covariance of sensor noise, where I_n denotes the n-by-n identity matrix. In the source level, a covariance component form was listed,

$$C_2 = \sum_{i=1}^{k} \gamma_i V_i, \tag{9.14}$$

where $\gamma \equiv [\gamma_1, \gamma_2, \ldots, \gamma_k]^T$ denotes a vector of k non-negative parameters that control the contribution of V_i. The components set, $V = \{V_1, V_2, \ldots, V_k\}$, is fixed and known. The formulation in Eq. (9.14) is extremely flexible, as a rich variety of candidate covariance bases can be easily combined in this framework (Mattout et al. 2006). The parameters γ are unknown, and they are estimated using restricted maximum likelihood (ReML) algorithm (Friston et al. 2007).

9.3.2 Priors from Anatomic Information

PEB model provides a natural structure to include priors from other modalities. This is based on the design of the covariance component form in Eq. (9.14). In the classical MN solution as defined in Eq. (9.2), covariance component is an identity matrix, i.e., $V_1 = I_d$ (Phillips et al. 2005). The Green function (Harrison et al. 2007) is employed to consider the local anatomic coherence,

$$G = \exp(\delta B) = [g_1, g_2, g_3, \ldots, g_\mathrm{d}], \quad (9.15)$$

where B is an adjacency matrix in the solution space, and it encodes the neighboring relationships among vertices of the cortical mesh. The source space has d vertices that are approximately uniformly distributed over the cortex surface. The jth column of G, g_j, defines the weights based on their distances to the center, the jth vertex. Then, we may consider LORETA-like prior as a more realistic model with two components modeling independent and coherent sources, respectively, $V = \{I, G\}$. In a recent approach (Friston et al. 2008), multiple sparse priors (MSP) from the columns of G are uniformly sampled to produce several hundred covariance components at the source level, $V = \{q_1 q_1^T, q_2 q_2^T, \ldots, q_k q_k^T\}$.

9.3.3 Network-Based Source Imaging

The extendable structure of covariance component provides large potential to contain other priors, such as the temporal coherent networks (TCNs) derived from the BOLD signal. Based on the PEB framework, we developed NEtwork-based SOurce Imaging (NESOI) (Lei et al. 2011). Different from the spatial activation pattern derived from statistic parameter mapping, the priors generated by independent component analysis (ICA) are free from all assumptions about the time courses contributing to signal changes. In addition, both transiently task-related and non-task-related components extracted by ICA can facilitate the EEG source imaging.

The novelty of NESOI includes the temporal coherent networks (TCNs) extracted from task fMRI. TCNs can involve cortical areas that are spatially distant. As thus, they differ from relations between spatially adjacent source information and local functional activation information. In NESOI, both EEG and fMRI recordings should be implemented on the same subject within the same paradigm, regardless of whether they are obtained simultaneously or separately. To construct TCNs, ICA was adopted to group brain areas that share response patterns (Beckmann et al. 2005). The spatial ICA decomposition of fMRI generates the spatial independent components and their corresponding time course. The spatial IC represents the brain activities, expressing the intensity distribution over all voxels. We re-scaled the IC to z scores, to standardize the relative contribution of each voxel. Voxel with absolute

z scores >3 is considered to be activated. Negative z scores indicate that the BOLD signals are modulated oppositely to the IC waveform (McKeown et al. 1998).

And then the activated voxel was projected to the EEG source space as matrix U ($d \times k$), where d is the number of nodes and k is the number of spatial ICs. The element U_{ij} is set to 1.0 if the absolute z score of the nearest neighbor voxel is larger than 3; otherwise U_{ij} is set to 0.0. Obviously, U is sparse matrix with a small number of nonzero elements. Because all the activated voxels in each IC would be expected to have similar temporal dynamics, we consider each IC as a TCN. In this condition, a simple way to define the covariance component from the ith IC is to set the diagonal terms with a binary of the ith column of U and the other terms with zero. Here, to consider the local coherence in source space, NESOI takes the covariance component V_i as

$$V_i = 1/n_i \Sigma g_j g_j \tag{9.16}$$

where g_j is the jth column of the Green function matrix in Eq. (9.15) and n_i is the total number of activated voxels in ith IC. Here, the covariance component with nonzero off-diagonal terms in Eq. (9.16) could model locally correlated sources. Obviously, the definition of our covariance component encodes a functional connectivity prior that is distinct from the neuronal-anatomical information utilized in other methods.

9.3.4 Resting-State Cortex Rhythms

NESOI is designed to reconstruct the source of event-related potential (ERP), and its spatial prior is the temporal coherent networks from the task-related fMRI with the same subject and the same paradigm. Recently, the brain in resting state is an attractive topic in neuroimaging studies. A natural extension of NESOI to resting-state EEG is resting-state NESOI (rsNESOI, Lei 2012). In rsNESOI, a template of resting-state networks (RSNs) was employed as spatial prior (Yeo et al. 2011). RSNs involve multiple brain regions that have similar BOLD activations during resting state, and they have been shown to be reproducible across large populations (Smith et al. 2009). We think the rhythmic EEG has a good degree of spatial concordance of the modules of large-scale functional networks. As the electrophysiological activity of cortex in macro-scale follows the same spatial pattern, resting-state networks may facilitate EEG source imaging (Lei et al. 2015).

The rsNESOI employs RSNs with fixed spatial template; hence it is free from any extra fMRI scanning. Eight large-scale brain networks are used to construct the covariance components of the EEG source reconstruction. Seven large-scale networks, i.e., visual, somatomotor, dorsal attention, ventral attention, limbic, fronto-parietal, and default networks, were identified based on the segmentation of resting-state fMRI (Yeo et al. 2011). We further considered the importance of the deep brain structures, including thalamus, caudate, hippocampus, amygdala, and olfactory.

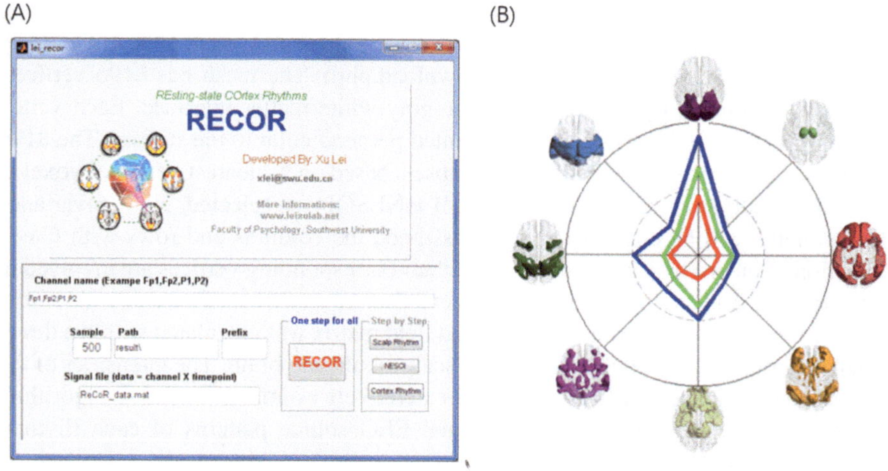

Fig. 9.4 Resting-state cortex rhythms (RECOR) is a toolbox to estimate the power of EEG rhythms in the eight large-scale brain networks. (**a**) The GUI interface of RECOR; (**b**) the power of each rhythm can be shown as radar map with eight brain networks above

Here the automated anatomical labeling (AAL) parcellation atlas (Tzourio-Mazoyer et al. 2002) was used to construct the eight large-scale networks. The main difference of the rsNESOI, when compared with the NESOI, is the utilization of RSNs instead of the various task-related functional co-activation patterns. In addition, as RSNs are fixed as some templates, the rsNESOI does not need any additional fMRI experiments. It is to say that a major novelty of rsNESOI is the employment of RSNs without any fMRI scanning (Lei 2012).

To define a pipeline for rhythmic EEG analysis, we developed a toolbox to estimate the power of EEG rhythms in a template of resting-state networks (Lei 2012): resting-state cortex rhythms (RECOR, Fig. 9.4). RECOR included three steps to calculate the power of EEG rhythms in each brain network. Firstly, the absolute power values of seven rhythms are calculated in each channel of the EEG signal (see Fig. 9.9). The rhythms include delta, theta, low alpha, high alpha, low beta, high beta, and gamma. The spectrum is estimated via Welch's method and the absolute power value was log-transformed. Secondly, EEG source reconstruction was implemented in each rhythm to estimate the cortical sources (Fig. 9.10). The default algorithm is LORETA-like prior with two components $V = \{I, G\}$. An alternative algorithm is rsNESOI (Lei et al. 2012), which was introduced in the above section. The last step of RECOR is averaging the solutions of source reconstruction across all vertices of a given large-scale brain network.

The result of resting-state cortex rhythms (RECOR). For three groups, we illustrated the estimated power of EEG alpha rhythms in the eight large-scale brain networks. Obviously health control group (blue line) has the largest alpha power in visual network.

In RECOR, the forward model is restricted to a high-density canonical cortical mesh, which was extracted from a structural MRI of a neurotypical male in FieldTrip software (http://fieldtrip.fcdonders.nl/download.php). The mesh has 8196 vertices, which were uniformly distributed on the gray-white matter interface. Each vertex node is assumed to have one dipole, oriented perpendicular to the surface. The 8196 vertices were further separated to eight subsets based on its nearest neighbor voxel in the large-scale brain network templates. If rsNESOI was selected, eight covariance components will be constructed. V_i is assigned the columns and rows with Green function of the mesh adjacency matrix, if their corresponding vertices are involved in ith network, and the other terms with zero (Lei et al. 2011). The electrodes were registered to the scalp surface, and the lead field matrix was calculated within a three-shell spherical head model including the scalp, skull, and brain. The intensities of the neural electric sources of EEG rhythms are iteratively estimated by ReML algorithm.

Rather than estimation of the punctual EEG source patterns of each rhythm, RECOR focused on the large-scale distribution of EEG source and calculated an averaged current density at each network. This is in line with the low spatial resolution of EEG recording. This averaging step may minimize the effects of poor performance of reconstruction method in estimating the deep brain structure. We recommend skip reporting precisely the location of the reconstructed source but focus on the output with more macro-scale in the eight brain networks. Note that RECOR is recommended to be used to analyze resting-state EEG, rather than task-related EEG or event-related potential. This source imaging of the spontaneous EEG may provide important findings in the understanding of brain functioning and variations of these functions during rest, sleep, cognitive task, maturation, or psychiatric diseases.

9.4 Limitation and Future Direction

9.4.1 Localization Accuracy

Simulation studies suggest that impressively high spatial localization accuracy in source imaging – matching that of fMRI – can be obtained with accurate electrode positions and realistic forward models (Brookes et al. 2010; Murzin et al. 2011). However, such high accuracy is hard to obtain in practice. There are many uncertainties such as skull and scalp conductivity, channel-to-MRI co-registration, brain anatomy, and channel localization (Steinsträter et al. 2010). Source reconstruction voxels are typically 5–10 mm^3, and spatial smoothing and normalization to a common brain space further reduce spatial precision. Thus, it is rare to see claims about fMRI level anatomical accuracy of source localization results, and caution should be applied when claiming of functional-anatomical dissociations of less than a few centimeters based on the results of EEG source imaging. Although it is possible to have higher spatial precision, particularly in hypothesis-driven research, the "effective" spatial precision used in most source reconstruction studies is on the order of several centimeters (Barnes et al. 2004). Developing a more realistic source reconstruction method is an important and active research topic, even though standard head models can provide fairly accurate results (Fuchs et al. 2002).

9.4.2 Combination with Other Modalities

There are many available modalities to serve as complementary techniques, such as MEG, magnetic resonance imaging (MRI), functional MRI (fMRI), positron emission tomography (PET), diffusion tensor imaging (DTI), transcranial magnetic stimulation (TMS), etc. (Lei et al. 2015). From the perspective of source analysis, the modality's specific contributions are at three levels: head model construction, spatial priors, and connectivity priors. In the above section, we had introduced the contribution of MRI and CT in head model construction. For spatial priors, EEG-correlated fMRI, temporally coherent networks, and resting-state fMRI are systematically introduced as priors for source analysis. In addition, both the fiber tracking (diffusion tensor imaging, DTI) and neuro-stimulation techniques (transcranial magnetic stimulation, TMS) have the potential to be involved as connectivity priors, because they can help to infer the neuroelectric connectivity in the source space.

Recently, the simultaneous MEG and EEG recording are utilized as a common practice. A long-lasting discussion concerning source analysis with MEG and EEG stated that MEG is superior to EEG for source localization accuracy. Simulation studies suggest that the localization accuracy of EEG can be as good as, or even superior to, that of MEG (Murzin et al. 2011; Steinsträter et al. 2010), in part because EEG is sensitive to radial and tangential dipoles, whereas MEG is maximally sensitive to tangential dipoles. However, improved accuracy for EEG source reconstruction requires many electrodes as well as precise electrode positions, head shapes, brain-forward models, and accurate estimates of tissue conductivities. Such precision in measurement is not typically done for EEG, whereas it is more often done with MEG. Thus, in practice, localization accuracy of MEG is often better than localization accuracy of EEG.

As EEG electrode technology improves and recording subject-specific electrode positions becomes more commonplace, the accuracy of EEG source reconstruction results will further improve. We conclude that combining EEG source imaging with other complementary modalities is a promising approach toward the study of brain networks in cognitive and clinical neurosciences.

9.5 Example Application

9.5.1 Experiment

Ten healthy subjects (6 females, age 20.9 ± 2.7 years) participated in the eyes-closed resting-state EEG study. All participants completed a questionnaire package, which included Pittsburgh Sleep Quality Index (PSQI), Self-Rating Depression Scale (SDS), and Self-Rating Anxiety Scale (SAS). After a detailed explanation of the study protocol, a written informed consent was obtained. Our study was approved by the Ethics Committee of Southwest University. In addition, all the related procedures were in accordance with the sixth revision of the Declaration of Helsinki. The dataset can be downloaded from the link (http://www.leixulab.net/data.asp).

EEG data was recorded about 3 min from the 64 electrodes mounted in an elastic cap (Brain Products, Munich, Germany) around 9:00 to 11:00 in the morning. The sampling frequency is set to 500 Hz. Before EEG recording, the impedance was kept below 5 kΩ. The pre-processing was conducted using EEGLAB (http://sccn.ucsd.edu/eeglab). The original EEG signal was filtered with band-pass between 0.1 and 45 Hz and referenced to common average. Then, data was fragmented into 2-s segments, and segments with ocular, muscular, and other types of artifacts were preliminarily identified and excluded. For all subjects, we selected the first 20 segments, i.e., 40 s of artifact-free EEG data.

We employed pwelch function to calculate the power spectrum of EEG recording. Here, both eyes-closed and eyes-open conditions of subject #10 were analyzed. A local maximum was observed in the frequency range of the alpha rhythm (8–12 Hz), which was the most prominent rhythm during eyes-closed condition (Fig. 9.5). In the following source localization, we will focus on this alpha rhythm. Three software toolboxes were compared: DIPFIT, sLORETA, and RECOR. Notice that the source analysis in frequency domain should base on cross-spectra, not the log-transformed scalp electric potential spectral powers shown here. We just use the power of alpha rhythm as mimic ERP distribution to show the main procedures of different toolboxes, though it is an incorrect usage for frequency domain source analysis (see Frei et al. 2001) (Code 9.1).

9.5.2 DIPFIT

We employed an EEGLAB plugin (Delorme and Makeig 2004) to localize equivalent current dipole for alpha rhythm: DIPFIT. Note DIPFIT was usually used for localizing independent component scalp maps, and its main functions are from the FieldTrip toolbox. As illustrated in Code 9.2, we introduced in detail its main steps for localizing the alpha rhythm.

To fit dipole models to the power of alpha rhythm, both the potential distribution and channel location files need to be loaded. To follow the dipole fitting example used in this tutorial, download a dataset (alphaRhythm.rar) from the link (http://www.leixulab.net/data.asp). This sample dataset contains a channel location file and an alpha power file.

As illustrated in Code 9.2, we first re-referenced the value to average. "dipfit_erpeeg" is the main function for dipole fitting. There are three steps to create equivalent current dipole. First is setting model and preferences. This involves choosing the model (spherical or boundary element) and excluding some channels

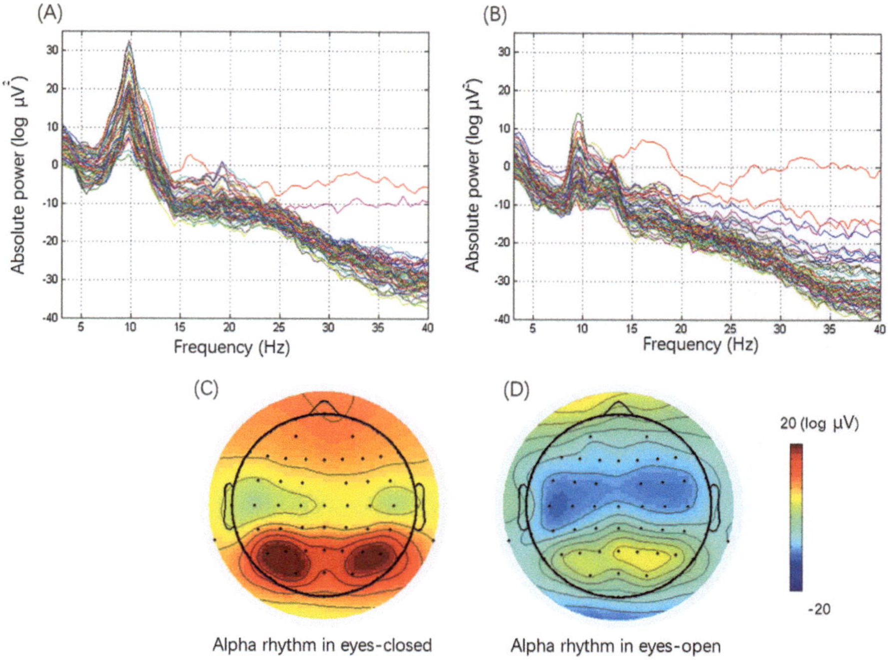

Fig. 9.5 Raw power spectra of resting-state EEG (**a**, **b**) and the topography (**c**, **d**) of alpha rhythm (8–12 Hz) for a single subject (#10). Here, the alpha rhythm predominates in power spectra and distributes around the occipital region

from the fitting procedure (e.g., eye channels). Here, we set the input values of "dipfit_erpeeg" as the scalp potential we want to localize, channel location, assumed dipoles model, coordination format, head model file, and MRI file. Second is grid scanning the solution space. This involves scanning all the coarse 3-D grids to determine an initiate searching point of equivalent dipoles. Third is non-linear optimization processing. This involves running an optimization searching to obtain the best position for each component. Note that the grid scanning and non-linear optimization are performed automatically and you may only find the procedures as output in the command window.

The next step is plotting the dipole in 3-D. This output was inset within a GUI interface. The Talairach coordinate and residual variance are listed in the middle left panel. There are some other display types that can be chosen. In this example, we also plot the dipole in the surface of the scalp. The result of executing the script is illustrated in Fig. 9.6.

As illustrated in Fig. 9.6, we may report our result as follows:

"Two symmetric dipoles were identified in the occipital regions; their Talairach locations are [53 −65 50] mm and [−53 −65 50] mm. Their momentations are [−3.4563 −0.0101 −1.3170] dB and [−4.9173 −0.8642 −0.3486] dB. From their momentations, it is obvious that the left dipole is larger than the right dipole (1.35:1). The residual variance was 27.46%."

```
eegFilePath='D:\\resting-state EEG\\';

%% Calculate the power spectrum for subject 10
for i=1:2 % eyes-open eyes-closed
    if i==1   % load the file
        subFile='sub10c.set';
    else
        subFile='sub10o.set';
    end
    EEG = pop_loadset('filename',subFile,'filepath',eegFilePath);
    signalRest=reshape(EEG.data,[61,100*240]);
    for j=1:61 % each channel
        [Pxx] = pwelch(signalRest(j,:),200,0,256,100);
        Pxx=20.*log10(Pxx);
        Px(:,j,i)=Pxx;
    end
end

%% Plot the spectrum and the topography in EC and EO
Fs=[0:128]./128*50;
figure
for i=1:2
    subplot(2,2,i)
    plot(Fs,Px(:,:,i));
    grid on;
    axis([3 40 -40 35]);
end

Pxm=squeeze(mean(Px(22:31,:,:))); % alpha band [8 12]
for i=1:2
    subplot(2,2,i+2)
    topoplot(Pxm(:,i), EEG.chanlocs,'maplimits',[-20 20]);
end
alphaRhythm=Pxm(:,1);    % alpha band for eyes-closed
save alphaRhythm.mat alphaRhythm;
chanlocs= EEG.chanlocs;   % channel location
save chanlocs.mat chanlocs;
```

Code 9.1 The MATLAB code to calculate the power spectrum and topography of resting-state EEG. The functions of pwelch and topoplot were used to estimate the spectrum and topography, respectively. The power of alpha band for eyes-closed condition and the channel localization were saved for the analysis of the following section.

9.5.3 sLORETA

We employed standardized low-resolution brain electromagnetic tomography (sLORETA) to estimate the cortical distribution of current source density of alpha

```
%% dipole fitting
load alphaRhythm.mat;  % eyes-closed
alphaRhythm=alphaRhythm-mean(alphaRhythm); % re-reference to average
load chanlocs.mat;
templateFile='D:\matlab tool\eeglab13_5_4b\plugins\dipfit2.3\standard_BESA\';
[ dipole,model] = dipfit_erpeeg(alphaRhythm, chanlocs,'dipoles',2,'settings',...
    {'coordformat','spherical','hdmfile', [templateFile,'standard_BESA.mat'],...
    'mrifile', [templateFile,'avg152t1.mat']});

%% plot the dipole in 3-D
dipplot(dipole, 'image', 'mri', 'sphere', 85, 'normlen', 'on','projimg', 'on', ...
    'projlines', 'on', 'gui', 'off');

%% plot the dipole plus the scalp map
dipolefortopoplot = [dipole(1).posxyz(:,1:2) dipole(1).momxyz(:,1:3)]/85;
figure
topoplot(alphaRhythm, chanlocs, 'dipole', dipolefortopoplot, 'dipnorm', 'on', ...
    'electrodes', 'off');
title([ 'Alpha rhythm, fit with two symmetric dipoles (RV '…
    num2str(dipole(1).rv*100,2) '%)']);
```

Code 9.2 The MATLAB code to estimate equivalent current dipole for alpha rhythm using DIPFIT plugin of EEGLAB. The functions of dipfit_erpeeg and dipplot were used to estimate and plot dipole, respectively

rhythm. A unique property of sLORETA is that, under ideal conditions, it localizes "test point sources" exactly. Both the electrode coordinates and the head model of sLORETA are based on the average MRI brain map of the Montreal Neurological Institute, i.e., the MNI152 (Mazziotta et al. 2001). The cortical gray matter constructs the solution space, with 6239 voxels and the spatial resolution of 5^3 mm^3.

Before run sLORETA, the data file and channel location file need to be translated into the sLORETA format. As illustrated in Code 9.3, we repeated our alpha rhythm power 10 times to fake an EEG wave and saved it as an ASCII file "Chapter9EEGdata.txt." This step is necessary because sLORETA is a localization method for wave. The channel location was saved to a separate file "Chapter9channel.txt."

First step, run the sLORETA setup program and a GUI interface opens (Fig. 9.7a). This requires a password, which should be obtained from the homepage: www.keyinst.unizh.ch/loreta. Then, click the main utilities, and click the button "Electrode names to coordinates." Based on the left file management panel, folders and files could be viewed. Open the folder that saved our alpha rhythm dataset. Select the file "Chapter9channel.txt" containing the list of electrode names. Note that electrode order must correspond to the recordings. Otherwise, this software will produce errors. Drag and drop this file to the input box of "File with electrode names." Automatically, the filename for the output is created as ∗.sxyz and then Run. Then, the Talairach electrode coordinates are created.

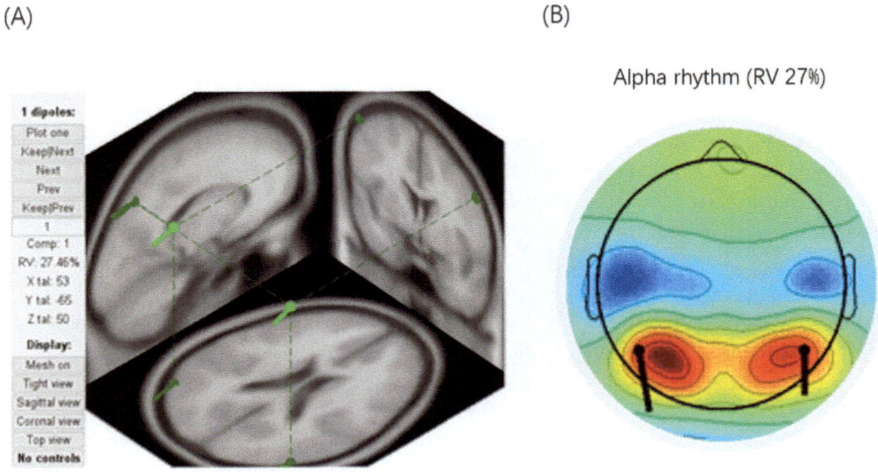

Fig. 9.6 The dipole fitted by DIPFIT in EEGLAB. (**a**) The dipole in 3-D with template MRI; (**b**) the dipole in the surface of the scalp. Two symmetric dipoles were reconstructed in the occipital region

```
% % Prepare data and channel files for LORETA
alphaSig=repmat(alphaRhythm',10,1);% repeat 10 times to fake an EEG wave
save -ascii Chapter9EEGdata.txt alphaSig;
fid = fopen('Chapter9channel.txt','w');
fprintf(fid,'%s\n',chanlocs.labels);
fclose(fid);
```

Code 9.3 The code for the prepared date and channel files for sLORETA. The one time point was repeated to 10 times, to fake an EEG wave with 10 time-bin recording in the scalp. Channel location was a file only listing the channel names.

Next step, click the tab for making the sLORETA transformation matrix. Drag and drop the Talairach coordinate file of electrode $*$.sxyz, and run the program. The output file is the sLORETA transformation matrix: Chapter9channel.spinv.

Then run sLORETA Viewer/Explorer, and click the File button. And then select electrode coordinate file "Chapter9channel.sxyz" and transformation matrix file "Chapter9channel.pinv." Then open EEG/ERP file "Chapter9EEGdata.txt." Then click the SamplingRateHz button and define it as 10 Hz. As we repeat our alpha rhythm 10 times, this setting means data has 1-second recording. We also reset the "Number of time frame per page" with 10. Click ViewInfo to check the information loaded up to now in the program. Then, click arbitrary point in the EEG/ERP signals window; a cursor appears. All other windows are locked to this cursor, displaying as follows: scalp maps of electric potentials, sLORETA in 3-D cortex, and sLORETA in Slice Viewer (Fig. 9.7b–d).

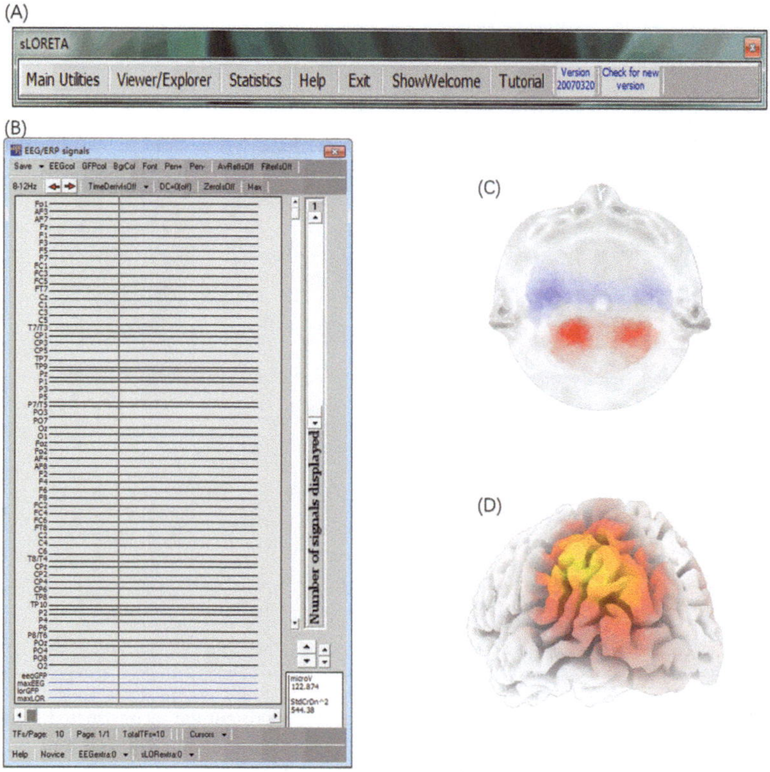

Fig. 9.7 The main IGU interface of sLORETA. (**a**) The main menu of sLORETA, (**b**) window of Viewer/Explorer, (**c**) scalp map of EEG/ERP signals, and (**d**) 3-D cortex view of sLORETA

Slice Viewer is the main output interface; here we introduce it in more detail. As illustrated in Fig. 9.8, the upper panel shows the axial, sagittal, and coronal slice of the whole brain. Use up menu to jump to other special locations. In the down left panel, the output text box lists some other parameters: the power value and the MNI coordinate. It also reports the best match number of Brodmann area.

You may save image as the report figure and report result of sLORETA as follows:"The distributed sources (cortex current potential) of sLORETA are localized around the occipital region. The MNI coordinate of central localization is [−50, −60 40], with Brodmann area being 40 in the inferior parietal lobule."

9.5.4 RECOR

Resting-state cortex rhythms (RECOR) is a toolbox to estimate the power of EEG rhythms in the eight large-scale brain networks (Lei et al. 2011; Lei 2012). This

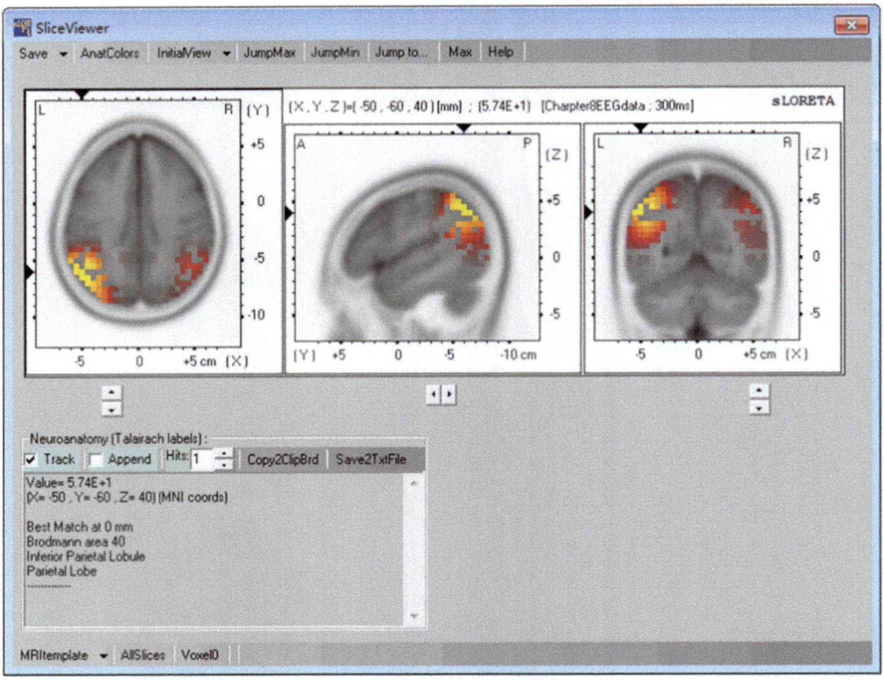

Fig. 9.8 Slice Viewer of sLORETA. The colored squares represent the reconstructed sources

toolbox is thoroughly tested and works on MATLAB R2008a and higher. RECOR can be downloaded from http://www.leixulab.net/recor.asp. After uncompressing the WinRAR file, add RECOR directories to the MATLAB search path.

Before run RECOR, we need to prepare our data to follow the RECOR format. EEG data should be saved as a ".mat" file, which contains only one variable: data. Data is a matrix, with the row being the number of channels and the column being the number of time points. In our example, we first use EEGLAB to load the existing dataset of "sub10c.set." Then we run command data = EEG.data in the command window of MATLAB. And data was saved as a single file, sub10c.mat, which is the input data for RECOR. After the RECOR path is set, input lei_recor in the command window. RECOR toolbox (Fig. 9.4a) opens in a new figure window. There is also an option to run RECOR using a batch script (Code 9.4). Batch script is very useful for running large datasets.

Here we have the following sets in the main GUI interface. For the channel name, the default input is the electrode name, separated by commas. Note that the electrode order must correspond to the recordings. Otherwise, this software will produce errors. Here we input the following: "Fp1, AF3, AF7, Fz, F1, F3, F5, F7, FC1, FC3, FC5, FT7, Cz, C1, C3, C5, T7, CP1, CP3, CP5, TP7, TP9, Pz, P1, P3, P5, P7, PO3, PO7, Oz, O1, Fpz, Fp2, AF4, AF8, F2, F4, F6, F8, FC2, FC4, FC6, FT8, C2,

```
srate=100;                      % sample rate 100 Hz
signal='sub10c.mat';             % signal file with variance "data"
path='recor result\';             % sample rate 100 Hz
prefix = '';                     % no prefix will be added to the output file
label={'Fp1','AF3','AF7','Fz','F1','F3','F5','F7','FC1','FC3','FC5',...

'FT7','Cz','C1','C3','C5','T7','CP1','CP3','CP5','TP7','TP9','Pz',...

'P1','P3','P5','P7','PO3','PO7','Oz','O1','Fpz','Fp2','AF4','AF8',...
    'F2','F4','F6','F8','FC2','FC4','FC6','FT8','C2','C4','C6','T8',...
    'CPz','CP2','CP4','CP6','TP8','TP10','P2','P4','P6','P8','POz',...
    'PO4','PO8','O2'};       % channel name of the 61 electrodes
inverseType='LORETA';   % reconstruction algorithm
%                   'NESOI'  NEtwork-based SOurce Imaging
%                   'LORETA'     LORETA-like model
template = 'RSN';          % output template is RSN
%                   'RSN'    Resting-state network (8 network)
%                   'BRM'    Brodmann areas (6 areas)
%%%%%%%%%%%%%%%%%%%%%%%%%%%%%%%%%%%
recorpath = which('lei_recor.m');
recorpath = recorpath(1:end-length('lei_recor.m'));
load(signal);
EEG.data=data;
EEG.srate=srate;
rhythmScalp=lei_step1_extractRhythm(EEG,label,prefix,path,recorpath);
rhythmCortex=lei_step2_NESOI(label,prefix,path,inverseType,recorpath);
rhythmNetwork=lei_step3_ReCor(prefix,path,template,recorpath);
```

Code 9.4 Batch analysis code for RECOR. The main input is the sample rate, signal file, output folder, and channel name. Inverse algorithm and output template have multiple choices

C4, C6, T8, CPz, CP2, CP4, CP6, TP8, TP10, P2, P4, P6, P8, POz, PO4, PO8, O2." The inputs for sample, path, and signal file are "100," "recor result\," and "sub10c.mat," separately.

The bottom right has four buttons; "RECOR" means all the steps are run at once. The analysis can also be run by selecting steps from "Scalp Rhythm," "LORETA," and "Cortex Rhythms" in order. Here we selected RECOR. RECOR will generate 9 figures and 12 files in current file path. In the following text, we introduce the main results.

First output is the power spectrum and topography in scalp surface (Fig. 9.9).The average power spectrum and topography of EEG rhythms are as follows: delta (2–4 Hz), theta (4–8 Hz), low alpha (8–10.5 Hz), high alpha (10.5–13 Hz), low beta (13–20 Hz), high beta (20–30 Hz), and gamma (30–40 Hz).

Second output is the current density in the cortex for each rhythm (Fig. 9.10). This is the inverse solution of LORETA or NESOI, based on your selection. Here, the current densities of both alpha1 and alpha2 are shown.

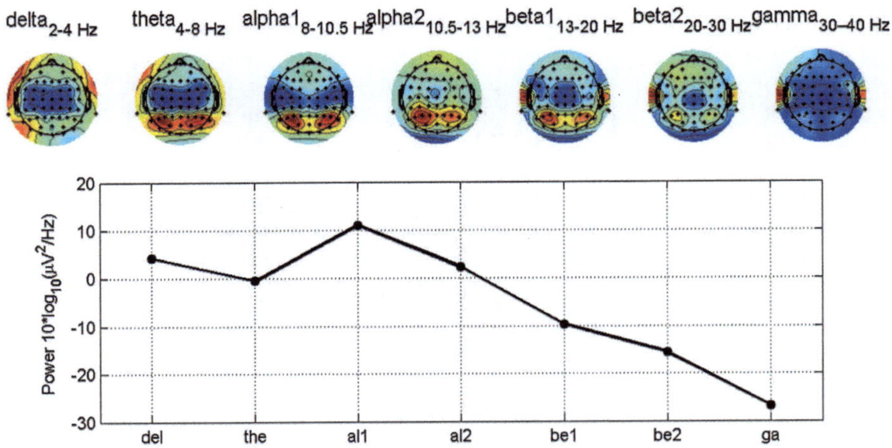

Fig. 9.9 Power spectrum and topography. Note that the alpha rhythm (8–12 Hz) in above Fig. 9.5 is separated to low (8–10.5 Hz) and high (10.5–13 Hz) alpha in RECOR

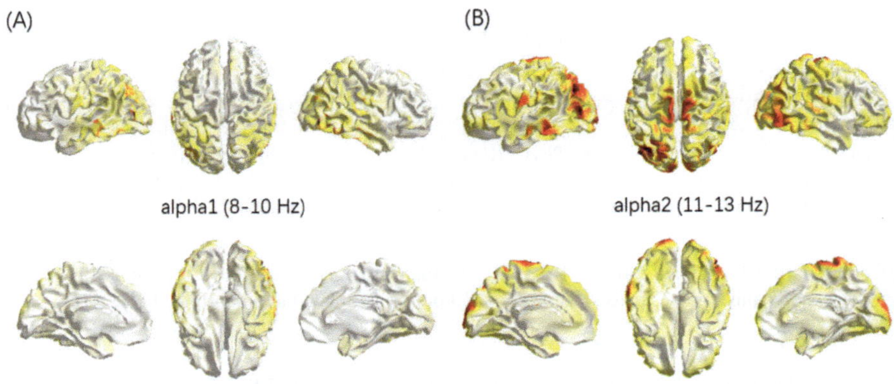

Fig. 9.10 Current densities in the cortex for alpha1 (8–10 Hz) and alpha2 (11–13 Hz) rhythms

Third output is the EEG spectral power density for each network (Fig. 9.11). There are eight panels corresponding to eight large-scale networks, each with seven power values of rhythms from delta to gamma.

The output folder has three mat files: ReCoR_rhythmScalp.mat, ReCoR_rhythmCortex.mat, and ReCoR_rhythmNetwork.mat. These files are the rhythm power in scalp surface, cortex, and large-scale brain network and are the main resources for further quantitative analysis. The corresponding dimensions of these files are [61 × 7], [8196 × 7], and [8 × 7]. All the second dimensions are 7, i.e., the seven rhythms. Here, we may be interested in finding the maximum activity of alpha rhythm; hence we load ReCoR_rhythmCortex.mat. In the third and fourth columns, we found the maximum (absolute) values as −190.44 and − 90.36,

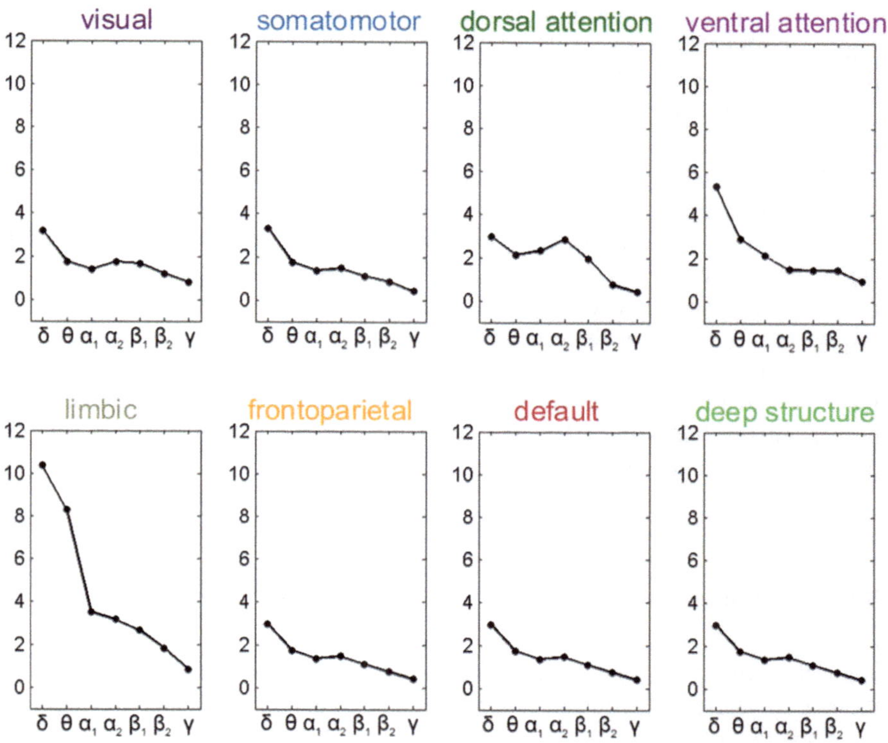

Fig. 9.11 EEG spectral power density of eight large-scale networks in rhythm from delta to gamma

respectively, and the corresponding subscript is 282. This means the 282nd dipole has the largest activity. We can further find its location in file of forward\ mesh8196.mat: [−48.23–81.07 5.11] mm, which is based on MNI coordinate.

The batch analysis in Code 9.3 has the same output as our prior setting in GUI. Here, you can find that the inverse type and template both have two options. The output networks template can be based on resting-state network template and the Brodmann areas.

9.5.5 *Comparing of the Three Methods*

In this tutorial we reconstruct the sources of alpha rhythm with three toolboxes: DIPFIT, sLORETA, and RECOR. DIPFIT is based on the equivalent dipole model, and the last two are based on the distributed source model. Note that all the source reconstructions are usually based on the time wave, not for power spectrum which is shown here. Source reconstruction in frequency domain should consider the phase and usually be based on cross-power spectrum. Our application is an incorrect usage for frequency domain source analysis (see Frei et al. 2001).

When comparing these three methods, we found the localizations as follows (mm): DIPFIT [53–65 50], sLORETA [−50, −60 40], and RECOR [−48.23–81.07 5.11]. Thinking the first one to be a Talairach coordinate, it is obvious that the sources of DIPFIT and sLORETA are close to each other, and the result of RECOR is far away. Both methods are based on distributed source model, and the main inter-dipole distance is 5 mm. However, there are many differences between them. First, the dipole is assumed to be radial to the cortex surface in RECOR, but the dipole in sLORETA has free direction. Second, dipoles in RECOR are mainly localized in the cortex surface, but dipoles in sLORETA are scattered all over the gray matter. In addition, RECOR has more dipoles (8196) than sLORETA (6239), resulting a more detailed subdivision of the whole brain. In this result, the maximum dipole of RECOR is far away from that of DIPFIT and sLORETA; this may imply that RECOR has the worst result. However, the source reconstruction is only its intermediate step for RECOR, and the final report of brain network is based on the average value within a large-scale network.

9.6 Conclusion

Algorithms for source analysis have been rapidly evolved into many directions over the last 20 years: from equivalent current dipole to distributed source model, from the prior of task-evoked activity (Henson et al. 2010) to the prior of task-free connectivity (Lei et al. 2011), and from single model source reconstruction to the multimodal fusion (Baillet et al. 2001a). There are some free and open-access toolboxes for EEG source analysis. Some main properties of these toolboxes include the support for MATLAB and the connectivity analysis. Another property is that some toolboxes are more specialized for component analysis and time series, such as EEGLAB. However, others specialize in scalp topographical analysis and clustering, such as CARTOOL. OpenMEEG can be used to make subject-specific BEM models (Gramfort et al. 2010). SPM supports dynamic causal model, which identifies effective connectivity in source space. eConnectome provides some causality analysis methods (He et al. 2011). Brainstorm (Tadel et al. 2011) and FieldTrip (Oostenveld et al. 2011) are popularly utilized for their strong function in EEG source analysis. We introduced only a few examples of the better-known toolboxes to guide the reader. All of the toolboxes introduced above have extensive online tutorials and user guides and are relatively easy to set up and work with.

Given its current advances and continuing development, EEG source analysis will increasingly be used for clinical and basic applications. The inexpensive EEG setups available in most clinical settings, the availability of computers, and the accessible open-source source analysis tools will enable broad applications of EEG source imaging. Furthermore, the capability of imaging dynamic brain activity from the whole brain makes EEG source analysis a desirable means for studying large-scale brain networks in humans.

While EEG source analysis can provide highly valuable information about brain networks and dynamics, it also has the advantage of being integrated with other modalities, such as fMRI, to combine the high temporal resolution of EEG with the high spatial resolution of fMRI (see Chap. 18 for more details). Efforts to increase the spatiotemporal resolution in imaging brain function are ongoing. It is envisioned that developments in designing better source reconstruction algorithms and combining EEG with other neural imaging or modulation techniques, such as MEG and fMRI, will be at the frontier of advances in functional neuroimaging.

References

Babiloni F, Babiloni C, Carducci F, Romani GL, Rossini PM, Angelone LM, Cincotti F. Multimodal integration of high-resolution EEG and functional magnetic resonance imaging data: a simulation study. NeuroImage. 2003;19:1–15.

Baillet S, Garnero L. A Bayesian approach to introducing anatomo-functional priors in the EEG/MEG inverse problem. IEEE Trans Biomed Eng. 1997;44:374–85.

Baillet S, Mosher J, Leahy R. Electromagnetic brain mapping. IEEE Signal Process Mag. 2001a;18:14–30.

Baillet S, Riera JJ, Marin G, Mangin JF, Aubert J, Garnero L. Evaluation of inverse methods and head models for EEG source localization using a human skull phantom. Phys Med Biol. 2001b;46:77–96.

Barnes GR, Hillebrand A, Fawcett IP, Singh KD. Realistic spatial sampling for MEG beam former images. Hum Brain Mapp. 2004;23:120–127.

Beckmann CF, DeLuca M, Devlin JT, Smith SM. Investigations into resting-state connectivity using independent component analysis. Philos Trans R Soc Lond Ser B Biol Sci. 2005;360:1001–13.

Benar CG, Gunn RN, Grova C, Champagne B, Gotman J. Statistical maps for EEG dipolar source localization. IEEE Trans Biomed Eng. 2005;52:401–13.

Brookes MJ, Zumer JM, Stevenson CM, Hale JR, Barnes GR, Vrba J, Morris PG. Investigating spatial specificity and data averaging in MEG. NeuroImage. 2010;49:525–38.

Dalal SS, Guggisberg AG, Edwards E, Sekihara K, Findlay AM, Canolty RT, Berger MS, Knight RT, Barbaro NM, Kirsch HE, Nagarajan SS. Five-dimensional neuroimaging: localization of the time-frequency dynamics of cortical activity. NeuroImage. 2008;40:1686–700.

Delorme A, Makeig S. EEGLAB: an open source toolbox for analysis of single-trial EEG dynamics including independent component analysis. J Neurosci Methods. 2004;134:9–21.

El Badia A, Ha-Duong T. An inverse source problem in potential analysis. Inverse Prob. 2000;16:651–63.

Fischl B, Sereno MI, Tootell RB, Dale AM. High-resolution intersubject averaging and a coordinate system for the cortical surface. Hum Brain Mapp. 1999;8:272–84.

Frei E, Gamma A, Pascual-Marqui R, Lehmann D, Hell D, Vollenweider FX. Localization of MDMA-induced brain activity in healthy volunteers using low resolution brain electromagnetic tomography (LORETA). Hum Brain Mapp. 2001;14:152–65.

Friston K, Harrison L, Daunizeau J, Kiebel S, Phillips C, Trujillo-Barreto N, Henson R, Flandin G, Mattout J. Multiple sparse priors for the M/EEG inverse problem. NeuroImage. 2008;39:1104–20.

Friston K, Mattout J, Trujillo-Barreto N, Ashburner J, Penny W. Variational free energy and the Laplace approximation. NeuroImage. 2007;34:220–34.

Fuchs M, Kastner J, Wagner M, Hawes S, Ebersole JS. A standardized boundary element method volume conductor model. Clin Neurophysiol. 2002;113:702–12.

Güllmar D, Haueisen J, Reichenbach JR. Influence of anisotropic electrical conductivity in white matter tissue on the EEG/MEG forward and inverse solution. A high-resolution whole head simulation study. NeuroImage. 2010;51:145–63.
Gorodnitsky IF, George JS, Rao BD. Neuromagnetic source imaging with FOCUSS: a recursive weighted minimum norm algorithm. Electroencephalogr Clin Neurophysiol. 1995;95:231–51.
Gramfort A, Papadopoulo T, Olivi E, Clerc M. OpenMEEG: opensource software for quasistatic bioelectromagnetics. Biomed Eng Online. 2010;9:45.
Gross J, Kujala J, Hämäläinen M, Timmermann L, Schnitzler A, Salmelin R. Dynamic imaging of coherent sources: Studying neural interactions in the human brain. Proc Natl Acad Sci U S A. 2001;98:694–9.
Hämäläinen MS, Ilmoniemi RJ. Interpreting magnetic fields of the brain: minimum norm estimates. Med Biol Eng Comput. 1994;32:35–42.
Hämäläinen MS, Sarvas J. Realistic conductivity geometry model of the human head for interpretation of neuromagnetic data. IEEE Trans Biomed Eng. 1989;36:165–71.
Harrison LM, Penny W, Ashburner J, Trujillo-Barreto N, Friston KJ. Diffusion-based spatial priors for imaging. NeuroImage. 2007;38:677–95.
He B, Dai Y, Astolfi L, Babiloni F, Yuan H, Yang L. eConnectome: A MATLAB toolbox for mapping and imaging of brain functional connectivity. J Neurosci Methods. 2011;195:261–9.
Helmholtz H. Ueber einige Gesetze der Vertheilung elektrischer Ströme in körperlichen Leitern mit Anwendung auf die thierisch-elektrischen Versuche. Annalen der Physik und Chemie. 1853;165:211–33.
Henson RN, Flandin G, Friston KJ, Mattout J. A Parametric Empirical Bayesian framework for fMRI-constrained MEG/EEG source reconstruction. Hum Brain Mapp. 2010;31:1512–31.
Lawson CL, Hanson RJ. Solving least squares problems. Englewood Cliffs: Prentice-Hall; 1974.
Lei X, Electromagnetic brain imaging based on standardized resting-state networks. 5th International Conference on Biomedical Engineering and Informatics (BMEI), Chongqing, China; 2012, p 40–4.
Lei X, Hu J, Yao D. Incorporating fMRI Functional Networks in EEG Source Imaging: A Bayesian Model Comparison Approach. Brain Topogr. 2012;25:27–38.
Lei X, Wu T, Valdes-Sosa P. Incorporating priors for EEG source imaging and connectivity analysis. Front Neurosci. 2015;9:284.
Lei X, Xu P, Luo C, Zhao J, Zhou D, Yao D. fMRI Functional Networks for EEG Source Imaging. Hum Brain Mapp. 2011;32:1141–60.
Lei X, Chen A, Xu P, Yao D. Gaussian Source Model based Iterative Algorithm for EEG source Imaging. Comput Biol Med. 2009;39(11):978–88.
Mattout J, Phillips C, Penny WD, Rugg MD, Friston KJ. MEG source localization under multiple constraints: an extended Bayesian framework. NeuroImage. 2006;30:753–67.
Mazziotta J, Toga A, Evans A, Fox P, Lancaster J, Zilles K, Woods R, Paus T, Simpson G, Pike B, Holmes C, Collins L, Thompson P, MacDonald D, Iacoboni M, Schormann T, Amunts K, Palomero-Gallagher N, Geyer S, Parsons L, Narr K, Kabani N, Le Goualher G, Feidler J, Smith K, Boomsma D, Hulshoff Pol H, Cannon T, Kawashima R, Mazoyer B. A four-dimensional probabilistic atlas of the human brain. J Am Med Inform Assoc. 2001;8:401–30.
McKeown MJ, Makeig S, Brown GG, Jung TP, Kindermann SS, Bell AJ, Sejnowski TJ. Analysis of fMRI data by blind separation into independent spatial components. Hum Brain Mapp. 1998;6:160–88.
Michel CM, Murray MM, Lantz G, Gonzalez S, Spinelli L, Grave de Peralta R. EEG source imaging. Clin Neurophysiol. 2004;115:2195–222.
Mitiche A, Grisell R, Aggarwal JK. On Smoothness of a Vector Field - Application to Optical-Flow. IEEE Trans Pattern Anal Mach Intell. 1988;10:943–9.
Mosher JC, Lewis PS, Leahy RM. Multiple dipole modeling and localization from spatio-temporal MEG data. IEEE Trans Biomed Eng. 1992;39:541–57.
Murzin V, Fuchs A, Kelso JA. Anatomically constrained minimum variance beamforming applied to EEG. Exp Brain Res. 2011;214:515–28.

Nolte G, Curio G. Current multipole expansion to estimate lateral extent of neuronal activity: a theoretical analysis. IEEE Trans Biomed Eng. 2000;47:1347–55.

Oostenveld R, Fries P, Maris E, Schoffelen JM. FieldTrip: Open Source Software for Advanced Analysis of MEG, EEG, and Invasive Electrophysiological Data. Comput Intell Neurosci. 2011;2011:1.

Pascual-Marqui RD. Standardized low-resolution brain electromagnetic tomography (sLORETA): technical details. Methods Find Exp Clin Pharmacol. 2002;24(Suppl D):5–12.

Pascual-Marqui RD, Michel CM, Lehmann D. Low resolution electromagnetic tomography: a new method for localizing electrical activity in the brain. Int J Psychophysiol. 1994;18:49–65.

Phillips C, Mattout J, Rugg MD, Maquet P, Friston KJ. An empirical Bayesian solution to the source reconstruction problem in EEG. NeuroImage. 2005;24:997–1011.

Phillips C, Rugg MD, Fristont KJ. Systematic regularization of linear inverse solutions of the EEG source localization problem. NeuroImage. 2002;17:287–301.

Phillips JW, Leahy RM, Mosher JC. MEG-based imaging of focal neuronal current sources. IEEE Trans Med Imaging. 1997;16:338–48.

Scherg M, Von Cramon D. Evoked dipole source potentials of the human auditory cortex. Electroencephalogr Clin Neurophysiol/Evoked Potentials Section. 1986;65:344–60.

Schmidt DM, George JS, Wood CC. Bayesian inference applied to the electromagnetic inverse problem. Hum Brain Mapp. 1999;7:195–212.

Sekihara K, Nagarajan SS, Poeppel D, Marantz A, Miyashita Y. Reconstructing spatio-temporal activities of neural sources using an MEG vector beamformer technique. IEEE Trans Biomed Eng. 2001;48:760–71.

Smith SM, Fox PT, Miller KL, Glahn DC, Fox PM, Mackay CE, Filippini N, Watkins KE, Toro R, Laird AR. Correspondence of the brain's functional architecture during activation and rest. Proc Natl Acad Sci. 2009;106:13040–5.

Steinsträter O, Sillekens S, Junghoefer M, Burger M, Wolters CH. Sensitivity of beamformer source analysis to deficiencies in forward modeling. Hum Brain Mapp. 2010;31:1907–27.

Tadel F, Baillet S, Mosher JC, Pantazis D, Leahy RM. Brainstorm: a user-friendly application for MEG/EEG analysis. Comput Intell Neurosci. 2011;2011:879716.

Tzourio-Mazoyer N, Landeau B, Papathanassiou D, Crivello F, Etard O, Delcroix N, Mazoyer B, Joliot M. Automated anatomical labeling of activations in SPM using a macroscopic anatomical parcellation of the MNI MRI single-subject brain. NeuroImage. 2002;15:273–89.

Valdes-Hernandez PA, von Ellenrieder N, Ojeda-Gonzalez A, Kochen S, Aleman-Gomez Y, Muravchik C, Valdes-Sosa PA. Approximate average head models for EEG source imaging. J Neurosci Methods. 2009;185:125–32.

van den Broek SP, Reinders F, Donderwinkel M, Peters MJ. Volume conduction effects in EEG and MEG. Electroencephalogr Clin Neurophysiol. 1998;106:522–34.

Van Veen BD, van Drongelen W, Yuchtman M, Suzuki A. Localization of brain electrical activity via linearly constrained minimum variance spatial filtering. IEEE Trans Biomed Eng. 1997;44:867–80.

Xu P, Tian Y, Chen H, Yao D. Lp norm iterative sparse solution for EEG source Localization. IEEE Trans Biomed Eng. 2007;54:400–9.

Yao D. The equivalent source technique and cortical imaging. Electroencephalogr Clin Neurophysiol. 1996;98:478–83.

Yao D. Electric potential produced by a dipole in a homogeneous conducting sphere. IEEE Trans Biomed Eng. 2000;47:964–6.

Yeo BT, Krienen FM, Sepulcre J, Sabuncu MR, Lashkari D, Hollinshead M, Roffman JL, Smoller JW, Zollei L, Polimeni JR, Fischl B, Liu H, Buckner RL. The organization of the human cerebral cortex estimated by intrinsic functional connectivity. J Neurophysiol. 2011;106:1125–65.

Chapter 10
Single-Trial Analysis

Li Hu

Abstract In modern neuroscience, accurate estimations of single-trial parameters in event-related brain responses have been set as a highly desirable goal, since the traditionally used across-trial averaging approach could lead to the loss of the information concerning across-trial variability of both phase-locked ERP and non-phase-locked ERS/ERD responses. In this chapter, we provided the technical details of single-trial analysis methods both in the time domain and the time-frequency domain to enhance the signal-to-noise ratio of event-related brain responses and estimate their single-trial parameters (e.g., latency and amplitude of ERP peaks, as well as latency, frequency, and magnitude of EEG oscillatory features). These methods included probabilistic independent component analysis and common spatial pattern for spatial filtering, continuous wavelet transform for time-frequency filtering, as well as multiple linear regression without/with a dispersion term for feature extraction. Finally, we emphasized the importance of single-trial analysis and discussed its promising applications in basic studies and clinical practice.

Keywords Spatial filtering · Wavelet filtering · Multiple linear regression · Variability · Single trial

10.1 Introduction

Human electroencephalogram (EEG) mostly reflects the synchronous changes of slow postsynaptic potentials occurring inside a variety of similarly oriented pyramidal neurons in the cerebrum (Nunez and Srinivasan 2006). Brisk sensory, motor or psychological events can elicit transient changes within the spontaneous EEG activity. Such changes are traditionally detected as event-related potentials (ERPs)

L. Hu (✉)
CAS Key Laboratory of Mental Health, Institute of Psychology, Chinese Academy of Sciences, Beijing, China

Department of Psychology, University of Chinese Academy of Sciences, Beijing, China
e-mail: huli@psych.ac.cn

L. Hu, Z. Zhang (eds.), *EEG Signal Processing and Feature Extraction*,
https://doi.org/10.1007/978-981-13-9113-2_10

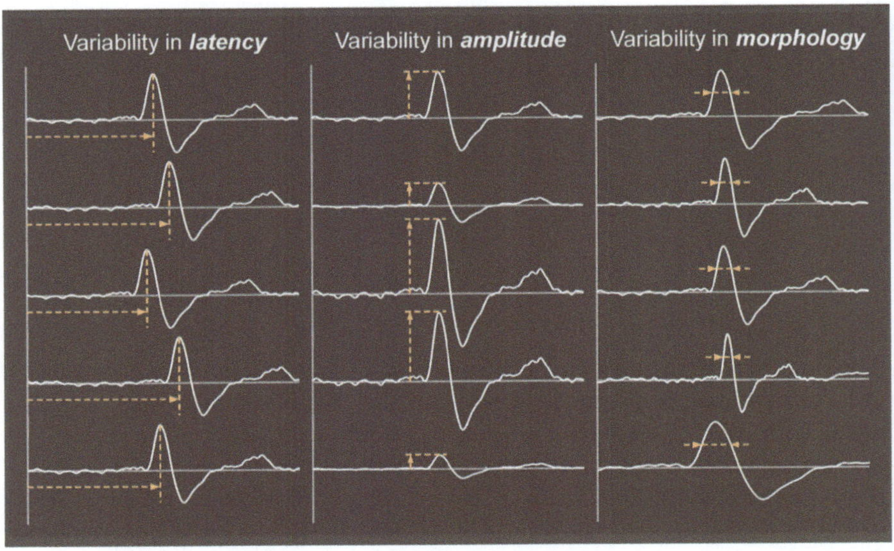

Fig. 10.1 Trial-by-trial variability of ERP latency, amplitude, and morphology

that are time-locked and phase-locked to the onset of stimulus (Pfurtscheller and Lopes da Silva 1999; Mouraux and Iannetti 2008). The same events can also induce non-phase-locked modulations of ongoing EEG oscillations, consisting of transient decreases (event-related desynchronization, ERD) or increases (event-related synchronization, ERS) of oscillatory power, which usually confine to a specific frequency band. The functional significance of ERD and ERS differs tremendously at the frequency bands of their occurrence. However, the magnitude of almost all these brain responses is often much smaller than the magnitude of the background EEG activities. Therefore, the identification and characterization of these brain responses rely on signal processing techniques to improve their signal-to-noise ratio (SNR).

The most widely used approach to detect event-related brain responses both in the time domain (Dawson 1951, 1954) and the time-frequency domain (Pfurtscheller and Lopes da Silva 1999) is the across-trial averaging. Notably, this across-trial averaging approach is based on the assumption that single-trial brain responses are stationary, i.e., the latency, amplitude, and morphology are invariant across different trials (Spencer 2005). However, this assumption is invalid in practice. The phase-locked ERP responses are composed of multiple waves whose latency, amplitude, and morphology could remarkably and independently vary from trial to trial (Spencer 2005; see Fig. 10.1). Likewise, the same situation holds for the non-phase-locked ERS and ERD responses. Consequently, such an across-trial averaging approach could lead to the loss of the information concerning across-trial variability of both phase-locked ERP and non-phase-locked ERS/ERD responses (Mouraux and Iannetti 2008). It should be noted that the across-trial variability usually includes physiologically relevant information, which could reflect the changes of stimulus features (e.g., duration, intensity, and location) (Iannetti et al. 2005b, 2006; Mayhew et al. 2006) and the fluctuations of mental states (e.g., alertness, expectation, and attention) (Legrain et al. 2002, 2003; Lee et al. 2009). Hence, to reliably estimate brain responses at the level of single trial, which include features both in the time and time-frequency domains, is particularly demanding and necessary, as we could explore such physiologically relevant information in a range of follow-up analyses. These analyses include within-subject comparisons between single-trial brain responses and behavioral indicators (e.g., the perceived intensity of sensation, reaction time) (Iannetti et al. 2005b), single-trial correlations with features of pre-stimulus EEG oscillations (Tu et al. 2016), and integration of simultaneously recorded EEG and functional magnetic resonance imaging (fMRI) data (Debener et al. 2006).

In modern neuroscience, accurate estimations of single-trial parameters in event-related brain responses have been set as a highly desirable goal. Advanced single-trial analysis techniques have been proposed to (1) greatly enhance the SNR of single-trial brain responses and (2) accurately estimate the single-trial parameters of brain responses. The use of these advanced techniques could provide new insights into the functional significance of the information processing underlying these brain responses (Mouraux and Iannetti 2008).

10.2 How to Do the Single-Trial Analysis?

How to enhance the SNR of event-related brain responses effectively and reliably is the main challenge in obtaining a reliable estimate of these responses at the single-trial level. Actually, various methods for the single-trial analysis of phase-locked ERPs have been proposed in previous literature (Quiroga 2000; Jung et al. 2001; Quiroga and Garcia 2003; Tang et al. 2005; Barbati et al. 2006, 2008; Mayhew et al. 2006; Tecchio et al. 2007; Porcaro et al. 2008, 2009; Hu et al. 2010; Mayhew et al. 2010; Porcaro et al. 2010; Hu et al. 2011b). Based on their rationales, these methods could be roughly divided into four main categories. The first primary category is the temporal filtering that removes or attenuates frequencies that do not contain the ERP signals. In this category, the simplest and most widely adopted strategy is the bandpass filtering, such as Wiener filtering (Walter 1968; Doyle 1975), time-varying Wiener filtering (de Weerd 1981; de Weerd and Kap 1981; Yu and McGillem 1983), and adaptive filtering (Hu et al. 2005; Lam et al. 2005). These strategies showed a considerable prospect in improving the SNR of ERP responses at the single-trial level. With regard to the second category, spatial filtering based on blind source separation (BSS) methods (e.g., independent component analysis (Ab Aziz and Ahmad 2006) and second-order blind identification (Tang et al. 2005)) has been demonstrated to be effective in isolating stimulus-related responses from background EEG activities (Makeig et al. 1997; Hyvarinen 1999; Bingham and Hyvarinen 2000; Hyvarinen and Oja 2000; Jung et al. 2001; Tang et al. 2005). The third category is the time-frequency filtering that was based, for instance, on a continuous or discrete wavelet transform (Quiroga 2000; Quiroga and Garcia 2003;

Mouraux and Plaghki 2004; Jongsma et al. 2006; Hu et al. 2010, 2011b). The time-frequency filtering can also be used to isolate stimulus-related, phase-locked responses from the background EEG activities and non-cerebral artifacts effectively. The last category of single-trial analysis is aimed at estimating single-trial ERP features. For instance, multiple linear regression that was proposed by Mayhew et al. (Mayhew et al. 2006) was used to estimate ERP latencies and amplitudes. This technique was further refined by including a dispersion term to improve the estimation accuracy of single-trial parameters through increasing the number of regressors to capture more dynamic information in ERPs (Hu et al. 2011a).

Importantly, most of the available single-trial analysis methods were aimed at estimating stimulus-evoked phase-locked responses in the time domain (i.e., ERP responses), but remained entirely blind to estimate stimulus-induced non-phase-locked modulations of ongoing EEG oscillations (i.e., ERD and ERS). Extending these approaches to explore the dynamic information of stimulus-induced non-phase-locked activity at the single-trial level is particularly necessary. Indeed, this analysis involves two crucial steps: feature isolation in the time-frequency domain and parameter estimation at the single-trial level. In a previous study, we showed that different time-frequency features (i.e., ERP, ERD, and ERS) elicited by transient sensory stimuli can be isolated and characterized using principal component analysis (PCA) decomposition with Varimax rotation (Hu et al. 2015). Time-frequency multiple linear regression approaches (TF-MLR without dispersion term and TF-MLRd with dispersion term) were developed to enhance the SNR of ERP/ERD/ERS in single trials, thus providing an unbiased estimation of their latency, frequency, and magnitude at single-trial level.

In the following sections, we provided the technical details of widely used single-trial analysis approaches both in the time domain and the time-frequency domain.

10.2.1 Single-Trial Analysis in the Time Domain

10.2.1.1 Spatial Filtering

Spatial filtering could isolate EEG/ERP components by incorporating information from multiple channels into a weighted combination. Recently, spatial filtering approaches have been extensively investigated, and the most generally adopted strategy is the ICA, which can decompose a multivariate signal into additive sub-components based on the assumption of mutual statistical independence of the non-Gaussian source signals (Makeig et al. 1997; Hyvarinen and Oja 2000). When applied to multichannel EEG or ERP data, ICA could decompose the signal into a series of independent components (ICs) equaling to the total number of these recording electrodes (Makeig et al. 1997; Mouraux and Iannetti 2008), which is referenced to as unconstrained ICA. ICA could be reliably used in isolating stimulus-related components from artifacts (e.g., eye blinks, eye movements, and muscle activity) and the background EEG activity, thus enhancing the SNR of event-related responses in single trials (Jung et al. 2001). Several novel algorithms have also been developed to improve the performance of the spatial filtering, including the probabilistic ICA (PICA) (Beckmann and Smith 2004; Mouraux and Iannetti 2009), the common spatial pattern (CSP) (Koles et al. 1990; Koles 1991), the second-order blind identification (SOBI) algorithm (Tang et al. 2005), the functional source separation (FSS) (Barbati et al. 2006, 2008; Tecchio et al. 2007; Porcaro et al. 2008, 2009, 2010), and the ICA with reference (James and Gibson 2003; Lu and Rajapakse 2006; Huang and Mi 2007). In the following section, we provided the technical details of both PICA and CSP.

Probabilistic Independent Component Analysis (PICA)

The PICA (aims to constrain the total number of estimated ICs to an effective number) could resolve the overestimation problem of the unconstrained ICA in the case that the real number of sources is less than the number of recording electrodes, thus providing a more accurate estimation of ICs (Beckmann and Smith 2004; Mouraux and Iannetti 2009). Theoretically, PICA is an effective combination of probabilistic principal component analysis (PPCA) (Tipping and Bishop 1999a, b; Minka 2001; Beckmann and Smith 2004) and an information maximization algorithm (Infomax ICA) (Bell and Sejnowski 1995; Delorme and Makeig 2004). Specifically, PPCA was applied to decompose an EEG signal into a designated number of principal components (PCs) and a Gaussian noise residual. The subspace of the PCs was then orthogonalized by Infomax ICA (Bell and Sejnowski 1995) to obtain the temporally independent ICs. More detailed processes were showed as follows.

PPCA decomposed the EEG/ERP data into a lower-dimensional source subspace and a noise subspace using the maximum-likelihood density estimation (Tipping and Bishop 1999b). The PPCA model was characterized by defining a high-dimensional vector of observations as a linear combination of low-dimensional PCs and an additive Gaussian noise:

$$\boldsymbol{x} = \boldsymbol{A}\boldsymbol{y} + \boldsymbol{\mu} + \boldsymbol{n} \tag{10.1}$$

where $\boldsymbol{x}$ is a d-dimensional vector of multichannel EEG data, $\boldsymbol{y}$ represents a k-dimensional ($k \leq d$) vector of PCs with distribution $\boldsymbol{y}\tilde{}\mathcal{N}(0, \boldsymbol{I}_k)$, $\boldsymbol{\mu}$ is the mean of $\boldsymbol{x}$, $\boldsymbol{A}$ represents the mixing matrix, and $\boldsymbol{n}$ is the Gaussian noise with distribution $\boldsymbol{n}\tilde{}\mathcal{N}(0, \sigma^2\boldsymbol{I}_d)$. The inverse of the mixing matrix $\boldsymbol{A}$ gives the unmixing matrix $\boldsymbol{W}_p$. Within this model, the distribution of the observed data $\boldsymbol{x}$ is

$$\boldsymbol{x}\tilde{}\mathcal{N}\left(\boldsymbol{\mu}, \boldsymbol{A}\boldsymbol{A}^T + \sigma^2\boldsymbol{I}_d\right). \tag{10.2}$$

Based on the maximum-likelihood density estimation (Tipping and Bishop 1999a, b; Minka 2001), we have

$$\boldsymbol{\mu}_{ML} = \frac{1}{N}\sum\nolimits_{i=1}^{N} \boldsymbol{x}_i \tag{10.3}$$

$$\boldsymbol{S} = \frac{1}{N}\sum\nolimits_{i=1}^{N} (\boldsymbol{x}_i - \boldsymbol{\mu})(\boldsymbol{x}_i - \boldsymbol{\mu})^T \tag{10.4}$$

$$\boldsymbol{A}_{ML} = \boldsymbol{U}_k\left(\boldsymbol{\Lambda}_k - \sigma^2\boldsymbol{I}_k\right)^{1/2}\boldsymbol{R} \tag{10.5}$$

$$\sigma^2_{ML} = \frac{1}{d-k}\sum\nolimits_{j=k+1}^{d} \lambda_j \tag{10.6}$$

where the column orthogonal matrix $\boldsymbol{U}_k$ contains the top k eigenvectors of $\boldsymbol{S}$, $\boldsymbol{\Lambda}_k$ denotes the diagonal matrix containing the top k eigenvalues of $\boldsymbol{S}$, λ_j is the $(j,j)^{th}$ entry of $\boldsymbol{\Lambda}_k$, and $\boldsymbol{R}$ represents an orthogonal matrix, which is generally taken as $\boldsymbol{R} = \boldsymbol{I}$. A more complicated expectation-maximization (EM) algorithm (Roweis 1998) can also be used for an efficient calculation of $\boldsymbol{R}$.

Please note that the number of sources that determine the valid subspace should be provided prior to the PC estimation, and this could be achieved using a Laplace approximation algorithm (Beckmann and Smith 2004). Once the PCs were identified by PPCA, multidimensional orthogonal vectors would be transformed into ICs with maximal statistical independence from each other. The orthogonal procedure could be performed on the obtained PCs using the Infomax ICA, which has been specified in previous literature (Bell and Sejnowski 1995; Delorme and Makeig 2004).

Common Spatial Pattern (CSP)

Different from the blind source separation algorithms, CSP is a mathematical tool to decompose two populations of multivariate signals into a set of spatial patterns, which maximize their differences in terms of variance (Muller-Gerking et al. 1999; Ramoser et al. 2000). This approach was first used to detect abnormalities during intraoperative neurophysiological monitoring in EEG analysis (Koles et al. 1990; Koles 1991) and has been proved to be a powerful strategy to discriminate different mental intentions in the research field of brain-computer interface (Blankertz et al. 2008). Additionally, this approach could enhance the SNR of single-trial EEG/ERP responses for the prediction of pain perception (Huang et al. 2013).

When adopting the CSP to the single-trial analysis, one population consisted of EEG activity before the stimuli (i.e., pre-stimulus EEG activity), and the other population consisted of EEG activity after the stimuli of the same trial (i.e., post-stimulus EEG activity). The pre- and post-stimulus EEG activities over all channels of the same trial could form two matrices $\boldsymbol{X}_{pre}, \boldsymbol{X}_{post} \in R^{N \times T}$, where N is the number of channels and T is the number of samples in each trial. To maximize the difference between the variability of $\boldsymbol{X}_{pre}$ and $\boldsymbol{X}_{post}$, CSP aims to estimate the generalized eigenvector or the projection vector **w** by solving the generalized eigenvalue problem as follows:

$$X_{post}X_{post}^{T}w = \lambda X_{pre}X_{pre}^{T}w \tag{10.7}$$

where $\langle\cdot\rangle$ is the averaging operator across trials for the same population and λ is the generalized eigenvalue. The matrix $W = [w_1, \cdots, w_N] \in R^{N \times N}$ is the spatial filter, where $w_1, \cdots, w_N$ are N eigenvectors estimated from Eq. (10.7), and $A = W^{-1} \in R^{N \times N}$ is the spatial pattern represented as a weighting of EEG channels. It should be noted that CSP provided an ordered list of spatial patterns according to the discriminative power between two populations. As a result, the spatial pattern with the maximal variance of X_{post} will capture the minimal variance of X_{pre}, and vice versa. Typically, only a few spatial patterns were sufficient to discriminate two populations (Muller-Gerking et al. 1999). These few spatial patterns, which could isolate the stimulus-evoked EEG responses (contained only in X_{post}) from the spontaneous EEG activity (contained in X_{post} as well as X_{pre}), worked as an effective spatial filter. In practice, a limited number of eigenvectors corresponding to the largest eigenvalues of X_{post} and lowest eigenvalues of X_{pre} could be selected to reconstruct the spatial-filtered single-trial EEG responses.

10.2.1.2 Time-Frequency Filtering

Time-frequency filtering (Quiroga 2000; Quian Quiroga and Garcia 2003; Mouraux and Plaghki 2004; Jongsma et al. 2006; Wang et al. 2007), which based on the time-frequency decomposition methods, such as windowed Fourier transform or wavelet transform, can isolate stimulus-induced and phase-locked brain responses from the background EEG activities and noise-related artifacts, thus enhancing the SNR of single-trial ERP waveforms. For wavelet transform, the discrete wavelet transform (DWT) and the continuous wavelet transform (CWT) are the commonly used approaches. The DWT-based time-frequency filtering could retain only wavelet coefficients correlating with the ERP signal and thereby generate a denoised average ERP waveform (Quiroga 2000; Quiroga and Garcia 2003; Jongsma et al. 2006). Different from the DWT that is only operated at specific scales and transitions (generally, at dyadic scales and transitions), we have developed a CWT-based time-frequency filtering (Hu et al. 2010), which can be performed at every possible time and frequency scale, thus providing a better performance in enhancing the single-trial SNR of stimulus-evoked brain responses (Tognola et al. 1998; Mouraux and Iannetti 2008).

Specifically, the CWT-based time-frequency filtering was achieved in the following three steps (see Fig. 10.2). First, single-trial EEG/ERP waveforms were decomposed into time-frequency representations using the CWT. Second, specific areas on the time-frequency plane corresponding to the ERP response were characterized and used as the wavelet filtering model. Third, time-domain ERP waveforms were reconstructed from the time-frequency domain to the time domain using an inverse continuous wavelet transform (ICWT) (Hu et al. 2010).

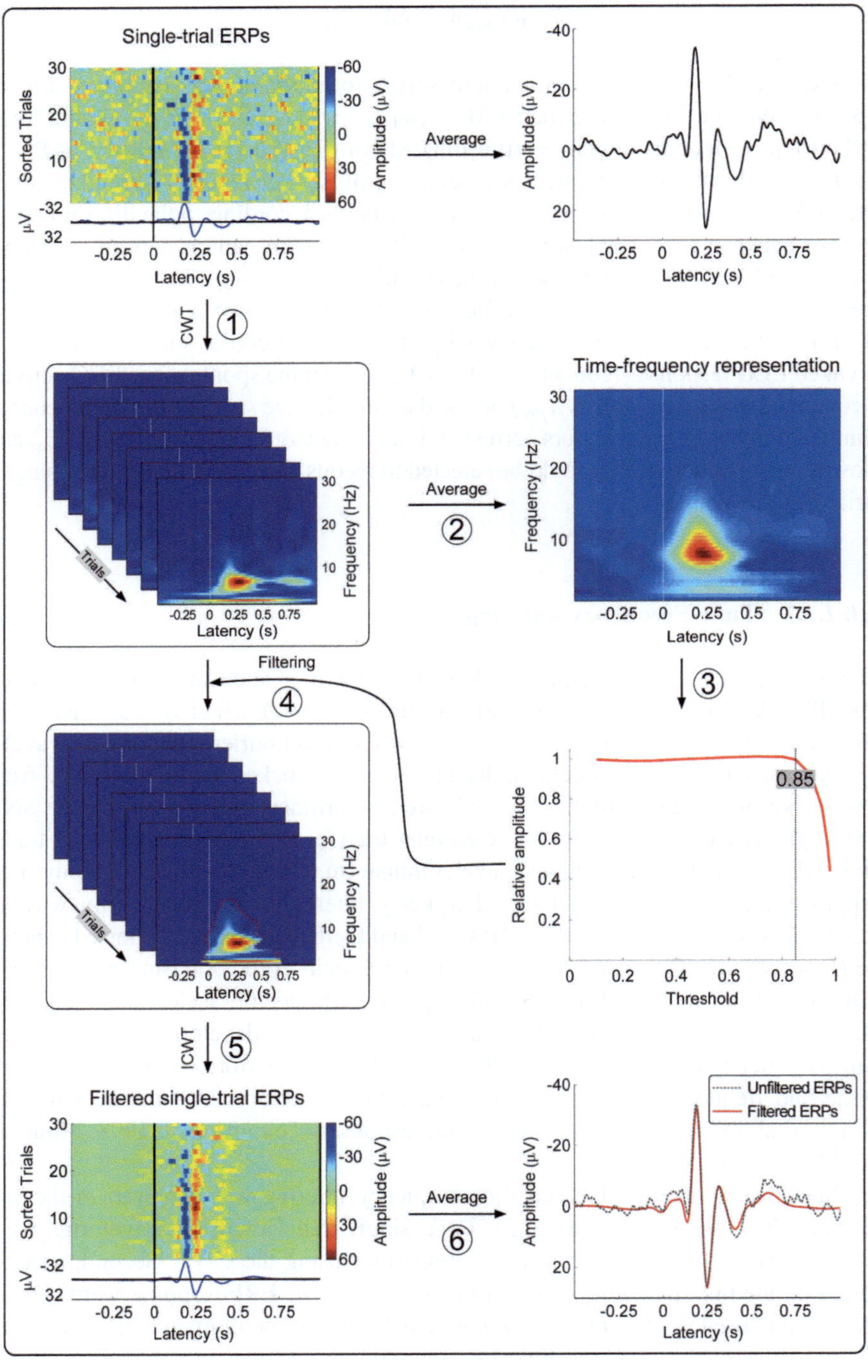

Fig. 10.2 The flowchart describing the procedures of time-frequency filtering based on continuous wavelet transform. (Reproduced from Hu et al. 2011a)

Continuous Wavelet Transform

The CWT is used to estimate the time-frequency representation of EEG or ERP signals that could offer an optimal compromise of time and frequency resolutions by adapting the window width as a function of the estimated frequency (Mouraux et al. 2003; Iannetti et al. 2008; Mouraux and Iannetti 2008). The CWT is defined as follows (Tognola et al. 1998):

$$WT(\tau, f) = \int_t x(t) \cdot \sqrt{f/f_0} \cdot \psi^*(f/f_0 \cdot (t - \tau))dt \tag{10.8}$$

$$\psi(t) = \frac{1}{\sqrt{\pi f_b}} e^{2i\pi f_o x} e^{-\frac{x^2}{f_b}} \tag{10.9}$$

where τ and f are the time and frequency indices, respectively, and $x(t)$ represents the original signal in the time (t) domain; $\psi(t)$ denotes the mother wavelet function with the central frequency f_0. The mother wavelet $\psi(t)$ is a complex Morlet wavelet (Eq. 10.9). Bandwidth parameters f_b and f_0 could be set to 0.05 and 6, respectively, which displays a good time-frequency resolution when detecting frequencies ranging from 1 to 30 Hz. The squared magnitude of $WT(\tau,f)$ is the scalogram or power spectrum.

Wavelet Filtering Model

In order to apply time-frequency filtering to enhance the SNR of ERP responses at single-trial level, a weighted, binary time-frequency template W_f was generated to identify the distribution of ERP responses evoked by the stimuli in the time-frequency plane. This template was used to filter out the contribution of non-stimulus-related background EEG activity and noise-related artifacts, thus enhancing the SNR of single-trial ERP responses.

For each single-trial EEG activity, time-frequency filtering was achieved by Eq. (10.10):

$$FWT_i(\tau, f) = W_f \cdot WT_i(\tau, f) \tag{10.10}$$

where WT_i is the time-frequency representation of trial i obtained by CWT, and FWT_i is the filtered time-frequency representation that is calculated by multiplying WT_i with the binary time-frequency template W_f.

Inverse Continuous Wavelet Transforms (ICWT)

The time-frequency filtered signal $y_i(t)$ was reconstructed into the time domain by applying the ICWT to the filtered time-frequency distribution $FWT_i(\tau,f)$ according to the following equation (Tognola et al. 1998):

$$y_i(t) = C_\psi \int_\tau \int_f FWT_i(\tau,f) \cdot \sqrt{f/f_0} \cdot \psi(f/f_0(t-\tau)) \cdot (f/f_0)^2 \cdot d\tau \cdot df \quad (10.11)$$

where C_ψ denotes a coefficient that depends on the Fourier transform of $\psi(t)$.

10.2.1.3 Parametric Approaches

Multiple Linear Regression (MLR)

The single-trial detection method based on multiple linear regression (MLR) was firstly proposed by Mayhew et al. (2006) to estimate latency and amplitude of single-trial ERPs automatically (Mayhew et al. 2006). This method has been successfully applied to detect single-trial parameters of the N2 and P2 waves of the LEP responses (Mayhew et al. 2006) and the N1 and P2 waves of the AEP responses (Mayhew et al. 2010). Recently, we have combined the wavelet filtering and MLR to enhance the SNR of single-trial ERP responses and estimate more accurately the single-trial parameters of even smaller ERP components (i.e., the N1 wave of the LEP responses) (Hu et al. 2010). Such a MLR approach is a procedure commonly used to analyze fMRI data (Friston et al. 1998), where not only the canonical hemodynamic response function (analogous to the average ERP in this case) but also its temporal derivative (to account for the temporal variability of the hemodynamic response) is fitted to the data in a general linear model (GLM) framework. Thus, the inclusion of the temporal derivative in the regressors allows capturing not only the amplitude variability but also the latency jitter of the single-trial ERP responses.

Taking N2 and P2 waves of the LEP responses as an example, the variability of latency and amplitude of both waves can be described as follows:

$$f(t) = k_N y_N(t + a_N) + k_P y_P(t + a_P) \quad (10.12)$$

where $f(t)$ denotes a single-trial ERP waveform that varies as a function of time. $f(t)$ can be modeled by the sum of the varied version of the N2 wave ($k_N y_N(t + a_N)$ and P2 wave ($k_P y_P(t + a_P)$. k_N and k_P are the weighted constants of the N2 wave and P2 wave, while a_N and a_P represent the latency jitter of the N2 wave and P2 wave, respectively. Please note that the N2 and P2 waves were modeled separately, since the N2 and P2 peaks of the LEP responses reflect the activities of different neural generators (Garcia-Larrea et al. 2003), and their amplitudes can be differentially modulated by several experimental factors (e.g., spatial attention and probability of perception) (Legrain et al. 2002; Lee et al. 2009). Using the Taylor expansion, the MLR model can be detailed as

$$f(t) \approx k_N y_N(t) + a_N k_N y'_N(t) + k_P y_P(t) + a_P k_P y'_P(t) \quad (10.13)$$

where $y_N(t)$ and $y_P(t)$ represent the averages of N2 and P2 waves. $y'_N(t)$ and $y'_P(t)$ denote the temporal derivatives of N2 and P2 waves, respectively. Briefly, the single-trial ERP waveform is approximated as the sum of the weighted averages of the N2 and P2 waves and their respective temporal derivatives.

Based on the fitted waveform, latency and amplitude of the ERP response in each single trial were estimated by calculating: for the N2 wave, the most negative peak if $k_N > 0$ (positive fit) or the most positive peak if $k_N < 0$ (negative fit), and for the P2 wave, the most positive peak if $k_P > 0$ (positive fit) or the most negative peak if $k_P < 0$ (negative fit), within a predefined time window centered at the N2 and P2 latencies in the average ERP waveform for each subject. Single-trial latencies were estimated from the latencies of the corresponding amplitudes.

Multiple Linear Regression with Dispersion Term (MLRd)

In addition to the trial-by-trial variability of ERP latency and amplitude, the morphology of the ERP waveform is not constant across trials (Jung et al. 2001; Casarotto et al. 2005; Mouraux and Iannetti 2008). The trial-by-trial variability of ERP morphology could be clearly identified in clinical ERP data, as many abnormal conditions (e.g., optic neuritis in multiple sclerosis) are characterized by the so-called "desynchronized" ERPs. In these cases, ERP waveforms capture reduced amplitude, longer latency, and, importantly, increased width of their waves (Pelosi et al. 1997; Orssaud 2003). When using ERPs as a diagnostic tool in clinical practice, these variations in ERP morphology (i.e., wave width) could be important parameters to be considered and quantified. In addition, to obtain a more accurate estimation of single-trial ERPs, it would be desirable to take into account not only the variability of ERP latency and amplitude (Mayhew et al. 2006; Hu et al. 2010) but also the variability of ERP morphology (Hu et al. 2011a).

To achieve this objective, we included a dispersion term (representing the variability of ERP morphology) in the MLR model, thus obtaining a dispersed version of MLR (MLRd):

$$f(t) = k_N y_N(s_N t + a_N) + k_P y_P(s_P t + a_P) \tag{10.14}$$

where s_N and s_P denote the time dispersion coefficients that determine the compression ratios of the width of N2 and P2 waves in the single-trial ERPs compared to those in the average ERPs, respectively.

However, it is difficult to estimate these single-trial parameters (i.e., s, a, and k), as the real function expressing ERP waveforms is unknown. For this reason, we adopted a nonparametric data-driven approach, named PCA, to define a basic set of regressors to fit the single-trial ERP waves. This method is similar to the procedure that has been used to model the hemodynamic response in some BOLD-fMRI studies (Friman et al. 2003; Hossein-Zadeh et al. 2003; Woolrich et al. 2004). Please note that PCA can transform the ERP data into several uncorrelated principal components (PCs) (Jolliffe 2002). When performing PCA on the variability matrix (see Hu et al. 2011a for the generation of the variability matrix), three main PCs for each ERP wave representing (1) the average ERP wave across trials, (2) the variability of ERP latency, and (3) the variability of ERP morphology, respectively, could be identified to model the single-trial N2 and P2 waves. To include the variability of ERP latency and morphology of each wave as regressors in the linear model, the MLRd procedure is mainly composed of four steps (Fig. 10.3, bottom panel), and the details of these steps have been specified in our previous publication (Hu et al. 2011a).

Please note that the decision of selecting the first three PCs of the variability matrices was based on the fact that the first three PCs were capable of explaining most of the variability of single-trial ERPs, namely, their latency, amplitude, and morphology (Spencer 2005). These first three PCs were compared with the three Taylor expansion series (i.e., a basic Gamma function, a temporal derivative of the Gamma function, and a morphological derivative of the Gamma function) and were demonstrated to be quite similar to the Taylor expansion series in previous literature (see Fig. 11 in (Hossein-Zadeh et al. 2003)). The inclusion of a dispersion term in the multiple linear regression is crucial to capture the physiological information reflected in the variability of ERP morphology. Importantly, we observed that including a regressor to capture the variability in ERP morphology could significantly improve the accuracy of single-trial parameter estimation when the noise level of the single-trial EEG activity is low (Hu et al. 2011a). This observation is valid since introducing the variability of ERP morphology as a regressor to model the noisy data might increase the possibility of fitting noise. Indeed, the performance of MLR and MLRd in extracting single-trial amplitudes was comparable when modeling the data with a high level of noise. Therefore, the use of MLRd is beneficial either (1) when the noise level of the raw data is low or (2) when the noise level of the data has been effectively reduced by additionally filtering the data (e.g., the wavelet filtering) (Hu et al. 2011a).

10.2.2 Single-Trial Analysis in the Time-Frequency Domain

As introduced previously, the techniques to perform single-trial analysis in the time domain can only be able to detect the phase-locked information (i.e., ERP responses) and be entirely blind to detect the non-phase-locked information (i.e., modulations of ongoing EEG oscillations, ERD and ERS). Therefore, advanced signal processing techniques are highly needed to explore the dynamic information of non-phase-locked information at the level of single trials. Different from the time-domain analysis, single-trial analysis in the time-frequency domain involves two important

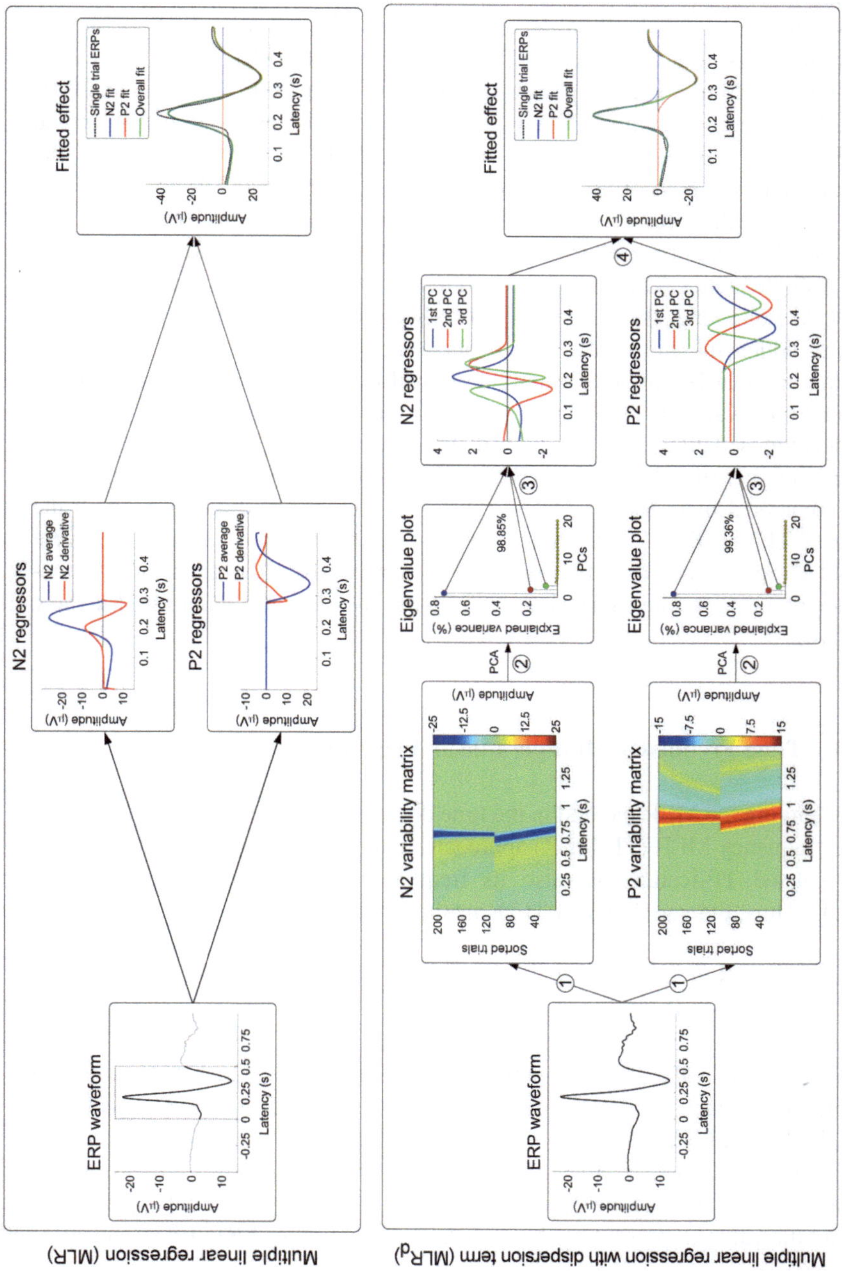

Fig. 10.3 Flowcharts describing the procedures of multiple linear regression (MLR, top) and multiple linear regression with dispersion term (MLRd, bottom). (Reproduced from Hu et al. 2011a)

steps: feature separation in the time-frequency domain and parameter estimation at the single-trial level (Hu et al. 2015).

As successfully implemented in several previous studies (Bernat et al. 2005, 2007; Mayhew et al. 2010), PCA decomposition with Varimax rotation is effective in separating physiologically distinct TF-features (i.e., ERP, ERD, and ERS) within the TFDs of single-trial EEG responses (Kayser and Tenke 2003; Bernat et al. 2005; Dien 2009). The procedure of PCA with Varimax rotation consists of five crucial steps (Hu et al. 2015): (1) data concentration, (2) PCA decomposition of the covariance matrix, (3) Varimax rotation, (4) rearrangement of PC vectors to TFDs, and (5) TFD thresholding. Firstly, the TFD of every single trial was rearranged as a vector, and all vectors from all single trials of all subjects were stacked sequentially to form a single matrix. Then, the generated matrix was decomposed into a set of PC using PCA. Following, the decomposed PCs were further rotated by the Varimax algorithm, which maximizes the sum of the variances of the squared loadings to ensure that the matrix can be optimally described by a linear combination of a few basic functions (Kaiser 1979; Richman 1986; Kayser and Tenke 2003). Fourthly, the three decomposed PCs that explained the maximal variance in the matrix (representing the stimulus-elicited ERP, ERD, and ERS) were rearranged into three-dimensional matrices (i.e., having the same number of dimensions with the single-trial TFDs). Finally, the time-frequency representations of each PC were thresholded using a cutoff at two standard deviations from the mean of all time-frequency points to isolate signal changes from the background EEG noise (Mayhew et al. 2010).

10.2.2.1 Time-Frequency Multiple Linear Regression (TF-MLR)

When extending the MLR method in the time domain to the time-frequency domain, the time-frequency MLR (TF-MLR) approach should take not only the latency jitter of the isolated TF-feature but also its frequency variability into account (see Fig. 10.4). Thus, the variability of TFDs in single trials can be modeled as follows:

$$\begin{aligned} \mathrm{F}(t,f) = & k_1\mathrm{F}_1[(t+a_1),(f+b_1)] + k_2\mathrm{F}_2[(t+a_2),(f+b_2)] \\ & + k_3\mathrm{F}_3[(t+a_3),(f+b_3)] + \varepsilon \end{aligned} \tag{10.15}$$

where $\mathrm{F}(t,f)$ denotes the TFD of a single-trial EEG response, which represents as a joint function of time t and frequency f. $\mathrm{F}_1(t,f)$, $\mathrm{F}_2(t,f)$, and $\mathrm{F}_3(t,f)$ are the averages of ERP, ERD, and ERS, respectively. $\mathrm{F}(t,f)$ can be modeled as the weighted sum of the ERP, ERD, and ERS, and the background noise ε, k_1, k_2, and k_3 are the weighted constants; a_1, a_2, and a_3 are the values representing the latency variability; and b_1, b_2, and b_3 are the values representing the frequency variability of ERP, ERD, and ERS, respectively.

Using the Taylor expansion, the TF-MLR model can be detailed as

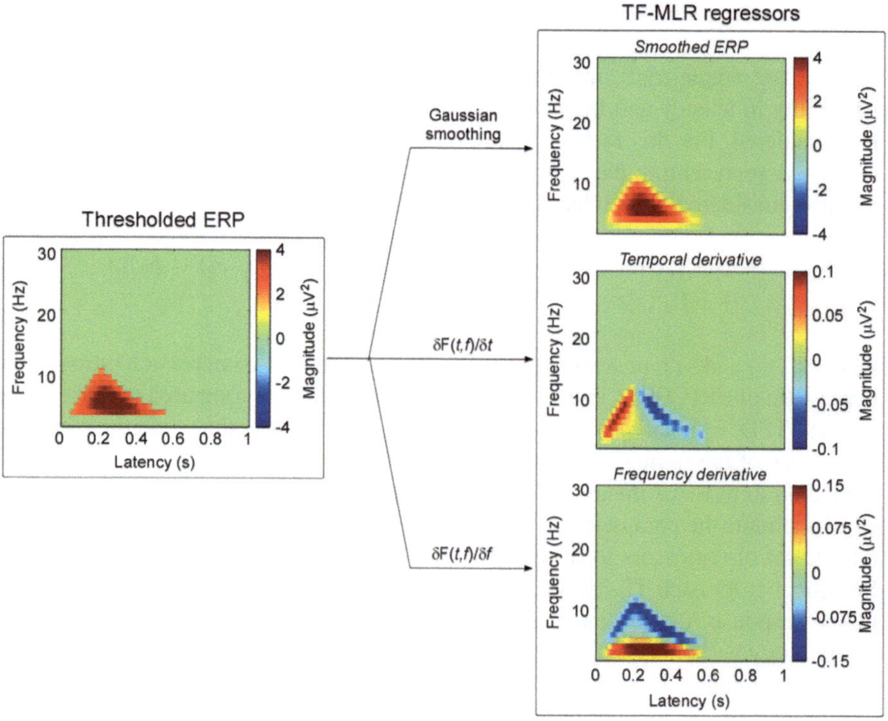

Fig. 10.4 The flowchart describing the procedure of time-frequency multiple linear regression (TF-MLR). (Reproduced from Hu et al. 2015)

$$\begin{aligned} \mathrm{F}(t,f) \approx\ & k_1\mathrm{F}_1(t,f) + k_1a_1\frac{\partial \mathrm{F}_1(t,f)}{\partial t} + k_1b_1\frac{\partial \mathrm{F}_1(t,f)}{\partial f} \\ & + k_2\mathrm{F}_2(t,f) + k_2a_2\frac{\partial \mathrm{F}_2(t,f)}{\partial t} + k_2b_2\frac{\partial \mathrm{F}_2(t,f)}{\partial f} \\ & + k_3\mathrm{F}_3(t,f) + k_3a_3\frac{\partial \mathrm{F}_3(t,f)}{\partial t} + k_3b_3\frac{\partial \mathrm{F}_3(t,f)}{\partial f} \end{aligned} \tag{10.16}$$

where $\frac{\partial \mathrm{F}_1(t,f)}{\partial t}$, $\frac{\partial \mathrm{F}_2(t,f)}{\partial t}$, and $\frac{\partial \mathrm{F}_3(t,f)}{\partial t}$ represent the temporal derivatives of ERP, ERD, and ERS; $\frac{\partial \mathrm{F}_1(t,f)}{\partial f}$, $\frac{\partial \mathrm{F}_2(t,f)}{\partial f}$, and $\frac{\partial \mathrm{F}_3(t,f)}{\partial f}$ denote the frequency derivatives of ERP, ERD, and ERS, respectively. A single-trial TFD can be modeled as the sum of a set of weighted bases (average, its temporal derivative and its frequency derivative). Single-trial parameters (latency, frequency, and magnitude) of each TF-feature were calculated based on the fitted single-trial TFD (Hu et al. 2015).

10.2.2.2 TF-MLR with Dispersion Term (TF-MLRd)

To estimate single-trial time-frequency responses more accurately, not only their variability in latency and frequency but also their variability in morphology should be considered. For this reason, in addition to the basic set of TF-MLR, two more regressors representing the scaling of the single-trial response in time or frequency are also considered, which leads to the TF-MLRd model (Fig. 10.5):

$$\begin{aligned}\mathrm{F}(t,f) = {} & k_1\mathrm{F}_1[(s_1t + a_1), (c_1f + b_1)] + k_2\mathrm{F}_2[(s_2t + a_2), (c_2f + b_2)] \\ & + k_3\mathrm{F}_3[(s_3t + a_3), (c_3f + b_3)] + \varepsilon\end{aligned} \tag{10.17}$$

where s_1, s_2, and s_3 represent the coefficients that code the compression ratios of the time width of ERP, ERD, and ERS of each single-trial TFD compared to those of the average TFD, respectively. c_1, c_2, and c_3 are the coefficients that code the compression ratios of the frequency width of ERP, ERD, and ERS of each single-trial TFD compared to those of the average TFD, respectively.

To estimate the parameters *k*, *s*, *c*, *a*, and *b* of each single-trial TFD, we generated a basic set of regressors to fit the ERP, ERD, and ERS using PCA (Jolliffe 2002). Specifically, for each TF-feature, we generated five PCs representing (1) the average of the response, (2) the latency variability, (3) the frequency variability, (4) the morphology variability in the time domain, and (5) the morphology variability in the frequency domain. The whole procedure for generating the TF-MLRd regressors is detailed in Fig. 10.5 and in our previous publication (Hu et al. 2015).

Whereas the TF-MLR approach that uses fewer regressors is more specific when detecting the stimulus-related responses (Friman et al. 2003), the TF-MLRd approach is able to capture the variability of the response morphology both in the time domain and the frequency domain. In other words, the TF-MLR provides a simple and robust approach to estimate single-trial parameters and is more appropriate when the SNR of EEG responses is relatively low. In contrast, the TF-MLRd approach is more sensitive in detecting response variability (Hu et al. 2011a) with the possibility of fitting some noise (Friman et al. 2003). For this reason, the TF-MLRd approach is more suited to accurately estimate single-trial responses when the SNR of EEG responses is relatively high (e.g., intracranial recordings, interictal spikes in epilepsy patients).

In addition, to improve the performance of single-trial estimation of TF-features, the SNR of the responses could be enhanced beforehand using other techniques (e.g., the use of spatial filtering or the use of tensor decomposition that could span the EEG

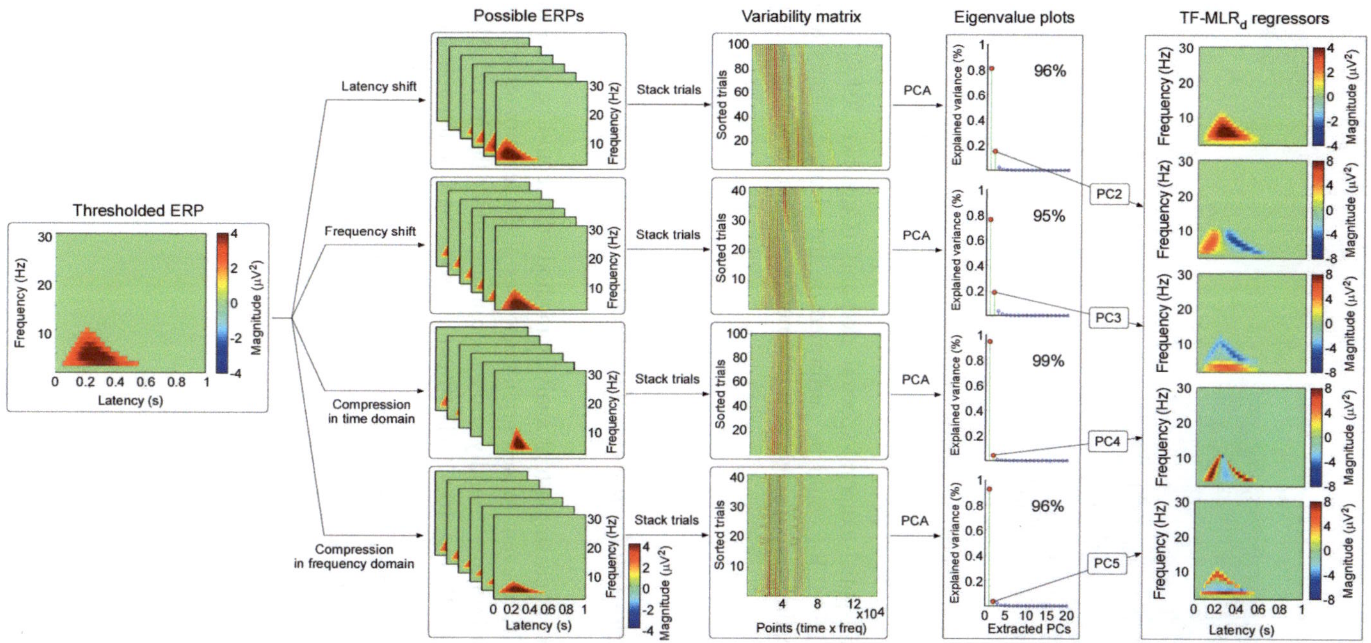

Fig. 10.5 The flowchart describing the procedure of time-frequency multiple linear regression with dispersion term (TF-MLRd). (Reproduced from Hu et al. 2015)

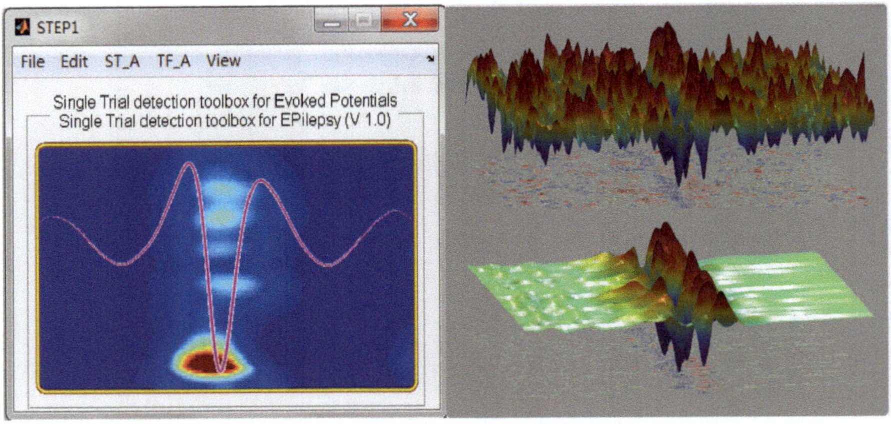

Fig. 10.6 Single-trial analysis toolbox (*left*) and a representative performance of single-trial analysis to enhance the signal-to-noise ratio. (*right*: top, raw EEG data; bottom, filtered EEG data)

signal into the spatial-temporal-spectral domain) (Cichocki 2013). Afterward, TF-MLR and TF-MLRd could be performed on the cleaned signals to obtain a more accurate estimation of single-trial parameters of brain responses at the level of single trials.

10.2.3 Single-Trial Analysis Toolbox

Most of the discussed single-trial analysis methods that could be used to greatly enhance the SNR of event-related brain responses at the level of single trials have been developed into user-friendly software running under the MATLAB environment, named STEP1 (Fig. 10.6). STEP1 could also be used to automatically extract single-trial parameters (e.g., latency and amplitude of ERP waves) of different features in event-related brain responses. The toolbox can be freely downloaded from the following websites: http://www.hulilab.com and http://www.iannettilab.net.

10.3 What're the Possible Applications of Single-Trial Analysis?

Single-trial analysis would help us explore the single-trial dynamics of stimulus-elicited brain responses, thus providing a rich source of dynamic information to be applied in a series of basic and clinical researches.

First, the single-trial dynamics of both phase-locked and non-phase-locked information would enable us to perform within-subject correlation analysis with, for example, stimulus parameters and behavioral variables, thus providing the possibility to improve our understanding of the functional significance of the brain responses (Mouraux et al. 2003; Mouraux and Plaghki 2004; Iannetti et al. 2008). This is particularly relevant in the field of cognitive neuroscience, since event-related brain responses exhibit high across-trial variability (Iannetti et al. 2005a; Purves and Boyd 1993). Please note that the across-trial variability is determined by both peripheral (e.g., stimulus intensity and duration) and central factors (e.g., fluctuations in vigilance, attentional focus, and task strategy) (Legrain et al. 2002, 2003; Lee et al. 2009). Importantly, since the within-subject correlation analysis would not be influenced by the between-subject variability (e.g., age, gender, and weight), the exploration of single-trial dynamics of brain responses would help increase the power of statistical analysis, thus providing new insights into brain functions.

Second, when the single-trial analysis was applied in simultaneous EEG and fMRI, the single-trial dynamics of both phase-locked and non-phase-locked information would enable us to better exploit the neurovascular coupling between EEG and fMRI signals. This analytic strategy would help us disclose brain functions with an optimal combination of temporal and spatial resolutions for two reasons. Firstly, EEG data collected simultaneously with fMRI are contaminated with substantial artifacts in the fMRI environment (Allen et al. 1998). For this reason, several algorithms have been developed to remove (or attenuate) these artifacts (Allen et al. 2000; Debener et al. 2007). However, event-related brain responses both in the time domain and the time-frequency domain would still suffer from reduced SNR (Debener et al. 2005; Iannetti et al. 2005a; Niazy et al. 2005). Therefore, single-trial analysis is highly needed to enhance the SNR of event-related brain responses at the level of single trials. Secondly, the single-trial parameters of event-related brain responses can be used to build a function of the predicted hemodynamic response, which is included as an additional regressor in the general linear model analysis (Iannetti et al. 2005a; Mayhew et al. 2010). In this case, voxels whose BOLD signal correlates in the time course with the across-trial variability of event-related brain responses would be identified, and the obtained results would reflect physiological information about the dynamic cortical processing (Bagshaw and Warbrick 2007; Goldman et al. 2009).

Third, the single-trial dynamics of both phase-locked responses and non-phase-locked responses would provide useful information, which could be used in machine learning to predict, for example, the subjective perception of pain (Huang et al. 2013). The strategy that combines single-trial analysis and machine learning techniques (e.g., binary classification and continuous regression) could be applied to predict subjective experience, thus satisfying the most requirement of various basic and clinical applications (Huang et al. 2013). In the field of pain neuroscience, this technique to decode subjective pain experience is profoundly important, as it could help optimize the diagnosis, monitoring, and treatment of pain for noncommunicative patients and patients with disorders of consciousness (Schnakers et al. 2010; Schulz et al. 2012).

Finally, single-trial analysis to explore dynamics of both phase-locked responses and non-phase-locked responses could also be used in some crucial clinical applications. For example, it is widely accepted that abnormalities of somatosensory-evoked potentials (SEPs) provide direct evidence for the functional impairment of the ascending somatosensory pathway (Cruccu et al. 2008). For this reason, SEPs have been widely used both in clinical diagnosis (Aminoff et al. 1988; Zeman and Yiannikas 1989; Yiannikas and Vucic 2008) and in intraoperative neurophysiological monitoring (Nuwer 1998; Luk et al. 2001; Deletis and Shils 2002; Minahan 2002; Hu et al. 2003; Devlin et al. 2006). The availability of single-trial analysis could provide dynamic information for the intraoperative monitoring to detect the temporary malfunctioning at an early stage, which could help prevent the possible irreversible spinal cord damage (Wiedemayer et al. 2002; Rossi et al. 2007).

References

Ab Aziz CB, Ahmad AH. The role of the thalamus in modulating pain. Malays J Med Sci. 2006;13 (2):11–8.

Allen PJ, Polizzi G, Krakow K, Fish DR, Lemieux L. Identification of EEG events in the MR scanner: the problem of pulse artifact and a method for its subtraction. NeuroImage. 1998;8 (3):229–39.

Allen PJ, Josephs O, Turner R. A method for removing imaging artifact from continuous EEG recorded during functional MRI. NeuroImage. 2000;12(2):230–9.

Aminoff MJ, Olney RK, Parry GJ, Raskin NH. Relative utility of different electrophysiologic techniques in the evaluation of brachial plexopathies. Neurology. 1988;38(4):546–50.

Bagshaw AP, Warbrick T. Single trial variability of EEG and fMRI responses to visual stimuli. NeuroImage. 2007;38(2):280–92.

Barbati G, Sigismondi R, Zappasodi F, Porcaro C, Graziadio S, Valente G, et al. Functional source separation from magnetoencephalographic signals. Hum Brain Mapp. 2006;27(12):925–34.

Barbati G, Porcaro C, Hadjipapas A, Adjamian P, Pizzella V, Romani GL, et al. Functional source separation applied to induced visual gamma activity. Hum Brain Mapp. 2008;29(2):131–41.

Beckmann CF, Smith SM. Probabilistic independent component analysis for functional magnetic resonance imaging. IEEE Trans Med Imaging. 2004;23(2):137–52.

Bell AJ, Sejnowski TJ. An information-maximization approach to blind separation and blind deconvolution. Neural Comput. 1995;7(6):1129–59.

Bernat EM, Williams WJ, Gehring WJ. Decomposing ERP time-frequency energy using PCA. Clin Neurophysiol. 2005;116(6):1314–34.

Bernat EM, Malone SM, Williams WJ, Patrick CJ, Iacono WG. Decomposing delta, theta, and alpha time-frequency ERP activity from a visual oddball task using PCA. Int J Psychophysiol. 2007;64(1):62–74.

Bingham E, Hyvarinen A. A fast fixed-point algorithm for independent component analysis of complex valued signals. Int J Neural Syst. 2000;10(1):1–8.

Blankertz B, Losch F, Krauledat M, Dornhege G, Curio G, Muller KR. The Berlin Brain--Computer Interface: accurate performance from first-session in BCI-naive subjects. IEEE Trans Biomed Eng. 2008;55(10):2452–62.

Casarotto S, Bianchi AM, Cerutti S, Chiarenza GA. Dynamic time warping in the analysis of event-related potentials. IEEE Eng Med Biol Mag. 2005;24(1):68–77.

Cichocki A. Tensor decompositions: a new concept in brain data analysis? CoRR. 2013:abs/1305.0395.

Cruccu G, Aminoff MJ, Curio G, Guerit JM, Kakigi R, Mauguiere F, et al. Recommendations for the clinical use of somatosensory-evoked potentials. Clin Neurophysiol. 2008;119(8):1705–19.

Dawson GD. A summation technique for detecting small signals in a large irregular background. J Physiol. 1951;115(1):2p–3p.

Dawson GD. A summation technique for the detection of small evoked potentials. Electroencephalogr Clin Neurophysiol. 1954;6(1):65–84.

de Weerd JP. A posteriori time-varying filtering of averaged evoked potentials. I Introduction and conceptual basis. Biol Cybern. 1981;41(3):211–22.

de Weerd JP, Kap JI. A posteriori time-varying filtering of averaged evoked potentials. II Mathematical and computational aspects. Biol Cybern. 1981;41(3):223–34.

Debener S, Ullsperger M, Siegel M, Fiehler K, von Cramon DY, Engel AK. Trial-by-trial coupling of concurrent electroencephalogram and functional magnetic resonance imaging identifies the dynamics of performance monitoring. J Neurosci. 2005;25(50):11730–7.

Debener S, Ullsperger M, Siegel M, Engel AK. Single-trial EEG-fMRI reveals the dynamics of cognitive function. Trends Cogn Sci. 2006;10(12):558–63.

Debener S, Strobel A, Sorger B, Peters J, Kranczioch C, Engel AK, et al. Improved quality of auditory event-related potentials recorded simultaneously with 3-T fMRI: removal of the ballistocardiogram artefact. NeuroImage. 2007;34(2):587–97.

Deletis V, Shils JL. Neurophysiology in neurosurgery: a modern intraoperative approach. San Diego: Academic; 2002.

Delorme A, Makeig S. EEGLAB: an open source toolbox for analysis of single-trial EEG dynamics including independent component analysis. J Neurosci Methods. 2004;134(1):9–21.

Devlin VJ, Anderson PA, Schwartz DM, Vaughan R. Intraoperative neurophysiologic monitoring: focus on cervical myelopathy and related issues. Spine J. 2006;6(6 Suppl):212S–24S.

Dien J. Evaluating two-step PCA of ERP data with Geomin, Infomax, Oblimin, Promax, and Varimax rotations. Psychophysiology. 2009;

Doyle DJ. Some comments on the use of Wiener filtering for the estimation of evoked potentials. Electroencephalogr Clin Neurophysiol. 1975;38(5):533–4.

Friman O, Borga M, Lundberg P, Knutsson H. Adaptive analysis of fMRI data. NeuroImage. 2003;19(3):837–45.

Friston KJ, Fletcher P, Josephs O, Holmes A, Rugg MD, Turner R. Event-related fMRI: characterizing differential responses. NeuroImage. 1998;7(1):30–40.

Garcia-Larrea L, Frot M, Valeriani M. Brain generators of laser-evoked potentials: from dipoles to functional significance. Neurophysiol Clin. 2003;33(6):279–92.

Goldman RI, Wei CY, Philiastides MG, Gerson AD, Friedman D, Brown TR, et al. Single-trial discrimination for integrating simultaneous EEG and fMRI: identifying cortical areas contributing to trial-to-trial variability in the auditory oddball task. NeuroImage. 2009;47(1):136–47.

Hossein-Zadeh GA, Ardekani BA, Soltanian-Zadeh H. A signal subspace approach for modeling the hemodynamic response function in fMRI. Magn Reson Imaging. 2003;21(8):835–43.

Hu Y, Luk KD, Lu WW, Leong JC. Application of time-frequency analysis to somatosensory evoked potential for intraoperative spinal cord monitoring. J Neurol Neurosurg Psychiatry. 2003;74(1):82–7.

Hu Y, Lam BS, Chang CQ, Chan FH, Lu WW, Luk KD. Adaptive signal enhancement of somatosensory evoked potential for spinal cord compression detection: an experimental study. Comput Biol Med. 2005;35(9):814–28.

Hu L, Mouraux A, Hu Y, Iannetti GD. A novel approach for enhancing the signal-to-noise ratio and detecting automatically event-related potentials (ERPs) in single trials. NeuroImage. 2010;50 (1):99–111.

Hu L, Liang M, Mouraux A, Wise RG, Hu Y, Iannetti GD. Taking into account latency, amplitude, and morphology: improved estimation of single-trial ERPs by wavelet filtering and multiple linear regression. J Neurophysiol. 2011a;106(6):3216–29.

Hu L, Zhang ZG, Hung YS, Luk KD, Iannetti GD, Hu Y. Single-trial detection of somatosensory evoked potentials by probabilistic independent component analysis and wavelet filtering. Clin Neurophysiol. 2011b;122(7):1429–39.

Hu L, Zhang ZG, Mouraux A, Iannetti GD. Multiple linear regression to estimate time-frequency electrophysiological responses in single trials. NeuroImage. 2015;111:442–53.

Huang DS, Mi JX. A new constrained independent component analysis method. IEEE Trans Neural Netw. 2007;18(5):1532–5.

Huang G, Xiao P, Hung YS, Iannetti GD, Zhang ZG, Hu L. A novel approach to predict subjective pain perception from single-trial laser-evoked potentials. NeuroImage. 2013;81:283–93.

Hyvarinen A. Fast and robust fixed-point algorithms for independent component analysis. IEEE Trans Neural Netw. 1999;10(3):626–34.

Hyvarinen A, Oja E. Independent component analysis: algorithms and applications. Neural Netw. 2000;13(4–5):411–30.

Iannetti GD, Niazy RK, Wise RG, Jezzard P, Brooks JC, Zambreanu L, et al. Simultaneous recording of laser-evoked brain potentials and continuous, high-field functional magnetic resonance imaging in humans. NeuroImage. 2005a;28(3):708–19.

Iannetti GD, Zambreanu L, Cruccu G, Tracey I. Operculoinsular cortex encodes pain intensity at the earliest stages of cortical processing as indicated by amplitude of laser-evoked potentials in humans. Neuroscience. 2005b;131(1):199–208.

Iannetti GD, Zambreanu L, Tracey I. Similar nociceptive afferents mediate psychophysical and electrophysiological responses to heat stimulation of glabrous and hairy skin in humans. J Physiol. 2006;577.(Pt 1:235–48.

Iannetti GD, Hughes NP, Lee MC, Mouraux A. Determinants of laser-evoked EEG responses: pain perception or stimulus saliency? J Neurophysiol. 2008;100(2):815–28.

James CJ, Gibson OJ. Temporally constrained ICA: an application to artifact rejection in electromagnetic brain signal analysis. IEEE Trans Biomed Eng. 2003;50(9):1108–16.

Jolliffe IT. Principal component analysis. 2nd ed. New York: Springer; 2002.

Jongsma MLA, Eichele T, Van Rijn CM, Coenen AML, Hugdahl K, Nordby H, et al. Tracking pattern learning with single-trial event-related potentials. Clin Neurophysiol. 2006;117 (9):1957–73.

Jung TP, Makeig S, Westerfield M, Townsend J, Courchesne E, Sejnowski TJ. Analysis and visualization of single-trial event-related potentials. Hum Brain Mapp. 2001;14(3):166–85.

Kaiser HF. Citation classic – varimax criterion for analytic rotation in factor-analysis. Cc/Soc Behav Sci. 1979;(24):20.

Kayser J, Tenke CE. Optimizing PCA methodology for ERP component identification and measurement: theoretical rationale and empirical evaluation. Clin Neurophysiol. 2003;114 (12):2307–25.

Koles ZJ. The quantitative extraction and topographic mapping of the abnormal components in the clinical EEG. Electroencephalogr Clin Neurophysiol. 1991;79(6):440–7.

Koles ZJ, Lazar MS, Zhou SZ. Spatial patterns underlying population differences in the background EEG. Brain Topogr. 1990;2(4):275–84.

Lam BS, Hu Y, Lu WW, Luk KD, Chang CQ, Qiu W, et al. Multi-adaptive filtering technique for surface somatosensory evoked potentials processing. Med Eng Phys. 2005;27(3):257–66.

Lee MC, Mouraux A, Iannetti GD. Characterizing the cortical activity through which pain emerges from nociception. J Neurosci. 2009;29(24):7909–16.

Legrain V, Guerit JM, Bruyer R, Plaghki L. Attentional modulation of the nociceptive processing into the human brain: selective spatial attention, probability of stimulus occurrence, and target detection effects on laser evoked potentials. Pain. 2002;99(1–2):21–39.

Legrain V, Bruyer R, Guerit JM, Plaghki L. Nociceptive processing in the human brain of infrequent task-relevant and task-irrelevant noxious stimuli. A study with event-related potentials evoked by CO2 laser radiant heat stimuli. Pain. 2003;103(3):237–48.

Lu W, Rajapakse JC. ICA with Reference. Neurocomputing. 2006;69:2244–57.

Luk KD, Hu Y, Lu WW, Wong YW. Effect of stimulus pulse duration on intraoperative somatosensory evoked potential (SEP) monitoring. J Spinal Disord. 2001;14(3):247–51.

Makeig S, Jung TP, Bell AJ, Ghahremani D, Sejnowski TJ. Blind separation of auditory event-related brain responses into independent components. Proc Natl Acad Sci U S A. 1997;94 (20):10979–84.

Mayhew SD, Iannetti GD, Woolrich MW, Wise RG. Automated single-trial measurement of amplitude and latency of laser-evoked potentials (LEPs) using multiple linear regression. Clin Neurophysiol. 2006;117(6):1331–44.

Mayhew SD, Dirckx SG, Niazy RK, Iannetti GD, Wise RG. EEG signatures of auditory activity correlate with simultaneously recorded fMRI responses in humans. NeuroImage. 2010;49 (1):849–64.

Minahan RE. Intraoperative neuromonitoring. Neurologist. 2002;8(4):209–26.

Minka TP. Automatic choice of dimensionality for PCA. Adv Neural Inf Proces Syst. 2001;13:598–604.

Mouraux A, Iannetti GD. Across-trial averaging of event-related EEG responses and beyond. Magn Reson Imaging. 2008;26(7):1041–54.

Mouraux A, Iannetti GD. Nociceptive laser-evoked brain potentials do not reflect nociceptive-specific neural activity. J Neurophysiol. 2009;101(6):3258–69.

Mouraux A, Plaghki L. Single-trial detection of human brain responses evoked by laser activation of Adelta-nociceptors using the wavelet transform of EEG epochs. Neurosci Lett. 2004;361 (1–3):241–4.

Mouraux A, Guerit JM, Plaghki L. Non-phase locked electroencephalogram (EEG) responses to CO2 laser skin stimulations may reflect central interactions between A partial partial differential- and C-fibre afferent volleys. Clin Neurophysiol. 2003;114(4):710–22.

Muller-Gerking J, Pfurtscheller G, Flyvbjerg H. Designing optimal spatial filters for single-trial EEG classification in a movement task. Clin Neurophysiol. 1999;110(5):787–98.

Niazy RK, Beckmann CF, Iannetti GD, Brady JM, Smith SM. Removal of FMRI environment artifacts from EEG data using optimal basis sets. NeuroImage. 2005;28(3):720–37.

Nunez PL, Srinivasan R. Electric fields of the brain: the neurophysics of EEG. 2nd ed. Oxford/New York: Oxford University Press; 2006.

Nuwer MR. Spinal cord monitoring with somatosensory techniques. J Clin Neurophysiol. 1998;15 (3):183–93.

Orssaud C. Leber's hereditary optic neuropathy. Orphanet Encyclopedia; 2003

Pelosi L, Geesken JM, Holly M, Hayward M, Blumhardt LD. Working memory impairment in early multiple sclerosis. Evidence from an event-related potential study of patients with clinically isolated myelopathy. Brain. 1997;120(Pt 11):2039–58.

Pfurtscheller G, Lopes da Silva FH. Event-related EEG/MEG synchronization and desynchronization: basic principles. Clin Neurophysiol. 1999;110(11):1842–57.

Porcaro C, Barbati G, Zappasodi F, Rossini PM, Tecchio F. Hand sensory-motor cortical network assessed by functional source separation. Hum Brain Mapp. 2008;29(1):70–81.

Porcaro C, Coppola G, Di Lorenzo G, Zappasodi F, Siracusano A, Pierelli F, et al. Hand somatosensory subcortical and cortical sources assessed by functional source separation: an EEG study. Hum Brain Mapp. 2009;30(2):660–74.

Porcaro C, Ostwald D, Bagshaw AP. Functional source separation improves the quality of single trial visual evoked potentials recorded during concurrent EEG-fMRI. NeuroImage. 2010;50 (1):112–23.

Purves AM, Boyd SG. Time-shifted averaging for laser evoked potentials. Electroencephalogr Clin Neurophysiol. 1993;88(2):118–22.

Quiroga RQ. Obtaining single stimulus evoked potentials with wavelet denoising. Physica D. 2000;145(3–4):278–92.

Quiroga RQ, Garcia H. Single-trial event-related potentials with wavelet denoising. Clin Neurophysiol. 2003;114(2):376–90.

Ramoser H, Muller-Gerking J, Pfurtscheller G. Optimal spatial filtering of single trial EEG during imagined hand movement. IEEE Trans Rehabil Eng. 2000;8(4):441–6.

Richman MB. Rotation of Principal Components. J Climatol. 1986;6(3):293–335.

Rossi L, Bianchi AM, Merzagora A, Gaggiani A, Cerutti S, Bracchi F. Single trial somatosensory evoked potential extraction with ARX filtering for a combined spinal cord intraoperative neuromonitoring technique. Biomed Eng Online. 2007;6:2.

Roweis S. EM algorithms for PCA and SPCA. Adv Neural Inf Proces Syst. 1998;10:626–32.

Schnakers C, Chatelle C, Majerus S, Gosseries O, De Val M, Laureys S. Assessment and detection of pain in noncommunicative severely brain-injured patients. Expert Rev Neurother. 2010;10 (11):1725–31.

Schulz E, Zherdin A, Tiemann L, Plant C, Ploner M. Decoding an individual's sensitivity to pain from the multivariate analysis of EEG data. Cereb Cortex. 2012;22(5):1118–23.

Spencer KM. Averaging, detection, and classification of single-trial ERPs. In: Handy TC, editor. Event-related potentials: a methods handbook. Cambridge, MA: MIT Press; 2005. p. 209–27.

Tang AC, Sutherland MT, McKinney CJ. Validation of SOBI components from high-density EEG. NeuroImage. 2005;25(2):539–53.

Tecchio F, Porcaro C, Barbati G, Zappasodi F. Functional source separation and hand cortical representation for a brain-computer interface feature extraction. J Physiol. 2007;580 (Pt.3):703–21.

Tipping ME, Bishop CM. Mixtures of probabilistic principal component analyzers. Neural Comput. 1999a;11(2):443–82.

Tipping ME, Bishop CM. Probabilistic principal component analysis. J R Stat Soc Ser B Stat Methodol. 1999b;61:611–22.

Tognola G, Grandori F, Ravazzani P. Wavelet analysis of click-evoked otoacoustic emissions. IEEE Trans Biomed Eng. 1998;45(6):686–97.

Tu Y, Zhang Z, Tan A, Peng W, Hung YS, Moayedi M, et al. Alpha and gamma oscillation amplitudes synergistically predict the perception of forthcoming nociceptive stimuli. Hum Brain Mapp. 2016;37(2):501–14.

Walter DO. A posteriori “Wiener filtering” of average evoked responses. Electroencephalogr Clin Neurophysiol Suppl. 1968;27:61+.

Wang Z, Maier A, Leopold DA, Logothetis NK, Liang H. Single-trial evoked potential estimation using wavelets. Comput Biol Med. 2007;37(4):463–73.

Wiedemayer H, Fauser B, Sandalcioglu IE, Schafer H, Stolke D. The impact of neurophysiological intraoperative monitoring on surgical decisions: a critical analysis of 423 cases. J Neurosurg. 2002;96(2):255–62.

Woolrich MW, Behrens TE, Smith SM. Constrained linear basis sets for HRF modelling using Variational Bayes. NeuroImage. 2004;21(4):1748–61.

Yiannikas C, Vucic S. Utility of somatosensory evoked potentials in chronic acquired demyelinating neuropathy. Muscle Nerve. 2008;38(5):1447–54.

Yu KB, McGillem CD. Optimum filters for estimating evoked potential waveforms. IEEE Trans Biomed Eng. 1983;30(11):730–7.

Zeman BD, Yiannikas C. Functional prognosis in stroke: use of somatosensory evoked potentials. J Neurol Neurosurg Psychiatry. 1989;52(2):242–7.

Chapter 11
Nonlinear Neural Dynamics

Yang Bai, Xiaoli Li, and Zhenhu Liang

Abstract Brain activities measured by EEG exhibit complex behavior with nonlinear dynamic properties. Therefore, methods derived from nonlinear theory could contribute to the understanding of the EEG dynamics and the underlying brain processes. Until now, a number of nonlinear dynamic methods have been proposed. These methods reveal various nonlinear properties of the EEG signals. Among them, "complexity" and "entropy" are the widely used concept in the EEG analysis. Moreover, entropy-based measures have been applied into clinical practice of monitoring depth of anesthesia. This study selects three classical types of nonlinear dynamic measures (total 12), introducing their basic theory and giving examples of applying them in real EEG analysis. All the MATLAB codes can be downloaded from https://pan.baidu.com/s/1ro2VEJWdsb5X6dCtueGUOA with password: hf4x. Although previous studies compared the performance of various nonlinear dynamic measures at different situations, it is still improper to determine which measure is the best one. The measure selection in your study should take a number of factors into

---

Electronic supplementary material The online version of this chapter (https://doi.org/10.1007/978-981-13-9113-2_11) contains supplementary material, which is available to authorized users.

Y. Bai (✉)
Department of Basic Medical Science, School of Medicine, Hangzhou Normal University, Hangzhou, Zhejiang, China
e-mail: baiyang@hznu.edu.cn

X. Li
State Key Laboratory of Cognitive Neuroscience and Learning, Beijing Normal University, Beijing, China
e-mail: xiaoli@bnu.edu.cn

Z. Liang
Institute of Electric Engineering, Yanshan University, Qinhuangdao, Hebei, China
e-mail: zhl@ysu.edu.cn

L. Hu, Z. Zhang (eds.), *EEG Signal Processing and Feature Extraction*,
https://doi.org/10.1007/978-981-13-9113-2_11

account, such as the parameter selection, robust to artifacts, compute consumption, the correlation of the nonlinear characteristics with the underlying neural process, and so on.

Keywords EEG · Complexity · Entropy · Nonlinear dynamics

11.1 Introduction

Detection of dynamic properties in complex systems is one of the most important issue in physical, medical, engineering, and biological sciences (Cao et al. 2004b). Especially, in neurophysiology, accurate assessing the underlying neuronal process may contribute to the diagnosis and treatment. Nonlinear dynamic analysis focuses on the revealing of the changes of dynamic properties in complex systems. The nonlinear dynamic theory considers the brain as a multiple dimensions' dynamic system. The neuronal activities represent properties of dynamic; thus, human brain could be simulated as a complex nonlinear dynamic system. The nonlinear and chaotic behavior characteristics of the neuronal systems make it reasonable to apply methods derived from theory of nonlinear dynamics to capture the neuronal activities. Since the nonlinear dynamics was firstly used in analyzing EEG of human sleeping, more and more researchers pay attention to expand nonlinear dynamic measures in EEG analyzing. For example, nonlinear measures based on the EEG in anesthesia gives an important reference guide in assessing anesthesia state of patients (Anier et al. 2010; Olofsen et al. 2008; Schultz et al. 2008).

During the last two decades, various nonlinear algorithms have been proposed for extracting EEG nonlinear characteristics to assess brain activities. Among them, the first important nature of the EEG lies in its dynamic "complexity," which can be characterized quantitatively by complexity analysis. The complexity analysis mainly represents the degree of randomness in time series. In the EEG analysis, the complexity measures the capacity of information in the signal fragment and then reflects the underlying activeness of the neurons. For example, the "complexity" of resting-state EEG with opening eyes will markedly be higher than that with closing eyes. The difference of capacity of information in EEG comes from the visual senses, which make the brain in a vigilant state with more information in the neuronal activities. When the eyes are closed, the brain comes to a relatively resting state with decreased active degree of neurons, which represents lower "complexity" in EEG signals. Lempel-Ziv complexity (LZC) was proposed by Lempel and Ziv (Lempel and Ziv 1976). As it is a nonlinear dynamic measure indicating the rate of appearance of new patterns in time series, only the activities adding or reducing the patterns of underlying system are considered by LZC. It does not matter whether the system is deterministic or stochastic (Li et al. 2008d). LZC as well as its derivatives has found numerous applications in characterizing the randomness of biological signals. In EEG analysis, LZC was applied to estimate levels of consciousness during anesthesia (Jouny and Bergey 2012) and to detect seizure onset (Zhang et al. 2001), depression (Li et al. 2008d), Parkinson's disease (Chen et al. 2010), schizophrenia (Sabeti et al. 2009), and Alzheimer's disease (Abasolo et al. 2006; Dauwels et al. 2011).

The "entropy," which was originally derived from thermodynamics, was firstly proposed by Clausius in 1865. When considered as a physical concept, it was used to describe the distribution probability of molecules of gaseous or fluid systems. In 1949, Claude E. Shannon introduced entropy to the information theory (Shannon and Weaver 1949). The Shannon entropy (ShEn) measures the predictability of later amplitude values of the signal based on the probability distribution of amplitude values former in the signal (Yoon et al. 2011; Bruhn et al. 2001). Up to now, the entropy theory undergoes development more than 100 years. It has been continuously improved with various definition derived. During its development, several entropy measures have been used in various neuron information processing. For example, in a study of the depth of anesthesia, several entropy measures have showed good performance, such as the spectral entropy (SpEn) (which includes response entropy (RE) and state entropy (SE)) (Klockars et al. 2012; Viertiö-Oja et al. 2004), wavelet entropy (WE) (Särkelä et al. 2007), Hilbert-Huang spectral entropy (HHSE) (Li et al. 2008b), phase space reconstruction-based entropy (sample entropy (SampEn), fuzzy entropy (FuzzyEn), and approximate entropy (ApEn) (Bruhn et al. 2000). Permutation entropy (PE) was proposed by Bandt and Pompe based on measuring the symbolic dynamics for time series analysis (Bandt and Pompe 2002a), and then it has been widely used in EEG signal analysis (Cao et al. 2004a; Li et al. 2007; Li et al. 2008a). Overall, all the entropy measures describe the capacity of information at different perspective, and the successful clinical application is to monitor the depth of anesthesia. Generally, the anesthetics will suppress the neural activities and result in a fading of consciousness. Therefore, an increasing of anesthetics would lead to a reducing of capacity of information interaction in the brain, which shows decreasing of entropy in EEG signals.

In addition to the complexity and entropy measures, many other nonlinear dynamic measures have also been proposed to analyze the characteristics of EEG, such as recurrence quantification analysis (RQA) methods (Webber and Zbilut 1994), Lyapunov exponents (Röschke et al. 1995), correlation dimension (Shayegh et al. 2014), and detrended fluctuation analysis (DFA) (Matic et al. 2015). In particular, the Hurst exponent (H), which measures the smoothness of a fractal time series derived from rescaled range analysis or DFA, has been successfully applied to evaluate the self-similarity and correlation properties of EEG recordings (Rajendra et al. 2005). It is an index describing the self-affine characteristics of irregular time series, which characterizes the persistence/anti-persistence properties of signals. The RQA measures is demonstrated as a suitable tool to characterize EEG signals because it does not require any assumption on stationarity, length, or noise of time series (Schinkel et al. 2009; Webber and Zbilut 1994). Due to its advantages, RQA has been used to reveal sleep stages (Song et al. 2004), condition of consciousness/unconsciousness (Klaus et al. 2010), and pre-ictal activity in epileptic subjects (Ouyang et al. 2008).

In this study, we select several nonlinear dynamic measures to introduce their algorithm and give examples of applying them in analyzing EEG signals during awake and anesthesia states. The measures include complexity methods of LZC as well as its derivatives; entropy methods of ShEn, ApEn, SampEn, PE, WE, HHSE, and HE; and indexes derived from RQA. All the MATLAB codes correlating with this study are also attached.

11.2 Complexity

Lempel-Ziv complexity was proposed by Lempel and Ziv (Lempel and Ziv 1976) and along with its derivatives has found numerous applications in characterizing the randomness of biological signals, especially in EEG analysis (Abasolo et al. 2006; Aboy et al. 2006; Huang et al. 2003; Zhang and Roy 2001).

11.2.1 Lempel-Ziv Complexity

LZC analysis is based on a coarse-graining, which transformed the signal $\{s(n)\}$ into a finite sequence with a few symbols (Zhang et al. 2001). The coarse-graining process conventionally converted the signal $s(i)$ into a binary $x(i)$. It determines how much information is in the original signal. Here, we give four types of LZC measures which depend on commonly used coarse-graining methods (mean, LZC_{mean}; median, LZC_{median}; midpoint, LZC_{mid-p}; k-means, $LZC_{k-means}$).

The coarse-graining approaches convert the original signal into 0–1 sequence through comparing the amplitude values with the threshold(s) (LZC_{mean}, LZC_{median}, and $LZC_{mid-point}$ only need one threshold T_d, while the $LZC_{k-means}$ needs two threshold values). Then, the complexity of the signal counts the number of different patterns in the sequence.

- LZC_{mean}

In this coarse-graining approach, the mean value of the amplitude values will be selected as T_d (Zhang et al. 2000). Then, the coarse-graining can be expressed as

$$x(i) = \begin{cases} 0, & if \;\; s(i) < T_d \\ 1, & otherwise \end{cases} \tag{11.1}$$

- LZC_{median}

For the LZC_{median}, the median value of the amplitude values will be chosen as T_d. After comparing each data point with the T_d ($s(i) - T_d$), we define the positive value as 1 and the others as 0 (Zhang et al. 2001).

- $LZC_{mid\ -\ p}$

The midpoint value of the amplitude values is determined by the smallest value s_{min} and the largest value s_{max},

$$T_d = \frac{s_{min} + s_{max}}{2} \tag{11.2}$$

Then, the subsequent step is exactly the same as explained in Eq. (11.1).

- $LZC_{k\ -\ means}$

The $LZC_{k\ -\ means}$ estimates the grouping of data points around centroids corresponding to points around which most of the data is agglomerated. In the initial iteration of the method, we set the two initial centroids as $z_1(1) = s_m + \varepsilon * s_m$ and $z_2(1) = s_m - \varepsilon * s_m$. The ε, which relates with the stationarity of the signal, actually determines the up and down thresholds. Here, we set it as 0.05 with experience following previous studies (Linde et al. 1980; Zhou et al. 2011). s_m is the mean of the amplitude values. Then, the distance between the centroid to each data point can be obtained,

$$\begin{cases} D_1^i = \|s(i) - z_1(1)\|^2 \\ D_2^i = \|s(i) - z_2(1)\|^2 \end{cases} \tag{11.3}$$

Then, the signal is converted into a binary sequence as follows:

$$x(i) = \begin{cases} 1 & if\ D_1^i < D_2^i \\ 0 & if\ D_1^i \geq D_2^i \end{cases} \tag{11.4}$$

11.2.2 Permutation Lempel-Ziv Complexity

Permutation approach considers the order relation within the values but not the absolute amplitude values. Previous studies have shown that the permutation approach has several advantages, such as simplicity, robustness, and low complexity in computation (Bandt and Pompe 2002b; Bandt 2005). Facilitated by these advantages, in addition to the PLZC, several other nonlinear methods such as permutation entropy (Bandt and Pompe 2002b; Bruzzo et al. 2008; Li et al. 2010, 2008c), permutation conditional mutual information (Li and Ouyang 2010), and permutation transfer entropy (Li and Li 2013) have been proposed to describe the chaotic systems.

PLZC was proposed as an improved LZC measure by using permutation process to enhance the robust to noise (Bai et al. 2015). The permutation scheme is illustrated in Fig. 11.1. The patterns are also referred as motifs (Olofsen et al. 2008). m is the

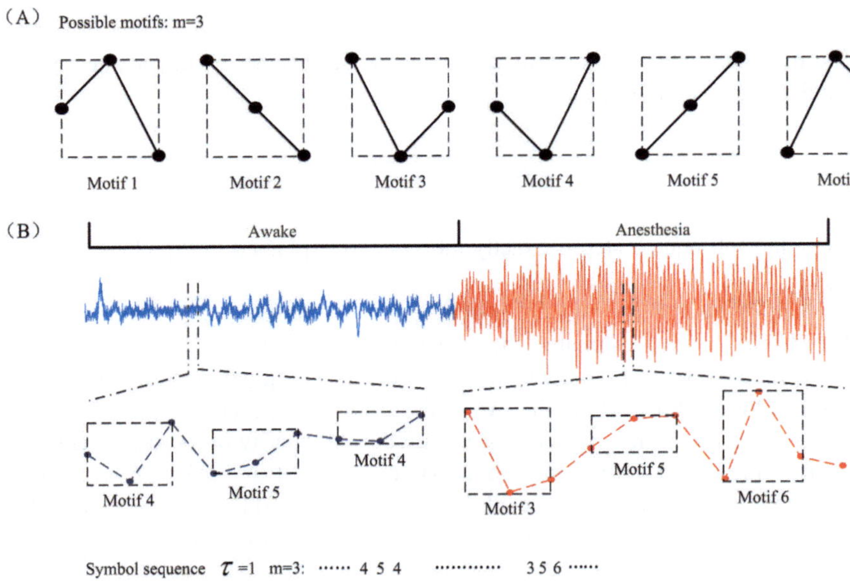

Fig. 11.1 Illustration of the permutation process with $m = 3$. (**a**) The possible motifs for $m = 3$. (**b**) Transform the signal into the symbol sequence using permutation

number of data points in each motif, so the total number of possible motifs is $m!$. $m = 3$ means three points in each motif and total six ($m!$) possible motifs can be obtained. By this way, the EEG signal could be transformed into a finite symbolic sequence with possible elements of 1 to 6. In order to ensure that all data points are mapped into one of the possible permutations, the equal values were arranged by the index of data involved. For example, if $x(i) = x(i + 1)$, it would regard as $x(i) < x(i + 1)$ for permutation. The most important parameters for this permutation procedure are m and τ.

There are generally two definitions for the complexity parsing. The primary LZC measure (Lempel and Ziv 1976) is most widely used in applications with two or more symbols. The detailed PLZC indexes could be calculated following the eight steps.

1. Step 1: Transform the signals into a finite sequence $\{x(n)\}$ with permutation procedure before mentioned. In this way, the signals will be replaced by no more than $m!$ kinds of symbols representing permutation patterns.
2. Step 2: Initialize the PLZC measure. Let $S$and Q denote the first and the second symbol, respectively, of the $\{x(n)\}$ and set the complexity factor $c(n) = 1$.
3. Step 3: Merge S and Q intoSQ. SQv denotes the string derived from SQ with operation of deleting the last character ofSQ. For example, S=$x(1)$, $x(2)$, ..., $x(i)$, Q=$x(i + 1)$, ..., $x(i + j - 1)$,$x(i + j)$, then SQ=$x(1)$, $x(2)$, ..., $x(i)$, $x(i + 1)$, . . ., $x(i + j - 1)$, $x(i + j)$ and SQv=$x(1)$, $x(2)$, ..., $x(i)$, $x(i + 1)$, . . ., $x(i + j - 1)$.

4. Step 4: Judge whether Q has reached the last symbol of the sequence. If reached, this measure will go on with normalization.
5. Step 5: List the subsequence of SQv and collect all of them in a vocabulary denoted as SQv^{sub}. Then, if Q belongs to SQv^{sub}, go to Step 6; otherwise, Q is a new sequence and go to step 7.
6. Step 6: Update Q by adding the next symbol to it and go to step 3.
7. Step 7: Update S as SQ and Q equal to the next symbol of $\{x(n)\}$. Meanwhile, the complexity factor $c(n)$is increased by one.
8. Step 8: Now $c(n)$ is the complexity of the symbol sequence $\{x(n)\}$ which denote the number of distinct patterns in the source sequence. The total number of subsequences present in $\{x(n)\}$ has an upper bound (Hu et al. 2006), denoted as $L(n)$:

$$L(n) = c(n)[log_{m!}\{c(n)\} + 1] \tag{11.5}$$

Then, PLZC can be indicated as a normalized $c(n)$, defined as

$$\text{PLZC} = \frac{c(n)[log_{m!}c(n) + 1]}{n} \tag{11.6}$$

where n denotes the total length of the symbol sequence. When n is very large, PLZC can be simplified as

$$\text{PLZC} = \frac{c(n)\{log_{m!}n\}}{n} \tag{11.7}$$

Figure 11.2 shows the results of the five LZC measures in describing the EEG signals during the awake and anesthesia states. All the measures show lower complexity of patterns during the anesthesia than the awake state. Among them, PLZC represents better stability especially during the awake state. Similar with the other four LZC measures, PLZC also evaluates the number of new patterns and their rate of occurrence along a given sequence. However, the coarse-graining approaches of the traditional LZC measures use the absolute values of the amplitude. Therefore, the coarse-graining approach makes the four LZC measures more sensitive to the artifacts in the signal. On the other hand, the thresholds used in the traditional LZC procedures have a distinct limitation. Since the binary sequence is simply constructed using a threshold value, it would result in a loss of frequency information in the original signal. However, the permutation in PLZC reflects the mutual relation of the signal points (described as motifs), and the variation of these motifs indicates the changes of the signal itself. Changes in both high and low frequency will alter these motifs, and variations in both will be present in the permutation. This property is particularly useful in characterizing the neuronal dynamic changes in the EEG signals, because brain activities are more likely to represent in EEG frequency domain. Hence, these led to better performance of the PLZC in distinguishing the awake and anesthesia states based on EEG.

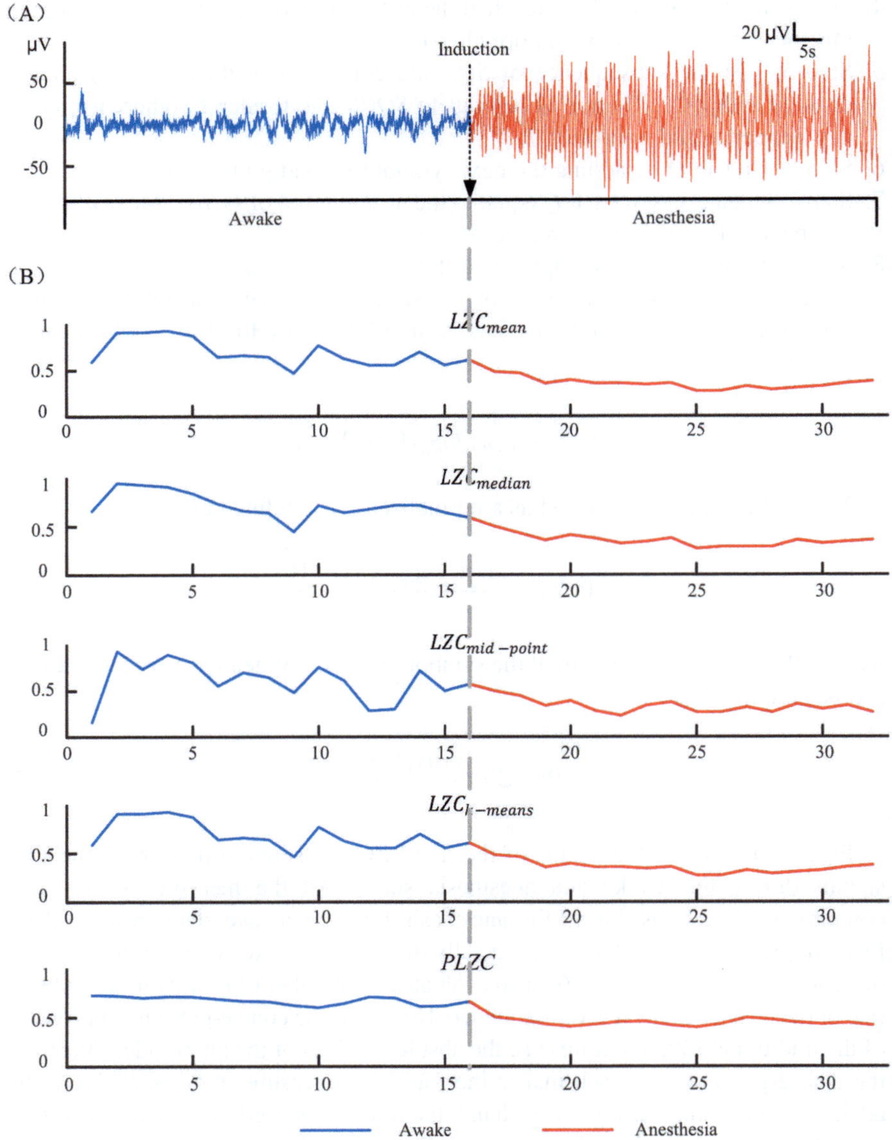

Fig. 11.2 Lempel-Ziv complexity methods in analyzing EEG signals. (**a**) EEG signals of awake (blue) and anesthesia (red). (**b**) Results of the LZC_{mean}, LZC_{median}, LZC_{mid-p}, $LZC_{k-means}$, and *PLZC* during awake and anesthesia states

11.3 Entropy

11.3.1 Entropy Based on Time Domain: ShEn, ApEn, SampEn, and PE

11.3.1.1 Shannon Entropy

Claude E. Shannon was the first to propose entropy to measure for information theory (Shannon 1922). Shannon entropy measures the probability density based on the probability distribution of amplitude values. It may be used globally, taking all data into account, or locally, around certain points. This measure can provide additional information about specific events, for example, outliers or rare events. Information entropy S_{en} of a random variable S that takes the values $s_1, s_2, \ldots, s_N$ is defined as

$$S_{en} = \sum_{i=1}^{n} p(s_i)\log_a \frac{1}{p(s_i)} = -\sum_{i=1}^{n} p(s_i)\log_a p(s_i), a > 1 \tag{11.8}$$

where $p(s_i)$ are probabilities of acceptance by the random variable S values s_i.

11.3.1.2 Approximate Entropy

ApEn is a modification of the Kolmogorov-Sinai entropy introduced by Pincus in 1991 (Pincus 1991). ApEn measures the predictability of future amplitude values of the signal based on the information of previous amplitude values. It can be used to the finite length signal, and it belongs to nonlinear dynamics which describes the unpredictability or randomness of the signal. Its computation involves embedding signal into phase space and estimates the rate of increment in the number of phase space pattern within a predefined value r when the embedding dimension of phase space is increasing from m to $m + 1$.

For a time series $x(i)(1 \leq i \leq N)$ of finite length N to reconstitute the $N\text{-}m + 1$ vectors $X_m(i)$ following the form:

$$\begin{aligned} X_m(i) &= \{x(i), x(i+1), \ldots x(i+m-1)\}, \\ i &= 1,2, \ldots N-m+1 \end{aligned} \tag{11.9}$$

where m is the embedding dimension.

Let $C_i^m(r)$ be the probability that any vector $X_m(j)$ is within r of $X_m(i)$; it can be defined as

$$C_i^m(\mathrm{r}) = \frac{1}{N-m+1}\sum_{j=1}^{N-m+1}\Theta\left(d_{ij}^m - r\right) \quad (11.10)$$
$$i,j = 1,2,\ldots N-m+1$$

where d is the distance between the vectors $X_m(i)$ and $X_m(j)$, defined as

$$d_{ij}^m = d\left[X_i^m, X_j^m\right] = \max(|x(i+k) - x(j+k)|) \quad (11.11)$$
$$k = 0,1,\ \ldots, m$$

and $\Theta$is the Heaviside function.

After that, define a parameter $\Phi^m(r)$:

$$\Phi^m(r) = (N-m+1)^{-1}\sum\nolimits_{i=1}^{N-m+1}\ln C_i^m(r) \quad (11.12)$$

Next, when the dimension changes to $m + 1$, it will repeat above process.

$$\Phi^{m+1}(r) = (N-m)^{-1}\sum\nolimits_{i=1}^{N-m}\ln C_i^{m+1}(r) \quad (11.13)$$

Finally, the approximate entropy is defined by:

$$ApEn(m,r,N) = \Phi^m(r) - \Phi^{m+1}(r) \quad (11.14)$$

The detail algorithm is shown in (Bruhn et al. 2000). And the ApEn index is easily influenced by data length (N), similar tolerance (r), and embedding dimension (m). According to Pincus (1991) and Bruhn et al. (2000), N is recommended as about 1000 of the standard deviation of the signal, r as 0.1~0.25, and m as 2~3.

11.3.1.3 Sample Entropy

The SampEn is proposed by Richman and Moorman (Richman and Moorman 2000) based on the ApEn but differs from it in three ways, which all remove bias existing in ApEn.

1. SampEn eliminates self-matches.
2. To avoid $\ln 0$ caused by removing self-matches, SampEn computes the addition operation of the total number of template well-matches prior to logarithmic operation.

3. In order to have equal number of patterns for both embedding dimension m and $m+1$, the time series reconstitution in SampEn have $N-m$ rows instead of $N-m+1$ in ApEn in embedding dimension m.

The first step is the same as ApEn. When the embedding dimension is m, the total number of template matches:

$$\mathrm{B}^m(r) = (N-m)^{-1}\sum_{i=1}^{N-m} C_i^m(r) \tag{11.15}$$

Similarly, the embedding dimension is $m+1$, and the total number of template matches:

$$\mathrm{A}^m(r) = (N-m)^{-1}\sum_{i=1}^{N-m} C_i^m(r) \tag{11.16}$$

Finally, the SampEn of the time series is estimated by:

$$\mathrm{AampEn}(r, m, N) = -\ln\frac{\mathrm{A}^m(r)}{\mathrm{B}^m(r)} \tag{11.17}$$

The SampEn is based on ApEn, so its parameters and their selection procedure are the same as ApEn.

11.3.1.4 Permutation Entropy

PE is a quantitative complexity measure that explores the local order structure of a dynamic time series. It transforms given time series into series of ordinal patterns, each describing the order relation between the present and a fixed number of equidistant past values at a given time (Bandt 2005). The advantages of the method are simplicity, robustness, and low computational complexity (Li et al. 2007).

Given a scalar time series, $\{x(i) : 1 \le i \le N\}$. Firstly, reconstruction time series:

$$\begin{array}{l} X_i = \{x(i), x(i+\tau), \ldots, x(i+(m-1)\tau)\} \\ i = 1,2, \ldots N-(m-1)\tau \end{array} \tag{11.18}$$

where τ is time delay, and m is the embedding dimension.

Then, rearrange X_i in an increasing order:

$$\{x(i+(j_1-1)\tau) \le x(i+(j_2-1)\tau) \le \ldots \le x(i+(j_m-1)\tau) \tag{11.19}$$

There are $m!$ permutations for m dimensions. Each vector X_i can be mapped to one of the $m!$ permutations.

Next, the probability of the jth permutation occurringp_j can be defined as

$$p_j = \frac{n_j}{\sum_{j=1}^{m!} n_j} \quad (11.20)$$

where n_j is the number of times the jth permutation is occurring.

The permutation entropy of the time series $\{x(i) : 1 \leq i \leq N\}$ is defined by:

$$H_x(m) = -\sum_{j=1}^{m!} p_j \ln p_j \quad (11.21)$$

when the time series is random, the $H_x(m)$ approaches its maximum value of $\ln(m!)$; when the time series is regular, the $H_x(m)$ approaches to zero.

Finally, normalizing $H_x(m)$ by diving $\ln(m!)$:

$$PE = \frac{H_x(m)}{\ln(m!)} \quad (11.22)$$

Figure 11.3 shows the indexes of ShEn, AE, SampEn, and PE in analyzing the EEG signals. It can be seen that the AE, SampEn, and PE could distinctly distinguish the two EEG states. The anesthesia state shows lower entropy than the awake state. Although the ShEn shows capability of estimating depth of anesthesia in previous studies, the ShEn indexes show indiscriminate between the awake and anesthesia state in this EEG data set. From the view of algorithm, ShEn simply measures the distribution of signal amplitude values, which would be easily contaminated by artifacts in the EEG signals, while AE, SampEn, and PE firstly reconstruct the signal and then investigate the ordinal relationship of the amplitude values. Thus, these measures are more sensitive to the changes of covert components in the EEG signal and are more robust to the artifacts.

11.3.2 Entropy Based on Time-Frequency Domain: WE and HHSE

Spectral entropy is a somewhat confusing but an important term in the nonlinear analysis. It uses spectral analysis to firstly transform the signal into time-frequency domain and then apply Shannon function to measure the distribution of spectral power. Although it uses Shannon entropy concept, spectral entropy reflects the power distribution of the transformations of EEG signal, which has been normalized to unit power so as to reduce the individual difference effects (Inouye et al. 1991). Therefore, the efficiency of spectral entropy extremely depends on the spectral

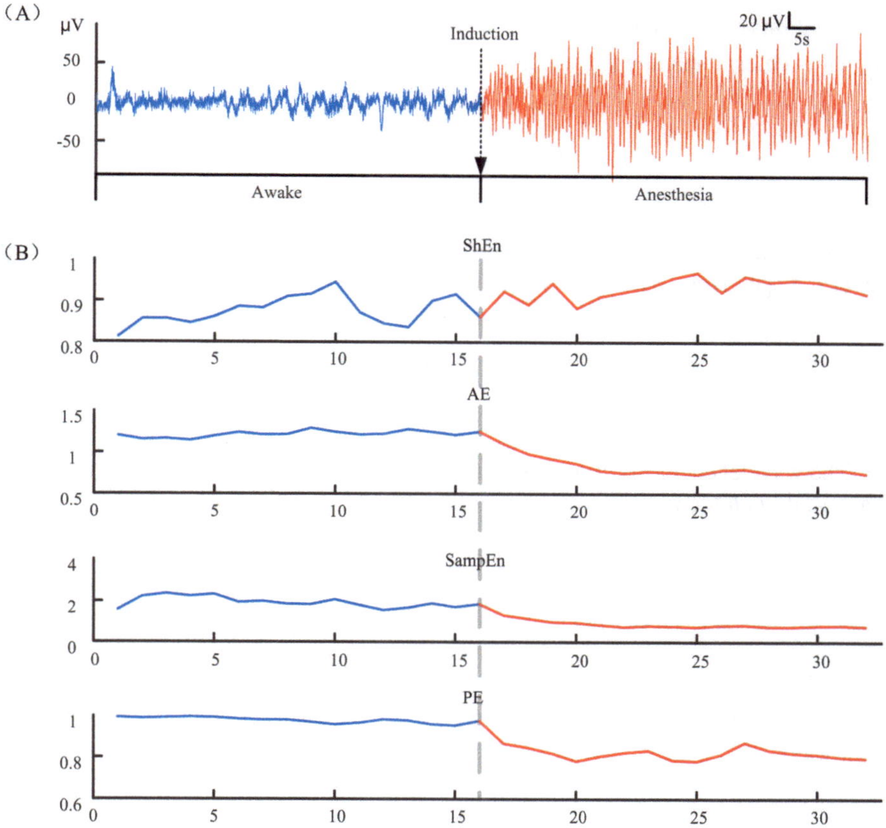

Fig. 11.3 Results of ShEn, AE, SampEn, and PE in analyzing EEG signals during awake (blue) and anesthesia (red) states

measurement. Normally, Fast Fourier transformation is frequently used to identify the power spectrum with time window moved stepwise. Besides, wavelet transform and Hilbert-Huang transform are the favorite approaches in signal processing area. Here, we present the wavelet entropy and Hilbert-Huang spectral entropy as examples.

11.3.2.1 Wavelet Entropy

In the recent years, WE is physiologically meaningful because it differentiates specific physiological brain states under spontaneous or stimulus-related conditions. When EEG signal behave more regularly and there is a symptom of synchronization in neuron activities, WE value reduces. Moreover, WE can recognize the time localizations of a dynamic process. To calculate the WE, wavelet energy $E_j^{(N_j)}$ of signal is determined at each scale$j$ as follows:

$$E_j^{(N_j)} = \sum_{k=m_j}^{m_j+N_j-1} d_j(k)^2 \tag{11.23}$$

where m_j and N_j are running index and the length of running windows within each level j, respectively. The total energy over all scales is obtained by:

$$E_{tot} = \sum_j E_j^{(N_j)} = \sum_i \sum_{k=m_j}^{m_j+N_j-1} d_j(k)^2 \tag{11.24}$$

The wavelet energy is divided by total energy to obtain relative wavelet energy at each scale j and length of running windowsN_j:

$$p_j^{(N_j)} = \frac{E_j^{(N_j)}}{E_{tot}} = \frac{E_j^{(N_j)}}{\sum_j E_j^{(N_j)}} = \frac{\sum_{k=m_j}^{m_j+N_j-1} d_j(k)^2}{\sum_j \sum_{k=m_j}^{m_j+N_j-1} d_j(k)^2} \tag{11.25}$$

The WE is calculated from entropy of $p_j^{(N_j)}$ distribution between scales as follows:

$$WE = -\sum_j p_j^{(N_j)} \log p_j^{(N_j)} \tag{11.26}$$

The detail of the algorithm used in this study can be seen in Särkelä et al. (2007). In addition, the values of WE are dependent on the wavelet basis function, the numbers of decomposed layers (n), and the data length (N). Among them, the wavelet basis function is the most important for the WE. For the lack of fixed criterion, it is very difficult to select an appropriate wavelet basis function in practical applications, and many studies are based on experiments.

11.3.2.2 Hilbert-Huang Spectral Entropy

To overcome some shortcoming of wavelet transform, Huang et al. in 1998 proposed empirical mode decomposition (EMD) in conjunction with Hilbert transform (Hilbert-Huang transform) method to process nonlinear and nonstationary signals (Huang et al. 1998). Now, it has been widely used to extract the features of EEG recordings (Liang et al. 2000; Rilling et al. 2003; Li 2006; Shalbaf et al. 2012). In particular, Li et al. had developed HHSE and successfully applied it to the anesthetic EEG signals (Li et al. 2008b). The HHSE is based on the transform, which applies the Shannon entropy concept to the Hilbert-Huang spectrum (Li et al. 2008b). For a given nonstationary signal $x(t)$, by the EMD method, decompose the signal into a series of intrinsic mode functions (IMFs), $C_n(1, 2, \ldots, N)$, where N is the number of IMFs. A signal $x(t)$ can be written by:

$$x(t) = \sum_{i=1}^{n-1} imf(t)_i + r_n(t) \tag{11.27}$$

Applying the Hilbert transform to the IMF components,

$$Z(t) = imf(t) + iH[imf(t)] = a(t)e^{i\int \omega(t)dt} \tag{11.28}$$

in which $a(t) = \sqrt{imf^2(t) + H^2[imf(t)]}$,$\omega(t) = \frac{d}{dt}[\arctan(H[imf(t)]/imf(t))]$,

$$h(\omega) = \int H(\omega, t)dt \tag{11.29}$$

where $\omega(t)$ and $a(t)$ are the instantaneous frequency and amplitude of IMF, respectively.

To simplify the representation, the Hilbert-Huang spectrum is denoted as a function of frequency (f) instead of angle frequency (ω). The marginal spectrum is normalized by:

$$\hat{h}(f) = h(f)/\sum h(f) \tag{11.30}$$

Next, apply the Shannon entropy concept to the Hilbert-Huang spectrum; Hilbert-Huang spectral entropy can be obtained by:

$$HHSE = -\sum_f \hat{h}(f)\log(\hat{h}(f)) \tag{11.31}$$

where HHSE is called Hilbert-Huang spectral entropy.

The HHSE values are mainly affected by frequency resolution and data length (N). To compute accurately, the frequency resolution is chosen as 0.1 Hz. The N directly influences the EMD. In general, that the N is too long or too small may induce the boundary effect, which can contaminate data and make power spectrum distortion.

Figure 11.4 shows results of WE and HHSE indexing the EEG signals. Both the WE and HHSE can show the trend of decreasing spectral entropy from the awake state to the anesthesia state. Comparing with WE, HHSE shows better consistency at each state and more distinct changes during the state transition. The wavelet transform has been suggested as a useful tool for presenting EEG signals in different time and frequency scales (Zoughi et al. 2012). In particular, entropy of signals in the wavelet domain (WE) indicates the variation of signal in each frequency scale

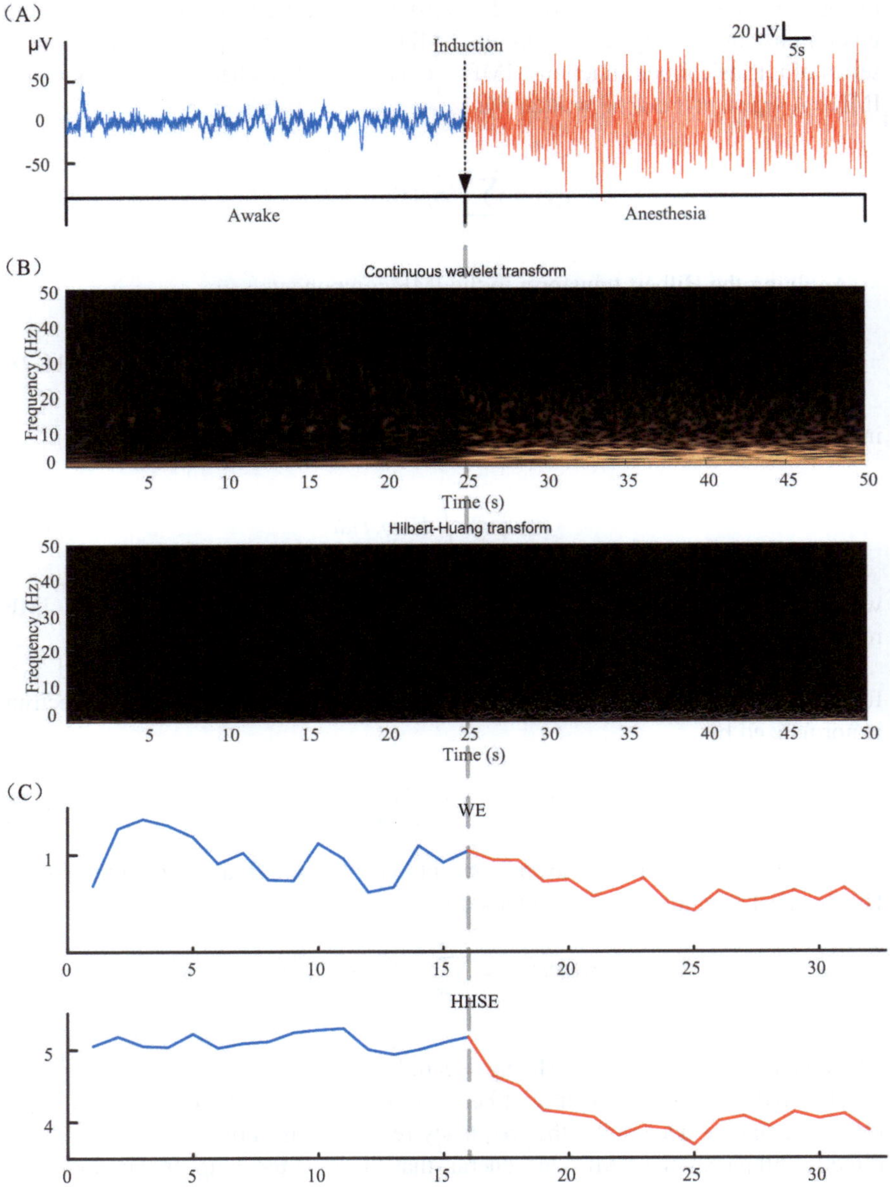

Fig. 11.4 Time-frequency spectrum and spectral entropy of the EEG signal at awake and anesthesia states. (**a**) The EEG signals. (**b**) Time-frequency spectrum measured by continuous wavelet transform and Hilbert-Huang transform. (**c**) Results of WE and HHSE indexing the awake (blue) and anesthesia (red) states

(Rosso et al. 2001). Although WE has been used to study the frequency contents in many studies, the wavelet transform algorithm presents some limitations. Wavelet transform can only deal with linear nonstationary signals in practical application. It is difficult to exactly localize a single oscillatory content contemporarily in time and frequency. In comparison with wavelet transform, the Hilbert-Huang transform could provide higher resolution simultaneously in time and frequency domains, which gives HHSE better ability of capturing the frequency changes in EEG signals. Moreover, the Hilbert-Huang transform can break down a complicated signal into several oscillatory modes without a basis function, such as sine or wavelet functions. Thus, it will not lead to a distortion of oscillatory modes in the instantaneous amplitude. Therefore, the advantages of the Hilbert-Huang transform make the HHSE more precise for estimating the complexity of the frequency components.

11.4 Hurst Exponent

Recently, studies found that the neural processing in human brain obey the power-law scaling relations (Worrell et al. 2002). The Hurst exponent concept is always employed to describe the dynamic change in power-law characteristics. It has been used to extract informative features from EEG recordings to indicate the effect of anesthetic drugs (Liang et al. 2012). The Hurst exponent analysis is based on spectrum. In this study, the harmonic wavelet was used to provide an unbiased and consistent estimation of the EEG power spectrum. The wavelet transform passes a filter $\psi(\cdot)$ over a time series $x(t)$:

$$\mathrm{W}_x(a,\tau) = \frac{1}{\sqrt{|a|}} \int x(t)\psi\left(\frac{t-\tau}{a}\right)dt \tag{11.32}$$

where $\psi(\cdot)$ means the basic or mother wavelet function. a and τ denote the scale factor and the translation of the origin, respectively. Then, the harmonic wavelet function can be given by:

$$\psi_{m,n}(t) = \frac{e^{jn2\pi t} - e^{jm2\pi t}}{j(n-m)2\pi t} \tag{11.33}$$

where m and n are the real-scale parameters, but not necessarily integers. Then, the wavelet transform can be denoted as

$$\mathrm{W}_s(n) = \sum_{n=0}^{N-1} x_n \psi_0^*\left(\frac{n-N}{s}\right) \tag{11.34}$$

where $*$ indicates the complex conjugate. Then, the wavelet spectrum of the x_n time series can be defined by replacing s using f:

$$S(f) = \bar{W}^2(f) \tag{11.35}$$

Then, the power-law could be analyzed by:

$$S(f) = \alpha f^{-\beta} \tag{11.36}$$

Considering the brain neural activities following the fractional Brownian motion random field model, the Hurst exponent H could be defined by:

$$\mathrm{H} = (\beta - 1)/2 \tag{11.37}$$

where $0 < \mathrm{H} < 1$, characterizing the persistence or anti-persistence of the x_n.

Figure 11.5 shows the Hurst exponent analysis for the EEG signal during awake and anesthesia states. In this study, the Hurst exponent was calculated using a 16 s time window with 75% overlap following a previous study (Yan et al. 2016). With a sliding time window, Hurst exponents could be calculated by each 16 s EEG recording. It shows that the Hurst exponents stay below 0.5 and rapidly increase to upper 0.5 after anesthetics given. Following the power-law concept, $\mathrm{H} < 0.5$ suggests anti-persistence behavior, while $0.5 < \mathrm{H} < 1$ suggests the persistence of the EEG recording. $\mathrm{H} = 0.5$ is a criticality state, which indicates a random fluctuation in the neural behaviors. The result suggests that the anesthetics could alter the neural oscillation model, which could be captured by power-law characteristics of EEG recordings. Thus, the Hurst exponents could be used to index the consciousness states in anesthesia. However, it should be noted that the power-law characteristics relay on a hypothesis that the brain neural behaviors obey the fractional Brownian motion random field model. It should be taken into consideration when using Hurst exponents for monitoring the altered states of a neural system.

11.5 Recurrence Plots

In the framework of nonlinear methods, the recurrence quantification analysis (RQA) is a suitable tool to characterize the dynamics of nonlinear systems through the recurrence plot, which provides a pictorial representation of recurrences in complex time series (Facchini et al. 2007; Mocenni and Stanley 2010). The recurrence plot is analyzed by RQA measures able to automatically extract and quantify recurrences.

Step 1: For a given time series $x_1, x_2, \cdots x_L$, the phase space vector X_i can be reconstructed by using the Takens' time delay method, $X_k = \left(x_k, x_{k+\tau}, \cdots, x_{k+(m-1)\tau}\right)$(Takens 1981), based on the observations x_k.

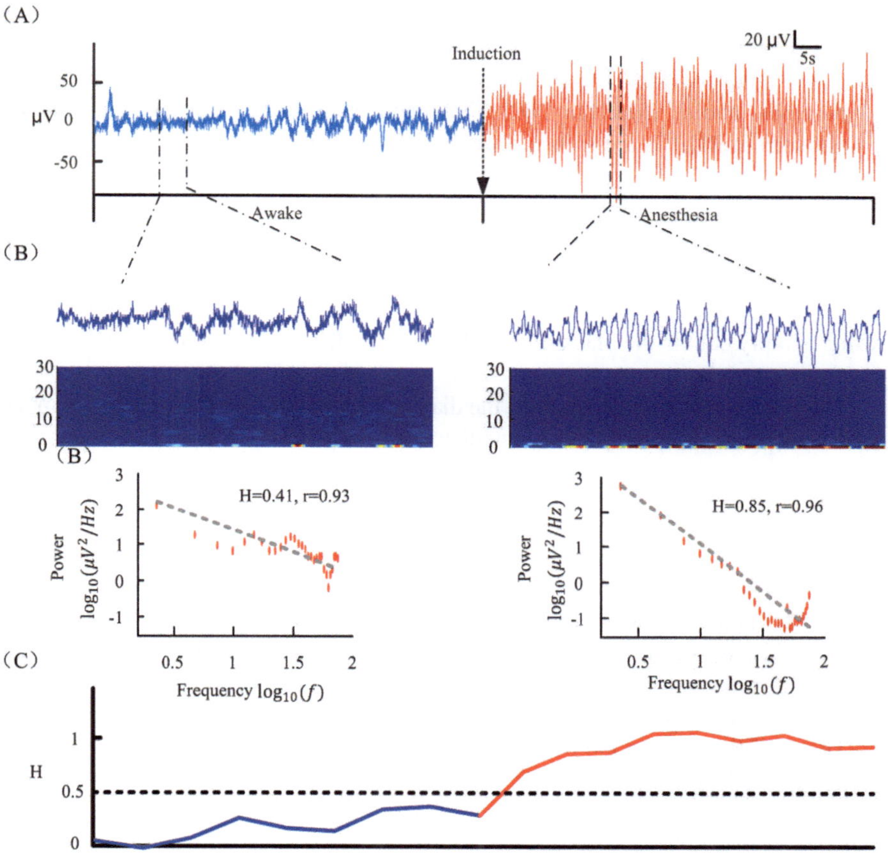

Fig. 11.5 Hurst exponent characteristic for power-law analysis of the EEG recording during the awake and anesthesia states. (**a**) EEG signals during the awake and anesthesia states. (**b**) Epochs of the EEG signal and corresponding harmonic wavelet spectrum, Hurst exponents, and linear correlation coefficient. (**c**) Hurst exponents for the EEG signal

Step 2: Determine the parameters m,τ, and r.

Step 3: Recurrence plotting (RP) is a new tool to visualize the time-dependent behavior of orbits X_i in a phase space. The key step of RP is to calculate the following N × N matrix:

$$R_{i,j} = \begin{cases} 1 : & \|X_i - Y_j\| \leq r \\ 0 : & otherwise \end{cases} \qquad i,j = 1, \cdots, N \tag{11.38}$$

where N is the point number of the times series for analysis and ||•||is the norm (the L_∞-norm is selected, because it is computationally faster and allows study of some features in RPs analytically). r is the cutoff distance defining an area centered at X_i.

Step 4: In order to further investigate the properties of RP, several measures of complexity quantify the small-scale structures in RP called recurrence quantification analysis (RQA). More details of RP and RQA can be found in several previous studies (Marwan 2003; Eckmann et al. 1987). Firstly, the simplest measure of RQA is the recurrence rate (RR), calculated by:

$$\mathrm{RR} = \frac{1}{N^2} \sum_{i,j=1}^{N} R_{i,j} \tag{11.39}$$

which simply counts the black dots in the RP. The RR is a measure of the density of recurrence points.

The ratio of recurrence points on the diagonal structures to all recurrence points is called determinism (DET) and is calculated by:

$$\mathrm{DET} = \sum_{l=l_{\min}}^{N} lP(l) \Big/ \sum_{i,j=1}^{N} R_{i,j} \tag{11.40}$$

where $P(l)$ is the frequency distribution of the lengths of the diagonal structures in the RP. $l_{\min}$ is the threshold, which excludes the diagonal lines formed by the tangential motion of a phase space trajectory. The DET is a determinism (or predictability) measure of a system.

The ENTR refers to the Shannon entropy of the frequency distribution of the diagonal line lengths; it is

$$\mathrm{ENTR} = -\sum_{l=l_{\min}}^{N} p(l) \ln p(l) \tag{11.41}$$

where ENTR is considered as a complexity measure of a deterministic structure in a dynamic system. The more complex the deterministic structure, the larger the ENTR value.

Figure 11.6 shows the estimators derived from the RQA. All the measures could distinctly represent the difference between the awake and anesthesia states as well as the decreasing trend during the state transition. In addition to the anesthesia, measures of RQA have been used in assessing pathological brain activities based on EEG (Ouyang et al. 2008). One of the advantages of the RQA method is that it can measure the dynamic characteristics of a short and nonstationary signal with noise (Li et al. 2004). In addition, RQA can deal with a linear and nonlinear time series to quantify the activity of a system irrespective of the number or dynamic nature of the individual sources (Marwan et al. 2007).

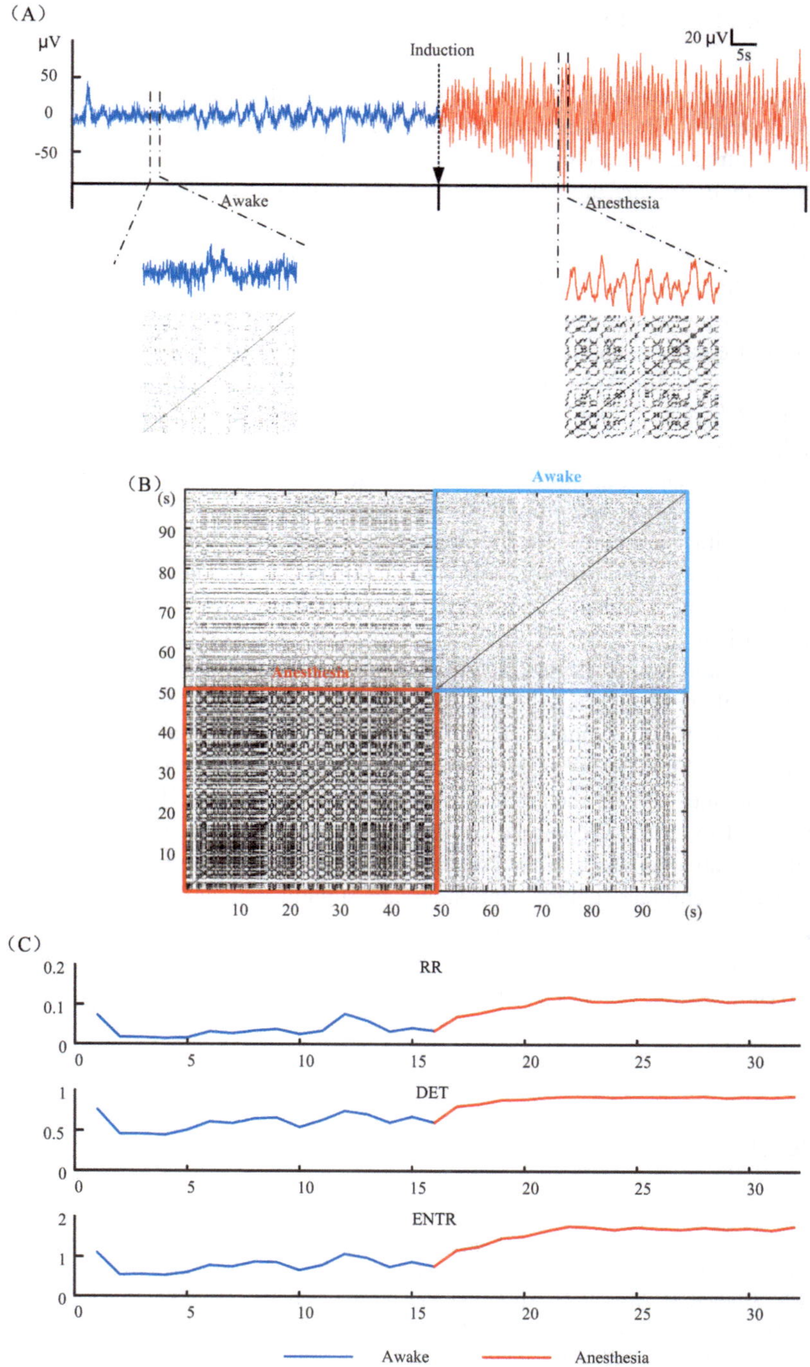

Fig. 11.6 (**a**) EEG signals during the awake and anesthesia states and corresponding recurrence plot patterns. (**b**) Recurrence plot patterns during awake (blue box) and anesthesia (red box) states. (**c**) The RR, DET, and ENTR index during awake (blue) and anesthesia (red) states

11.6 Conclusions

This study includes total 12 nonlinear dynamic measures which are frequently used in EEG studies. These measures could be simply divided into three classes: complexity, entropy, and pictorial analysis. The complexity measures the information contents by extracting the covert components in the signals. The entropy measures the distribution of characteristics of the signal based on Shannon information theory. Therefore, a number of entropy measures have been proposed based on statistic of different characteristics, which reflect various properties of the signals. In this study, there includes time domain-based measures (such as ApEn, SampEn, PE, and RQA) and time-frequency domain-based measures (such as the WE and HHSE). Among the 12 measures, the ApEn and SampEn are all based on the phase space analytical methods (Chen et al. 2009), while the PE and PLZC are based on the ordinal patterns (Bandt 2005).

All the measures but ShEn in this study show good ability in distinguishing the awake and anesthesia states based on EEG signals. However, all the results are based on the specific parameters for the measures, which are selected arbitrarily in this study. In practice, the parameter selection determines the efficiency of the measures. The details of the parameter selection of these measures could be found in previous studies (Liang et al. 2015). On the other hand, each of the measures has its advantages and limits. Some researchers have made efforts in comparing the performance of these measures (Bein 2006; Sleigh et al. 2001; Sleigh et al. 2005). However, from our point of view, there is no measure that could be simply considered as the best one. Generally, it just to say that there is a relatively more appropriate measure for the application in a specific situation. The parameter selection, robust to artifacts, compute consumption, and the requirements of the signals are all the valuable issues in the comparison of the measures.

Overall, nonlinear dynamic measures are suitable for describing the nonlinear property of the EEG signals. However, the comprehensive consideration of the signal properties as well as the relationship of the EEG characteristics with the underlying neural process is necessary in the selection of nonlinear dynamic measures.

References

Abasolo D, Hornero R, Gomez C, Garcia M, Lopez M. Analysis of EEG background activity in Alzheimer's disease patients with Lempel-Ziv complexity and central tendency measure. Med Eng Phys. 2006;28(4):315–22. https://doi.org/10.1016/j.medengphy.2005.07.004.

Aboy M, Hornero R, Abasolo D, Alvarez D. Interpretation of the Lempel-Ziv complexity measure in the context of biomedical signal analysis. IEEE Trans Biomed Eng. 2006;53(11):2282–8. https://doi.org/10.1109/TBME.2006.883696.

Anier A, Lipping T, Jantti V, Puumala P, Huotari A Entropy of the EEG in transition to burst suppression in deep anesthesia: surrogate analysis. In: Engineering in medicine and biology society (EMBC), 2010 annual international conference of the IEEE, 2010. IEEE, pp. 2790–3.

Bai Y, Liang Z, Li X. A permutation Lempel-Ziv complexity measure for EEG analysis. Biomed Signal Process Control. 2015;19:102–14.

Bandt C. Ordinal time series analysis. Ecol Model. 2005;182(3):229–38.

Bandt C, Pompe B. Permutation entropy: a natural complexity measure for time series. Phys Rev Lett. 2002a;88(17):174102.

Bandt C, Pompe B. Permutation entropy: a natural complexity measure for time series. Phys Rev Lett. 2002b;88(17):174102.

Bein B. Entropy. Best Pract Res Clin Anaesthesiol. 2006;20(1):101–9.

Bruhn J, Lehmann LE, Röpcke H, Bouillon TW, Hoeft A. Shannon entropy applied to the measurement of the electroencephalographic effects of desflurane. Anesthesiology. 2001;95 (1):30–5.

Bruhn J, Röpcke H, Hoeft A. Approximate entropy as an electroencephalographic measure of anesthetic drug effect during desflurane anesthesia. Anesthesiology. 2000;92(3):715–26.

Bruzzo AA, Gesierich B, Santi M, Tassinari CA, Birbaumer N, Rubboli G. Permutation entropy to detect vigilance changes and preictal states from scalp EEG in epileptic patients. A preliminary study. Neurol Sci. 2008;29(1):3–9. https://doi.org/10.1007/s10072-008-0851-3.

Cao Y, Tung W, Gao J, Protopopescu V, Hively L. Detecting dynamical changes in time series using the permutation entropy. Phys Rev E. 2004a;70(4; Part 2):46217.

Cao Y, Tung WW, Gao JB, Protopopescu VA, Hively LM. Detecting dynamical changes in time series using the permutation entropy. Phys Rev E Stat Nonlinear Soft Matter Phys. 2004b;70 (4. Pt 2):046217.

Chen CC, Hsu YT, Chan HL, Chiou SM, Tu PH, Lee ST, Tsai CH, Lu CS, Brown P. Complexity of subthalamic 13-35 Hz oscillatory activity directly correlates with clinical impairment in patients with Parkinson's disease. Exp Neurol. 2010;224(1):234–40. https://doi.org/10.1016/j.expneurol.2010.03.015.

Chen W, Zhuang J, Yu W, Wang Z. Measuring complexity using FuzzyEn, ApEn, and SampEn. Med Eng Phys. 2009;31(1):61–8.

Dauwels J, Srinivasan K, Ramasubba Reddy M, Musha T, Vialatte FB, Latchoumane C, Jeong J, Cichocki A. Slowing and loss of complexity in Alzheimer's EEG: two sides of the same coin? Int J Alzheimers Dis. 2011;2011:539621. https://doi.org/10.4061/2011/539621.

Eckmann JP, Kamphorst SO, Ruelle D. Recurrence plots of dynamical systems. EPL (Europhys Lett). 1987;4:973.

Facchini A, Mocenni C, Marwan N, Vicino A, Tiezzi E. Nonlinear time series analysis of dissolved oxygen in the Orbetello lagoon (Italy). Ecol Model. 2007;203(3–4):339–48.

Hu J, Gao J, Príncipe JC. Analysis of biomedical signals by the Lempel-Ziv complexity: the effect of finite data size. IEEE Trans Biomed Eng. 2006;53(12):2606–9.

Huang L, Yu P, Ju F, Cheng J. Prediction of response to incision using the mutual information of electroencephalograms during anaesthesia. Med Eng Phys. 2003;25(4):321–7.

Huang NE, Shen Z, Long SR, Wu MC, Shih HH, Zheng Q, Yen NC, Tung CC, Liu HH. The empirical mode decomposition and the Hilbert spectrum for nonlinear and non-stationary time series analysis. Proc Roy Soc Lond A. 1998;454(1971):903–95.

Inouye T, Shinosaki K, Sakamoto H, Toi S, Ukai S, Iyama A, Katsuda Y, Hirano M. Quantification of EEG irregularity by use of the entropy of the power spectrum. Electroencephalogr Clin Neurophysiol. 1991;79(3):204–10.

Jouny CC, Bergey GK. Characterization of early partial seizure onset: frequency, complexity and entropy. Clin Neurophysiol. 2012;123(4):658–69. https://doi.org/10.1016/j.clinph.2011.08.003.

Klaus B, Gerhard S, Matthias E, Andreas R, Kochs EF, Walter Z, Hans-Ulrich D. Anaesthesia monitoring by recurrence quantification analysis of EEG data. PLoS One. 2010;5(1):e8876.

Klockars JG, Hiller A, Munte S, van Gils MJ, Taivainen T. Spectral entropy as a measure of hypnosis and hypnotic drug effect of total intravenous anesthesia in children during slow induction and maintenance. Anesthesiology. 2012;116(2):340–51. https://doi.org/10.1097/ALN.0b013e3182410b5e.

Lempel A, Ziv J. On the complexity of finite sequences. IEEE Trans Inf Theory. 1976;22(1):75–81.
Li D, Li X, Liang Z, Voss LJ, Sleigh JW. Multiscale permutation entropy analysis of EEG recordings during sevoflurane anesthesia. J Neural Eng. 2010;7(4):046010.
Li X. Temporal structure of neuronal population oscillations with empirical model decomposition. Phys Lett A. 2006;356(3):237–41.
Li X, Cui S, Voss LJ. Using permutation entropy to measure the electroencephalographic effects of sevoflurane. Anesthesiology. 2008a;109(3):448.
Li X, Li D, Liang Z, Voss LJ, Sleigh JW. Analysis of depth of anesthesia with Hilbert–Huang spectral entropy. Clin Neurophysiol. 2008b;119(11):2465–75. https://doi.org/10.1016/j.clinph.2008.08.006.
Li X, Ouyang G, Richards DA. Predictability analysis of absence seizures with permutation entropy. Epilepsy Res. 2007;77(1):70.
Li X, Ouyang G, Yao X, Guan X. Dynamical characteristics of pre-epileptic seizures in rats with recurrence quantification analysis. Phys Lett A. 2004;333(1–2):164–71.
Li XL, Cui SY, Voss LJ. Using permutation entropy to measure the electroencephalographic effects of sevoflurane. Anesthesiology. 2008c;109(3):448–56. https://doi.org/10.1097/Aln.0b013e318182a91b.
Li XL, Ouyang GX. Estimating coupling direction between neuronal populations with permutation conditional mutual information. NeuroImage. 2010;52(2):497–507. https://doi.org/10.1016/j.neuroimage.2010.05.003.
Li Y, Tong S, Liu D, Gai Y, Wang X, Wang J, Qiu Y, Zhu Y. Abnormal EEG complexity in patients with schizophrenia and depression. Clin Neurophysiol. 2008d;119(6):1232–41. https://doi.org/10.1016/j.clinph.2008.01.104.
Li Z, Li X. Estimating temporal causal interaction between spike trains with permutation and transfer entropy. PLoS One. 2013;8(8):e70894.
Liang H, Lin Z, McCallum R. Artifact reduction in electrogastrogram based on empirical mode decomposition method. Med Biol Eng Comput. 2000;38(1):35–41.
Liang Z, Li D, Ouyang G, Wang Y, Voss LJ, Sleigh JW, Li X. Multiscale rescaled range analysis of EEG recordings in sevoflurane anesthesia. Clin Neurophysiol. 2012;123(4):681–8.
Liang Z, Wang Y, Sun X, Li D, Voss LJ, Sleigh JW, Hagihira S, Li X. EEG entropy measures in anesthesia. Front Comput Neurosci. 2015;9:16.
Linde Y, Buzo A, Gray RM. Algorithm for vector Quantizer design. Ieee T Commun. 1980;28(1):84–95. https://doi.org/10.1109/Tcom.1980.1094577.
Marwan N. Encounters with neighbours: current developments of concepts based on recurrence plots and their applications. Norbert Marwan; 2003.
Marwan N, Romano MC, Thiel M, Kurths J. Recurrence plots for the analysis of complex systems. Phys Rep. 2007;438(5):237–329.
Matic V, Cherian PJ, Koolen N, Ansari AH, Naulaers G, Govaert P, Huffel SV, Vos MD, Vanhatalo S (2015) Objective differentiation of neonatal EEG background grades using detrended fluctuation analysis. Front Hum Neurosci 9:189.
Mocenni C, Stanley HE. Identifying the dynamics of complex spatio-temporal systems by spatial recurrence properties. Proc Natl Acad Sci U S A. 2010;107(18):8097–102.
Olofsen E, Sleigh JW, Dahan A. Permutation entropy of the electroencephalogram: a measure of anaesthetic drug effect. Br J Anaesth. 2008;101(6):810–21. https://doi.org/10.1093/bja/aen290.
Ouyang G, Li X, Dang C, Richards DA. Using recurrence plot for determinism analysis of EEG recordings in genetic absence epilepsy rats. Clin Neurophysiol. 2008;119(8):1747–55.
Pincus SM. Approximate entropy as a measure of system complexity. Proc Natl Acad Sci. 1991;88(6):2297–301.
Röschke J, Fell J, Beckmann P. Nonlinear analysis of sleep eeg in depression: calculation of the largest lyapunov exponent. Eur Arch Psychiatry Clin Neurosci. 1995;245(1):27–35.
Rajendra AU, Faust O, Kannathal N, Chua TL, Laxminarayan S. Non-linear analysis of EEG signals at various sleep stages. Comput Methods Programs Biomed. 2005;80(1):37–45.

Richman JS, Moorman JR. Physiological time-series analysis using approximate entropy and sample entropy. Am J Phys Heart Circ Phys. 2000;278(6):H2039–49.

Rilling G, Flandrin P, Gonçalvés P. On empirical mode decomposition and its algorithms. In: IEEE-EURASIP workshop on nonlinear signal and image processing, NSIP-03, Grado (I), 2003.

Rosso OA, Blanco S, Yordanova J, Kolev V, Figliola A, Schurmann M, Basar E. Wavelet entropy: a new tool for analysis of short duration brain electrical signals. J Neurosci Methods. 2001;105 (1):65–76.

Särkelä MOK, Ermes MJ, van Gils MJ, Yli-Hankala AM, Jäntti VH, Vakkuri AP. Quantification of epileptiform electroencephalographic activity during sevoflurane mask induction. Anesthesiology. 2007;107(6):928–38.

Sabeti M, Katebi S, Boostani R. Entropy and complexity measures for EEG signal classification of schizophrenic and control participants. Artif Intell Med. 2009;47(3):263–74.

Schinkel S, Marwan N, Kurths J. Brain signal analysis based on recurrences. J Physiol Paris. 2009;103(6):315–23.

Schultz A, Siedenberg M, Grouven U, Kneif T, Schultz B. Comparison of Narcotrend index, Bispectral index, spectral and entropy parameters during induction of propofol-remifentanil anaesthesia. J Clin Monit Comput. 2008;22(2):103–11. https://doi.org/10.1007/s10877-008-9111-6.

Shalbaf R, Behnam H, Sleigh JW, Voss LJ. Using the Hilbert–Huang transform to measure the electroencephalographic effect of propofol. Physiol Meas. 2012;33(2):271–85. https://doi.org/10.1088/0967-3334/33/2/271.

Shannon CE. A mathematical theory of communication, Bell System. New York: American Telephone and Telegraph Co.; 1922.

Shannon CE, Weaver W. The mathematical theory of communication. Urbana: University of Illinois Press; 1949.

Shayegh F, Sadri S, Amirfattahi R, Ansari-Asl K. A model-based method for computation of correlation dimension, Lyapunov exponents and synchronization from depth-EEG signals. Comput Methods Programs Biomed. 2014;113(1):323–37.

Sleigh J, Voss L, Barnard J. What are electroencephalogram entropies really measuring? Int Congr Ser. 2005;1283:231–4. https://doi.org/10.1016/j.ics.2005.06.048.

Sleigh JW, Olofsen E, Dahan A, De Goede J, Steyn-Ross DA (2001) Entropies of the EEG: the effects of general anaesthesia.

Song IH, Lee DS, Kim SI. Recurrence quantification analysis of sleep electoencephalogram in sleep apnea syndrome in humans. Neurosci Lett. 2004;366(2):148–53.

Takens F. Dynamical systems and turbulence. Lecture Notes in Mathematics. 1981;898(9):366.

Viertiö-Oja H, Maja V, Särkelä M, Talja P, Tenkanen N, Tolvanen-Laakso H, Paloheimo M, Vakkuri A, Yli-Hankala A, Meriläinen P. Description of the Entropy™ algorithm as applied in the Datex-Ohmeda S/5™ Entropy Module. Acta Anaesthesiol Scand. 2004;48(2):154–61.

Webber JC, Zbilut JP. Dynamical assessment of physiological systems and states using recurrence plot strategies. J Appl Physiol. 1994;76(2):965.

Worrell GA, Cranstoun SD, Echauz J, Litt B. Evidence for self-organized criticality in human epileptic hippocampus. Neuroreport. 2002;13(16):2017.

Yan J, Wang Y, Ouyang G, Yu T, Li Y, Sik A, Li X. Analysis of electrocorticogram in epilepsy patients in terms of criticality. Nonlinear Dyn. 2016;83(4):1909–17.

Yoon YG, Kim TH, Jeong DW, Park SH. Monitoring the depth of anesthesia from rat EEG using modified Shannon entropy analysis. Conf Proc IEEE Eng Med Biol Soc. 2011;2011:4386–9. https://doi.org/10.1109/IEMBS.2011.6091088.

Zhang HX, Zhu YS, Wang ZM. Complexity measure and complexity rate information based detection of ventricular tachycardia and fibrillation. Med Biol Eng Comput. 2000;38(5):553–7.

Zhang X-S, Roy RJ, Jensen EW. EEG complexity as a measure of depth of anesthesia for patients. IEEE Trans Biomed Eng. 2001;48(12):1424–33.

Zhang XS, Roy RJ. Derived fuzzy knowledge model for estimating the depth of anesthesia. IEEE Trans Biomed Eng. 2001;48(3):312–23. https://doi.org/10.1109/10.914794.

Zhou S, Zhang Z, Gu J Interpretation of coarse-graining of Lempel-Ziv complexity measure in ECG signal analysis. Conference proceedings: annual international conference of the IEEE Engineering in Medicine and Biology Society IEEE Engineering in Medicine and Biology Society conference 2011: pp. 2716–2719; 2011. https://doi.org/10.1109/IEMBS.2011.6090745.

Zoughi T, Boostani R, Deypir M. A wavelet-based estimating depth of anesthesia. Eng Appl Artif Intell. 2012;25(8):1710–22. https://doi.org/10.1016/j.engappai.2011.10.006.

Chapter 12
Connectivity Analysis

Huibin Jia

Abstract The EEG signals could be used to assess the communication between brain regions. Various techniques have been developed in order to quantify the EEG connectivity of scalp-level EEG signals or source-level activities. Briefly speaking, four kinds of EEG connectivity measures are evaluated in literatures, including coherence-based measures, phase synchronization-based measures, generalized synchronization-based measures, and granger causality-based measures. All measures have their own advantages and disadvantages. Here, we illustrated the common sources problem in EEG analysis, the measures in EEG connectivity analysis, how to conduct EEG connectivity analysis using resting-state EEG signals and event-related EEG signals, and source-level connectivity. Moreover, we provided two examples of EEG connectivity, along with the EEG datasets and MATLAB codes, which are focused on the EEG connectivity of resting-state signals and event-related signals, respectively.

Keywords Functional connectivity · Source localization · Synchronization · Granger causality

12.1 Introduction

Each part of the human brain has its own function in human behaviors; thus, dozens of brain areas are commonly involved in even very easy tasks. Moreover, it has been well recognized that even during task-free resting state, the brain regions in default

---

Electronic supplementary material The online version of this chapter (https://doi.org/10.1007/978-981-13-9113-2_12) contains supplementary material, which is available to authorized users.

H. Jia (✉)
Key Laboratory of Child Development and Learning Science of Ministry of Education, School of Biological Sciences & Medical Engineering, Southeast University, Nanjing, Jiangsu, China

L. Hu, Z. Zhang (eds.), *EEG Signal Processing and Feature Extraction*,
https://doi.org/10.1007/978-981-13-9113-2_12

mode network (DMN) are still active (Buckner et al. 2008). Efficient communication between function-specific brain regions is essential for healthy cognitive functioning (Abrams et al. 2013). In macroscopic scale, this connectivity can be quantified by modern neuroimaging techniques. For example, we could detect and investigate the anatomical connections using MRI and DTI (Rasero et al. 2017; Ercan et al. 2016). On the other hand, the EEG technique is applied to quantify the statistical dependency, i.e., temporal coupling/dependency, between the EEG signals of spatially remote brain regions (O'Reilly et al. 2017).

In order to make full use of the rich temporal-spatial-spectral information of EEG signals and overcome the inherent limitations (e.g., low spatial resolution and reference dependent) of EEG techniques, researchers have developed many different measures in EEG connectivity. According to the ability whether the measures could detect the causal relationship between brain regions, they could be divided into two categories: functional connectivity (FC) and effective connectivity (EC). The measures in FC could assess the linear or nonlinear statistical dependence between the signals without providing any causal information, whereas the measures in EC could assess the causal relationship between signals. According to the logic of how to define statistical dependence, they could be divided into coherence-based measures, phase synchronization-based measures, generalized synchronization-based measures, and granger causality-based measures (Niso et al. 2013). A large number of studies have been conducted in the past decades using these EEG connectivity measures. These studies have enriched our understanding of how the human brain operates in normal or pathological conditions and found that several factors (e.g., age, gender, task operation, cognitive processing, and neuropsychological diseases) could modulate the connectivity between different brain regions (Abrams et al. 2013; Duann and Ide 2009; Huster et al. 2014; Heise et al. 2014).

Below, we will illustrate the common sources problem in EEG analysis, the measures in EEG connectivity analysis, how to conduct EEG connectivity analysis using resting-state EEG signals and event-related EEG signals, and source-level connectivity.

12.2 Common Sources Problem

The common sources problem in EEG analysis refers to two kinds of problems: the active reference electrode(s) and the volume conduction.

A very important operation in EEG preprocessing is selecting appropriate active reference electrode(s). Several different types of reference, including the nose tip reference (NT), the vertex reference (the electrode Cz), the linked mastoids reference (LM), the common average reference (CAR), and the left mastoid reference (L), are currently used in literatures (Yao et al. 2007). It has been shown in many studies that the selection of reference electrode(s) can affect estimation of certain EEG measures, such as EEG spectral power, amplitude and latency of ERP component, and some measures in EEG connectivity (Yao et al. 2007; Yao et al. 2005; Nunez et al. 1997).

Volume conduction is the phenomenon that electromagnetic fields generated at a location in the brain will propagate through tissues (brain, skull, skin, etc.) and will be recorded by several electrodes (van den Broek et al. 1998). Volume conduction is a double-edged sword. Without it, noninvasive EEG would not be possible; because of it, EEG connectivity measurements can be confounded (Cohen 2015). In order to overcome the disadvantage of volume conduction and recover the true connectivity between brain regions, we could use either of the following two approaches (Cohen 2015). In the first approach, we could apply spatial filters (e.g., the surface Laplacian transform and source localization), which could strongly attenuate volume conduction and therefore permit the valid interpretation of standard connectivity analysis methods, to scalp EEG data, and then estimate the measures in connectivity analysis (Srinivasan et al. 2007; Sun et al. 2014). In the second approach, the EEG connectivity measures which are insensitive to the volume conduction are applied to the scalp EEG signals (Vinck et al. 2011; Stam et al. 2007; Nolte et al. 2004). Note that these two different approaches are used in combination in some studies, i.e., computing the volume conduction insensitive measures on spatial filtered signals.

12.3 The Measures in EEG Connectivity Analysis

Below, four kinds of EEG connectivity measures are illustrated: coherence-based measures, phase synchronization-based measures, generalized synchronization-based measures, and granger causality-based measures. Note that the simplest measure in EEG connectivity is Pearson's correlation coefficient between EEG signals of two electrodes/brain regions. Although this measure is very simple and fast to compute, it is rarely used in literatures due to its inherent limitations (e.g., failing to use the rich spectral information of EEG signals, and difficulty in interpreting the negative correlation coefficients). Thus, this measure is not explained below.

12.3.1 Coherence-Based Measures

The magnitude-squared coherence (or, simply, the coherence) could assess the linear relationship between two signals at each frequency bin evaluated (Niso et al. 2013). Suppose $X(t)$ and $Y(t)$ represent the EEG signals over electrodes (or brain regions) X and Y, respectively. Firstly, the time domain signals $X(t)$ and $Y(t)$ are converted to frequency domain using fast Fourier transform (FFT) or discrete/continuous wavelet transform (DWT/CWT). Then, for each frequency bin f, the individual spectral power density $S_{xx}(f)$ and $S_{yy}(f)$ and their cross spectral power density $S_{xy}(f)$ are estimated. The coherency function $K_{xy}(f)$ is calculated as the ratio between cross spectral power density and individual spectral power density of two signals.

$$K_{xy}(f) = \frac{S_{xy}(f)}{\sqrt{S_{xx}(f)S_{yy}(f)}} \tag{12.1}$$

Lastly, the coherence at frequency bin f is computed as

$$\mathrm{COH}_{xy}(f) = \left|K_{xy}(f)\right|^2 \tag{12.2}$$

The range of coherence value is 0–1. $\mathrm{COH}_{xy}(f) = 0$ means that there is no linear dependence between $X(t)$ and $Y(t)$ at frequency f. Larger coherence value suggests higher statistical dependence between the two signals, and vice versa.

The coherence has been widely used in EEG connectivity studies. However, it has some limitations. Firstly, it could only assess the linear dependence between two signals thus fails to detect their nonlinear relationship. Secondly, it is significantly influenced by the amplitudes of signals. Thirdly, it could not dissociate volume conduction from true interactions between brain regions.

To overcome some limitations of the coherence, Nolte et al. (2004) proposed that the imaginary part of the coherency function (ImC) can serve as an alternative EEG connectivity measure, which could eliminate the influence of volume conduction based on the hypothesis that volume conduction of uncorrelated sources cannot create a nonzero ImC. Note that this ImC also has a crucial disadvantage. As has been shown in Stam et al. (2007), this measure is strongly influenced by the phase of the coherency function, since it is most effective in detecting connectivity with a phase lag corresponding to a quarter cycle (i.e., $\pi/4$), and breaks down if the two sources of interest are in phase or in phase opposition (Stam et al. 2007).

These two coherence-related measures could not provide any causal information about the EEG signals; thus, they are regarded as FC measures. In the family of coherence-based measures, researchers developed the Phase Slope Index (PSI), which is an EC measure that could assess the causal dependence between signals. The basic idea behind PSI is that interactions between neural processes require some time, and if the speed at which different waves travel is similar, then the phase difference between the sender and recipient increases with frequency, and we expect a positive slope of the phase spectrum (Nolte et al. 2007).

In Nolte et al. (2007), the PSI is defined as

$$\tilde{\Psi}_{xy} = \Im\left(\sum_{f \in F} K_{xy}^{*}(f) K_{xy}(f + \delta f)\right), \tag{12.3}$$

where $K_{xy}(f)$ is the coherency function, δf is the frequency resolution of FFT, F is the set of frequencies over which the slope is summed, and $\Im(\cdot)$ is the imagery part.

It is convenient to normalize $\tilde{\Psi}_{xy}$ by an estimate of its standard deviation, i.e.,

$$\Psi_{xy} = \tilde{\Psi}_{xy}/\mathrm{std}\left(\tilde{\Psi}_{xy}\right) \tag{12.4}$$

The $\mathrm{std}\left(\tilde{\Psi}_{xy}\right)$ can be estimated by the jackknife method. Normally, a Ψ_{xy} value larger than 2 indicates statistically significant time delay between $X(t)$ and $Y(t)$ in the frequency range considered, i.e., EEG signal $X(t)$ could significantly influence $Y(t)$. Nolte et al. (2007) proved that this method outperformed traditional EC measures, i.e., the Granger causality, in the detection of directionality.

12.3.2 Phase Synchronization-Based Measures

The coherence-based measures could not separate the amplitude and phase of EEG signals during assessing connectivity, whereas the phase synchronization-based measures illustrated here could separate the phase component and amplitude component of EEG signals for a given frequency or frequency band.

The phase synchronization has been widely used to explore the nonlinear dynamics in chaotic systems (Yu 2013). If the phases of two oscillatory activities synchronize with a constant phase difference, we could say that phase synchronization exists. Suppose $\phi_x(t)$ and $\phi_y(t)$ are the phase time series of EEG signals $X(t)$ and $Y(t)$ at time point t, respectively, and $\Delta\phi_{xy}(t)$ is their phase difference or relative phase at time point t. If, for each time point, the following equation holds,

$$\Delta\phi_{xy}(t) = \left|\phi_x(t) - \phi_y(t)\right| \leq \text{constant} \tag{12.5}$$

then phase synchronization is detected. For practical analysis, the relative phase difference should be wrapped to the interval $[0, \ 2\pi)$ through the following equation:

$$\Delta\phi_{rel}(t) = \Delta\phi_{xy}(t)\mathrm{mod}2\pi \tag{12.6}$$

where mod means the remainder after division.

In practical EEG connectivity analysis, firstly we need to obtain the instantaneous phase time series of the EEG signal at each electrode (or brain region) through classical frequency domain transform methods, such as combining band-pass filtering and the Hilbert transform, short-time Fourier transform (STFT), and DWT/CWT. Below we will show how to extract instantaneous phase time series through band-pass filtering and the Hilbert transform and compute phase synchronization-related measures.

Before the Hilbert transform, a band-pass filtering is needed since the Hilbert transform requires a narrowband signal for the phase and amplitude components of the signal to be straightforward to separate. One of the most commonly used filters is the finite impulse response (FIR) filter. It is suggested that the filter order of the FIR filter is at least three cycles of the lower-frequency component of the interested frequency band.

Suppose the band-pass filtered signals at two electrodes (or brain regions) are $X(t)$ and $Y(t)$, respectively. Through the Hilbert transform, we can obtain their analytic signals $X_{an}(t)$ and $Y_{an}(t)$, as

$$X_{\text{an}}(t) = X(t) + iX_{\text{H}}(t) \tag{12.7}$$

$$\mathrm{Y}_{\text{an}}(\mathrm{t}) = \mathrm{Y}(\mathrm{t}) + \mathrm{iY}_{\text{H}}(\mathrm{t}) \tag{12.8}$$

where $X_H(t)$ and $Y_H(t)$ are the Hilbert transform of band-pass filtered signals $X(t)$ and $Y(t)$, respectively. Namely,

$$X_{\text{H}}(t) = \frac{1}{\pi}\text{P.V.}\int_{-\infty}^{+\infty} \frac{X(\tau)}{t-\tau}\text{d}\tau \tag{12.9}$$

$$Y_{\text{H}}(t) = \frac{1}{\pi}\text{P.V.}\int_{-\infty}^{+\infty} \frac{Y(\tau)}{t-\tau}\text{d}\tau \tag{12.10}$$

where P. V. is the Cauchy principal value.

Using the analytic signal $X_{an}(t)$, we can calculate the instantaneous amplitude $A_x(t)$ and instantaneous phase $\phi_x(t)$ (similar for $Y_{an}(t)$).

$$A_{\text{x}}(t) = \sqrt{X(t)^2 + X_{\text{H}}(t)^2} \tag{12.11}$$

$$\phi_{\text{x}}(t) = \tan^{-1}\frac{X_{\text{H}}(t)}{X(t)} \tag{12.12}$$

After the instantaneous phase time series $\phi_x(t)$ and $\phi_y(t)$ are computed, we can compute their phase difference $\Delta\phi_{rel}(t)$ using Eqs. (12.5) and (12.6).

Lachaux et al. (1999) proposed that the phase synchronization between EEG signals can be quantified using the phase difference, resulting in an index called the Phase Locking Value (PLV) (Lachaux et al. 1999). Suppose $\Delta\phi_{rel}(t)$ is the phase difference between $X(t)$ and $Y(t)$, then their PLV is evaluated as

$$\text{PLV}_{\text{XY}} = \left|\frac{1}{N}\sum_{n=1}^{N} \text{e}^{i\Delta\phi_{\text{rel}}(t_n)}\right| \tag{12.13}$$

where N is the length (i.e., number of points) of $\Delta\phi_{rel}(t)$.

This PLV actually assesses the distribution of phase difference time series over $[0, 2\pi)$. A larger PLV indicates that the phase difference time series occupies a small portion of the unit circle of $[0, 2\pi)$.

The range of PLV value is [0, 1]. If the PLV is 1, then the phase difference time series is constant across the entire time series. If the PLV is 0, then the phase difference is uniformly distributed over $[0, 2\pi)$. Although this measure has been widely in traditional EEG connectivity, it has an important limitation, i.e., sensitive to the common sources problem.

The phase lag index (PLI) proposed by Stam et al. (2007) can provide reliable estimate of phase synchronization that is robust against the presence of common sources. A basic idea underlying EEG connectivity analysis is that the true interactions between neural oscillations (as opposed to volume conduction effects) need some period; thus, we could observe a certain time lag between these two neural signals with true connectivity (Stam et al. 2007). A lagged interaction with fixed time, as opposed to instantaneous interaction which may be caused by the common sources problem, should produce a constant phase difference between two neural signals at a value different from 0 and π. This is also the basic hypothesis of the ImC illustrated above. Irrespective of the magnitude of the phase leads and lags, the PLI discards phase distributions that centered on 0 mod π and assesses the asymmetry of phase difference distribution. If the number of time points with phase difference $\Delta\phi_{\text{rel}}(t)$ in the interval $(-\pi, 0)$ is different from that with phase difference $\Delta\phi_{\text{rel}}(\text{t})$ in the interval $(0, \pi)$, then asymmetry of phase difference distribution exists, suggesting that true connectivity is present between the two neural signals. Larger asymmetry suggests larger EEG connectivity, whereas symmetric distribution (i.e., the number of time points with phase difference $\Delta\phi_{\text{rel}}(t)$ in the interval $(-\pi, 0)$ is equal to that with phase difference $\Delta\phi_{\text{rel}}(\text{t})$ in the interval $(0, \pi)$) indicates no true coupling. Formally, the PLI is computed using the following equation:

$$\text{PLI} = \left| \frac{1}{N} \sum_{n=1}^{N} \text{sign}(\Delta\phi_{\text{rel}}(t_n)) \right| \tag{12.14}$$

The range of PLI is between 0 and 1. A PLI of zero indicates either no coupling or coupling with a phase difference centered around 0 mod π. A PLI of 1 indicates perfect phase locking at a value different from 0 mod π.

Stam et al. (2007) showed that compared to the ImC, the PLI performed better in detecting the true connectivity and was more robust to the common sources problem. However, the discontinuity of this measure is an important disadvantage, since small perturbations turn phase lags into leads and vice versa, a problem that may become serious for small-magnitude synchronization effects (Vinck et al. 2011). In order to overcome this limitation, Vinck et al. (2011) introduced a related index, namely, the weighted phase lag index (WPLI). In WPLI, the contribution of the observed phase leads and lags is weighted by the magnitude of the imaginary component of the cross-spectrum. Compared to the PLI, the WPLI has at least two advantages, in terms of reduced sensitivity to additional, uncorrelated noise sources and increased statistical power to detect changes in phase synchronization. Assuming S is the cross-spectrum between $X(t)$ and $Y(t)$, the WPLI can be calculated as

$$\mathrm{WPLI} = \frac{|\langle \Im(S) \rangle|}{\langle |\Im(S)| \rangle} = \frac{|\langle |\Im(S)| \mathrm{sign}(\Im(S)) \rangle|}{\langle |\Im(S)| \rangle} \tag{12.15}$$

The range of WPLI is between 0 and 1. Higher WPLI value indicates higher coupling between neural oscillations, and vice versa.

The three phase synchronization-based measures (i.e., PLV, PLI, and WPLI) could not assess the causal relationship between the EEG signals of two electrodes (or brain regions).

12.3.3 Generalized Synchronization-Based Measures

Apart from the coherency and phase synchronization-based measures, some researchers argued that the EEG connectivity could be evaluated by the generalized synchronization-based measures (Stam and van Dijk 2002). For two dynamical systems X and Y, if the state of the response system Y is a function of the state of the driving system X, i.e., $Y = F(X)$, the generalized synchronization exists. Although several algorithms have been introduced to detect the generalized synchronization between physiological signals, the most commonly used measure in brain connectivity analysis is the synchronization likelihood (SL) proposed by Stam and van Dijk (2002), which could avoid the bias of other generalized synchronization-based measures and provide a normalized estimation of dynamical interaction between multi-electrode EEG signals (Stam and van Dijk 2002).

Assuming the number of electrodes or brain regions is M and the total number of time point is N in our EEG signals, the EEG data is $x_{k,\ i}$, where k is electrode or region index ($k = 1, \ldots, M$) and i is the time point ($i = 1, \ldots, N$). From this multi-electrodes or multi-regions EEG signal, the embedded vectors $X_{k,\ i}$ are constructed as following:

$$X_{k,i} = \left(x_{k,i}, x_{k,i+\tau}, x_{k,i+2\tau}, \ldots, x_{k,i+(d-1)\tau}\right) \tag{12.16}$$

where τ and d are the time lag and embedding dimension, respectively.

Then, for electrode (or brain region) k and time point i, $P^{\varepsilon}_{k,i}$, which could assess the probability that embedded vectors are closer to each other than a given distance ε, is defined:

$$P^{\varepsilon}_{k,i} = \frac{1}{2(w_2 - w_1)} \sum_{\substack{j=1 \\ w_1 < |i-j| < w_2}}^{N} \theta\left(\varepsilon - \left|X_{k,i} - X_{k,j}\right|\right) \tag{12.17}$$

where θ is the Heaviside step function ($\theta(x) = 1$ when $x \leq 0$, otherwise 0), w_1 is the Theiler window which is used to avoid the autocorrelation effects and should be at least of the order of autocorrelation time, and w_2 is a window that sharpens the time resolution of the synchronization measure and is chosen such that $w_1 \ll w_2 \ll N$.

For each electrode (or brain region) and each time point, a critical distance $\varepsilon_{k,\,i}$ is determined according to $P_{k,i}^{\varepsilon_{k,i}} = P_{\text{ref}}$, where $P_{\text{ref}} \ll 1$. Then, for each pair of time point (i,j) within the window being considered ($w_1 < |i - j| < w_2$), we could compute the number of channels $H_{i,\,j}$ where the distance between $X_{k,\,i}$ and $X_{k,\,j}$ is smaller than the critical distance $\varepsilon_{k,\,i}$. This $H_{i,\,j}$ is calculated as:

$$H_{i,j} = \sum_{k=1}^{M} \theta\left(\varepsilon_{k,i} - \left|X_{k,i} - X_{k,j}\right|\right) \tag{12.18}$$

The synchronization likelihood $\mathrm{SL}_{k,\,i,\,j}$, where k denotes a given electrode (or brain region) i and j denote two distinct time points, is evaluated using the following criterion:

$$\text{if } \left|X_{k,i} - X_{k,j}\right| < \varepsilon_{k,i}, \text{then } \mathrm{SL}_{k,i,j} = \frac{\mathrm{H}_{i,j} - 1}{M - 1}. \text{ Otherwise } \mathrm{SL}_{k,i,j} = 0. \tag{12.19}$$

Through averaging over all the j, we could obtain the synchronization likelihood $\mathrm{SL}_{k,\,i}$, which describes how strongly electrode k at time i is synchronized to all the other $M - 1$ electrodes:

$$\mathrm{SL}_{k,i} = \frac{1}{2(w_2 - w_1)} \sum_{\substack{j=1 \\ w_1 < |j-i| < w_2}}^{N} \mathrm{SL}_{k,i,j} \tag{12.20}$$

The range of SL value is between P_{ref} and 1. If the calculated SL is P_{ref}, then all the M electrodes (or brain regions) are unrelated with each other. If the calculated SL is 1, we could say that a maximum synchronization between these M electrodes (or brain regions) is revealed. Note that this measure could not provide any causal information about the M time series.

12.3.4 Granger Causality-Based Measures

All of the measures illustrated above could not assess the causal relationship or direction of information flow between different neural oscillations, except the PSI. In order to assess the causal dependence or directed information flow between brain regions, researchers have developed several methods that have been applied to the fMRI, EEG, and MEG connectivity, such as the dynamic causal modeling, structural equation modeling, transfer entropy, and the Granger causality methods (Friston et al. 2003; Astolfi et al. 2004; Vicente et al. 2011; Gao et al. 2011).

The theory behind Granger causality (GC) was first developed by Wiener (1956), and later reformulated and formalized by Granger (1969) through linear stochastic autoregressive models (Wiener 1956; Granger 1969). Granger defined the causality between time series based on the following two assumptions: (1) the cause must happen prior to its effect; (2) the cause has information that could improve the prediction of the effect, above and beyond the information contained in the past of the effect. The basic idea of GC is that if time series *X*(*t*) is the GC of another time series *Y*(*t*), then predictions of the further value of *Y*(*t*) based on its own past values and the past value of X(*t*) are better than predictions of the further value of *Y*(*t*) only based on its own past values. Below, we will illustrate the GC using mathematical statement.

Assuming we want to test whether time series *X*(*t*) is the GC of another time series *Y*(*t*), we need to construct the first autoregressive (AR) model with *Y*(*t*) as dependent variable and its past values as independent variables:

$$Y(t) = a_0 + a_1 Y(t-1) + a_2 Y(t-2) + \ldots + a_p Y(t-p) + \varepsilon_{Y|Y^*} \tag{12.21}$$

Then, the second AR model is constructed with *Y*(*t*) as dependent variable, too. Different from the first AR model, both the past values of *X*(*t*) and Y(*t*) are used as independent variables.

$$\begin{aligned} Y(t) = b_0 + b_1 Y(t-1) + b_2 Y(t-2) + \ldots + b_p X(t-p) + c_1 X(t-1) \\ + c_2 X(t-2) + \ldots + c_p X(t-p) + \varepsilon_{Y|X^*,Y^*} \end{aligned} \tag{12.22}$$

In the above two AR models, $a_i(i = 1,2,\ldots,p)$, $b_i(i = 1,2,\ldots,p)$, and $c_i(i = 1,2,\ldots, p)$ are the model parameters, which can be estimated through least-squares methods, lattice algorithms (e.g., Vieira-Morf), or state-space methods (e.g., Kalman filtering). The p in the AR models is model order, which needs to be determined before AR model fitting. Commonly, the model order is determined by selecting a model order that could minimize one or more information criteria evaluated over a range of model orders. The information criteria that have been used in literatures include Akaike information criterion (AIC), Bayesian information criterion (BIC), Akaike's final prediction error criterion (FPE), and Hannan-Quinn criterion (HQ) (Lutkepohl 2005). Since the model order is the only one parameter needed to be defined by user, it is crucial that a proper model order must be selected. In practice, the model order should be selected by examining the results of multiple information criteria. Moreover, it should be noted that the model order selection can be a time-consuming iterative process. In the above two AR models, $\varepsilon_{Y|Y^*}$ and $\varepsilon_{Y|X^*,Y^*}$ are the residuals associated to the models.

After the AR models have been fitted, the variances of residuals were computed over time:

$$V_{Y|Y^*} = \text{var}\left(\varepsilon_{Y|Y^*}\right) \tag{12.23}$$

$$V_{Y|X^*,Y^*} = \mathrm{var}\left(\varepsilon_{Y|X^*,Y^*}\right) \tag{12.24}$$

where var(·) denotes variance over time.

If *X*(*t*) Granger-causes *Y*(*t*), then the variance of residuals of the second AR model (i.e., $V_{Y|X^*,Y^*}$) should be smaller than that of the first AR model (i.e., $V_{Y|Y^*}$). From the variances of residuals of two AR models, we could calculate a classical time-domain GC measure, which is defined as:

$$\mathrm{GC}_{X\to Y} = \ln\left(\frac{V_{Y|Y^*}}{V_{Y|X^*,Y^*}}\right) \tag{12.25}$$

The range of this $\mathrm{GC}_{X\to Y}$ measure is $[0, +\infty)$. If $\mathrm{GC}_{X\to Y}$ equals 0, the past information of *X*(*t*) could not improve the prediction of *Y*(*t*). If $\mathrm{GC}_{X\to Y}$ is larger than 0, *X*(*t*) Granger-causes *Y*(*t*).

It should be noted that the traditional GC tests are designed to test the causal relationship between pairs of processes, which may introduce misleading results since their relationship may be affected by another or more processes. In order to solve this problem, multivariate Granger causality analysis has been developed, which is performed by fitting a vector autoregressive (VAR) model to multi-electrodes or multi-regions EEG signals. Consider a VAR model of order p with dimension M (i.e., the number of electrodes or brain regions considered in GC analysis is M):

$$\begin{pmatrix} X_1(\mathrm{t}) \\ \cdot \\ \cdot \\ \cdot \\ X_M(\mathrm{t}) \end{pmatrix} = \sum_{\tau=1}^{p} \mathrm{A}_\tau \begin{pmatrix} X_1(t-\tau) \\ \cdot \\ \cdot \\ \cdot \\ X_M(t-\tau) \end{pmatrix} + \begin{pmatrix} \varepsilon_1(\mathrm{t}) \\ \cdot \\ \cdot \\ \cdot \\ \varepsilon_M(\mathrm{t}) \end{pmatrix} \tag{12.26}$$

where A_τ ($\tau = 1, \ldots, p$) are $M \times M$ coefficient matrices.

If the coefficient matrices of VAR model are Fourier transformed, we could compute more complex frequency-domain GC measures, such as the partial directed coherence (PDC) and directed transfer function (DTF) (Baccalá and Sameshima 2001; Kamiński and Blinowska 1991).

Note that the VAR modeling has two crucial assumptions of the signals being investigated, which are stationary and stability. A stochastic process *X*(*t*) is weakly stationary if its mean and covariance do not change over time. A stable process will not "blow up" (i.e., diverge to infinity). An important fact is that stability implies stationarity; thus, it is sufficient to test the stability to ensure that a VAR process is both stable and stationary. It should be noted that the EEG signals, especially those containing prominent evoked potentials, are usually nonstationary time series, exhibiting large fluctuations in both the mean and variance over time. Many methods have been proposed for fitting VAR models to nonstationary series. These methods include segmentation-based adaptive VAR (AMVAR) approaches, the Kalman filtering approach, and nonparametric methods based on minimum-phase spectral matrix factorization (Ding et al. 2000; Sommerlade et al. 2009; Dhamala et al. 2008). Apart from these methods, some preprocessing procedures could improve the stationarity of EEG signals, such as filtering, differencing, and detrending. If EEG data contains low-frequency drift or pronounced artifacts within certain frequency bands, the signal filtering can dramatically improve data stationarity. Since the phase difference in frequency domain are key element in information flow modeling, it is critical to apply a zero-phase (acausal) filter, which does not introduce phase distortion. The differencing can also improve the stationarity of signals. A first-order differencing for time series $X(t)$ is calculated as $\nabla X^1(t) = X(t) - X(t-1)$. This operation can extend to nth order differencing: $\nabla^n X(t) = \nabla^{n-1} X(\mathrm{t}) - \nabla^{n-1} X(t-1)$. Note that differencing is a high-pass filter and may influence the magnitude of frequency-domain GC measures. If linear trend is present in the signals, we may need to linearly detrend the signals using a least-squares fit, which will act as a high-pass filtering.

12.4 Summary

Above, the classical measures in EEG connectivity are explained in four sections: coherence-based measures, phase synchronization-based measures, generalized synchronization-based measures, and Granger causality-based measures. The pipeline of estimating these measures has been shown in Fig.12.1. Note that many other

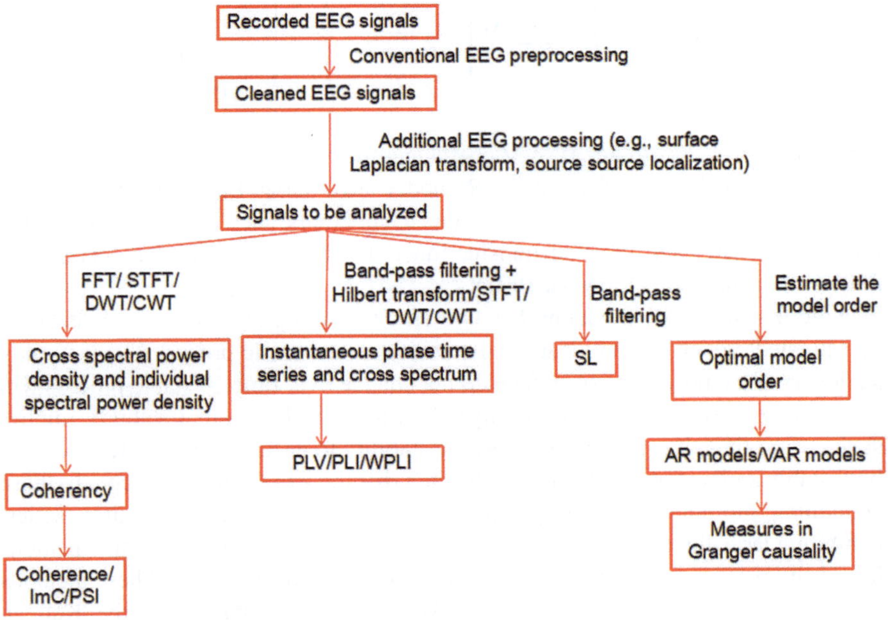

Fig. 12.1 The pipeline of estimating four classical kinds of functional connectivity measures

measures still exist in literatures, such as the directionality phase indexes (DPI), mutual information (MI), and transfer entropy (TE). In this chapter, we only explained how to compute the most frequently used measures in the field of EEG connectivity.

According to whether they can provide any information about the causal relationship or the direction of information flow, they can be divided into FC measures and EC measures. The FC measures include coherence, ImC, PLV, PLI, WPLI, and SL, whereas the EC measures include PSI and Granger causality-based measures.

Some of these measures, including the ImC, PSI, PLI, and WPLI, were designed to overcome the common sources problem. Before using other measures influenced by this problem, it is suggested that we should apply spatial filters (e.g., the surface Laplacian transform and the source localization) to the scalp EEG signals, which could strongly attenuate volume conduction.

Specifically, the surface Laplacian transform could estimate the current source density (CSD), which is a reference-free measure and is the second spatial derivative of the electric field. The CSD could enhance and separate focal activity and remove low spatial frequency assumed to originate from deeper sources; thus, you cannot use the surface Laplacian transform when you expect deep or widely distributed coherent sources. This method should be conducted on time-domain EEG signals, not frequency-domain data. In order to conduct the surface Laplacian transform, we can use the CSD toolbox (http://psychophysiology.cpmc.columbia.edu/software/csdtoolbox/).

The CSD is surface- or electrode-based measures and thus cannot provide the electric activity of each cerebral region. In order to obtain the source activity of each brain region, source localization should be conducted before measure computation. More information about the techniques in source localization can be seen in Chap. 7. Some excellent free-available toolboxes exist, such as the LORETA software (http://www.uzh.ch/keyinst/loreta.htm), the fieldtrip software (http://www.fieldtriptoolbox.org/), and the Brainstorm software (https://neuroimage.usc.edu/brainstorm/).

12.5 Computation of Connectivity Measures in Resting-State and Event-Related EEG Signals

All these measures shown above can be applied to both resting-state and event-related EEG signals. However, there are some differences between the two types of EEG datasets.

12.5.1 EEG Connectivity for Resting-State EEG Signals

In resting-state EEG studies, the subjects are asked to keep eye-closed or eye-open for a certain length of time (such as 5 min). In these studies, we are not interested in the timing information of human cognitive processing. This is the fundamental difference between resting-state and event-related experimental design.

Before the measures in EEG connectivity are evaluated, the resting-state EEG data should be properly preprocessed. In my opinion, the continuous EEG data should be segmented into dozens of epochs with equal length. The rationales for this operation can be summarized as following: Firstly, although the continuous EEG signals have been preprocessed, they may still contain artifacts that cannot be corrected by typical preprocessing operations. In this situation, we could delete those artifacts contaminated epochs by setting a threshold (such as ±75 μV). Secondly, the measures using Fourier transform and Granger causality-based measures require stationarity of the signal. It has been widely recognized that the resting-state EEG can be regarded as quasi-stationary only on the order of a few seconds. Several epoch lengths have been used in previous studies, such as 1 sec, 2 sec, and 5 sec. Moreover, due to details of the implementation of the FFT algorithm, to increase computational speed, it is also desirable that the epoch length (i.e., point number) is a power of 2 (e.g., 64, 128, and 256).

After the continuous EEG data have been segmented into artifact-free epochs, the phase synchronization-based measures, generalized synchronization-based measures, and Granger causality-based measures were computed for each epoch, each frequency band, and each electrode pair (or pair of brain regions). Then, for each frequency band and each electrode pair (or pair of brain regions), the measures are averaged across all the epochs. However, this is not the case for coherence-based measures. During the computation of these measures (e.g., coherence, ImC, and PSI), the FFT is conducted on each epoch; then for each pair of electrodes, the power spectral density of each electrode and their cross spectral density are computed. Then, the power spectral densities of electrodes and cross spectral densities of pairs are averaged across epochs. Lastly, the coherence-based measures are evaluated based on these epoch-averaged power spectral densities and cross spectral densities.

12.5.2 EEG Connectivity for Event-Related EEG Signals

As has been explained above, a fundamental difference between resting-state and event-related experimental design is that we need to extract the timing information of human cognitive processing in event-related EEG signals.

Before the measures in EEG connectivity are evaluated, the event-related EEG signals should be properly preprocessed and should be segmented into epochs according to the experimental conditions. For each condition, those artifact-contaminated epochs should be deleted.

When the interested measures are those based coherency and phase synchronization, we should compute the analytic signal for each frequency bin and each time point. These analytic signals can be calculated through combining band-pass filtering and the Hilbert transform. They could also be extracted through classical time-frequency techniques, such as STFT and CWT/DWT. Similar to ERD/ERS analysis, the epoch length should be relatively larger than typical ERP analysis, such as 3 sec (1 sec before stimuli onset and 2 sec after stimuli onset).

When we want to compute the coherence-based measures in event-related signals, the power spectral densities of electrodes and cross spectral densities of electrode pairs are computed for each frequency bin, each time point, and each epoch. Then, for each time-frequency point, the power spectral densities and cross spectral densities are averaged across epochs. Lastly, the coherence-based measures are evaluated, which will result in a time-frequency representation of coherence-based measures for each pair of electrodes. You can see that the procedures are very similar compared to those for resting-state EEG signals. The only difference is that for event-related EEG signals we should not only compute the measures for each frequency but also for each time point (latency).

When we want to compute the phase synchronization-based measures in event-related signals, we should extract the phase information for each frequency bin, each time point, each epoch, and each electrode through combining band-pass filtering and Hilbert transform and typical time-frequency analysis techniques (e.g., STFT and CWT/DWT). When the measure to be computed is the WPLI, we should also compute the imagery part of cross spectral densities for each pair of electrodes. Note that, for resting-state EEG signals, the consistency of phase difference or that of the sign of phase difference is evaluated across all the time points within a given epoch. However, for event-related EEG signals, this consistency is evaluated across all the epochs for each time-frequency point, respectively, which will result in a time-frequency representation of phase synchronization-based measures for each pair of electrodes.

In Granger causality-related measures, data stationarity is a necessary precondition for accurate VAR estimation. However, it is well known that EEG signals containing ERP waveforms are highly nonstationary. In neuroscience researches, many methods have been proposed in order to solve this problem. Among these methods, the simplest method is called segmentation-based adaptive VAR modeling, which is rather similar to the concept behind the STFT or other windowing techniques (Ding et al. 2000). Namely, a window with length N is extracted from the time-domain multi-electrodes EEG dataset, and the VAR model is fitted to the data in this window. Then, the window is incremented by a short step. The VAR model is fitted to this new time window. These procedures will continue until the last point of the sliding window reaches the end of the epoch.

12.6 Source-Level Connectivity

EEG connectivity can be conducted both on scalp potentials and source-localized activities. Performing EEG connectivity on the source signals may make the results more interpretable, since we are investigating the connectivity between certain brain regions, not that between electrodes over the scalp. In order to perform source-level connectivity, we need to compute the source-level activities (i.e., current source densities) of all the voxels of the grey matters or those of dipoles in human brain, depending on whether distributed source analysis (e.g., LORETA\sLORETA \eLORETA, MNE, Beamforming) or discrete source analysis (e.g., BESA) is conducted. Then, the activities of user-defined regions of interest (ROIs) are extracted according to classical brain atlas, such as the Automated Anatomical Labeling (AAL) and Brodmann areas. Lastly, we could compute the connectivity measures between ROIs using the methods illustrated above.

In source-level connectivity, the following points should be mentioned. Firstly, the results of source-level connectivity are heavily related with the quality and accuracy of source activities, which depend on the signal-to-noise ratio (SNR) of scalp EEG signals and source localization techniques. Secondly, the accuracy of source activities is significantly influenced by the depth of brain structures. Thus, it is recommended by some researchers that we should not investigate the connectivity between deep brain structures. Thirdly, the source activities obtained from some source localization techniques (e.g., LORETA\sLORETA\eLORETA and MNE) are highly correlated between adjacent voxels. In these cases, using measures which could overcome the common sources problem (e.g., PLI, WPLI) should produce more reasonable results.

12.7 Statistical Test in EEG Connectivity

Although the basic principles of statistical test in EEG connectivity is the same for the resting-state EEG signals and event-related EEG signals, there are some differences in actual operations.

For resting-state EEG signals, the connectivity measures will be computed for each frequency band, each pair of electrodes, and each EEG dataset. The statistical test should be conducted on each frequency band, each pair of electrodes, respectively, which is named as mass-univariate testing. In typical resting-state EEG analysis, several frequency bands are commonly defined, such as delta band, theta band, alpha band, beta band, and gamma band. Assuming the number of electrodes is N, the number of electrode pairs is C_N^2 (i.e., $\frac{N(N-1)}{2}$). This will make the multiple comparisons problem very serious in connectivity analysis. Traditionally, this problem can be corrected by controlling the family-wise error rate (FWE) or the false discovery rate (FDR). The advantage of this approach is that it does not require interpretation of any abstract organizational or topological properties. Its main disadvantage though is the inherent massive number of multiple comparisons that must be performed. However, with such a large number of multiple comparisons, together with a potentially low contrast-to-noise ratio, this approach may not offer sufficient power. An alternative approach is the network-based statistic (NBS), which is a validated statistical method to deal with the multiple comparisons problem on a graph (Zalesky et al. 2010).

Briefly speaking, this NBS method consists of the following steps. Firstly, univariate testing (such as t-test and ANOVA) is conducted on all the connections or pairs. Secondly, the pairs exceeding a given threshold (such as $p < 0.01$) are admitted to a set of supra-threshold pairs. Thirdly, "connected graph components" are identified among these supra-threshold pairs. A connected graph component is defined as a set of supra-threshold pairs for which a path can be found between any two electrodes (or brain regions). The size of each component is assessed by the total number of pairs it comprised. Fourthly, the null distribution of size of connected component is empirically derived using a permutation approach with 5000 permutations. The component with largest size is recorded for each permutation, which yields an empirical null distribution for the size of the largest component size. Lastly, the FWER corrected p-value for an originally identified component of given size is then estimated as the proportion of permutations for which the largest component is of the same size or greater. The NBS toolbox, which could conduct NBS, can be downloaded from https://www.nitrc.org/projects/nbs.

For event-related EEG signals, the situation is more complex, since not only the connectivity measures will be computed for each frequency bin and each pair of electrodes but also will be computed for each time point. Typically, the statistical tests and the multiple comparisons correction afterward are conducted for each time-frequency matrix, respectively, using similar methods as the typical ERD/ERS analysis in time-frequency analysis. Since the number of electrode pairs is very large, it is usually not practical to test for all the time-frequency matrices. Thus, pairs of interest should be selected based the research hypothesis.

For event-related EEG signals, we can also use another approach. Firstly, we need to define a time-frequency region of interest (ROI) based on the research hypothesis or observation of the time-frequency representation. Then, the values within this time-frequency ROI are averaged. Through these procedures, the complexity of statistical tests is largely reduced. Then, we could statistically test each pair of interest.

12.8 Example: Connectivity Analysis for Resting-State EEG Signals

Basic Information of the Demo Datasets In the experiment, both eyes-closed and eyes-open resting-state EEG data were recorded about 5 min from the 64 scalp tin electrodes mounted in an elastic cap (Brain Products, Munich, Germany), with the sampling frequency of 500 Hz around 9:00 to 12:00 in the morning. Subject was introduced to eyes-closed first, and then eyes-open. The EEG datasets consist of 20 four-minute EEG recordings (eye-closed and eye-open condition), obtained from 10 volunteers in the Sleep and Neuroimaging Center of Southwest University, China. In preprocessing, the continuous EEG data were down-sampled to 250 Hz and digitally filtered within the 0.1–45 Hz frequency band using a Chebyshev II-type

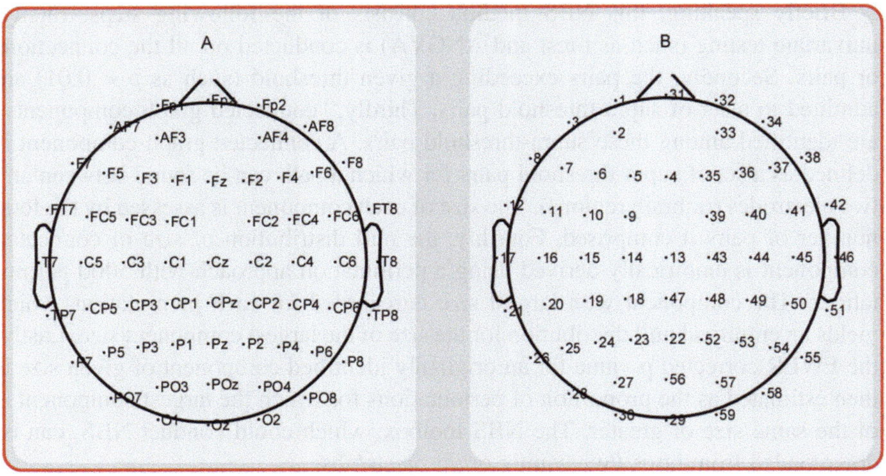

Fig. 12.2 The spatial location of each electrode. The name and number of each electrode is shown in panel A and B, respectively

filter. The filtered EEG recordings were re-referenced to average reference and then segmented to 2 sec. The segmentations with ocular, muscular, and other types of artifact were identified and excluded. We only retained the first 120 segmentations, constituting a four-minute EEG recording for each subject.

The preprocessed EEG signals contain the electrodes TP9 and TP10 which were located over left mastoid and right mastoid, respectively. Since the signals collected through these two electrodes may not contain the EEG signals produced by the human brain, these two electrodes were removed before further analysis (In EEGLAB, this could be done through "Edit >> Select data"). After these two electrodes have been removed, the signals of the remaining electrodes were re-referenced to their common average. The name, number, and spatial location of each electrode are shown in Fig.12.2.

In order to solve the common sources problem, the surface Laplacian transform, which was based on the second spatial derivative of the recorded surface potentials, was conducted through CSD toolbox. The application of surface Laplacian transform in the CSD toolbox has the following steps:

1. Obtain the label of each electrode. In the original preprocessed EEG datasets, the number of electrodes is 61. Thus, the number of electrodes is 59 after removing the electrodes TP9 and TP10. If your preprocessed EEG datasets have been imported into the EEGLAB, using the following command, you can obtain the labels: *electrodes = {EEG.chanlocs.labels}'*. Note that the labels of electrodes should be a column vector in CSD toolbox; thus, transposition is conducted.
2. Get the EEG montage. Since the scalp surface Laplacian estimates are based on the second spatial derivative of the recorded surface potentials, we need to get the spatial locations of the electrodes (i.e., the EEG montage). In the CSD toolbox, the head is represented as a unit sphere (i.e., a sphere with a radius of 1) and all the

electrodes located on the surface of this sphere. If the Cartesian coordinates are used, the x-y plane is marked by the great circle combining the locations Fpz, T7, Oz, and T8, with the x-axis running through T7 (−1.0) and T8 (+1.0), the y-axis running through Oz (−1.0) and Fpz (+1.0), and the z-axis running through the origin of the x-y plane (0) and Cz (+1.0). For spherical coordinates, the angle theta denotes the rotation from the x-axis toward the y-axis (positive poles, respectively), whereas phi denotes the angular displacement from the x-y plane toward the positive pole of the z-axis. The CSD toolbox contains a look-up file (i.e., "10–5-System_Mastoids_EGI129.csd") which has 330 standard scalp sites based on the classical 10-20 system and 129 scalp locations of a first-generation geodesic sensor net (GSN). Getting the EEG montage or electrodes, locations should use the function "ExtractMontage" in CSD toolbox. Supposing the variable "electrodes" contains the labels of electrodes, the following command can help us obtain the EEG montage: *Montage = ExtractMontage('10–5-System_Mastoids_EGI129.csd',electrodes)*. To verify the integrity and correctness of the identified EEG montage, we could use the function "MapMontage" in CSD toolbox by entering "MapMontage(Montage)" in the MATLAB command window, which will produce a topographical plot of the EEG montage.

3. Generate the transformation matrices "G" and "H." The surface Laplacian transform in CSD toolbox is based on two "electrodes-by-electrodes" transformation matrices "G" and "H". The "G" and "H" can be derived by the following command: *[G,H] = GetGH(Montage)*. Then, we could save these two matrices for further usage using the MATLAB function "save."
4. Load the EEG datasets into EEGLAB, apply the surface Laplacian transform, and save the CSD estimates. For each EEG dataset, we should firstly import the EEG data. If your EEG datasets are in EEGLAB format, you could use the "pop_loadset" function to import the EEG data. In EEGLAB, the imported EEG data, which can be found in the field "data" of variable "EEG" (i.e., *EEG.data*), is an "electrodes-by-samples" matrix or an "electrodes-by-samples-by-epochs" matrix, depending on whether the EEG data is continuous data or segmented data. The function "CSD" can be used to apply the surface Laplacian transform in the CSD toolbox through the following command: *X = CSD (D, G, H)*, where "*D*" is the EEG signals to be transformed, "*G*" and "*H*" are the transformation matrices computed in the last step, and "*X*" is the CSD estimates of "*D*." Note that the "*D*" should be an "electrodes-by-samples" matrix. Thus, if EEG signals are three-dimensions segmented EEG signals (i.e., electrodes-by-samples-by-epochs), you should conduct the CSD estimates for each epoch, respectively. The size of the CSD estimates is exactly the same of the original EEG signals. For convenience, the CSD estimates "*X*" should be saved in EEGLAB format. Since in EEGLAB, the EEG signals are saved in the field "data" of variable "EEG" (i.e., *EEG.data*), we could replace the EEG signals by CSD estimates through MATLAB command: *EEG.data = X*. Then, the updated variable "EEG" can be saved using the function "pop_saveset" in EEGLAB. Note that the saved files contain the CSD estimates, not the EEG signals.

In this example, the following EEG connectivity measures are computed: coherence, PLV, PLI, and WPLI. In order to compute the coherence between electrodes, the following steps should be performed for each subject:

1. Load the CSD estimates. In our example datasets, the CSD estimates for each subject is a 59 × 200 × 120 matrix.
2. Perform FFT on the CSD of each electrode and each epoch using the MATLAB function "fft." An example is *B* = *fft(A, NFFT, N)*. The input variables "*A*," "NFFT," and "*N*" are the CSD estimates of a given electrode and a given epoch, the next power of 2 from length of "*A*," and the length of "*A*," respectively. The output variable "*B*" is the FFT of vector "*A*." Here, the "*N*" is 200, and "NFFT" is 256.
3. Compute the coherence between each pair of electrodes for each frequency bin through Eq. (12.1) and (12.2). Note that the coherence is computed for each frequency bin, respectively. Thus, in order to get the coherence of a given frequency band (such as delta band, theta band, alpha band, beta band, and gamma band), we need to average the coherence of frequency bins within this frequency band.

In order to compute the PLV and PLI, the following steps should be performed for each subject and each frequency band:

1. Load the CSD estimates.
2. Band-pass filtering on the CSD estimates. A band-pass filtering, such as 8~13 Hz (i.e., the alpha band), is needed. In this example, the function "eegfilt" in EEGLAB is used. Note that the input data of this EEGLAB function should be a two-dimension data; thus, the CSD estimates should be reshaped into a two-dimension data (electrodes-by-samples) before band-pass filtering.
3. Get the instantaneous phase time series of band-pass filtered CSD estimates through the Hilbert transform. Supposing the band-pass filtered CSD estimates is "*X*," the phase time series "*Y*" is computed using the following command: *Y* = *angle(hilbert(X))*. Since the calculation of the Hilbert transform requires integration over infinite time, 10% of the calculated instantaneous phase values are discarded on each side of the time series. Then, the remaining instantaneous phase time series is reshaped into a three-dimension matrix (electrodes-by-samples-by-epochs).
4. Compute the PLV and PLI for each pair of electrodes and each epoch using Eq. (12.13) and (12.14). Then, you should average the PLV and PLI values across epochs.

In order to compute the WPLI, the following steps should be performed for each subject and each frequency band:

1. Load the CSD estimates and conduct band-pass filtering on the CSD estimates.
2. Compute analytic signal through the Hilbert transform, and discard the 10% of the calculated analytic signal on each side. Then, the remaining analytic signal is reshaped into a three-dimension matrix (electrodes-by-samples-by-epochs).

3. Compute the WPLI for each pair of electrodes and each epoch using Eq. (12.15), and then average the WPLI values across epochs.

Two example scripts, which could compute the CSD estimates and EEG connectivity measures (i.e., coherence, PLV, PLI, and WPLI), respectively, have been provided by us.

After these EEG connectivity measures have been computed across all the subjects and two conditions, the NBS toolbox is used to test the conditional effect (i.e., comparing the measures in EC condition and those in EO condition). Note that the band of interest is alpha band (8~13 Hz) in our example. The grand-average EEG connectivity matrices across all the subjects for each measure (coherence/PLV/PLI/WPLI) and each condition (EC/EO), along with the results of NBS, have been shown in Fig.12.3. From this figure, we could find that the PLI and WPLI could significantly attenuate the value of connectivity between adjacent electrodes, which may be caused by the common sources problem. Moreover, the NBS showed that dozens of pairs of electrodes have significant conditional effect with the measures of EC condition significantly larger than those of EO condition.

12.9 Example: Connectivity Analysis for Event-Related EEG Signals

Here, the datasets in folders "S" and "L" under folder "Data_AVSP_20" will be used as demo datasets. Firstly, the EEG datasets were re-sampled to 250 Hz in order to reduce the computational efforts. Secondly, the CSD were estimated through the CSD toolbox for each subject and each condition. Thirdly, EEG connectivity measures (i.e., PLV, PLI, and WPLI) were computed for each subject and each condition.

For the CSD dataset of each subject and each condition, the procedures of connectivity measures computation can be summarized as the following steps:

1. Load the CSD dataset, which was saved as EEGLAB format, through the "pop_loadset" function in EEGLAB.
2. Get the complex time-frequency representation through STFT. The MATLAB function "sub_stft" was used to conduct the STFT. The following parameters were used: window size 0.3 sec, the time points evaluated −1:0.004:1.996 sec (i.e., the same as the original time axis of EEG data/CSD), the frequency bins evaluated 1:1:30 Hz. The CSD for each dataset is an "electrodes-by-samples-by-epochs" matrix. Through STFT, we could obtain a complex time-frequency matrix with dimensions "frequency bins-by-time points-by-epochs-by-electrodes."
3. Compute the phase synchronization-based measures (i.e., PLV, PLI and WPLI). In order to compute the PLV and PLI, the phase information for each frequency bin, each time point, each epoch, and each electrode were derived from the

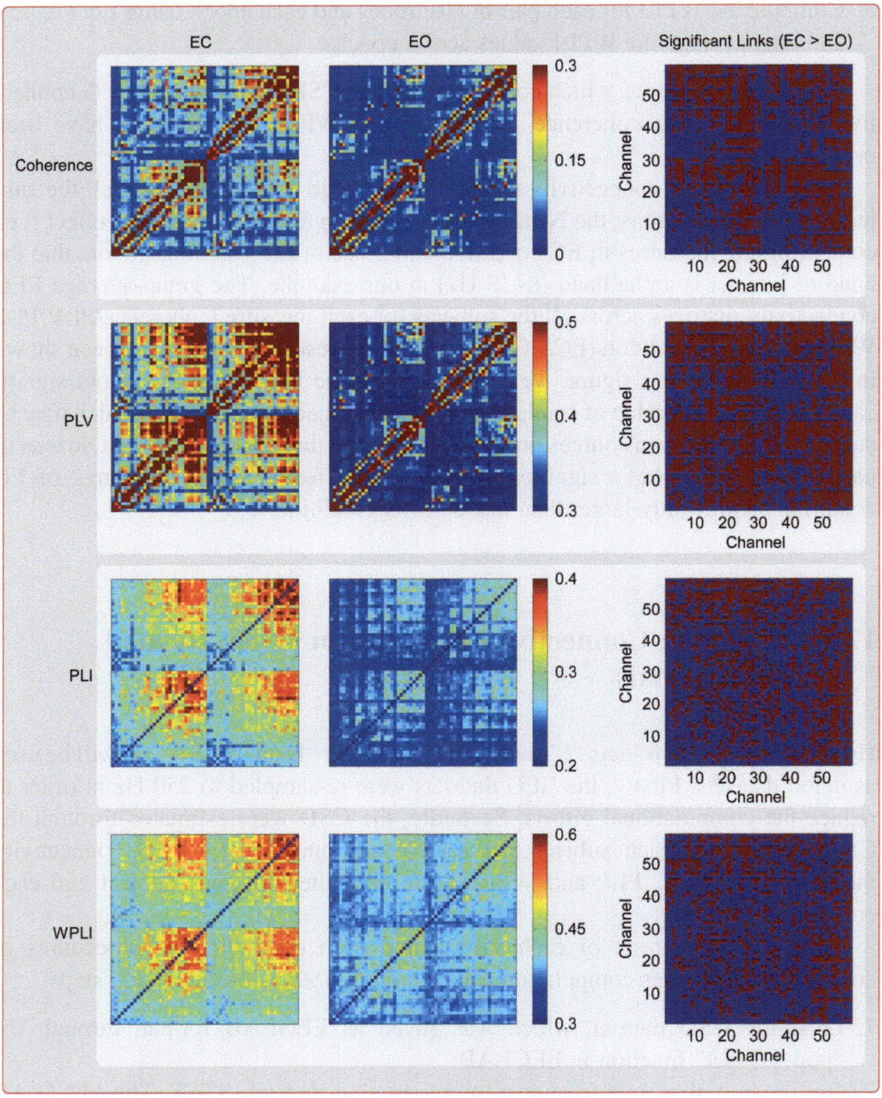

Fig. 12.3 The group-level coherence, PLV, PLI, and WPLI of alpha band under EC condition and EO condition and links with significant conditional effect (EC > EO)

complex time-frequency matrix computed in the last step using the MATLAB function "angle." Suppose the phase information of electrode i and j are stored in variables Phase_i and Phase_j, respectively, with dimensions "frequency bins-by-time points-by-epochs." Then, we computed the relative phase (i.e., phase difference) between Phase_i and Phase_j: relative_phase $= \text{Phase}_i - \text{Phase}_j$. The PLV between electrode i and j is computed using the following MATLAB command:

*PLV = abs(sum(exp(1i*relative_phase),3)/size(relative_phase,3))*. The PLI between electrode i and j is computed using the following MATLAB command: *PLI = abs(mean(sign((abs(relative_phase)- pi).*relative_phase),3))*. In order to compute the WPLI between electrode i and j, we need their complex time-frequency representation S_i and S_j. Then, the imagery part of their cross-spectrum is computed: *crossspec_imag = imag(S_i.* conj(S_j))*. Then, the WPLI between electrode i and j is computed using the following MATLAB command: *WPLI = abs(mean(crossspec_imag,3))./mean(abs(crossspec_imag),3)*.

Although the connectivity measures of all the pairs of electrodes can be computed, we may need to define electrodes of interest and pairs of interest before further statistical tests. In the current example, the electrodes of interest are C3 and C4. The grand-average connectivity measures (i.e., PLV, PLI, and WPLI) across all subjects in condition "L" and "S" are shown in Fig.12.4. The following conclusions can be

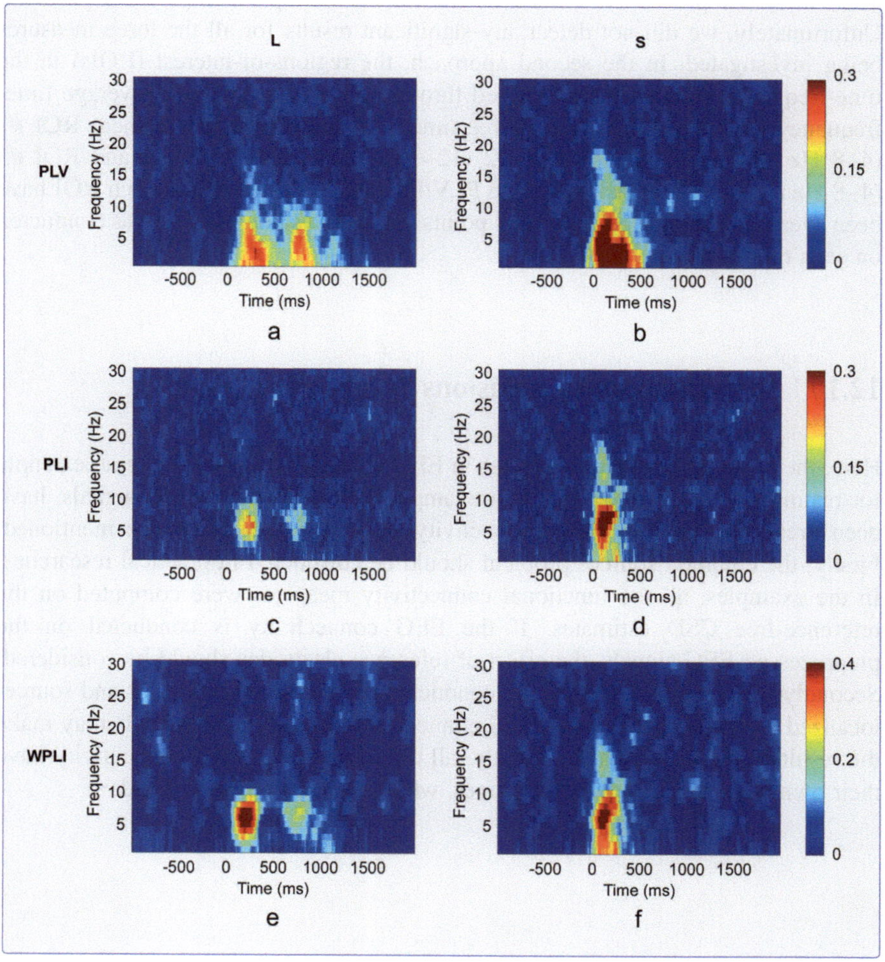

Fig. 12.4 The group-level PLV, PLI, and WPLI under condition "L" and condition "S"

seen from this figure. For condition "L," the PLV between electrode C3 and C4 can be clearly seen in two time-frequency regions: 1~8 Hz & 100~400 ms and 1~8 Hz & 700~900 ms, whereas for condition "S," the PLV between electrode C3 and C4 can be clearly seen in time-frequency regions 1~16 Hz & 50~500 ms. Note that the PLI/WPLI could attenuate the common sources problem through discarding the phase synchronization caused by consistent phase lag 0 or π across epochs. For condition "L," the PLI/WPLI between electrode C3 and C4 can be clearly seen in two time-frequency regions: 4~8 Hz and 100~300 ms and 4~8 Hz and 600~800 ms, whereas for condition "S," the PLI/WPLI between electrode C3 and C4 can be clearly seen in time-frequency regions 1~16 Hz and 50~300 ms.

After these EEG connectivity measures (i.e., PLV, PLI, and WPLI) have been computed across all the subjects and two conditions, we statistically tested whether there were significant conditional effects. Here, two approaches were used. In the first approach, the paired samples t-test was conducted on each time-frequency point. Then the FDR procedure is used to control the multiple comparisons problem. Unfortunately, we did not detect any significant results for all the three measures being investigated. In the second approach, the regions-of-interest (ROIs) in the time-frequency plane were predefined through observing the grand-average time-frequency representations. Here, three time-frequency ROIs are defined: ROI #1 (4~8 Hz and 100~300 ms), ROI #2 (12~16 Hz and 100~250 ms), and ROI #3 (4~8 Hz and 600~800 ms). After the PLV/PLI/WPLI values within each ROI have been averaged across time-frequency points, the paired samples t-test was conducted on each time-frequency ROI.

12.10 Remarks and Conclusions

Here, the basic theories and measures in EEG connectivity, along with an example for resting-state EEG signals and an example for event-related EEG signals, have been presented. For the EEG connectivity, several points should be mentioned. Firstly, the common sources problem should be considered in practical researches. In the examples, all the functional connectivity measures were computed on the reference-free CSD estimates. If the EEG connectivity is conducted on the preprocessed EEG signals, the effect of reference electrodes should be considered. Secondly, EEG connectivity can be conducted both on scalp potentials and source-localized activities. Performing EEG connectivity on the source signals may make the results more interpretable. Thirdly, all the measures in EEG connectivity have their own advantages and disadvantages, which should be kept in mind.

References

Abrams DA, Lynch CJ, Cheng KM, Phillips J, Supekar K, Ryali S, Uddin LQ, Menon V. Underconnectivity between voice-selective cortex and reward circuitry in children with autism. Proc Natl Acad Sci U S A. 2013;110(29):12060–5.

Astolfi L, Cincotti F, Mattia D, Salinari S, Babiloni C, Basilisco A, Rossini PM, Ding L, Ni Y, He B, Marciani MG, Babiloni F. Estimation of the effective and functional human cortical connectivity with structural equation modeling and directed transfer function applied to high-resolution EEG. Magn Reson Imaging. 2004;22(10):1457–70.

Baccalá LA, Sameshima K. Partial directed coherence: a new concept in neural structure determination. Biol Cybern. 2001;84(6):463–74.

Buckner RL, Andrewshanna JR, Schacter DL. The brain's default network: anatomy, function, and relevance to disease. Ann N Y Acad Sci. 2008;1124(1):1–38.

Cohen MX. Effects of time lag and frequency matching on phase-based connectivity. J Neurosci Methods. 2015;250:137–46.

Dhamala M, Rangarajan G, Ding M. Analyzing information flow in brain networks with nonparametric Granger causality. NeuroImage. 2008;41(2):354–62.

Ding M, Bressler SL, Yang W, Liang H. Short-window spectral analysis of cortical event-related potentials by adaptive multivariate autoregressive modeling: data preprocessing, model validation, and variability assessment. Biol Cybern. 2000;83(1):35–45.

Duann JR, Ide JX. Functional connectivity delineates distinct roles of the inferior frontal cortex and presupplementary motor area in stop signal inhibition. J Neurosci. 2009;29(32):10171–9.

Ercan ES, Suren S, Bacanlı A, Yazıcı KU, Callı C, Ardic UA, Aygunes D, Kosova B, Ozyurt O, Aydın C, ROhde LA. Altered structural connectivity is related to attention deficit/hyperactivity subtypes: a DTI study. Psychiatry Res Neuroimaging. 2016;256:57–64.

Friston KJ, Harrison LM, Penny WD. Dynamic causal modeling. NeuroImage. 2003;19 (4):1273–302.

Gao Q, Duan X, Chen H. Evaluation of effective connectivity of motor areas during motor imagery and execution using conditional Granger causality. NeuroImage. 2011;54(2):1280–8.

Granger CWJ. Investigating causal relations by econometric models and cross-spectral methods. Econometrica. 1969;37(3):424–38.

Heise V, Filippini N, Trachtenberg AJ, Suri S, Ebmeier KP, Mackay CE. Apolipoprotein E genotype, gender and age modulate connectivity of the hippocampus in healthy adults. NeuroImage. 2014;98(7):23–30.

Huster RJ, Plis SM, Lavallee CF, Calhoun VD, Herrmann CS. Functional and effective connectivity of stopping. NeuroImage. 2014;94(6):120–8.

Kamiński MJ, Blinowska KJ. A new method of the description of the information ow in the brain structures. Biol Cybern. 1991;65(3):203–10.

Lachaux JP, Rodriguez E, Martinerie J, Varela FJ. Measuring phase synchrony in brain signals. Hum Brain Mapp. 1999;8(4):194–208.

Lutkepohl H. New introduction to multiple time series analysis. Berlin: Springer; 2005.

Niso G, Bruña R, Pereda E, Gutiérrez R, Bajo R, Maestú F, del-Pozo F. HERMES: towards an integrated toolbox to characterize functional and effective brain connectivity. Neuroinformatics. 2013;11(4):405–34.

Nolte G, Bai O, Wheaton L, Mari Z, Vorbach S, Hallett M. Identifying true brain interaction from EEG data using the imaginary part of coherency. Clin Neurophysiol. 2004;115(10):2292–307.

Nolte G, Ziehe A, Nikulin VV, Schlögl A, Krämer N, Brismar T, Müller KR. Robustly estimating the flow direction of information in complex physical systems. Phys Rev Lett. 2007;100 (23):234101.

Nunez PL, Srinivasan R, Westdorp AF, Wijesinghe RS, Tucker DM, Silberstein RB, Cadusch PJ. EEG coherency : I: statistics, reference electrode, volume conduction, Laplacians, cortical imaging, and interpretation at multiple scales. Electroencephalogr Clin Neurophysiol. 1997;103 (5):499–515.

O'Reilly C, Lewis JD, Elsabbagh M. Is functional brain connectivity atypical in autism? A systematic review of EEG and MEG studies. PLoS One. 2017;12(5):e0175870.

Rasero J, Amoroso N, La Rocca M, Tangaro S, Bellotti R, Stramaglia S. Multivariate regression analysis of structural MRI connectivity matrices in Alzheimer's disease. PLoS One. 2017;12 (11):e0187281.

Sommerlade L, Henschel K, Wohlmuth J, Jachan M, Amtage F, Hellwig B, Lücking CH, Timmer J, Schelter B. Time-variant estimation of directed influences during Parkinsonian tremor. J Physiol-Paris. 2009;103(6):348–52.

Srinivasan R, Winter WR, Ding J, Nunez PL. EEG and MEG coherence: measures of functional connectivity at distinct spatial scales of neocortical dynamics. J Neurosci Methods. 2007;166 (1):41–52.

Stam CJ, Nolte G, Daffertshofer A. Phase lag index: assessment of functional connectivity from multi channel EEG and MEG with diminished bias from common sources. Hum Brain Mapp. 2007;28(11):1178–93.

Stam CJ, van Dijk BW. Synchronization likelihood: an unbiased measure of generalized synchronization in multivariate data sets. Physica D. 2002;163(3):236–51.

Sun Y, Lim J, Kwok K, Bezerianos A. Functional cortical connectivity analysis of mental fatigue unmasks hemispheric asymmetry and changes in small-world networks. Brain Cogn. 2014;85 (1):220–30.

van den Broek SP, Reinders F, Donderwinkel M, Peters MJ. Volume conduction effects in EEG and MEG. Electroencephalogr Clin Neurophysiol. 1998;106(6):522–34.

Vicente R, Wibral M, Lindner M, Pipa G. Transfer entropy--a model-free measure of effective connectivity for the neurosciences. J Comput Neurosci. 2011;30(1):45–67.

Vinck M, Oostenveld R, van Wingerden M, Battaglia F, Pennartz CMA. An improved index of phase-synchronization for electrophysiological data in the presence of volume-conduction, noise and sample-size bias. NeuroImage. 2011;55(4):1548–65.

Wiener N. The theory of prediction. In: Beckenbach EF, editor. Modern mathematics for the engineer. New York: McGraw-Hill; 1956.

Yao D, Wang L, Arendt-Nielsen L, Chen ACN. The effect of reference choices on the spatio-temporal analysis of brain evoked potentials: the use of infinite reference. Comput Biol Med. 2007;37(11):1529–38.

Yao D, Wang L, Oostenveld R, Nielsen KD, Arendtnielsen L, Chen AC. A comparative study of different references for EEG spectral mapping: the issue of the neutral reference and the use of the infinity reference. Physiol Meas. 2005;26(3):173–84.

Yu D. Additional brain functional network in adults with attention-deficit/hyperactivity disorder: a phase synchrony analysis. PLoS One. 2013;8(8):e54516.

Zalesky A, Fornito A, Bullmore ET. Network-based statistic: identifying differences in brain networks. NeuroImage. 2010;53(4):1197–207.

Chapter 13
Spatial Complex Brain Network

Dong Wen, Zhenhao Wei, Yanhong Zhou, Yanbo Sun, Fengnian Li, and Jiewei Li

Abstract This chapter introduces the research status of spatial complex brain networks from the perspective of graph theory and complex networks. Firstly, we review the theoretical concepts of graph theory and complex networks, and combined them with spatial complex brain networks. We focused on a variety of important network topological properties, and then introduced them based on structural connections, functional connections, and cost-effective connections. Three different types of brain networks were established, and further studies on the relationship between structural brain networks and functional brain networks, as well as brain network research based on computational models. Finally, we discussed the future research directions of spatial complex brain networks.

Keywords Spatial complex brain · Graph theory · Complex networks · Topological property

D. Wen (✉) · Z. Wei · Y. Sun
School of Information Science and Engineering, Yanshan University, Qinhuangdao, Hebei, China
e-mail: xjwd@ysu.edu.cn

Y. Zhou
School of Mathematics and Information Science and Technology, Hebei Normal University of Science and Technology, Qinhuangdao, Hebei, China

F. Li
Yanshan University Library, Yanshan University, Qinhuangdao, Hebei, China

J. Li
Department of Electrical and Electronic Engineering, The University of Hong Kong, Hong Kong, China

L. Hu, Z. Zhang (eds.), *EEG Signal Processing and Feature Extraction*,
https://doi.org/10.1007/978-981-13-9113-2_13

13.1 Introduction

The human brain is one of the most complex and functionally efficient organs in nature. It is considered to be a complex system. Its ingenious and perfect structure and functional connection mode make the brain have powerful information differentiation and integration functions (Murre and Sturdy 1995). The brain is made up of a large number of neurons with sparse connections between them, which operate at multiple organizational levels, and each level has its own temporal and spatial scale. Therefore, it is possible to analyze from multiple levels: neurons; local circuits; specific regions of the brain; large-scale organization of the cortex; the entire brain. Early neuroscience research focused on the localization of the function of each brain region, while modern perspectives tend to use complex network methods to analyze the structural and dynamic behavior of different levels of neural networks (Mesulam 1998).

We can divide the brain network according to different scales while considering temporal and spatial properties (Betzela and Bassett 2017). Complex network theory is a pioneering representation by the small world network (Watts and Strogatz 1998) and the scale-free network (Barabási and Albert 1999). It has developed into a multidisciplinary research frontier for many years, and it is also a hot spot in the field of brain science research. Existing brain imaging technologies based on EEG, MEG, fMRI, and DTI also provide powerful tools for the study of spatial complex brain networks. For example, the study of EEG signals in patients with epilepsy found that the topological properties of the brain's functional network show dynamic behaviors that vary with disease status (Schindler et al. 2008). An fMRI-based study found that the small world features of functional brain networks in patients with Alzheimer's disease are degraded relative to normal controls (Supekar et al. 2008). In addition to brain diseases, the researchers divided the subjects into high IQ groups and general IQ groups through IQ scores. Based on the DTI data of the subjects, the structural brain network was analyzed. It was found that the global efficiency of the structural brain network of the high IQ group was significantly higher than that of the general IQ group (Li et al. 2009).

At present, there are many analytical methods that can describe spatial complex brain networks. Among them, graph theory analysis methods can describe and evaluate brain functional networks and structural networks. As a branch of scientific computing, graph theory quantifies the brain network topology by defining brain network nodes and edges, and the introduction of attributes, such as "small world", effective network and degree distribution, provide a new basis for the structural segmentation and functional integration of the brain, which help us to understand the spatially complex networks of the brain more deeply. Below we will introduce the knowledge of graph theory and complex brain networks.

For ease of understanding, the main content of this chapter is shown in Fig. 13.1.

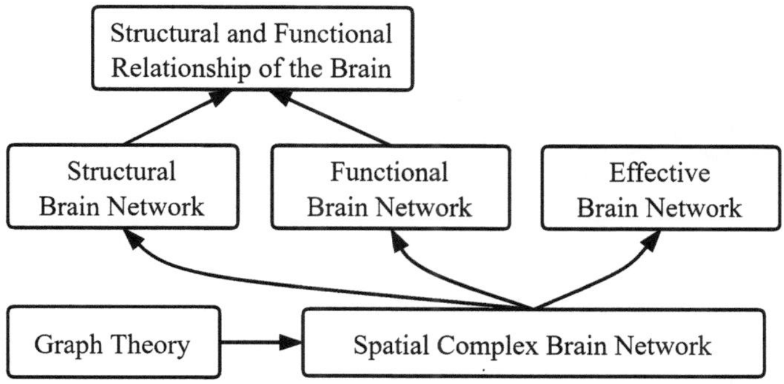

Fig. 13.1 The main content of this chapter

13.2 Graph Theory and Complex Networks

In the field of brain network research, the best practices for constructing and analyzing brain networks are still evolving. At present, people realize that brain networks are basically multi-scale physical spaces. Multi-scales include the representation of networks on a spatial scale, characterization on time scales, and at different topological scales. The spatial scale of a network is the granularity at which its nodes and edges are defined, ranging from single cells, synaptic granularity, brain regions, and large-scale fiber bundles. The time scale of the network refers to the measured time accuracy from sub-milliseconds to the entire life span to evolutionary changes in different species.

Complex brain network research is also known as brain network research based on graph theory analysis. This section introduces graph theory and complex networks. Graph theory is one of the branches of mathematics developed in the eighteenth century. It is a hotspot of complex brain network research because it can provide powerful network analysis methods to study the computational model of brain networks. A complex network is a network with complex topology characteristics. The two most famous complex networks are small world networks and scale-free networks. Small world networks have both large clustering coefficients and small feature path lengths. At present, the network has been widely confirmed to have small world characteristics, but it is controversial to have scale-free characteristics.

In graph theory, a complex network is represented by a graph, which consists of a series of nodes and edges between a series of nodes. According to whether the side has direction, the network can be divided into undirected networks and directed networks.

If the sides of the network have different weights, the network may be referred to as an undirected weighted network (as shown in Fig. 13.2 a), if the edges between

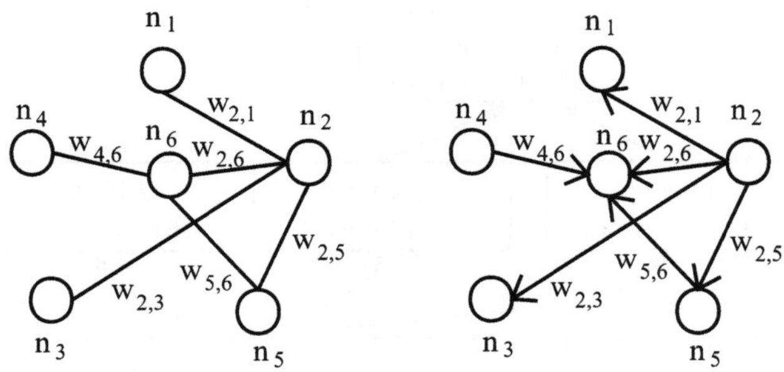

(a) undirected weighted network (b) directed weighted network

Fig. 13.2 According to whether there is direction between nodes, a network can be classified as an undirected weighted network or a directed weighted network

nodes have both directions and weights, the network may be referred to as a directed weighted network (as shown in Fig. 13.2 b). The degree of a node is defined as the number of edges that the node is connected to. The nodes in the directed network are divided into out-degree and in-degree. The out-degree refers to the number of edges from the node to other nodes, and the in-degree refers to the number of edges pointing from other nodes to the node. In a weighted network, the concept corresponding to degree is the node strength, defined as the weight sum of the edges connected to the node.

For brain networks, they can be depicted from different spatial scales, such as macro, micro, or large scales. However, due to limitations in technology and computing power, macro- or large-scale networks are currently studied. In a macro network, the electrodes of the EEG are usually used as nodes, and the definition of edges is the connection of functions or structures between nodes. Once the nodes and edges of the network are determined, the correlation matrix can be obtained, and the graph theory features are further analyzed.

The important features of the network topology mainly include the characteristic path length, the clustering coefficient, and the betweenness centrality.

The characteristic path length is defined as the shortest length of the path between two nodes in the network. In undirected and unprivileged networks, the distance between nodes is defined as the number of shortest paths connecting the two nodes. The characteristic path length of the network is the average of the characteristic path lengths between any two nodes in the network, and measures the global transmission capability of the network. When there is a part of the non-uniform part in the network, the value of the characteristic path length is infinite. To avoid this problem, efficiency can be used to characterize the connection characteristics of the network. Efficiency is defined as the reciprocal of the harmonic mean of the distance between all nodes in the network. The greater the efficiency, the less expensive it is for information or energy to be exchanged on the network.

The clustering coefficient is used to indicate the closeness of the neighbor nodes of the nodes in the network. The clustering coefficient is defined as the ratio of the number of edges actually present between a node and the k nodes connected to it to the total number of edges between these nodes. The clustering coefficient of the whole network is the average of the clustering coefficients of all nodes in the network, which can measure the local information transmission capability of the network. If k_i is the average network degree and two neurons i and j are connected to another one, then there is a probability that i and j are also connected with each other. This probability can be quantified using the clustering coefficient cci of a node i, which is defined by the fraction between the total number of links l_i between the vertices within its neighborhood and the number of links that could possibly exist between them:

$$\frac{k_i(k_i - 1)}{2} \tag{13.1}$$

The clustering coefficient is computed as:

$$\mathrm{cc}_i = \frac{2l_i}{k_i(k_i - 1)} \tag{13.2}$$

The importance of each node or edge in the network can be expressed in betweenness. The number of shortest paths through a node or edge is called the node's or the side's betweenness. A node with a high betweenness becomes a hub. Nodes or edges with high betweenness are usually important to maintain the effectiveness of the entire network communication. The importance of a node or edge to the network can be assessed by calculating the efficiency of the "damaged" network after the node or edge is removed. Suppose $\sigma_{st}(v)$ is the shortest path number from node s to node t through node v, and σ_{st} is the shortest path number from node s to node t, then the betweenness is computed as:

$$C_{\mathrm{B}}(v) = \sum_{s \neq v \neq t} \frac{\sigma_{st}(v)}{\sigma_{st}} \tag{13.3}$$

Based on betweenness, we can define betweenness centrality:

$$B_i = \sum_{i \neq k} \frac{\sigma(i, j, k)}{\sigma(j, k)} \tag{13.4}$$

In addition to the above indicators, other indicators such as grading can also be estimated, which are used to quantify the topology of the brain network. Next, we will introduce a complex brain network connected to the brain nerve.

13.3 Spatial Complex Brain Network

The brain neural connection network can be divided into structural brain networks, functional brain networks, and effective brain networks. The structural brain network consists of anatomical connections between neural units, which reflects the physiological structure of the brain.

The functional network describes the statistical connection between nodes of the brain network and is an undirected network. For effective brain networks, it describes the interaction or information flow between nodes of the brain network, which is a directed network.

There are currently two ideas for brain network research: data-driven research and computational model-based research. The former is based on experimentally measured data reflecting brain structural connectivity, or data reflecting functional connectivity of the brain. It calculates the connection relationship of pre-defined brain regions or nodes, and then constructs a network for analysis. The latter is based on a specific neural computational model. These two ideas are generally used to study various functions and diseases of the brain, such as brain development and aging, cognitive mechanisms of the brain, network mechanisms, and applications of mental or neurological diseases.

Building a brain network can be divided into three steps, that is, defining a node, defining and determining the strength of the connection between the nodes, selecting an appropriate threshold, and establishing a connection between nodes having a connection strength greater than a threshold. Figure 13.3 shows an example of EEG network illustrated as matrices (with weighted correlation coefficients or binary values). In the study of EEG, one channel or one source can be regarded as a node, while the connection strength between two nodes can be calculated as correlation, coherence, or other statistical relationships. Note that, the brain network can be illustrated by many ways, and Fig. 13.3 is just a simple and straightforward means.

13.3.1 Structural Brain Network

Structural connections between neurons, including electrical and chemical connections between axons and dendrites, are the material basis for functional connectivity of the brain. The human brain cannot obtain its complete neuronal structural connection network by physiological anatomy. At present, the research on the structural brain network of human living organisms is based on imaging techniques that reflect structural connections, and the role of EEG in structural brain networks is not prominent.

This section focuses on the determination of node definitions and connection relationships for structural brain networks.

Nodes of structural brain networks can be defined at different spatial scales, from individual neurons to local circuits, to brain regions or cortical regions with specific

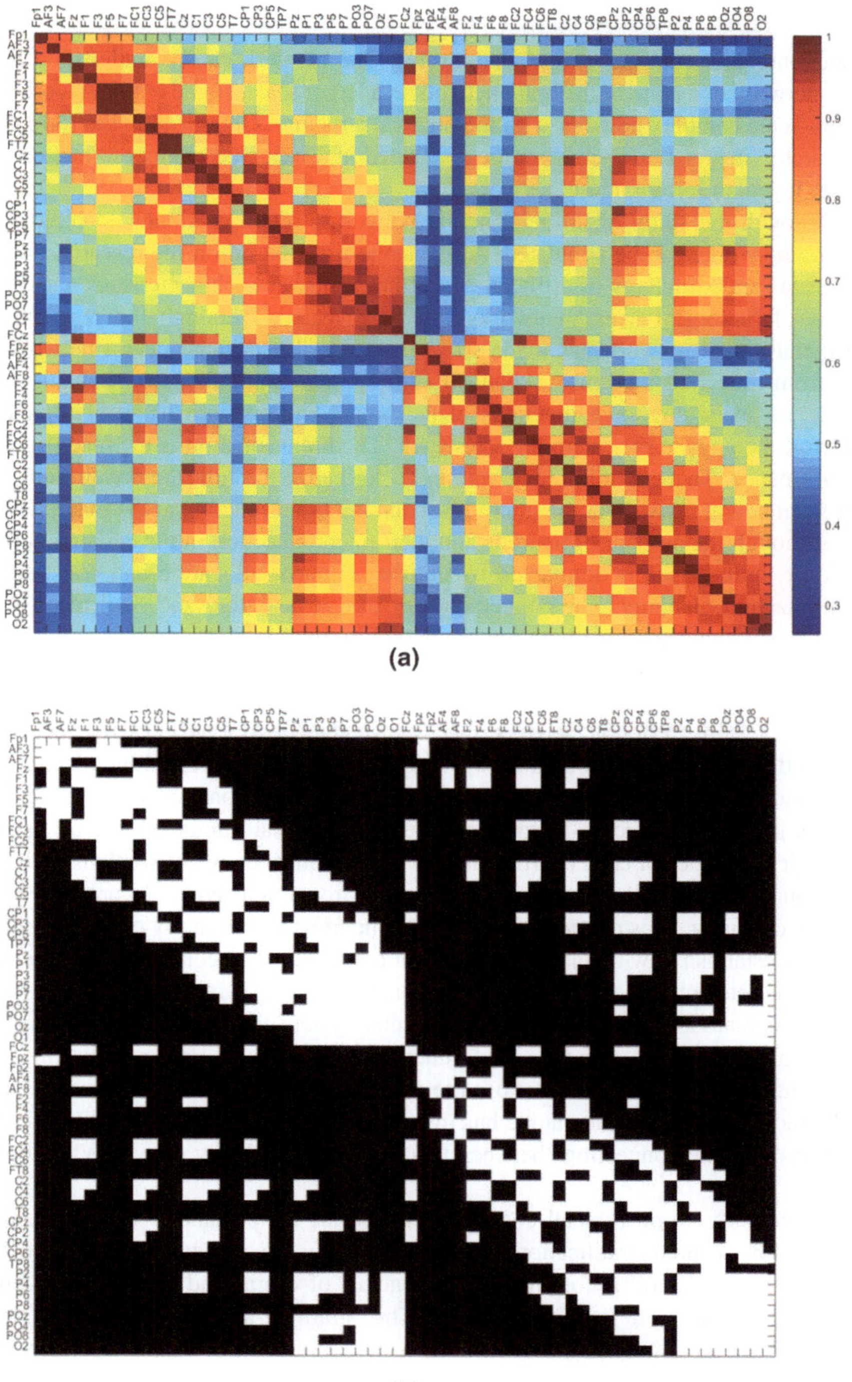

Fig. 13.3 An example of EEG-based brain networks. (**a**) The correlation matrix of 60 EEG channels. One EEG channel is one node, and the connectivity between two channels is calculated as the Pearson correlation coefficient between EEG of two channels. (**b**) The binary correlation matrix of (**a**) after thresholding at 0.85. White color means two channels are functionally connected, while black color means they are not

functions, and local nerves on small scales. The network can be understood as a sub-network or a node of the brain network at a larger scale, so the brain network is a multi-scale complex network with hierarchical structure. Organized in the form of a hierarchy, neurons on each layer are connected in a weighted manner to neurons in other layers to form a neural network.

Through the study of artificial neural networks, activities similar to biological neural networks can be simulated. At the same time, based on the research on the activities and structures of artificial neural networks, new characteristics and attributes of biological neural networks can be revealed. Although the research of artificial neural network has made some progress, the artificial neural network is far from reflecting the real mechanism of the brain, especially the mechanism of some advanced cognitive functions of the brain. The research of artificial neural network belongs to the micro level research. The brain function is the macroscopic performance. There is still a long way to go to realize the full interpretation of the artificial neural network at the micro level and the brain function at the macro level.

Although the number of neurons and their interconnection in the cerebral cortex is large, brain tissue is dominated by optimal resource allocation and constrained minimization planning. The connectivity of the cortical zone city will result in a small interconnection cost between the neural pivot points. This makes communication more efficient. During the evolution process, there has been a large increase in the number of neurons, and there is less direct connection between them. The nervous system (a neural network that is instantaneously connected by selective interaction) forms a distributed network, and the network is very dilute. There are indications that the human cerebral cortex is a dense, interconnected cortical region that is globally connected to form a large-scale cortical circuit. However, as the intrinsic properties of each neuron change over time (driven by interactions with the environment and with other neural groups), neuronal cell dynamics are also constantly changing. This dynamic nature allows neurons to constantly change connections or establish new connections based on computing and communication needs. Thus, the interaction between neurons becomes highly dynamic.

The nematode is an excellent experimental system for understanding the relationship between the structure and function of the entire nervous system. All potential of this system comes from the synthesis of both anatomical and electrophysiological data. The nematode has tight neural tissue composed of 302 different neurons, and the connections between them can be completely drawn (White et al. 1976).

In the early work, a small connection matrix of the primate visual cortex was obtained by manual stratification, in order to classify the connections as ascending, descending, or lateral according to their pattern of origin and termination of the cortex (Felleman and Van Essen 1991). This stratified analysis requires detailed anatomical information about the direction of the classification, which is used to place the brain regions at the upper, middle, and lower levels in the layered map, using an algorithm similar to simulated annealing. Several hierarchical models were estimated, modified, and selected under the cost function associated with anatomical constraints. The results of these methods fully reflect how the visual cortical areas are organized together. However, the exact hierarchical model has not yet been determined.

Watts et al. used complex network methods to study the nervous system, and clearly stated that neural networks have small world characteristics. They applied the quantization method of the complex network to study the neural network. Each neuron is a node in the network. The synaptic connection acts as the edge of the network, and a directed network composed of 282 nodes and 2462 edges is established. The study found that the topology of this network is neither a random network nor a regular network, but a small world network (Watts and Strogatz 1998).

13.3.2 Functional Brain Network

A functional connection describes the statistical relationship between functional signals between nodes in a certain period of time, but does not reflect the causal relationship between nodes. The functional brain network is constructed based on functional signals and can be constructed on a microscopic scale by point-distribution relationships between individual neurons. It can be constructed on a meso-scale by a local field potential that reflects the activity of the neuron cluster. It can be constructed on the macro scale by EEG, MEG, etc. between specific functional brain regions. At present, functional brain networks are more studied on a macro scale.

For multi-channel EEG signals, researchers often define the area covered by the electrode corresponding to each EEG channel as a node, then quantify the relationship between the signals of each channel of the EEG, and use the strength of this relationship as the functional connection strength between the corresponding brain regions, thereby constructing a functional brain network. The amplitude of the resting state is relatively large. According to the body of the subject and the state of the subject in the experiment, signals (such as alpha waves) in the corresponding frequency bands in each channel of the EEC are extracted first. The relationship between the waveforms of the various channels in this band is then quantified by a selected measure, and the network constructed from these quantitative relationships is referred to as a functional brain network within this band.

EEG-based brain network research can be divided into the following four steps.

1. Preprocessing of EEG signals: de-artifacts, noise removal, extraction of waveforms in a specific frequency band, and the like.
2. Quantify the relationship between every two of EEG channels. Phase synchronization, cross correlation, mutual information, and likelihood synchronization can be used.
3. Determine thresholds and build a cortical neural functional network. It is necessary to determine a suitable threshold to construct a brain network: whether there is a connection edge between two nodes depends on the strength of the

relationship between the corresponding EEG signal channels. When the relationship strength is greater than the threshold, the element corresponding to the adjacency matrix is set to 1. Otherwise, the element corresponding to the adjacency matrix is 0. The diagonal elements of the adjacency matrix are also set to 0 to avoid self-joining edges in the network. The network thus constructed is a 0–1 binary-valued unprivileged network. If the relationship strength between nodes with connected edges is assigned as the weight of the connected edge, and the element corresponding to the weight matrix is the connection strength, a weight brain network is generated. There are a number of methods for determining thresholds, such as studies using alternative data methods to determine thresholds, a simple and commonly used alternative data method is phase location chaos. The method generated by this method has the same power spectrum as the original data, but does not have the linear characteristics of the original data (Hurtado et al. 2004). For the selection of the threshold, a method has recently been proposed to automatically construct a brain function network from the EEG data according to the positional relationship of the vertices in the network. Nicolas Langer studied the variation of small world clustering coefficients, path lengths, and number of edges under different threshold increments (Langer et al. 2013). In the experiment of studying the differences of network characteristics of functional brain networks, the validity of the network characteristics is verified, and the network characteristics and the network with different weights of effective division can be accurately characterized (Gang Li et al. 2018).
4. Using the existing complex network measures to analyze the established functional brain network, such as the characteristic path length, efficiency, clustering coefficient and degree distribution, etc. Analyzing the physiological significance that these characteristic parameters may reflect.

The earliest revealing of the function of connecting small worlds is made by Stephan et al. (Stephan et al. 2000), who used invasive neurological tracer to obtain a functional network of monkey neuron connection data, which opened the prelude to the application of complex networks to study brain function networks. Using the teachings based on EEG recording of five healthy subjects under no-task conditions, Stam established a functional network of the human brain. Each channel acts as a node and gets 126 nodes. Then, one can calculate the synchronization likelihood value between any two channels in different frequency ranges. When the bit is greater than a given threshold, it defines the edge connection between the two nodes, thus obtaining five undirected networks with dilute sulfur. Finally, it is concluded that the brain functional network constructed under this experimental condition has a topology similar to that of the regular network in the α and β bands, while the low- and high-frequency networks in other bands have small world characteristics (Stam 2004). In 2006, Micheloyannis et al. applied the same method to analyze the EEG of 28 sensors of 14 subjects under working memory tasks and obtained different results: the functional network of the brain showed small world characteristics in all frequency ranges (Micheloyannis et al. 2006a). Bassett et al. analyzed the ERP data of different tasks in different subjects, and obtained the stability of small world characteristics, but the detailed features of brain functional connections in different frequency bands are different in the position of core nodes (Bassett et al. 2006). Pachou studied the relationship between functional network topology and response time in a resting-state functional network, and evaluated the correlation between network measurements and response times in each frequency band, respectively. The results show that humans may be associated with brain efficiency in integrating information from distributed brain regions. Longer response times are associated with reduced small world characteristics, further demonstrating the association between resting state functional brain networks and cognitive function (Micheloyannis et al. 2006a).

Small world characteristics are widespread. Jlenia studied the topological structure of the EEG-derived brain network when working memory was performed at different stages of different operations. Through analysis, it was found that in all EEG frequencies, the coding phase showed a significantly higher small world topology in the storage and retrieval of EEG networks (Toppi et al. 2018). In addition to the importance of discovering small world characteristics in EEG, it has also been found in other signals. For example, studies have explored age-related scale-free and small world network characteristics of ECG signals in male subjects at different sleep stages, and calculated averaging, clustering coefficients, and feature path lengths. And the results show that small world attributes can be found in all sleep stages of different age groups (Zhu et al. 2018).

Most of the brain network analysis has small world characteristics, which is not accidental. Many research results have suggested that the network structure with small world characteristics is the optimal choice for the synchronization of neural activity in different brain regions. It is the two organizational principles of brain function--the concentrated expression of information exchange properties of functional differentiation and functional integration. It is the brain adaptive results for stimulus signals. This feature of the brain facilitates the efficient transfer of information between multiple systems in a timely manner, facilitating the efficient processing of external stimulus information, enabling efficient interaction of information between different functional partitions.

Many researchers study the relationship between functional networks and disease. Micheloyannis et al. used EEG to find that the local clustering coefficient of brain function network in patients with schizophrenia was lower than that of normal people under resting state and working memory task, while the average shortest path length only increased significantly at rest (Micheloyannis, Pachou et al. 2006b). Pachou et al. analyzed the differences in functional connections between schizophrenic patients and normal subjects under memory tasks (Pachou et al. 2008). Studies by Rubinov et al. have shown that the small world characteristics of brain function networks in patients with schizophrenia are not obvious and the network topology is closer to random networks, validating the results of Micheloyannis (Rubinov et al. 2009).

Cai et al. reconstructed the phase space of EEG by embedding EEG time series into dimension by using the algorithm based on the distance set of phase points. The vector space distance of the phase space is used to obtain the closeness between the vectors. Describe the regression matrix of the phase space, and transform the obtained matrix to the connection matrix of the complex network. The differences in the characteristics of epilepsy activity and inactivity, mean value, clustering coefficient and average shortest distance were analyzed, and the small world characteristics of brain network were verified (Cai et al. 2011). Fang verified that the complex network characteristic index of alcohol addiction patients is significantly different from that of normal people. It defines the concept of brain neural network information entropy and neural network standard information entropy, and finds that the brain neural network information entropy of encephalopathy patients is obvious less than normal (Fang and Jiang 2007). When assessing the severity of Alzheimer's disease (AD), Chen considered the weighted undirected network of EEG signals and whole brain network measurements from different frequency bands to study the functional brain network and modify the alpha band. The network topology found that the topological reorganization network of functional brain networks can be used as a marker to assess the severity of AD symptoms (Chen et al. 2019). And in the study of schizophrenia (SZ), the mutual information is calculated for the EEG data, and then the corresponding characteristic indicators are calculated according to the graph theory. The results show that compared with the normal control, the SZ functional brain network has a smaller clustering coefficient, a larger average feature path length, and lower overall efficiency and local efficiency (Yin et al. 2017).

There are also studies that analyze the functional brain network using the minimum spanning tree approach (Fraga). When studying the status of functional connectivity in developmental dyslexia, the EEG quiescent state data was used and the phase-lag index (PLI) was used to calculate the weighted connection matrix for multiple bands. A minimum spanning tree (MST) map representing the sub-network with the greatest connectivity is obtained from the connection matrix and then analyzed. Differences in theta bands between the two graph indicators were found to indicate a reduction in communication between the network integration and network nodes of patients with dyslexia compared to the control group (González et al. 2016). In addition, Rene L used the minimum spanning tree for analysis in the study of differences in network measurements between cognitively normal Parkinson's disease (PD) patients and matched healthy controls, and found that there is a significant correlation between multiple network characteristics and cognitive performance of PD (Utianski et al. 2016).

Figure 13.4 shows the difference of EEG functional connectivity between the meditation state and resting state across different frequency bands. The results are obtained by the coherence method. Red lines indicate the significantly increased coherence in meditation than at rest, while blue lines indicate the significantly decreased coherence in meditation. The overall results show significant decrease in brain connectivity in most of the frequency bands, while only sporadic brain regions have increased connectivity mainly in the alpha1 and gamma bands

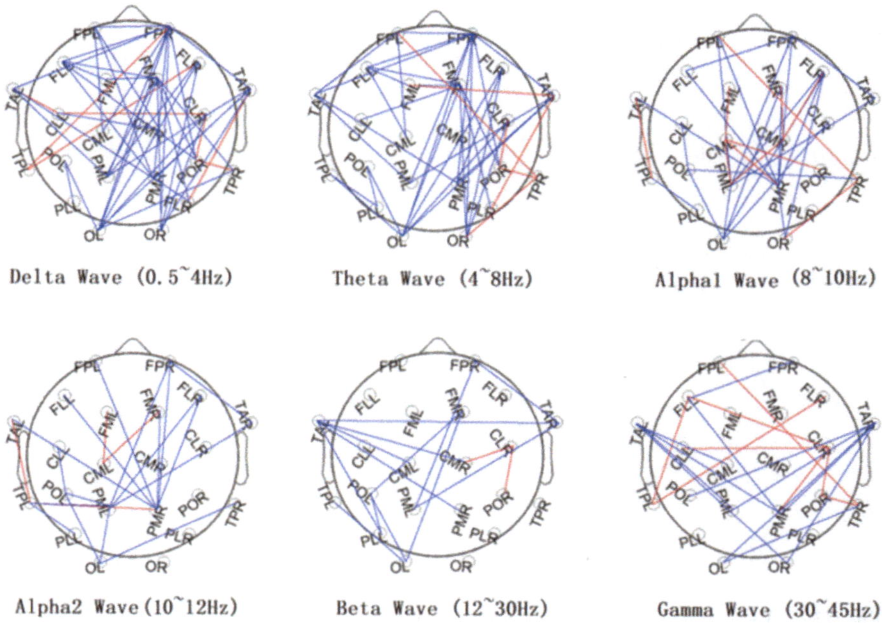

Fig. 13.4 An example of EEG-based functional networks. Networks are constructed based on coherence between EEG channels in six frequency bands. Red line indicates the significantly increased value during meditation than at rest, while blue line indicates decreased value during meditation

13.3.3 Effective Brain Networks

In brain network research, directional brain function connections are called causal connections. This connection describes the statistical causal relationship between nodes and can reflect the direction of information propagation between nodes. A network built by a causal connection is called a causal brain network. The study of the causal brain network is identical to the functional brain network in signal preprocessing and node definition, the difference being in the measure of the relationship between the quantized nodes. The strength of the causal link is generally quantified using causality analysis. At present, many methods for analyzing causality in the frequency domain are mostly based on Granger causality, such as directed transfer function (DTF) and partial directed coherence (PDC). After quantifying the causal relationship, the thresholds need to be further selected to construct a network of causalities. At this time, the method of estimating the significance level of the PDC value can be used, and the estimated significance threshold can be taken as the threshold: if the PDC is slightly larger than the significance threshold, it is considered that there is a significant causal relationship between the sequences corresponding to this PDC value, and the corresponding adjacency matrix element value is set to 1. Otherwise, there is no causal relationship, and the corresponding adjacency matrix element value is set to 0. If the PDC value is used as a weight, a

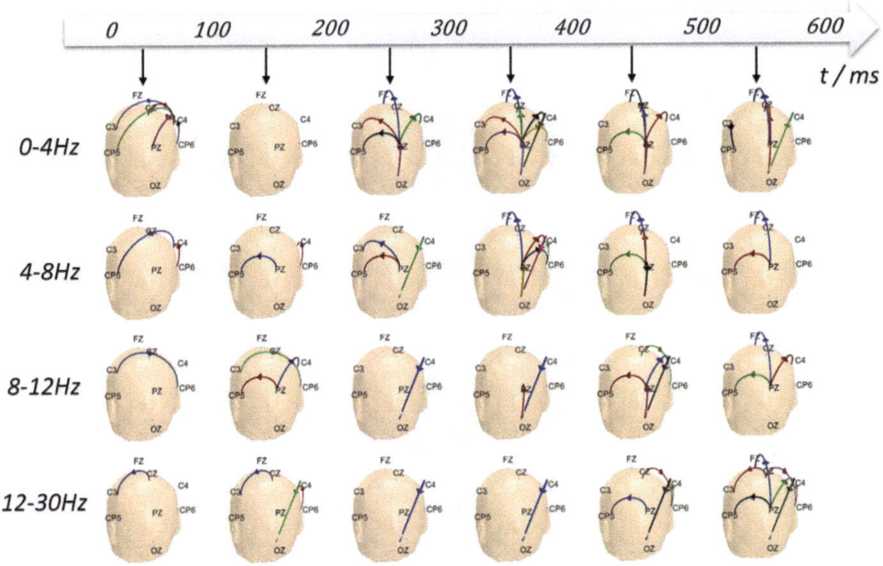

Fig. 13.5 An example of EEG-based effective networks. EEG data were recorded in a visual oddball experiment. Effective brain connectivity was obtained by the time-varying DTF method in different time intervals and frequency bands

weighted causality network can be further established. Then the complex network analysis method can be used to study the constructed causal brain network. The study of the causal brain network can also be carried out on a single frequency band. For example, Mohan compared the pathogenic brain network of healthy controls and tinnitus patients from the alpha frequency band (Mohan et al. 2018). For directed networks, degrees, feature path lengths, clustering coefficient, and other measures have their corresponding definitions relative to undirected networks (Park and Kim 2006). In addition, there are some proprietary feature parameters, such as link reciprocity, which measures the probability of simultaneous forward and reverse join edges between nodes.

Figure 13.5 shows an example of the EEG-based effective networks based on the DTF method. The results indicate the connectivity difference between target responses and standard responses in a visual oddball experiment

13.3.4 Structural and Functional Relationship of the Brain

The relationship between structural and functional connections has always been a concern for neuroscientists. From the perspective of the network, the function of a single brain network node is determined by its connection relationship with other nodes in the network, and the functions between nodes with similar connection patterns are similar (Passingham et al. 2002), while the structure of the brain nerve is similar. The connection gives the physical limits of these functional connections. There are three main ideas for the study of the relationship between structural connections and functional connections based on the network perspective:

1. Observing the data reflecting the brain structure and the data reflecting the brain function of the same subject, and then comparing the relationship between the structural network of the same research object and its functional network (Honey et al. 2009). The advantage of such a study is that the data used are real experimental results, but, limited by the current level of brain imaging technology, can only study the relationship between brain structure and brain function at the mesoscale and large scale.
2. Based on the brain computing model instead of experimental observation data, this research idea often defines a neurodynamic equation as a brain node, and the coupling relationship between dynamic equations as a structural connection between nodes. Then one or several state variables selected of each node (that is, the dynamic equation) are observed., and the data of state variables observed is used to estimate the functional connection relationship between the nodes. Finally, the relationship between functional connections and structural connections is obtained by studying the simulation (Ponten et al. 2009). The advantage of this method is that the structural connection relationship of the brain network can be completely controlled by the researcher according to the hypothesis, and the relationship between the structure and function of the brain network of any connection relationship can be explored at various scales without being limited by the brain imaging technology. Its limitation is that the network studied is a hypothetical simulation network. Without the combination of real experimental data, it is easy to produce pseudo-results without physiological significance.
3. The comprehensive experimental data and the computational model, that is, the structural connection relationship based on experimental observations, determine the coupling relationship of the established computational models, and then study the relationship between brain structure and brain function based on the computational model. This method integrates the advantages of the first two research ideas, and the results obtained are more convincing (Ponten et al. 2009).

Structure and function are the main core issues of complex network research. The research on the relationship between complex network structure and function is mainly focused on whether the same dynamic process shows different characteristics on different structure networks.

In early work, many papers pointed out the interaction between neuronal tissue and neuronal function, emphasizing that the nervous system connection structure is related to brain function. There are studies that treat nodes in complex networks as neurons. They examined the characteristics of this dynamic process on random, regular, and small world networks and found that it can produce fast responses on random networks but cannot produce sustained oscillations. The rule network can produce continuous volatility, but it cannot produce rapid response. On the small world network between the random network and the rule network, it can respond quickly and continuously. Therefore, their work proves the relationship between network structure and function from the perspective of whether the network can respond quickly and produce continuous oscillation (Lago-Fernández et al. 2000). Honey et al. studied the relationship between structure and function by studying the cortical structure network of macaques and found that there is a large degree of overlap between the two network connections (Honey et al. 2007).

Nonetheless, in order to understand the working mode and mechanism inside the complex system more deeply, it is very meaningful to study the evolution mechanism and model of the complex network to reveal the evolution mechanism of the actual system. Studies have found that the synaptic connection strength is also related to spatial association, that is, strong connections are more spatially aggregated than weak connections, forming a strong connection framework (Song et al. 2005). In order to deal with this problem, researchers proposed a spatial growth algorithm based on space constraints to simulate the evolution of brain networks. The process of establishing a spatial evolution network module is as follows: when the network is initiated, one or two nodes are established at a specific position in the two-dimensional space; then, considering that the new node and the probability of the edge between the old nodes is related to the spatial distance between the two nodes, each step can randomly adds a new node to any position in the two-dimensional space. If the new node fails to establish a connection with the old node, the new node is removed from the network. This produces an alternative network with an average shortest path and clustering coefficient similar to the actual cortical connection, and the results of the model also fit well with the results of the actual brain structure network in other aspects.

Undoubtedly, the brain has considerable complexity in its evolution history, structure, function, and coding of brain neurons. At this stage, people have already had a certain understanding of the dynamic characteristics of few neurons, and preliminary research has been made on the time coding methods of brain neuron discharge and their mutual response mechanisms. However, the brain contains many dynamic and spiritual factors and has a very rich information content, so it is still difficult to analyze as a unique complex adaptive system. However, through the study of complexity science, with the help of complex network theory and some related concepts, we can at least recognize some of the characteristics of complexity in the brain and achieve some scientific exploration on the memory and learning of the brain, the emergence of creative thinking, the formation of stream of consciousness, etc.

13.4 Summary and Discussion

So far, the research status of complex networks is still more empirical research than theoretical research. Computer numerical simulation methods are more than mathematical analysis and the proposed network model can only control one or two topological features, and the structure and mechanism of brain networks are still to be explored.

When exploring the unknown network structure, the topology structure of the complex network is like a black box. We can only obtain the unknown parameters of the topology structure and reconstruct the system structure through the output information of the network. How to build a network connection that is more consistent with the actual system from the data in the sea and establish a network model that conforms to the actual system working mechanism is a very important direction for future research.

Most of the previous studies have obtained undirected and unweighted graphs based on the connection model of mutual relations, and the interactions in the actual network sometimes do not occur continuously, and some can only occur under certain conditions. Establishing an undirected and unweighted network connection tends to ignore such conditional constraints, and the threshold for artificially given interactions is required when establishing an undirected and unweighted network. However, the threshold selection does not have a quantitative or statistical test for the quality of the criteria. Different thresholds have certain influence on the topology characteristics of the network. For example, the neural network studied by Onnela considers the neural connection strength to use different thresholds to generate networks with different structural parameters. Faust et al. pointed out that the influence of different thresholds on the nature of the network may be related to the selective edge-cut caused by the threshold selection, or it may also be indirectly caused by changes in network density due to different thresholds.

One of the major challenge of using neuroimaging to study brain functional connectivity is that it is difficult to draw clear causal inferences, because the functional connections between different brain regions in the nervous system change in real time. It may not be caused by the effect of one nervous system on another. Therefore, describing the effective connection of causal relationships in nervous system events and establishing a directed weighted neural network to reveal the new topological attributes of neural networks is the future direction of brain network research.

Cognitive neuroscience emphasizes multidisciplinary, multilevel intersections. The research levels include molecules, synapses, neurons, brain regions, brain region circuit, whole brain and overall behaviors, and even the environment, society, etc. In order to determine and analyze the relationship between the brain at different levels of time scale and spatial scale and the emergence mechanism of system complexity, the multilevel and interdisciplinary comprehensive research on human brain cognitive function and its neural mechanism has become the mainstream of contemporary scientific development one.

In the field of application of complex networks, most of the research still stays in the structural analysis and qualitative discussion of real network systems, but the causes of some complex phenomena appearing in real network systems still lack research. In addition, the research on the control of dynamic processes in real network systems needs to be strengthened. Although biological networks have accumulated a lot of research results as a large type of complex networks, the analysis of the network also reveals some basic structures and dynamic characteristics of the brain, but the research on the brain network is still in its infancy and the actual physiological significance of many indicators in the network is one of the problems that have been plaguing us all the time. How to analyze and understand brain data to obtain the relationship among the large-scale complex structures, dynamic activities, and cognitive functions of the brain will be a new direction in the future study of the brain.

From a clinical perspective, the conclusion of the study is still a long way from being truly clinically diagnosed. Firstly, the research conclusions of network science are mostly qualitative results. Secondly, different researchers use different experimental data and different refactoring techniques, which may result in inconsistent or even contradictory results. So in order to make a real contribution to the research of brain science, it is necessary for network science to strengthen and deepen the study of complex network theory, including network topology and dynamic behavior, to explore and deepen the study of network parameters and new statistical laws, and to conduct rigorous theoretical reasoning, from qualitative transfer to quantitative research, the judgment basis of cognitive activities, and reliable and effective disease diagnosis markers are given from the perspective of network science.

The study of complex networks is both increasing and rapidly developing. Therefore, many theories and methods are still being explored and developed. At present, the research on complex brain networks has just started. The analysis methods of complex networks need to be further expanded. It is the future research direction to find effective analysis methods that are more suitable for human brain function. In the future, with the analysis of the brain's massive data from different angles and different dimensions, it is bound to have a more comprehensive understanding of the brain mechanisms of diseases and genetics.

References

Barabási AL, Albert R. Emergence of scaling in random network. Science. 1999;286 (5439):509–12.

Bassett DS, Meyer-Lindenberg A, Achard S, et al. Adaptive reconfiguration of fractal small world human brain functional networks. Proc Natl Acad Sci U S A. 2006;103(51):19518–23.

Betzela RF, Bassett DS. Multi-scale brain networks. NeuroImage. 2017;160:73–83.

Cai SM, Hong L, Wei ZQ, et al. Regression analysis of EEG signals based on complex networks. Journal of University of Science and Technology of China. 2011;41(4):331–7.

Chen J, Liu C, Peng CK. Topological reorganization of EEG functional network is associated with the severity and cognitive impairment in Alzheimer's disease. PHYSICA A-STATISTICAL MECHANICS AND ITS APPLICATIONS. 2019;513:588–97.

Fang X, Jiang Z. Brain functional network analysis based on electroencephalogram. Acta Phys Sin. 2007;56(12):7330–9.

Felleman DJ, Van Essen DC. Distributed hierarchical processing in the primate cerebral cortex. Cereb Cortex. 1991;1(1):1–47.

Fraga González G, Van der Molen MJW, Žarić G, et al. Graph analysis of EEG resting state functional networks in dyslexic readers. Clin Neurophysiol. 2016;127(9):3165–75.

Li G, Li B, Jiang Y. A new method for automatically modelling brain functional networks. BIOMEDICAL SIGNAL PROCESSING AND CONTROL. 2018;45:70–9.

Honey CJ, Sporns O, Cammoun L, et al. Predicting human resting-state functional connectivity from structural connectivity. Proc Natl Acad Sci U S A. 2009;106(6):2035–40.
Honey CJ, Kötter R, Breakspear M, et al. Network structure of cerebral cortex shapes functional connectivity on multiple time scales. Proc Natl Acad Sci U S A. 2007;104(24):10240–5.
Hurtado JM, Rubchinsky LL, Sigvardt KA. Statistical method for detection of phase-locking episodes in neural oscillation. J Neurophysiol. 2004;96(4):1883–98.
White JG, Southgate E, Thomson JN, et al. The structure of the ventral nerve cord of Caenorhabditis elegans. Philos Trans R Soc Lond Ser B Biol Sci. 1976;275(983):327–48.
Lago-FernÃndez LF, Huerta R, Corbacho F. Fast response and temporal coherent oscillations in small world networks. Phys Rev Lett. 2000;84(12):2758–61.
Li Y, Liu Y, Li J, et al. Brain anstomical network and intelligence. PLoS Comput Biol. 2009;5(5): e1000395.
Mesulam M. From sensation to cognition. Brain. 1998;121:1013–52.
Micheloyannis S, Pachou E, Stam CJ, et al. Small-world networks and disturbed functional connectivity in schizophrenia. Schizophr Res. 2006a;87(123):60–6.
Micheloyannis S, Pachou E, Stam CJ, et al. Using graph theoretical analysis of multi channel EEG to evaluate the neural efficiency hypothesis. Neurosci Lett. 2006b;402(3):273–7.
Mohan A, Davidson C, De Ridder D, et al. Effective connectivity analysis of inter- and intramodular hubs in phantom sound perception – identifying the core distress network. Brain Imaging Behav. 2018:1–19. https://doi.org/10.1007/s11682-018-9989-7.
Murre JM, Sturdy DP. The connectivity of the brain: multi-level quantitative analysis. Biol Cybern. 1995;73(6):529–45.
Langer N, Pedroni A, Jäncke L. The problem of thresholding in small-world network analysis. PLoS One. 2013;8(1):e53199.
Pachou E, Vourkas M, Simos P, et al. Working memory in schizophrenia: an EEG study using power apectrum and coherence analysis to estimate cortical activation and network behavior. Brain Topogr. 2008;21(2):128–37.
Park SM, Kim BJ. Dynamic behaviors in directed networks. Phys Rev E Stat Nonlinear Soft Matter Phys. 2006;74(2 Pt 2):026114.
Passingham RE, Stephan KE, Kotter R. The anatomical basis of functional localization in cortex. Nat Rev Neurosci. 2002;3(8):606–16.
Ponten SC, Daffertshofer A, Hillebrand A, et al. The relationship between structural and functional connectivity:graph theoretical analysis of an EEG neural mass model. NeuroImage. 2009;52 (3):985–94.
Rubinov M, Knock SA, Stam CJ, et al. Small-world properties of nonlinear brain activity in schizophrenia. Hum Brain Mapp. 2009;30(2):403–16.
Schindler KA, Bialonski S, Horstmann MT, et al. Evolving functional network properties and synchronizability during human epileptic seizures. Chaos. 2008;18(3):033119.
Song S, Sjöström PJ, Reigl M, et al. Highly nonrandom features of synaptic connectivity in local cortical circuits. PLoS Biol. 2005;3(3):e68.
Stam CJ. Functional connectivity patterns of human magnetoencephalographic recordings: a small world network? Neurosci Lett. 2004;355(1/2):25–8.
Stephan KE, Hilgetag CC, Burns GA. Computational analysis of functional connectivity between areas of primate cerebral cortex. Philos Trans R Soc Lond Ser B Biol Sci. 2000;355 (1393):111–26.
Supekar K, Menon V, Rubin D, et al. Network analysis of intrinsic functional brain connectivity in AD. PLoS Comput Biol. 2008;4(6):e1000100.
Toppi J, Astolfi L, Risetti M, et al. Different topological properties of EEG-derived networks describe working memory phases as revealed by graph theoretical analysis. Frontiers in Human Neurosience. 2018;11:637.
Utianski RL, Caviness JN, van Straaten EC, et al. Graph theory network function in Parkinson's disease assessed with electroencephalography. Clin Neurophysiol. 2016;127(5):2228–36.

Watts DJ, Strogatz SH. Collective dynamics of small-world networks. Nature. 1998;393 (6684):440–2.

Yin ZL, Li J, Zhang Y. Functional brain network analysis of schizophrenic patients with positive and negative syndrome based on mutual information of EEG time series. BIOMEDICAL SIGNAL PROCESSING AND CONTROL. 2017;31:331–8.

Zhu G, Wang C, Liu F, et al. Age-related network topological difference based on the sleep ECG signal. Physiol Meas. 2018;39(8):084009.

Chapter 14
Temporal Complex Network Analysis

Zhongke Gao, Yuxuan Yang, and Qing Cai

Abstract Characterizing dynamical processes in a time-dependent complex system from observed time series is of great significance in many fields. Traditional time series analysis methods have difficulty in coping with some specific burdens, resulted from the increase of complexity of systems. Complex network, emerged in the last decade, provided a solution for dealing with these burdens. In this chapter, we introduce the basic concepts of complex network analysis of time series and some typical methods from univariate time series, namely, recurrence network, visibility graph, and horizontal visibility graph methods. In addition, the complex network analysis of multivariate time series is still a hot topic especially in the nowadays Big Data time. In this chapter, we still provide a case of multiscale complex network from multivariate time series and its application. Lastly, we introduce an expository example of a complex network-based study, to reveal the research steps in a complex network analysis for multivariate EEG signals where two different complex network methods are given.

Keywords Complex network · Time series · Recurrence network · Visibility graph · EEG signals

Characterizing dynamical processes in a time-dependent complex system from observed time series of just one or more variables is an extremely significant problem in many fields, ranging from physics and chemistry to economy and social

---

Electronic supplementary material The online version of this chapter (https://doi.org/10.1007/978-981-13-9113-2_14) contains supplementary material, which is available to authorized users.

Z. Gao (✉) · Y. Yang · Q. Cai
School of Electrical and Information Engineering, Tianjin University, Tianjin, China
e-mail: zhongkegao@tju.edu.cn

L. Hu, Z. Zhang (eds.), *EEG Signal Processing and Feature Extraction*,
https://doi.org/10.1007/978-981-13-9113-2_14

science. A variety of time series analysis methods have been proposed to fulfill this challenging task, including recurrence plot, multiscale entropy, and fractal analysis. Time series analysis has been broadly adopted in scientific research and engineering applications. Many theoretical developments for time series analysis have significantly contributed to the understanding of complex systems.

However, when system complexity increases, it becomes difficult to describe the dynamical behavior from time series, and traditional time series analysis methods have difficulty in coping with the specific burdens of this increased complexity. During the last decade, a new multidisciplinary methodology using complex network has emerged for characterizing complex systems. Charting the interactions among system components, abstracted as nodes and edges, has allowed us to represent a complex system as a complex network and then assess the system in terms of network theory. Numerous applications have already proven that complex network analysis of time series has great potential for characterizing important properties of complex dynamical systems.

Given the increasing interest and broad application, a tutorial introduction to complex network analysis of time series is provided in this chapter. We focus on introducing basic concepts, some typical methods, and its related applications.

14.1 What Is Complex Network?

Complex network research is originated from graph theory, which is a branch of discrete mathematics. However, differing from graph theory, network analysis mainly focuses on the real-life networks which are complicated and large. Complex network analysis can characterize the properties of a complex system by quantifying the topology of its network representation. The local and global properties (statistical measures) in network analysis are beneficial to the understanding of complex interrelations and information flow between different components in systems (Rubinov and Sporns 2010). The publication on the research of small-world network and scale-free network caused the emergence and spring in the study of complex networks. The initial research mainly focused on some small networks. Recently, more and more concerns have been given to large-scale complex systems with thousands or millions of nodes.

Actually, there exist various networks in real life, e.g., the Internet and electric power grids. In addition, networks can also be modeled in the abstract space, like the social network (Boccalettia et al. 2006). A network is actually a collection of nodes and edges between nodes (Rubinov and Sporns 2010). Commonly used network measures include degree, clustering coefficient, and characteristic path length. Specifically, the degree of a node is equivalent to the number of edges connected to that node. The degrees of all nodes in the network constitute the degree distribution, which can mark the development and resilience of the network. Clustering coefficient equals to the fraction of the node's neighbors that are also neighbors of each other. Clustering coefficient enables to locally depict the fraction of triangles around an individual node. Characteristic path length, also known as average shortest path length, measures the typical separation between two nodes and is the mean of geodesic lengths over all couples of nodes.

14.2 Typical Methods for Complex Network Analysis of Time Series

Constructing a network from a time series is an essential problem to be solved. A variety of novel methodologies have been proposed to map a univariate/multivariate time series into a complex network. These methods have been applied to address interdisciplinary challenges and have already proven great potential for characterizing important properties of complex dynamical systems. The literature on complex network analysis of time series is growing at a very fast rate due to its wide applications in a large variety of research fields (Gao et al. 2016b). Some classical methods for complex network analysis of time series will be introduced as follows.

14.2.1 Recurrence Network

An important network approach for analyzing time series is the recurrence network (RN), which has undergone an explosive growth in recent years. The idea of recurrence network was originated from the recurrence plot (RP), introduced to visualize the recurrences of dynamical systems.

The process of RP is as follows:

1). For a time series $\{u_i\}$, by choosing an embedding dimension d and time delay τ, one can construct a trajectory $\vec{x}(i) = \left(u_i, u_{i+\tau}, \ldots, u_{i+(d-1)\tau}\right), i = 1,2, \ldots, N$ in the d-dimensional space by the method of phase space reconstruction. Figure 14.1 gives a simple description of phase space reconstruction where A is one of the points in the trajectory.

2). Then the RP with size $N \times N$ can be obtained from the matrix:

$$R_{i,j} = \Theta\left(\varepsilon - \left\|\vec{x}(i) - \vec{x}(j)\right\|\right), i \neq j \tag{14.1}$$

where Θ is the Heaviside function, and ε is a predefined threshold. If the states at time i, j are similar (namely the distance between $\vec{x}(i)$ and $\vec{x}(j)$ is smaller than ε),

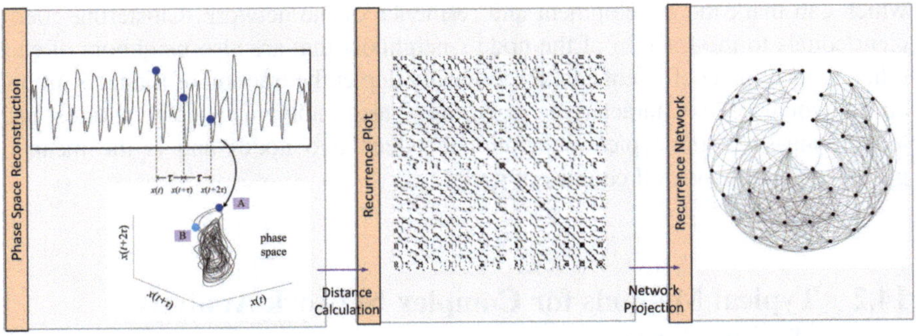

Fig. 14.1 The schematic diagram of the process of recurrence network

then $R_{i,\,j} = 1$ and one plots a dot at the point (i, j) to represent the appearance of recurrence. Thus, the RP can be plotted as shown in Fig. 14.1.

If the individual phase space vector $(\vec{x}\,(i))$ is served as a node i and the occurrence of a recurrence between $(\vec{x}\,(i))$ and $(\vec{x}\,(j))$ is used to indicate of the existence of one edge between node i, j, then one can project the RP into a RN as shown in Fig. 14.1.

Through the tool of complex network analysis, RN has been successfully applied to various research fields, including analyzing paleoclimate regime transitions from observational data (Marwan and Kurths 2015).

14.2.2 Visibility Graph and Horizontal Visibility Graph

The (horizontal) visibility graph theory has been proved to be computationally efficient and an analytically tractable method, which allows probing the dynamics underlying real complex systems from time series (Lacasa et al. 2008; Luque et al. 2009). The visibility graph method proceeds by mapping time series into graphs according to a specific geometric criterion, using complex network techniques to characterize time series (Gao et al. 2017). More recently, we extended the visibility graph theory to develop limited penetrable visibility graph theory (Gao et al. 2016a).

The process of visibility graph and limited penetrable visibility graph are as follows:

1) For continuous 10 data points of the illustrated time series, we display them in the form of vertical bars. Two arbitrary data values (t_a, y_a) and (t_b, y_b) will have visibility and consequently will become two connected nodes of the associated graph, if any other data (t_c, y_c) placed between them fulfills:

$$y_c < y_b + (y_a - y_b)\frac{t_b - t_c}{t_b - t_a} \tag{14.2}$$

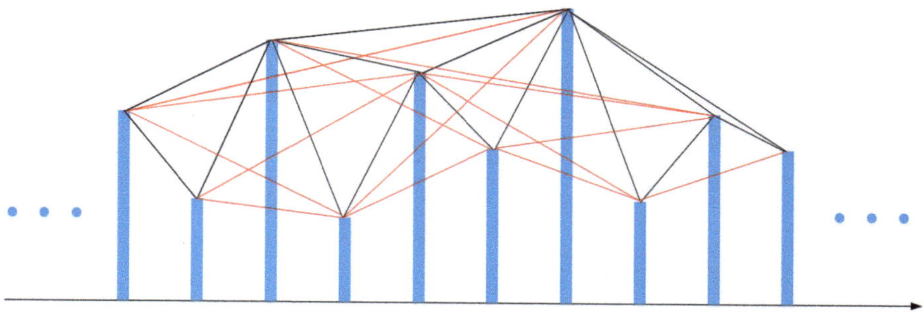

Fig. 14.2 Example of a time series and its corresponding VG (black lines); LPVG (black and red lines) with the limited penetrable distance L being 1, where every node corresponds to time series data in the same order. The visibility lines between data points define the links connecting nodes in the graph

2) The limited penetrable visibility graph is a development of visibility graph. In particular, if we set the limited penetrable distance to be L, a connection between two nodes exists if the number of in-between nodes that block the visibility line is no more than L.

Then, the LPVG can be constructed as shown in Fig. 14.2.

The process of horizontal visibility graph and limited penetrable horizontal visibility graph are as follows:

1) A connection between two nodes $y(i)$ and $y(j)$ exists (black lines in Fig. 14.3) if the following criterion is fulfilled:

$$y(i), y(j) > y(k) \text{ for all } k \text{ such that } i < k < j \tag{14.3}$$

2) The limited penetrable horizontal visibility graph is a development of the HVG. If we set the limited penetrable distance to L, a connection between two nodes exists if the number of in-between nodes that block the horizontal line is no more than L.

Then, the LPHVG can be constructed as shown in Fig. 14.3.

14.2.3 Complex Network Analysis of Multivariate Time Series

In addition, the complex network analysis of multivariate time series is still a hot topic especially in the nowadays Big Data time. Here, we provide a case of multiscale complex network and its application (Gao et al. 2015). In the literature, a novel measure named clustering coefficient entropy was also proposed to characterize the network property. The basic procedure for constructing a multiscale complex network from multivariate time series is as follows:

(1) For a multivariate signal that contains p sub-signals of equal length L ($\{x_{k,i}\}_{i=1}^{L}$, $k = 1,2, \ldots, p$), a coarse-grained process was first performed to define temporal scales, and further multivariate coarse-grained signals were obtained as follows:

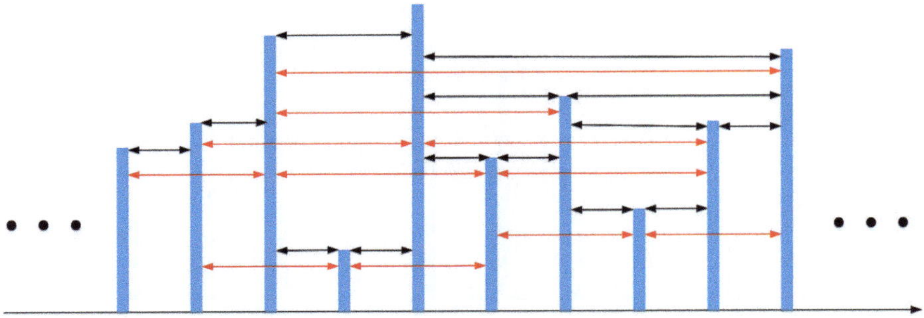

Fig. 14.3 Example of a time series and its corresponding HVG (black lines); LPHVG (black and red lines) with the limited penetrable distance L being 1, where every node corresponds to time series data in the same order. The horizontal visibility lines between data points define the links connecting nodes in the graph

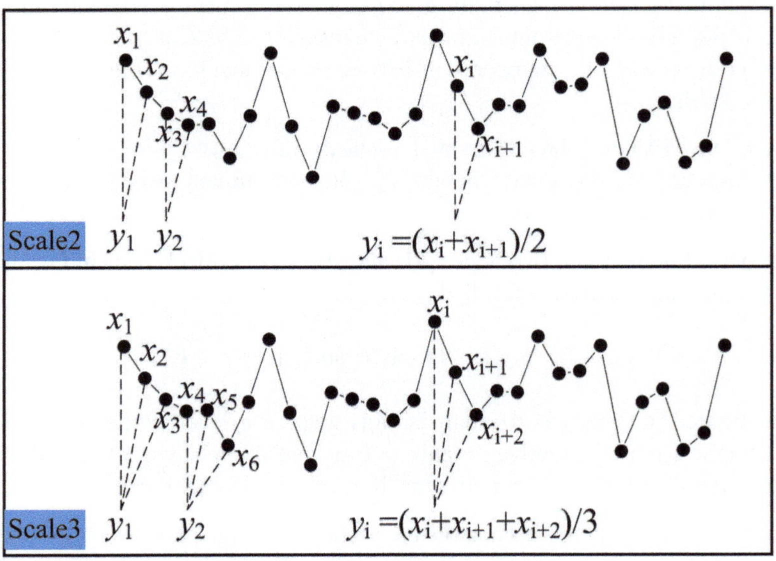

Fig. 14.4 The schematic diagram of the coarse-grain process for one of the channel k

$$y_{k,j}^{s} = \frac{1}{s} \sum_{i=(j-1)s+1}^{js} x_{k,i} \tag{14.4}$$

where s is the scale factor and $1 \leq j \leq \left\lfloor \frac{L}{s} \right\rfloor$, $k = 1, 2, \ldots, p$. Figure 14.4 gives the schematic diagram of the coarse-grain process for one of the channel k.

(2) Then multivariate embedding theory was used to construct a complex network from each obtained $y_{k,j}^{s}$, i.e., constructing a complex network at a different scale factor s. Specifically, the multivariate phase space reconstruction was performed on the $\{x_{k,i}\}_{i=1}^{L}$,$k = 1,2, \ldots, p$ as follows:

$$X_m(i) = \left[x_{1,i}, \ldots, x_{1,i+(m_1-1)\tau_1}, x_{2,i}, \ldots, x_{2,i+(m_2-1)\tau_2}, \ldots x_{p,i}, \ldots, x_{p,i+(m_p-1)\tau_p}\right] \tag{14.5}$$

where $\tau = [\tau_1, \tau_2, \ldots, \tau_p]$ and $M = [m_1, m_2, \ldots, m_p] \in R^p$ is the vector of time delay and vector of embedding dimension, respectively, and $X_m(i) \in R^m \left(m = \sum_{k=1}^{p} m_k\right)$.

(3) Then multiscale complex networks can be inferred from multivariate signals $\{x_{k,i}\}_{i=1}^{L}, k = 1,2, \ldots, p$ by the following steps:

(a) $(L - n)$ composite delay vectors $X_m(i) \in R^m$ was produced, where $n = \max\{M\} \times \max\{\tau\}$ and $i = 1, 2, \ldots, L - n$.

(b) The phase space distance between any two vectors $X_m(i)$ and $X_m(j)$, $j \neq i$ was defined in terms of maximum norm

$$d[X_m(i), X_m(j)] = \max_{l=1,\ldots,m}\{|x(i+l\text{-}1)\text{-}x(j+l\text{-}1)|\} \tag{14.6}$$

(c) Each phase space vector was regarded as a node and the phase space distance was used to determine the edges for constructing a complex network. By choosing a threshold, the adjacency matrix A of the complex network was obtained: An edge between node i and j exists ($A_{ij} = 1$) if the phase space distance between them is smaller than the threshold; while node i and j are not connected ($A_{ij} = 0$) otherwise. The topology of the derived complex network at different scales is determined entirely by the adjacency matrix A.

(d) Finally, the multiscale complex networks can be obtained by performing steps a)-c) on each coarse-grained multivariate signals. The threshold can be determined by the percentage (i.e., 15%) of total variation $T_r(s)$, where S is the covariance matrix of the multivariate signals.

The proposed multiscale complex network was applied to analyze multi-channel measurements from gas-liquid flows and the results suggested that multiscale complex networks allow quantitatively revealing the nonlinear flow behavior governing the transitions of flow patterns from the perspective of multiscale analysis and complex network analysis.

14.3 Complex Network Analysis of Time Series: Two Methods

In the following, we will introduce an expository example of a complex network-based study, to reveal the research steps in a complex network analysis for multivariate EEG signals. The EEG signals, collected from the subjects with the state of eyes-open and eyes-closed will be analyzed by two different complex network methods, respectively.

Opening and closing the eyes are fundamental behaviors for directing attention to the external versus internal world. However, it remains indistinct whether the resting state of eyes-open relative to eyes-closed are associated with different topological organizations of functional brain networks. Studies on resting-state functional networks from an electrophysiological perspective can take advantage of high temporal resolution. Considering that electroencephalograph (EEG) is low-cost and portable, many EEG-based brain network studies have received wide attention. We here give two kinds of networks from resting-state EEG signals to research the difference of topological properties of brain networks when the eyes are open versus closed.

14.3.1 EEG Acquisition and Preprocessing

The used dataset was created by Sleep and NeuroImaging Center and was collected from 10 healthy volunteers (five males and five females, age: 21 ± 0.8). Both eyes-closed and eyes-open resting-state EEG data were recorded about five minutes from the 64 scalp tin electrodes mounted in an elastic cap (Brain Products, Munich, Germany), with the sampling frequency of 500 Hz around 9:00 to 12:00 in the morning. The impedance of all electrodes was kept below 5 kΩ. Subject was introduced to eyes-closed first, and then eyes-open. The preprocessing was conducted using MATLAB scripts supported by EEGLAB (http://sccn.ucsd.edu/eeglab). The continuous EEG data were down-sampled to 100 Hz and digitally filtered within the 0.1–45 Hz frequency band using a Chebyshev II-type filter. The filtered EEG recordings were re-referenced to average reference. The segmentations with ocular, muscular, and other types of artifact were identified and excluded. We only retained the first 120 segmentations, constituting a four-minute EEG recording for each subject. After preprocessing, 61-channels EEG signals can be obtained in the state of eyes-closed and eyes-open, respectively, for each volunteer.

14.3.2 Weighted Recurrence Network Analysis

We first introduce the results of weighted recurrence networks method. (Marwan et al. 2009)

(1) We construct the weighted recurrence network for each epoch and the process can be described as follows. For a multi-channel signal $\{x_{k,l}\}_{l=1}^{L}, k = 1,2, \ldots, p$ which contains p sub-signal of equal length L, we first perform the phase-space reconstruction by using a suitable dimension m and a proper time delay τ (determined by FNN and C-C method, respectively) as follows (Kennel et al. 1992; Kim et al. 1992):

$$x\rightharpoonup_k(t) = (x_{k,t}, x_{k,t+\tau}, \ldots, x_{k,t+(m-1)\tau}), t = 1,2,...,N \tag{14.7}$$

where N is the number of vector points in the reconstructed phase space trajectory.

(2) For any generated phase space trajectories $\vec{x}_m(i)$ originated from any sub-signal x_m, a $N \times N$ recurrence plot (RP) can be achieved:

$$R_{i,j}^{\vec{x}_m}\left(\varepsilon^{\vec{x}_m}\right) = \Theta\left(\varepsilon^{\vec{x}_m} - \left\|\vec{x}_m(i) - \vec{x}_m(j)\right\|\right),\ i = 1, \ldots, N\ j = 1, \ldots, N \tag{14.8}$$

Thus, for a multi-channel signal $\{x_{k,l}\}_{l=1}^{L}, k = 1,2, \ldots, p$ containing p sub-signal, the number of obtained recurrence plots is p.

(3) For pair-wise RPs, in order to look for the times when both of them recur simultaneously, we can obtain the joint recurrence plot (JRP) as follows:

$$JRP_{i,j}^{\vec{x}_m,\vec{x}_n}\left(\varepsilon^{\vec{x}_m}, \varepsilon^{\vec{x}_n}\right) = R_{i,j}^{\vec{x}_m}\left(\varepsilon^{\vec{x}_m}\right) R_{i,j}^{\vec{x}_n}\left(\varepsilon^{\vec{x}_n}\right) \tag{14.9}$$

Thereafter, we can use joint recurrence rate (JRR) to quantify the density of recurrence points in each JRP as follows $JRR(\vec{x}_m, \vec{x}_n) = \frac{1}{N^2} \sum_{i,j=1}^{N} JRP_{i,j}^{\vec{x}_m,\vec{x}_n}$.

(4) Then we can characterize the synchronization of pair-wise time series as follows, where RR represents the recurrence rate of each recurrence plot which is 0.1. Thus, for a multi-channel signal $\{x_{k,l}\}_{l=1}^{L}, k = 1,2, \ldots, p$ containing p sub-signal, we can obtain a synchronization matrix $S(\vec{x}_m, \vec{x}_n)$ of the size $p \times p$. Finally, we can infer a multivariate weighted recurrence network by regarding each sub-signal as a node and deeming the synchronization $S(\vec{x}_m, \vec{x}_n)$ as the weight of the edge connecting node m and node n. The diagram of the method is shown in Fig. 14.5.

(5) As the generated network is a fully connected network in which it is difficult to depict its inherent topological structure and property, the integrals of the network measures over the sparsity range (corresponding to the areas of the network measure curve within the sparsity range) are taken into account. Specially, the sparsity ranges from 10% to 35% with step 1% (Korgaonkar et al. 2014; Zhang et al. 2011) and then we calculate the integrals of the weighted global efficiency (representing the functional integration of networks) and local efficiency (indicating the efficiency of local communication) (Rubinov and Sporns, 2010). Specifically, the weighted global efficiency is defined as:

$$E^w = \frac{1}{n} \sum\nolimits_{i \in N} \frac{\sum_{j \in N, j \neq i} \left(d_{ij}^w\right)^{-1}}{n-1} \tag{14.10}$$

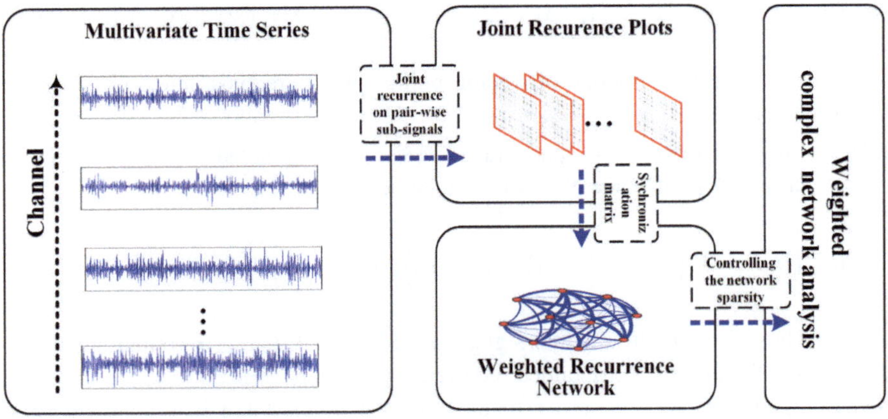

Fig. 14.5 The diagram of the weighted recurrence network method

where d_{ij}^{w} is the shortest weighted path length between nodes i and j, and the weighted local efficiency can be calculated by:

$$E_{loc}^{w} = \frac{1}{2n}\sum\nolimits_{i \in N} \frac{\sum_{j,h \in N, j \neq i}\left(w_{ij}w_{ih}\left[d_{jh}^{w}(N_i)\right]^{-1}\right)^{1/3}}{k_i(k_i - 1)} \tag{14.11}$$

where $d_{jh}^{w}(N_i)$ is the length of the shortest path between nodes j and h that contains only neighbors of node i and k_i is the degree of node i. All the above measures are calculated by using the Brain Connectivity Toolbox in Matlab (http://www.brain-connectivity-toolbox.net).

The results of the weighted recurrence network are as follows:

The acquired EEG signals were first divided into several same-length epochs of 24s without overlapping for each subject. Thus, for each subject, there were ten epochs for the state of eyes-open and eyes-closed, respectively.

Then, weighted global efficiency and local efficiency were calculated for each epoch and the average results for each subject are shown in Figs. 14.6 and 14.7. In addition, t-test was performed between the two states for each subject and the obtained p-values are also shown in Figs. 14.6 and 14.7. We can see that the *p*-values are smaller than 0.05 indicating the existence of statistical significance between the two states. As can be also seen, both weighted global efficiency and local efficiency for the state of eyes-closed are higher than the state of eyes-open for all subjects. These results suggest a decrease in specialized information processing with a decrease in integrated information processing in the state of eyes-open. All these results indicate that the proposed method has a capacity to explore the variety of brain dynamical properties from the state of eyes-closed to eyes-open.

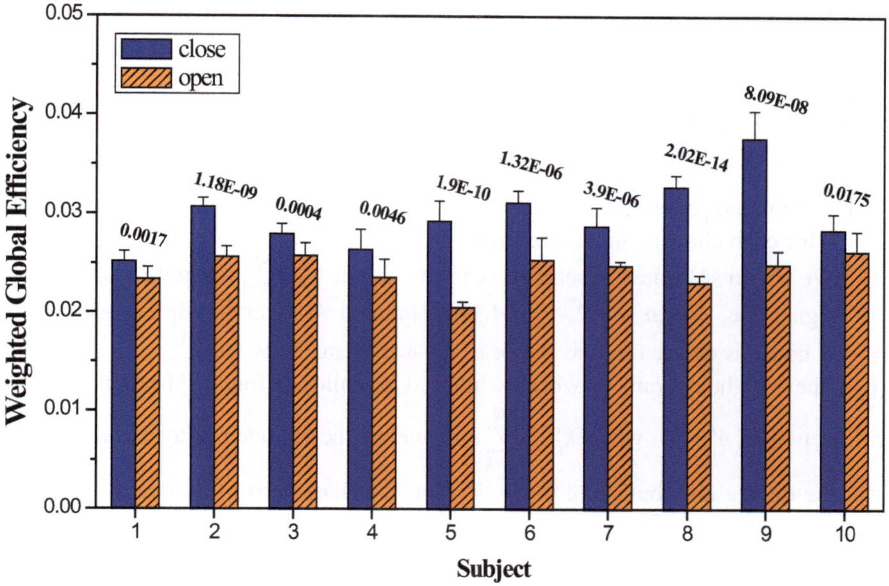

Fig. 14.6 The average weighted global efficiency and *p*-value of each subject with the state of eyes-open and eyes-closed, respectively

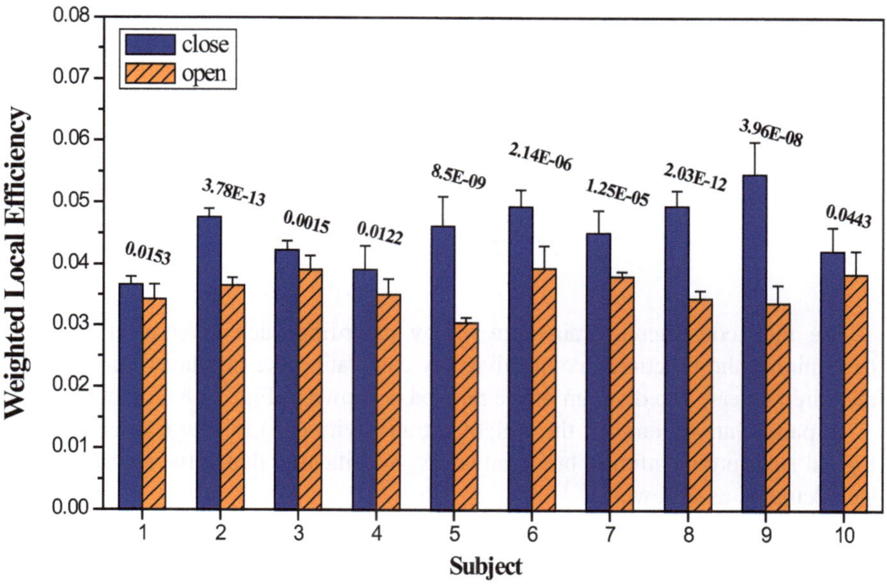

Fig. 14.7 The average weighted local efficiency and *p*-value of each subject with the state of eyes-open and eyes-closed, respectively

14.3.3 Multiplex Limited Penetrable Visibility Graphs

We construct the multiplex limited penetrable visibility graph for each epoch (a multi-channel EEG signal $\{x_{\alpha,i}\}_{i=1}^{N}, \alpha = 1,2, \ldots, M$), and the MLPVG method (Lacasa et al. 2015) can be implemented by the following steps:

(1) The limited penetrable visibility graph with the limited penetrable distance being 1 for each channel signal is constructed.
(2) We obtain M limited penetrable visibility graphs $\{A^{\alpha}\}_{\alpha=1}^{M}$ for multi-channel EEG signal $\{x_{\alpha,i}\}_{i=1}^{N}, \alpha = 1,2, \ldots, M$.The obtained M-layer multiplex network on N nodes is defined by the vector of adjacency matrices $\{A^{\alpha}\}_{\alpha=1}^{M}$.
(3) The weighted brain network is inferred as follows: for an M-layer multiplex network $\{A^{\alpha}\}_{\alpha=1}^{M}$, where $k_i^{\alpha} = \sum_{j=1}^{N} a_{ij}$ is the degree of node i at layer α and $p(k^{\alpha})$ is the degree distribution of the α-th layer, we characterize the interlayer correlation between two layers α and β by using the mutual information of the corresponding degree sequences $\{k_i^{\alpha}\}$ and $\left\{k_i^{\beta}\right\}$, i=1, ..., N. We compute the quantities:

$$p\left(k^{\alpha}, k^{\beta}\right) = \frac{N_{k^{\alpha},k^{\beta}}}{N} \tag{14.12}$$

where $N_{k^{\alpha},k^{\beta}}$ is the number of nodes having degree equal to k^{α} and k^{β}, respectively, on layer α and on layer β. The interlayer correlation of layers α and β is:

$$I_{\alpha,\beta} = \sum_{k^{\alpha}} \sum_{k^{\beta}} p\left(k^{\alpha}, k^{\beta}\right) \log \frac{p\left(k^{\alpha}, k^{\beta}\right)}{p(k^{\alpha}) p\left(k^{\beta}\right)} \tag{14.13}$$

We then construct a brain network by regarding each layer as a node and determining the functional connectivity by calculating the interlayer correlation of all paired layers. The diagram of the method is shown in Fig. 14.8 (Xia et al. 2013).

In particular, we employ the weighted transitivity (T^W) to characterize the topological structure of inferred brain networks. Specifically, this network measure can be calculated as follows:

$$T^W = \frac{\sum_{i \in N} \sum_{j,h \neq i \in N} \left(w_{ij} w_{hj} w_{hi}\right)^{1/3}}{\sum_{i \in N} k_i (k_i - 1)} \tag{14.14}$$

where N is the set of all nodes in the network, and n is the number of nodes. w_{ij}, w_{ih} and w_{jh} represent the weight between node i and j, i and h, j and h, respectively. The t-test is employed to obtain the p-values of weighted transitivity between eyes-closed

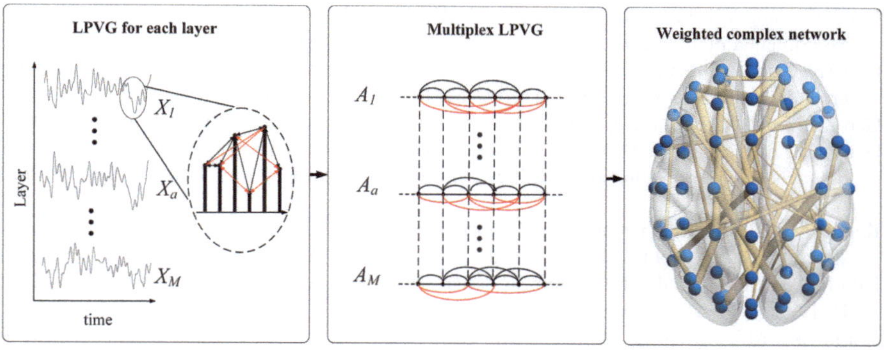

Fig. 14.8 The diagram of the multiplex limited penetrable visibility graph method

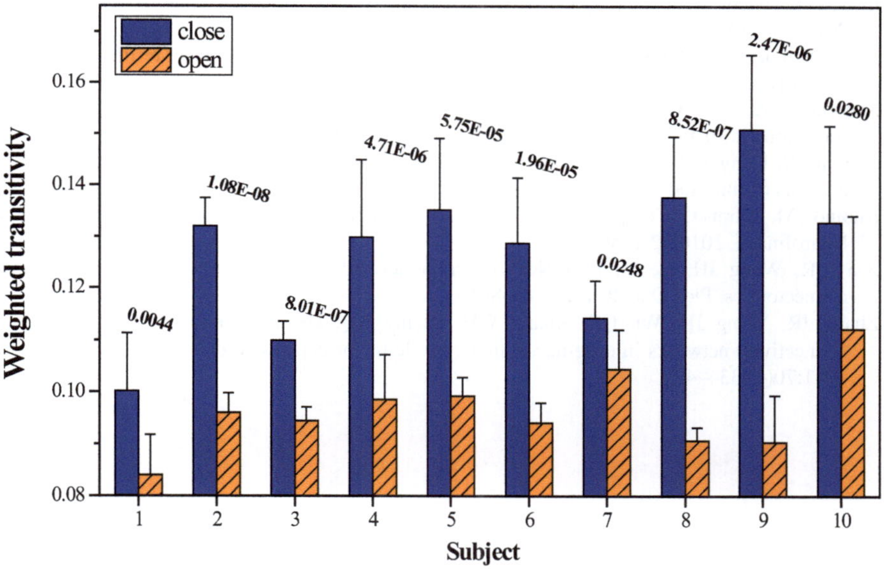

Fig. 14.9 The weighted transitivity and p-value of each subject with the state of eyes-open and eyes-closed, respectively

and eyes-open states. We show the results in Fig. 14.9. For different subjects, the p-values are all much smaller than 0.05 indicating statistical significance. The measure clearly enables to classify eyes-closed and eyes-open states.

References

Boccalettia S, Latora V, Moreno Y, Chavez M, Hwang DU. Complex networks: structure and dynamics. Phys Rep. 2006;424(4):175–308.

Gao ZK, Yang YX, Fang PC, Zou Y, Xia CY, Du M. Multiscale complex network for analyzing experimental multivariate time series. EPL. 2015;109:30005.

Gao ZK, Cai Q, Yang YX, Dang WD, Zhang SS. Multiscale limited penetrable horizontal visibility graph for analyzing nonlinear time series. Sci Rep. 2016a;6:35622.

Gao ZK, Small M, Kurths J. Complex network analysis of time series. EPL. 2016b;116(5):50001.

Gao ZK, Cai Q, Yang YX, Dong N, Zhang SS. Visibility graph from adaptive optimal kernel Time-Frequency Representation for classification of epileptiform EEG. Int J Neural Syst. 2017;27 (4):1750005.

Kennel MB, Brown R, Abarbanel HD. Determining embedding dimension for phase-space reconstruction using a geometrical construction. Phys Rev A. 1992;45(6):3403.

Kim HS, Eykholt R, Salas JD. Nonlinear dynamics, delay times, and embedding windows. Phys D. 1992;127(1–2):48–60.

Korgaonkar MS, Fornito A, Williams LM, Grieve SM. Abnormal structural networks characterize major depressive disorder: a connectome analysis. Biol Psychiatry. 2014;76(7):567–74.

Lacasa L, Luque B, Ballesteros F, Luque J, Nuño JC. From time series to complex networks: the visibility graph. PNAS. 2008;105(13):4972–5.

Lacasa L, Nicosia V, Latora V. Network structure of multivariate time series. Sci Rep. 2015;5:15508.

Luque B, Lacasa L, Ballesteros F, Luque J. Horizontal visibility graphs: exact results for random time series. Phys Rev E. 2009;80(4):046103.

Marwan N, Kurths J. Complex network based techniques to identify extreme events and (sudden) transitions in spatio-temporal systems. Chaos. 2015;25:097609.

Marwan N, Donges JF, Zou Y, Donner RV, Kurths J. Complex network approach for recurrence analysis of time series. Phys Lett A. 2009;373(46):4246–54.

Rubinov M, Sporns O. Complex network measures of brain connectivity: Uses and interpretations. NeuroImage. 2010;52:1059–69.

Xia MR, Wang JH, He Y. BrainNet viewer: a network visualization tool for human brain connectomics. Plos One. 2013;8(7):e68910.

Zhang JR, Wang JH, Wu QZ, Kuang WH, Huang XQ, He Y, Gong QY. Disrupted brain connectivity networks in drug-naive, first-episode major depressive disorder. Biol Psychiatry. 2011;70(4):334–42.

Chapter 15
Machine Learning

Yiheng Tu

Abstract Machine learning and pattern recognition have been widely applied in EEG analysis. They provide new approaches to decode and characterize task-related brain states and extract them from non-informative high-dimensional EEG data. Given the growth in the interest and breadth of application, we introduce how to apply machine learning techniques in EEG analysis. First, we give an overview of machine learning analysis and introduce several basic concepts. Then, we propose a scientific question of discriminating EEG data under eyes-open and eyes-closed resting-state conditions, and provide a step-by-step tutorial including extracting features, training features, feature selection and dimension reduction, selecting a classifier, testing the classifier, evaluating results, and pattern expression. We also discuss perspective, particularly the deep learning algorithms, for future study. In the last section of this chapter, we give detailed MATLAB codes for implementing machine learning analysis for classifying eyes-open and eyes-closed EEG data.

Keywords Machine learning · Classification · Feature · Training · Testing

With millisecond-level temporal resolution, electroencephalographic (EEG) has been widely used to study the neural dynamics of perception and cognition in humans (Hu et al. 2011; Zhang et al. 2012; Tu et al. 2014, 2016). When combined with an experiment that involves the delivery of external stimuli or a subject performing certain tasks, EEG can provide powerful methodology for building a link between the brain and behavior. However, to what extent the "brain decoding"

Electronic supplementary material The online version of this chapter (https://doi.org/10.1007/978-981-13-9113-2_15) contains supplementary material, which is available to authorized users.

Y. Tu (✉)
Department of Psychiatry, Massachusetts General Hospital and Harvard Medical School, Charlestown, MA, USA

L. Hu, Z. Zhang (eds.), *EEG Signal Processing and Feature Extraction*,
https://doi.org/10.1007/978-981-13-9113-2_15

can be practically realized using EEG is not clear. For example, how accurate can EEG detect one's behavioral state (e.g., eyes-open vs. eyes-closed)? Can we monitor individual's cognitive or perceptual level in real time? Besides the development of recording systems and experimental design, emerging analytical approaches seem to be the key to answer these questions.

There has been growing interest in the use of machine learning techniques to analyze EEG (Blankertz et al. 2011; Makeig et al. 2012; Müller et al. 2008). Accumulating studies have provided evidence that machine learning is capable of extracting meaningful information from high-dimensional and noisy EEG data. Given the growth in interest and breadth of application, in this chapter we provide a tutorial introduction to machine learning approaches for EEG. We focus on introducing the methodology and its related applications, and perspective, particularly the deep learning algorithms, for future study.

15.1 Machine Learning Analysis: An Overview

Neuroimaging experiments are designed to differentiate brain states, so that EEG data collected from the experiment can be modeled with cognitive or perceptual responses. Conventional analytical approaches typically use regression to build a relationship between hypothesis-based EEG features, such as event related potentials (ERPs) and behavioral scores, or seek to determine which feature in time or frequency domain is involved in the processing of cognitive or perceptual states by analyzing each sample separately. We call such approaches "univariate analysis". Theoretically, if the brain responses at a particular time point differ between two states, it is possible to use that activity to decode an individual's cognitive or perceptual responses. For example, the power of alpha band oscillation is strongly related to the eyes-closed and eyes-open conditions. However, it is often difficult to find an individual feature of interest that offers a sufficiently large difference between two conditions, and that may fail to make decisions.

A machine learning classifier is a function that takes the values of various features of brain activity as independent variables or predictors from a condition, and predicts the class that the condition belongs to. There are several basic concepts for machine learning needed to be introduced:

Class: the category to which an object belongs. In a class, a set of patterns share common attributes and usually originate from the same source.

Pattern: a collection of features of an object, along with the class information for that object.

Sample: any given pattern of an object is referred to as a sample.

Feature: a set of variables believed to carry discriminating and characterizing information about an object.

Feature vector: a collection of K such features for one sample, ordered in some way into a K-dimensional vector.
Feature space: the K-dimensional space in which the feature vector lies.

In an EEG experiment, the features could be extracted from time, frequency, or spatial domains, and the class could be cognitive or perceptual responses. We can denote a sample (e.g., trial or subject) as the vector $X \in R^{N \times K}$ and its class label as $y \in R^{N \times 1}$. A classifier can build up the relationships between features and class labels using the training data, by learning a number of parameters and estimating the weights of each feature. Therefore, given a sample X, the classifier is formally a function f that predicts label $y = f(X)$. The function f can be either classification, where the output is a discrete number corresponding to a limited number of classes, or regression, where the output is a continuous variable.

The machine learning classifier requires training by which the classifier learns the function f between features and their corresponding class labels. After training, the classifier can be applied to the test data to determine whether the features contain the discriminative information between classes of samples. If the trained classifier truly captured the relationships between features and labels, it would be able to predict the classes of samples in the test data that had not been previously seen. Normally, we assume that the training and test data are independently drawn from a "sample distribution." As shown in Fig. 15.1, we denote the training and test data as $X_{training} \in R^{N \times K}$ and $X_{test} \in R^{L \times K}$, respectively, where samples are the rows and features are the columns, and the class labels by column vectors $y_{training}$ and y_{test}. For classification, the most commonly used measure to evaluate the performance of a classifier is prediction error, or accuracy, which is the proportion of samples in the test data being correctly classified. For regression, we can use mean square error (MSE) $= \frac{1}{L}\sum_{i=1}^{L}\left(y_i - \tilde{y}_i\right)^2$ or prediction-outcome correlation $= corr\left(y, \tilde{y}\right)$ between the real labels y and the predicted labels $\tilde{y}$.

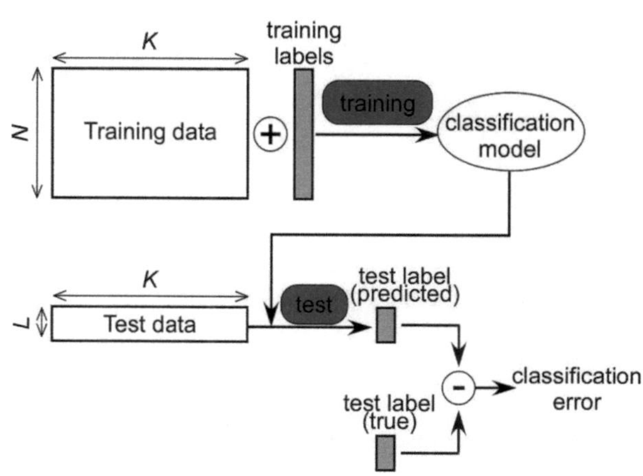

Fig. 15.1 Machine learning classifier framework. The classifier is learned from the training data and their class labels, and then used to predict labels for test data. The predicted labels are compared to the true labels and the accuracy of the classifier can be computed

15.2 Machine Learning Analysis: EEG Features

To perform machine learning analysis in EEG studies, we need to decide which EEG features can be used to discriminate between experimental conditions. For spontaneous EEG data, we usually study the signal in the frequency domain through characterizing rhythms by frequency and amplitude. For example, the alpha band-limited power for eyes-closed resting-state EEG is significantly higher than eyes-open condition (Fig. 15.2). Other features, including bispectrum (detect phase coupling between different frequency signals) (Gajraj et al. 1998), entropy (describe the distribution of EEG signal) (Wang et al. 2014), coherence (similarity of the frequency components across EEG channels) (Srinivasan et al. 2007), and connectivity (similarity of the time series across EEG sensors) (Schoffelen and Gross 2009), also can be used as features for machine learning analysis. For evoked potentials (EPs) and event-related potentials (ERPs), we can use the magnitude and latency of peaks in time domain and in time-frequency domain as features.

There are several features that could be applied to both spontaneous EEG and EPs/ERPs. For example, the joint time-frequency representation of EEG illustrates how frequency features vary with time (Hu et al. 2014) (Fig. 15.3). The power (e.g., power spectral density)/phase value (e.g, phase-locked value) of each time-frequency point, or a summarized value from a certain time frequency region, could be a feature. The EEG/ERP topography, which is the interpolated map of one EEG feature from multichannel EEG, representing the distribution of the feature on the scalp, could also be a feature (Van de Ville et al. 2010). The joint space-time-frequency features of multichannel EEG/ERP capture information from all three domains of EEG signal (Tu et al. 2014), but increase the computational complexity of the machine learning classifier.

In summary, selecting a proper collection of EEG features is crucial in machine learning analysis. Researchers can base their decisions on hypothesis or data driven

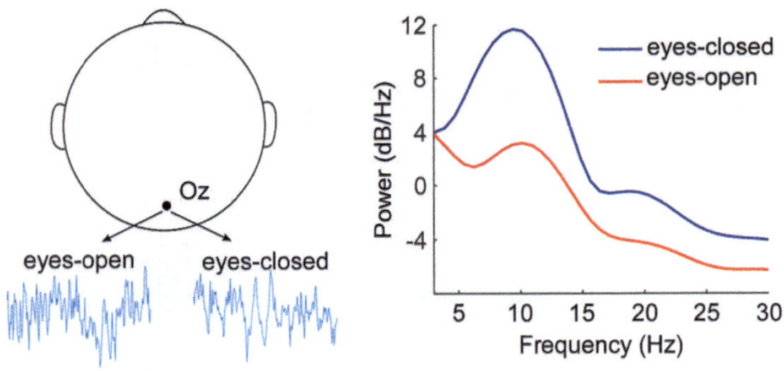

Fig. 15.2 Waveform and power spectrum density for eyes-open and eyes-closed resting-state EEG

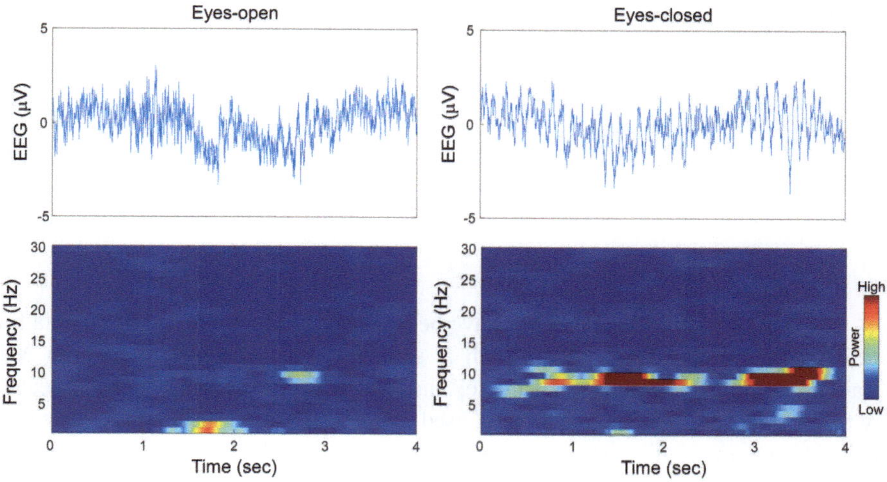

Fig. 15.3 EEG time series (upper) and time-frequency representations (lower)

search. We will provide an example of using alpha band limited power as a feature to classify eyes-open and eyes-closed conditions at the end of this chapter.

15.3 Machine Learning Analysis: Training

The goal of a machine learning classifier is to estimate the correct class corresponding to a give feature vector based on some prior knowledge obtained through training. Training is the process by which a classifier learns the mapping relationship between feature vectors and their corresponding class labels. This relationship between feature vectors and the corresponding class labels forms a decision boundary in the feature space that separates patterns of different classes from each other.

Before training the classifier, we need to organize the samples (feature vectors and class labels) into a training set (data used to train a classifier; with known class labels), a testing set (data used to evaluate the performance of a classifier; typically collected at the same time or carved of the training data; with known class labels), and in some situations, a field data set (unknown data to be classified by a trained classifier; with unknown class labels). There is an important tradeoff between having many noise samples (e.g., single trials) or fewer cleaner ones (e.g., averaged trials) for training the classifier. Generally, having more samples helps the classifier to obtain good estimates of their parameters in training and increases the power of the test for the significance of the accuracy in testing.

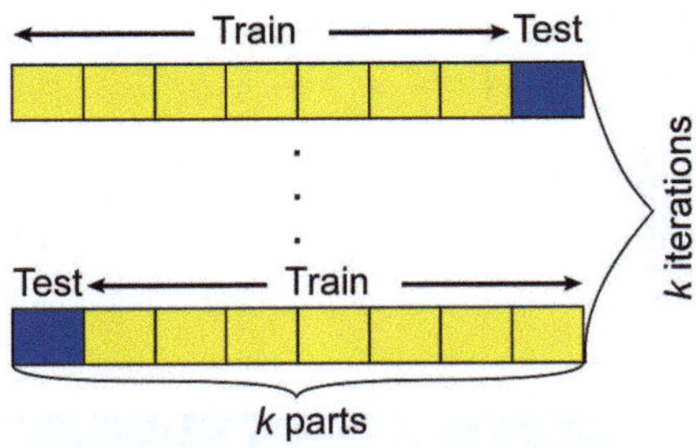

Fig. 15.4 Cross-validation. The data is split multiple times into a train set (with training samples), used to train the classifier, and a test set (with test samples), used to compute predictive power

The cross-validation approaches have been developed to make use of all data for training and testing (Fig. 15.4). With N samples, we can divide them into N-1 training samples and one test sample, and the same procedure can be repeated N times to ensure that each sample is used as the test sample once. This variant is called "leave-one-out" cross-validation (LOOCV) (Tu et al. 2016). Although in each iteration the classifier trained is technically different, one can expect them to be similar since they share much training data. Besides LOOCV, other variants of cross-validation, such as K-fold cross-validation, where K is the number of parts into which the dataset is divided (e.g., 5-fold or 10-fold), have been widely used in EEG studies. Compared to LOOCV, K-fold cross-validation can be computationally efficient and the classifier trained might not be over-fitted. It is worth mentioning that the training data in each fold must contain samples of all classes, otherwise the classifier for that fold will not be able to predict the absent ones.

The MATLAB function "crossvalind" can help us to generate indices for a K-fold cross-validation of N samples. Indices contain equal (or approximately equal) proportions of the integers 1 through K that define a partition of the N samples into K subsets.

```
clear all; close all;

Indices = crossvalind('Kfold',N,K)
```

To build a better classifier through training, researchers need to consider the training performance, which is the ability of a classifier to correctly identify the classes of training data, and the generalization performance, which is the ability of the classifier to identify the classes of (unseen) test data. A good machine learning classifier has the decision boundary that provides the best possible generalization performance, and not the one that provides perfect training performance. In some cases, perfect separation of training data is typically caused by noisy data, and finding a decision boundary that provides perfect training data classification would amount to learning the noise in the data. We call these cases, which learn noise or random error instead of the true underlying relationship, as overfitting.

15.4 Machine Learning Analysis: Feature Selection and Dimension Reduction

The performance of machine learning analysis in EEG is often hindered by the "curse-of-dimensionality"; where the EEG data are naturally high-dimensional and span multiple spaces but the number of samples is limited (Mwangi et al. 2014). For example, 64-channel time-frequency maps (1000 samples in the time domain and 100 bins in the frequency domain) of EEG have 6,400,000 features (if one time-frequency point is one feature). Feature selection or dimension reduction is critical in identifying a small set of discriminative features from high-dimensional EEG data for higher classification accuracy and better classifier interpretability (Tu et al. 2015).

Feature selection or dimension reduction can be realized with or without class labels (supervised or unsupervised). Principle component analysis (PCA) is a conventional dimension reduction approach that projects high-dimensional data along a direction for which the variance of the data is maximized in a linear subspace spanned by a small number of latent components (Fig. 15.5) (Jolliffe 2011). Mathematically, PCA uses orthogonal transformation to convert a set of observed data of

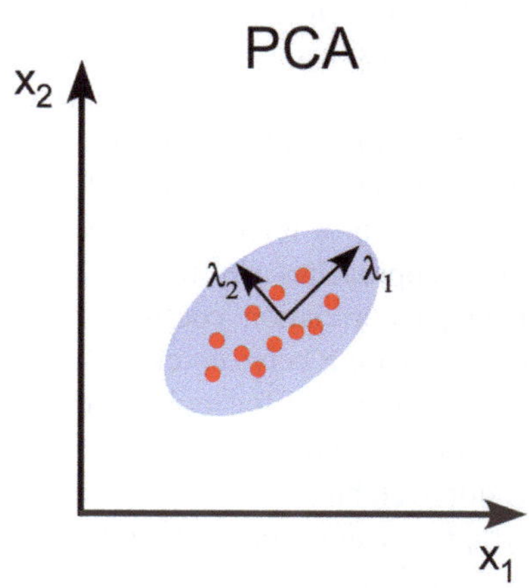

Fig. 15.5 Principle component analysis. The projection direction λ_1 captures the maximal variance of the data samples, while its orthogonal direction λ_2 has the minimal variance of the data samples

correlated variables into a set of values of linearly uncorrelated variables called principal components (PCs). By omitting a PC of low variance from the data, we lose only a small amount of information. Suppose we only keep L (which is smaller than the number of variables) PCs, then the new data will only have L columns but contain most of the information. Normally, PCA can be used in EEG analyses to reduce the dimensionality along time, frequency, channels, trials, and subjects, since EEG exhibits redundancy across samples in these domains (Hu et al. 2015).

To implement PCA in MATLAB, we can use the built-in function "pca" as follows:

```
clear all; close all;

[coef,score,latent] = pca(X)
```

The function returns the principal component coefficients in "coef," principal component scores (representations of X in the principal component space) in "score," and the principle component variances (eigenvalues of the covariance matrix of X) in "latent," for the n-by-k matrix X. Rows of X correspond to samples and columns correspond to features. Each column of "coef" contains coefficients for one principal component, and the columns are in descending order of component variance. Rows of "score" correspond to samples, and columns correspond to components.

However, unsupervised methods do not make use of class information to reduce the dimensionality and cannot guarantee that classes are well separated in low-dimensional space. In contrast, supervised methods exploit class information to ensure that high-dimensional data can be mapped into a low-dimensional space where different classes are well separated (Mwangi et al. 2014). Here, we give a brief introduction of three categories of supervised feature selection methods. The most direct approaches use univariate statistical techniques, including t-test, ANOVA, and Pearson correlation coefficient to rank features according to their relevance in discriminating between class differences. These univariate approaches normally do not perform well for multidimensional EEG data. Wrapper techniques use a classification or regression objective function to rank features according to their classification weights in the model. Recursive feature elimination (RFE) is the most popular wrapper feature selection approach (Shen et al. 2007; Gysels et al. 2005). It begins with all features in the training set, and features are iteratively eliminated from the set until an optimal number of features are found. Recently, the embedded methods that select features by enforcing certain "penalties" on a machine learning model, thus yielding a small subset of relevant features, have become increasingly popular. For example, least absolute shrinkage and selection operator (LASSO) minimizes the mean square error (MSE) with sparsity-enhancing L_1 regularization on regression coefficients, and it can shrink small regression coefficients to zero to realize feature selection (Tibshirani 1996). When dealing with strongly correlated predictors (e.g., adjacent EEG time points), LASSO arbitrarily selects one variable from a group of highly-correlated variables, which degrades the interpretability of the prediction model. Therefore, an elastic net approach was developed and applied in neuroimaging studies. Elastic net enhances LASSO by combining L_1 regularization (sparsity-enforcing) and L_2 regularization (grouping constraint) (Zou and Hastie 2005).

15.5 Machine Learning Analysis: Choosing a Classifier

After preparing samples and selecting the most informative features, we are ready to select a classifier (Lotte et al. 2007). Classifiers that learn the function f between features and class labels are categorized into discriminative and generative models. The goal of a discriminative model is to directly learn a prediction function with a given parametric form by setting its parameters. The most natural choice for a simple function is to have the classification depend on a linear combination of the features with:

$$\mathrm{X}w = x_1 w_1 + \ldots + x_K w_K \tag{15.1}$$

The classifier determines class A if $\mathrm{X}w > 0$ or class B if $\mathrm{X}w < 0$. Therefore, learning a linear classifier is equivalent to learning a line separating points into two classes as far as possible. We call this line a decision boundary. Linear discriminant analysis (LDA) and linear support vector machine (SVM) are commonly used approaches to learn the decision boundary (Fig. 15.6) (Subasi and Ismail Gursoy 2010).

LDA finds a projection direction to a line such that all samples from different classes are well separated. SVM constructs a hyperplane in a high-dimensional space so that an optimal decision boundary (which maximizes the margin between the boundaries of different classes) can be achieved in this hyperplane. The main novelty of SVM is its ability to use a kernel function to map the original feature space (where classes are not linearly separable) into a higher dimensional space (where the classes are linearly separable) and find the support vectors in that higher dimensional space. There are several parameter settings of SVM that need be tuned to perform a good prediction. For a linear SVM, a cost parameter (C), which tells the SVM optimization how much you want to avoid misclassifying each training sample, needs to be defined before training the classifier. For large values of C, the optimization will choose a smaller-margin hyperplane if that hyperplane does a better job of getting all the training points classified correctly. Conversely, a very small value of C will cause the optimizer to look for a

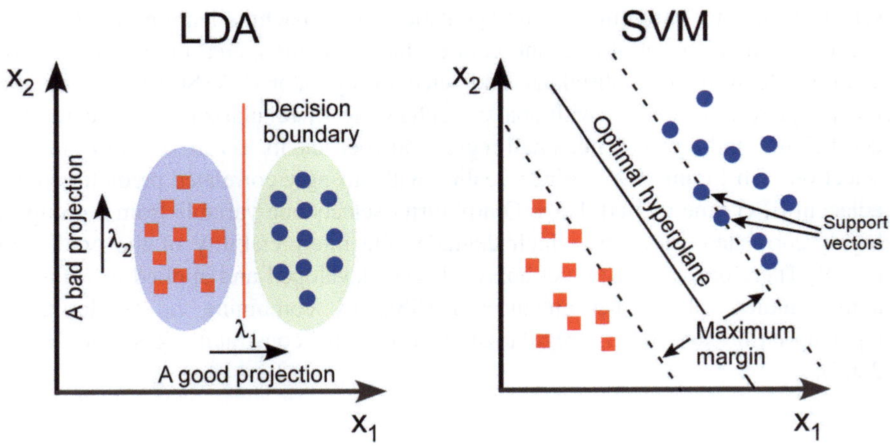

Fig. 15.6 Classifiers based on discriminative models learn a decision boundary separating points into two classes as far as possible

larger-margin separating hyperplane, even if that hyperplane misclassifies more points. For nonlinear versions of SVM, a grid-search via inner cross-validation would be helpful to define an optimal combination of parameters.

In contrast, the generative models, for example, Gaussian Naïve Bayes (GNB) classifier, learn a statistical model that can generate a sample belonging to a particular class and then make a prediction (Huang et al. 2013). What is modeled is the distribution of class-conditioned feature values, $p(X| y = A)$ and $p(X| y = B)$, which is then inverted via Bayes rule to classify by deciding which probability is largest. In practice, Naïve Bayes has been shown to provide good performance and only requires a small amount of training data to estimate the parameters necessary for classification.

The implementation of classifiers can be achieved using either the MATLAB built-in function "classify" or other toolboxes. To use the function "classify," we can run the following MATLAB codes,

```
clear all; close all;

class = classify(sample,training,group,'type')
```

The function classifies each row of the data in sample into one of the groups in training. Sample and training must be matrices with the same number of columns.

Group is a grouping variable for training and each element defines the group to which the corresponding row of training belongs. Specify *type* inside single quotes to decide the type of classifier, including "linear," "diaglinear," "quadratic," "diagquadratic," and "mahalanobis." The detailed description of function "classify" can be found in the MATLAB Documentation.

To perform SVM classifier, it is recommended to use LIBSVM toolbox (https://www.csie.ntu.edu.tw/~cjlin/libsvm/) which is an integrated software for support vector classification, regression, and distribution estimation. There are several steps for using the LIBSVM-MATLAB. First, after downloading the toolbox and unzipping the file, we need to set path for the toolbox in the MATLAB interface by "Home->set path ->add with subfolders->the path of your LIBSVM toolbox". Second, we need to select a compiler and configure the compiler, using the code:

```
mex -setup
```

and you will see the following contents in your MATLAB command window:

```
Please choose your compiler for building external interface (MEX) files

Would you like mex to locate installed compilers [y]/n?
```

This is asking you to locate the compilers installed on your computer and we can type "y" to select a compiler. The selection of compilers is based on the existing compilers on your computer and all supported compilers for LIBSVM can be found in the official website of LIBSVM.

```
Please choose your compiler for building external interface (MEX) files

Would you like mex to locate installed compilers [y]/n? y

Select a compiler:

[1] Microsoft Visual C++ 2015 Professional (C)

[0] None
```

Alternatively, we can select the proper compiler and run the file "make.m" in the LIBSVM-MATLAB folder using the code:

```
make
```

If you do not see any error, the LIBSVM-MATLAB toolbox is successfully installed. We can try the SVM classifier using the sample data in LIBSVM folder by the following MATLAB codes:

```
libsvmread('heart_scale');

model = svmtrain(heart_scale_label,heart_scale_inst);

[predict_label,accuracy] = svmpre-
dict(heart_scale_label,heart_scale_inst,model);
```

After implementing the above codes, if you see the following result in the MATLAB command window, the LIBSVM is ready for future classification tasks.

```
Accuracy = 86.6667% (234/270) (classification)
```

Although some classifiers naturally permit the use of more than two classes, most are by nature binary classifiers. Generally, multiclass classification (e.g., L classes) is realized using popular binary classifiers by different strategies:

One-vs.-All: train a binary classifier for each class to distinguish that class from all other classes, so there are L total classifiers. Combine binary classifiers by returning classifier with highest confidence/score (e.g., the highest posterior probability of a Naïve Bayes classifier).

One-vs.-One: train a binary classifier for each pair of classes so there are $L(L-1)/2$ total classifiers. Combine binary classifiers by finding the most frequent classification label.

15.6 Machine Learning Analysis: Evaluating Results

Once training is complete, the classifier's generalization performance needs to be evaluated on the test data or field data. The most direct measure for the evaluation of a machine learning classifier is the accuracy, which is the portion of correctly classified samples in all samples. It should be noted that accuracy is not suitable for unbalanced datasets. More importantly, accuracy cannot reflect the cost of errors because not all errors are equally costly. For example, two types of error in diagnosis of a disease:

1. Sick people incorrectly identified as healthy; the cost may include death;
2. Healthy people incorrectly identified as sick; the cost might be the side effects and the cost of administering drugs that are in fact not needed.

Therefore, we need other performance measures for a binary classification. Table 15.1 summarizes four typical measures. The positive and negative represent rejecting and not rejecting the null hypothesis, respectively. In a typical medical diagnosis, true positive (TP) represents sick people correctly diagnosed as sick; false positive (FP) represents healthy people incorrectly identified as sick; true negative (TN) represents healthy people correctly identified as healthy; false negative (FN) represents sick people incorrectly identified as healthy.

Another two measures, including sensitivity, which measures the proportion of positives that are correctly identified as such ($Sensitivity = \frac{TP}{TP+FN}$), and specificity, which measures the proportion of negatives that are correctly identified as such ($Specificity = \frac{TN}{TN+FP}$), are also important in practice.

For most systems, it is not possible to realize both a maximum sensitivity and maximum specificity using the same configuration (e.g., set of parameters). In this case, one has to decide which of the two is more important. A useful tool for investigating this issue is the receiver operating characteristic (ROC) plot (Fig. 15.7). The ROC curve is a fundamental tool for diagnostic test evaluation. In a ROC curve, the sensitivity is plotted as a function of the (1-specificity) for different criterion values of one parameter. The area under the ROC curve (AUC) is a measure of how well a parameter can distinguish between two diagnostic groups.

Providing an absolute number of accuracy, sensitivity, or specificity is not enough. A statistically significant classification result is the one where we can reject the null hypothesis that the classifier decides at random. If the classifier does not

Table 15.1 Performance measures for a classifier

		Actual value	
		Positive	Negative
Predicted outcome	Positive	True positive (TP)	False positive (FP)
		Correct outcome	Type I error
	Negative	False negative (FN)	True negative (TN)
		Type II error	Correct outcome

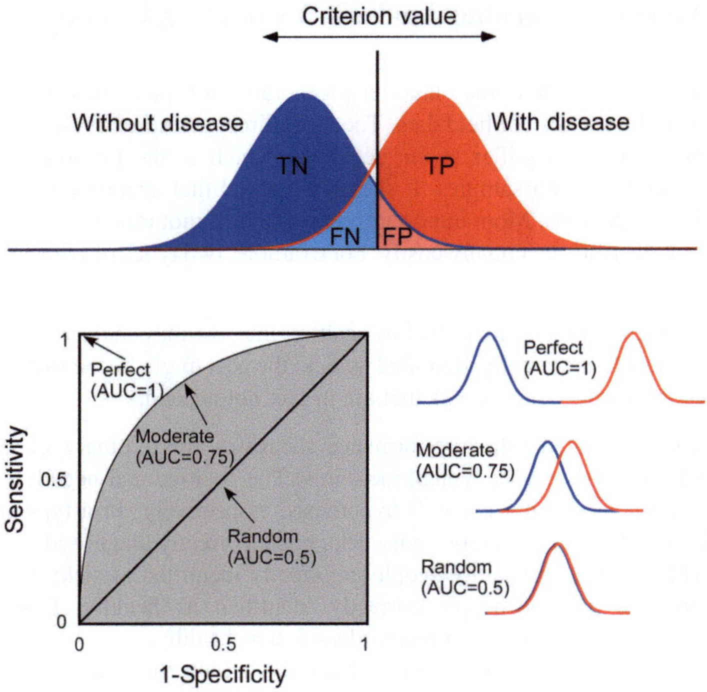

Fig. 15.7 Receiver operating characteristic (ROC)

have any information about the class that is being predicted using EEG data, we would expect that the classifier could only guess the class labels at random. More formally, if the test samples were drawn independently and the sample sizes of the two classes are equal, then we would expect an accuracy of 50% on average, based on the binomial distribution. Therefore, a statistical significance needs to be calculated by determining the probability that the obtained accuracy might vary to a value at least as high as 50% when the null hypothesis is satisfied. For example, we can use a one-sample t-test against 50% to show that the classification accuracy is significantly higher than chance level, or we can use a paired t test to compare the performance of different feature sets or classifiers.

However, if samples of EEG data do not satisfy independence, a non-parametric permutation test is recommended for assessing the significance (Maris and Oostenveld 2007). In permutation testing, we randomly permute the class labels of the data prior to training. Cross-validation is then performed on the permuted dataset, and the procedure is repeated 1000 or more times. If the classifier trained on real data labels has an accuracy that exceeded the 95% confidence interval generated from the accuracies of the classifiers trained on randomly relabeled data labels, the classification/prediction model is considered to be performing well.

15.7 Machine Learning Analysis: Pattern Expression

Once we know the class information is presented in the data, it is natural to ask where this information resides (e.g., which time interval in our data contains the most discriminative information). This can be addressed by two aspects: determining which features are being selected and determining how their weights affect the classification.

When using feature selection in each cross-validation fold, features being selected may be different and the overlap between features selected in all folds may be small (<1/3) because whatever features are selected contain sufficient information for classification. Features in EEG data might be redundant since many of them are intrinsically correlated and encode the same things. There might be a small subset present in all folds that could be necessary for the classification.

The use of linear classifiers makes it easier to know which features strongly affect classification, since each feature affects the decision only through the magnitude of its weight (linear classifiers do not consider the interactions between features). For intra-subject classification (training a classification model on each subject), we can assess the statistical significance of each feature in the classification by comparing the weight of each feature across all subjects against zero, using a one-sample t test. For inter-subject classification (training a classification model across subjects), we can use non-parametric statistical approaches. Besides the permutation test we introduced, bootstrap tests are widely used in neuroimaging to determine the significance of weight for each feature. In a bootstrap test, we construct 1000 or more bootstrap samples (with replacement) consisting of paired features and class labels, and run machine learning analysis on each. A one sample t-test is performed for each feature based on the proportion of weights below or above zero, and subjected to multiple comparison corrections.

15.8 Machine Learning Perspective: Deep Learning Algorithms

Deep learning is a subset of machine learning that is composed of multiple processing layers to learn representations of data with multiple levels of abstraction. Because deep learning can learn and understand complex representations directly from the raw signal, as well as automatically extract advanced features for classification, it has been applied to computer vision, speech recognition, and natural language processing. It significantly outperforms traditional machine learning algorithms with more training data and greater availability of graphics processing units (GPUs) for complex computation.

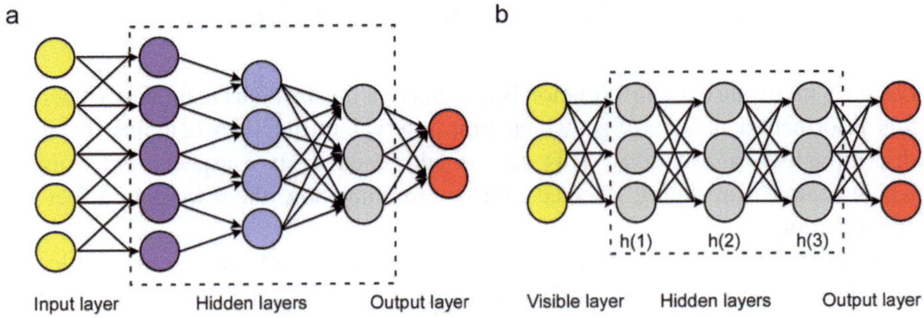

Fig. 15.8 Frameworks of deep learning algorithms. (**a**) convolutional neural network (CNN); (**b**) deep belief network (DBN)

A major limitation of current EEG analyses is that unknown and potentially relevant features may not be included. Although numerous successful applications of deep learning have been developed to large-scale image, video and text data, they remain relatively unexplored in the neuroimaging field. One reason for this may be that the number of samples in most neuroimaging datasets is very limited, thus making it difficult to train large-scale networks with millions of parameters. Nevertheless, recent studies have started to explore the potential of deep learning algorithms for EEG-based brain decoding. In this section, we will briefly introduce two deep learning algorithms including convolutional neural network (CNN) and deep belief network (DBN), as well as their applications in EEG analysis.

CNN is a class of deep, feed-forward artificial neural networks inspired by the connectivity pattern between neurons in the animal visual cortex. The basic framework of CNN is shown in Fig. 15.8a. A typical CNN consists of an input and an output layer, as well as multiple hidden layers. The hidden layers consist of convolutional layers, pooling layers, fully connected layers, and normalization layers. CNNs can learn local nonlinear features through convolutions and represent high level features as compositions of low-level features through multiple layers of processing. In EEG analysis, CNN is normally designed to extract a wide range of features from raw data, using two convolutional layers to handle the large number of input signals (e.g., time points, electrodes). The first convolution is used across time to process temporal information and the second convolution is used across electrodes to handle relationships between electrodes. In this way, the framework can be regarded as an integration of spatial block and temporal block. Since CNN has good performance in decoding and is easier for iterative training, it has been successfully applied in EEG studies. Mirowski et al. applied CNNs for epileptic seizure prediction in EEG and intercranial EEG (Mirowski et al. 2009). Cecotti and

Graser used a special CNN for classification of steady-state visual evoked potentials (SSVEPs, brain oscillation induced by visual stimuli) (Cecotti and Graeser 2008). The network integrated the Fourier transform between convolutional layers, which transformed the data from the time domain to a time-frequency representation. In Tabar and Halici, the authors aim to use deep learning methods to improve the classification performance of EEG motor imagery signals (Tabar and Halici 2017). They combined CNN and stacked autoencoders (SAE) to classify EEG motor imagery signals. A new form of input was introduced to combine time, frequency and spatial information extracted from EEG signals and it was used in CNN with one 1D convolutional and one max-pooling layer.

DBN is a generative model with connections between the multiple layers of latent variables ("hidden units"; Fig. 15.8b). DBNs can be viewed as a composition of unsupervised networks such as restricted Boltzmann machines (RBMs). The RBMs are stacked in such a way that the output layer of the RBM at layer *l-1* is the input to the RBM at layer *l*. A typical DBN can have two or more RBMs stacked in this way. The purpose of the RBM is to learn a probability distribution over a set of inputs. The high-dimensional data can be fed into the hidden layer through the visible layer, and the hidden layer can identify the characteristics of data based on weights of connections between layers. The best way to train a DBN is one layer at a time, in a layer-wise manner. Once every layer of the DBN has been trained, the output of the final RBM can be used as the input to a classifier for decoding. Therefore, one of the most important applications of DBN in EEG analysis is dimension reduction. For example, in a paper by Li et al., the authors proposed a DBN based model for affective state recognition from EEG signals (Li et al. 2013). Specifically, signals from each EEG channel are firstly processed with a DBN to effectively extract critical information from over thousands of features. The extracted low-dimensional characteristics are then utilized in the learning to avoid the small sample problem. Similar application was introduced by Zheng et al., who trained a DBN with differential entropy features extracted from multichannel EEG as input to classify two emotional categories (positive and negative) (Zheng et al. 2014).

In general, there are several advantages to using deep learning approaches for EEG analyses. First, deep learning approaches learn features directly from the raw data through several layers and take into account higher-order features and relationships between those features. Second, deep learning approaches can be applied to unlabeled data by using unsupervised methods, and thus have wider applications for abundant unlabeled EEG data.

15.9 Machine Learning Analysis: An Example

In this section, we demonstrate how to perform machine learning analysis on a toy example. The signals were selected at the Oz channel of the EEG data from one subject in two separate sessions (eyes-open [EO] and eyes-closed [EC]). There are 60 trials for EO and EC conditions, respectively, as training samples and another 60 trials with unknown labels as test samples. The task of this example is training a classifier to separate EO and EC conditions from training samples and identifying labels for test samples.

In this example, we summarize the following concepts as:

Classes: EC and EO
Samples: EEG trials
Feature: alpha power; in the previous section we showed that alpha power in EC is significantly higher than alpha power in EO
Feature vector: alpha power at Oz ($K = 1$)
Feature space: alpha power (1-dimensional space)
Pattern: alpha of EC > alpha of EO

We provided MATLAB codes including (1) extract features for training and test samples; (2) train an LDA classifier and evaluate the training performance using 10-fold cross validation; (3) evaluate the generalizability of the LDA classifier on test samples.

In the first step, we load MATLAB file "data_classification.mat" into the Workspace. The file contains (1) ec (60 EEG trials for eyes-closed condition; each trial has 2000 time points); (2) eo (60 EEG trials for eyes-open condition; each trial has 2000 time points); (3) test_samples (60 EEG trials for testing; each trial has 2000 time points); (4) test_labels (60 binary numbers, 1 or 0, representing the class labels for the test samples). It should be noted that the sampling rate of the EEG data is 200, and the recording length for each trial is 10 sec.

```
clear all; close all;

load data_classification.mat
```

In the second step, we calculate power spectral density (PSD) and extract alpha-band power for training and test samples, respectively.

```
%% parameters

fs_test = 200; % sampling rate

N_Train = size(ec,2); % number of traing trials

N_Test = size(test_samples,2); % number of test trials


%% PSD

nfft = 256; % Point of FFT

for n = 1:N_Train

    [P_ec(:,n),f] = pwelch(detrend(ec(:,n)),[],[],nfft,fs_test); % calculate PSD for
ec condition

    [P_eo(:,n),f] = pwelch(detrend(eo(:,n)),[],[],nfft,fs_test); % calculate PSD for
eo condition

end

for n=1:N_Test

    [P_test(:,n),f] = pwelch(detrend(test_samples(:,n)),[],[],nfft,fs_test); % calcu-
late PSD for test samples

end


%% feature extraction

alpha_idx = find((f<=12)&(f>=8));  % frequency index of alpha band power

a_ec_train = mean(P_ec(alpha_idx,:)); % extract alpha band power from eo

a_eo_train = mean(P_eo(alpha_idx,:)); % extract alpha band power from ec

a_test = mean(P_test(alpha_idx,:)); % extract alpha band power from test data
```

After running the codes above, we can find three vectors (dimension: 1×60) in the MATLAB Workspace: a_ec_train, a_eo_train, and a_test, representing the extracted features for training samples in eyes-closed condition, eyes-open condition, and for test samples.

In the third step, we train an LDA classifier and evaluate performance of the classifier using 10-fold cross validation. We define the class label for eyes-open condition as "1," and eyes-closed condition as "0."

```
all_samples = [a_eo_train,a_ec_train]'; % all samples

all_labels = [ones(size(a_eo_train,2),1);zeros(size(a_ec_train,2),1)]; % labels of all samples: 1 for eo; 0 for ec

K = 10; % K-fold CV

indices = crossvalind('Kfold',all_labels,K); % generate indices for CV

for k = 1:K % K iterations

   cv_test_idx = find(indices == k); % indices for test samples in one trial of validation

   cv_train_idx = find(indices ~= k); % indices for training samples in one trial of validation

   cv_classout = classify(all_samples(cv_test_idx,:),all_samples(cv_train_idx,:),all_labels(cv_train_idx));

   cv_acc(k) = mean(cv_classout==all_labels(cv_test_idx)); % compute accuracy

   TP = sum((cv_classout==all_labels(cv_test_idx))&(cv_classout==1)); % compute true positive

   TN = sum((cv_classout==all_labels(cv_test_idx))&(cv_classout==0)); % compute true negative

   FP = sum((cv_classout~=all_labels(cv_test_idx))&(cv_classout==1)); % compute false positive

   FN = sum((cv_classout~=all_labels(cv_test_idx))&(cv_classout==0)); % compute false negative

   cv_sensitivity(k) = TP/(TP+FN); % compute specificity

   cv_specificity(k) = TN/(TN+FP); % compute sensitivity

end

cv_acc_avg = mean(cv_acc); % averaged accuracy

cv_sensitivity_avg = mean(cv_sensitivity); % averaged sensitivity

cv_specificity_avg = mean(cv_specificity); % averaged specificity
```

After running the codes above, we can find three performance measures in the MATLAB Workspace. The averaged accuracy (cv_acc_avg), sensitivity (cv_sensitivity_avg), and specificity (cv_specificity_avg) of the trained classifier are 84.2%, 100%, and 68.3%, respectively.

In the last step, we apply the trained classifier to identify class labels for test samples and evaluate the performance.

```
%% test on test data

% Concatenate training/test data and specify the labels

train_samples = [a_eo_train';a_ec_train']; % training samples

train_labels = [ones(N_Train,1);zeros(N_Train,1)]; % labels of training samples: 1 for eo; 0 for ec

test_samples = [a_test']; % test samples

classout = classify(test_samples,train_samples,train_labels,'linear');

TP_test = sum((classout==test_labels)&(classout==1));

TN_test = sum((classout==test_labels)&(classout==0));

FP_test = sum((classout~=test_labels)&(classout==1));

FN_test = sum((classout~=test_labels)&(classout==0));

test_acc = sum(classout==test_labels)/N_Test; % compute accuracy

test_sensitivity = TP_test/(TP_test+FN_test); % compute specificity

test_specificity = TN_test/(TN_test+FP_test); % compute sensitivity
```

After running the codes above, we can find three performance measures in the MATLAB Workspace. The accuracy (test_acc), sensitivity (test_sensitivity), and specificity (test_specificity) of discriminating eyes-open and eyes-closed conditions in the test samples are 83.3%, 100%, and 66.7%, respectively.

References

Blankertz B, Lemm S, Treder M, Haufe S, Müller KR. Single-trial analysis and classification of ERP components – a tutorial. Neuroimage. 2011;56:814–25.

Cecotti H, Graeser A. Convolutional neural network with embedded Fourier transform for EEG classification. In: 19th international conference on pattern recognition. 2008. pp. 1–4.

Gajraj RJ, Doi M, Mantzaridis H, Kenny GN. Analysis of the EEG bispectrum, auditory evoked potentials and the EEG power spectrum during repeated transitions from consciousness to unconsciousness. Br J Anaesth. 1998;80:46–52.

Gysels E, Renevey P, Celka P. SVM-based recursive feature elimination to compare phase synchronization computed from broadband and narrowband EEG signals in brain-computer interfaces. Signal Process. 2005;85:2178–89.

Hu L, et al. Single-trial detection of somatosensory evoked potentials by probabilistic independent component analysis and wavelet filtering. Clin Neurophysiol. 2011;122:1429–39.

Hu L, Xiao P, Zhang ZG, Mouraux A, Iannetti GD. Single-trial time-frequency analysis of electrocortical signals: baseline correction and beyond. Neuroimage. 2014;84:876–87.

Hu L, Zhang ZG, Mouraux A, Iannetti GD. Multiple linear regression to estimate time-frequency electrophysiological responses in single trials. Neuroimage. 2015;111:442–53.

Huang G, et al. A novel approach to predict subjective pain perception from single-trial laser-evoked potentials. Neuroimage. 2013;81:283–93.

Jolliffe I. International encyclopedia of statistical science 1094–1096. Berlin Heidelberg: Springer; 2011.

Li K, Li X, Zhang Y, Zhang A Affective state recognition from EEG with deep belief networks. In: IEEE international conference on bioinformatics and biomedicine. 2013. pp. 305–310.

Lotte F, Congedo M, Lécuyer A, Lamarche F, Arnaldi B. A review of classification algorithms for EEG-based brain–computer interfaces. J Neural Eng. 2007;4:R1–R13.

Makeig S, et al. Evolving signal processing for brain-computer interfaces. Proc IEEE. 2012;100:1567–84.

Maris E, Oostenveld R. Nonparametric statistical testing of EEG- and MEG-data. J Neurosci Methods. 2007;164:177–90.

Mirowski P, Madhavan D, LeCun Y, Kuzniecky R. Classification of patterns of EEG synchronization for seizure prediction. Clin Neurophysiol. 2009;120:1927–40.

Müller KR, et al. Machine learning for real-time single-trial EEG-analysis: from brain–computer interfacing to mental state monitoring. J Neurosci Methods. 2008;167:82–90.

Mwangi B, Tian TS, Soares JC. A review of feature reduction techniques in Neuroimaging. Neuroinformatics. 2014;12:229–44.

Schoffelen JM, Gross J. Source connectivity analysis with MEG and EEG. Hum Brain Mapp. 2009;30:1857–65.

Shen KQ, Ong CJ, Li XP, Zheng H, Wilder-Smith EP. Feature selection method for multilevel mental fatigue EEG classification. IEEE Trans Biomed Eng. 2007;54:1231–7.

Srinivasan R, Winter WR, Ding J, Nunez PL. EEG and MEG coherence: measures of functional connectivity at distinct spatial scales of neocortical dynamics. J Neurosci Methods. 2007;166:41–52.

Subasi A, Ismail Gursoy M. EEG signal classification using PCA, ICA, LDA and support vector machines. Expert Syst Appl. 2010;37:8659–66.

Tabar YR, Halici U. A novel deep learning approach for classification of EEG motor imagery signals. J Neural Eng. 2017;14:016003.

Tibshirani R. Regression Shrinkage and Selection via the Lasso. J R Stat Soc Ser B (Methodological). 1996;58:267–88.

Tu Y, et al. An automated and fast approach to detect single-trial visual evoked potentials with application to brain–computer interface. Clin Neurophysiol. 2014;125:2372–83.

Tu Y, Hung YS, Hu L, Zhang Z. PCA-SIR: a new nonlinear supervised dimension reduction method with application to pain prediction from EEG. In: 7th International IEEE/EMBS Conference on Neural Engineering (NER). 2015. pp. 1004–1007.

Tu Y, et al. Alpha and gamma oscillation amplitudes synergistically predict the perception of forthcoming nociceptive stimuli. Hum Brain Mapp. 2016;37:501–14.

Van de Ville D, Britz J, Michel CM. EEG microstate sequences in healthy humans at rest reveal scale-free dynamics. Proc Natl Acad Sci U S A. 2010;107:18179–84.

Wang XW, Nie D, Lu BL. Emotional state classification from EEG data using machine learning approach. Neurocomputing. 2014;129:94–106.

Zhang ZG, Hu L, Hung YS, Mouraux A, Iannetti GD. Gamma-band oscillations in the primary somatosensory cortex – a direct and obligatory correlate of subjective pain intensity. J Neurosci. 2012;32:7429–38.

Zheng W-L, Zhu J-Y, Peng Y, Lu B-L EEG-based emotion classification using deep belief networks. In: IEEE International Conference on Multimedia and Expo (ICME). 2014. pp. 1–6.

Zou H, Hastie T. Regularization and variable selection via the elastic net. J R Stat Soc Ser B (Statistical Methodology). 2005;67:301–20.

Chapter 16
Deep Learning

Zhongke Gao and Xinmin Wang

Abstract Brain-computer interface (BCI) technologies enable direct communications between humans and computers by analyzing EEG signals. One of the challenges with traditional methods in classification tasks is receiving unsatisfactory recognition effects from EEG signals. In recent years, deep learning has drawn a great deal of attentions in diverse research fields, and could provide a novel solution for learning robust representations from EEG signals. In this chapter, we firstly introduce the basic concepts of deep learning techniques and two commonly used structures in time series analysis, namely, convolutional neural network and recurrent neural network. Then, we provide the applications of these two DL models to focus on the eye state detection task, which both achieve excellent recognition effects and are expected to be useful for broader applications in BCI systems.

Keywords Fatigue detection · EEG analysis · Brain-computer interface · Deep learning

Machine learning techniques allow to extract effective information from EEG signals, which play a vital role in different EEG-based classification research tasks. And machine learning methods have been applied to many control applications. For example, ErrP signals are decoded from a human operator in real time to control robots to perform a binary object selection task (Salazar-Gomez et al. 2017). Such systems may be further upgraded for more examples. The core point of EEG-based recognition systems is to develop practical computational algorithms, which are increasingly recognized as novel tools for rehabilitation therapy. In spite of much progress has been conducted, there is still considerable improvement room for the accuracy of information extraction from EEG signals. Hence, a direction from the area of machine learning attracts great interests of researchers, which is deep learning.

Z. Gao (✉) · X. Wang
School of Electrical and Information Engineering, Tianjin University, Tianjin, China
e-mail: zhongkegao@tju.edu.cn

L. Hu, Z. Zhang (eds.), *EEG Signal Processing and Feature Extraction*,
https://doi.org/10.1007/978-981-13-9113-2_16

16.1 What Is Deep Learning?

Machine learning is a subbranch of AI that focuses on teaching computers how to learn without the need to be programmed for specific tasks. The core idea of ML methods is to develop algorithms that learn from and make prediction on data. Deep learning is a particular subset of ML methods using artificial neural networks (ANNs), which are slightly inspired by the structure of neurons located on the human brain. Informally, the word *deep* refers to the presence of many layers in the artificial neural network, but this definition of deep has changed over time (Gulli and Pal 2017).

Recently, deep learning techniques have shown prominent abilities in so many fields, such as image classification, speech recognition, and time series prediction. It significantly improves previous state-of-the-art results achieved over dozens of years, which is due to the availability of more training data (such as ImageNet for images) and the relatively low-cost availability of GPUs for very efficient numerical computation. Based on these existing conditions, the research on network design is a very important part of deep learning techniques. Moreover, various studies have started to investigate the potential of CNNs and their variations for EEG signals decoding, including motor imagery classification, fatigue driving evaluation, and emotion recognition. Note that EEG signals are different with 2D images and speech signals, which is truly challenging to design a proper network for EEG-based classification tasks.

A deep learning framework is a function that takes the values of various features of EEG signals (or raw signals) as the network input, and predicts the class of the samples. The experiments are carried out to collect the EEG signals of subjects with cognitive or perceptual responses. Let vector $\mathrm{X} \in \mathrm{R}^{E \times T}$ and its class label as $y \in R^N$, where E denotes the cap electrode plates, T denotes the signal sampling points, and N denotes the classes of response states. A DL framework can build up the relationships between input samples and labels using the training data, and then can predict the label of a given sample.

16.2 Typical Deep Learning Methods for Time Series

Learning effective representations from EEG signals is a challenging problem to be investigated. Various novel methods have been proposed to build up a proper framework for classification tasks. These methods have shown great potentials for learning effective features from EEG signals. The commonly used methods include convolutional neural networks and recurrent neural networks.

16.2.1 Convolutional Neural Networks

In the following sections, we first explain the basic idea of CNNs, and then introduce architecture choices for EEG analysis. Finally, we describe the implementation details of CNN training.

Generally, CNNs have shown great superiority in many learning tasks, such as images and audio signals. These signals often have an inherent hierarchical structure. For example, images typically consist of edges that together form simple shapes which form larger and more complex shapes. CNNs can learn local non-linear features through convolutions and nonlinearities and represent higher-level features as compositions of low-level features through multiple layers of processing. In addition, many CNNs employ pooling layers to create a coarser inter-mediate feature representation to increase translation-invariant.

To decode EEG signals, CNNs are designed to extract a wide range of features from signal sequences with few prior knowledge, which can reach competitive performances. It has been proved that standard CNNs can be used as a general-purpose tool for brain-signal decoding tasks. Besides, keeping the architecture generic also increases the further applications. To deal with raw EEG signals, the CNNs should typically have two convolutional layers to better handle the large number of input channels, where one input channel per electrode compared to three input channels in RGB images. The first convolution is used across time to precede temporal information and the second convolution is used across electrodes to handle electrode relations. Therefore, the DL framework can be regarded as an integration of spatial block and temporal block. Many tricks are proposed to improve the performance of DL framework on spatial and temporal dimensions (Schirrmeister et al. 2017).

Meanwhile, many attempts are conducted to combine DL methods with the existing analysis methods. The EEG signals are converted into new characteristics by feature extraction methods, which can be fed into DL frameworks with more concrete information. Recently, by modifying the filter-bank common spatial patterns methods, EEG signals are turned into new temporal representations and a convolutional neural network architecture is introduced for motor imagery EEG data classification. The framework outperforms the existing results on the MI dataset (S. Sakhavi et al. 2018). Concretely, the first two layers of these frameworks should perform a temporal and a spatial convolution. Besides, they embed all the computational steps in a single network, and thus all steps can be optimized jointly. Also, due to having several pooling regions with one trial, these frameworks can learn a temporal structure of EEG signals, which is proved to help classification.

As for design choices, we give details for some of these aspects. (1) Batch normalization standardizes intermediate outputs of the network to zero mean and unit variance for a batch of training samples. This is meant to facilitate the optimization by keeping the inputs of layers closer to a normal distribution during training (S. Ioffe and Szegedy 2015). (2) Dropout randomly sets some inputs for a layer to zero in each training update. It is meant to prevent co-adaption of different units and can be seen as analogous to training an ensemble of networks. There are many other ways to improve accuracies and we do not list all here.

16.2.2 Recurrent Neural Networks

A recurrent neural network (RNN) is an extension of a conventional feedforward neural network which can handle a variable-length input. The RNNs can handle the variable-length sequence by having a recurrent hidden state whose activation at each time is dependent on that of the previous time. Standard RNNs are hard to train due to the well-known vanishing or exploding gradient problems. To address these problems, long short-term memory (LSTM) is proposed as the gated recurrent network architectures (Golmohammadi et al. 2017). The most commonly used architecture is described in (Graves and Schmidhuber 2005) as follows:

$$i_t = \sigma\left(U^i x_t + W^i s_{t-1} + p^i \cdot c_{t-1} + b^i\right) \tag{16.1}$$

$$f_t = \sigma\left(U^f x_t + W^f s_{t-1} + p^f \cdot c_{t-1} + b^f\right) \tag{16.2}$$

$$c_t = f_t \cdot c_{t-1} + i_t \cdot g(U^c x_t + W^c s_{t-1} + b^c) \tag{16.3}$$

$$o_t = \sigma(U^o x_t + W^o s_{t-1} + p^o \cdot c_t + b^o) \tag{16.4}$$

$$s_t = o_t \cdot g(c_t) \tag{16.5}$$

where i_t, f_t, c_t, o_t, and s_t are the input gate, forget gate, cell state, output gate and block output at time instance t, respectively; x_t is the input at time t; U^* and W^* are the weight matrices applied on input and recurrent hidden units, respectively; $\sigma(.)$ and g (.) are the sigmoid and tangent activation functions, respectively; p^* and b^* are the peep-hole connections and biases, respectively; and $\cdot$ means element-wise product. More specifically, a memory block of LSTMs is shown in Fig. 16.1.

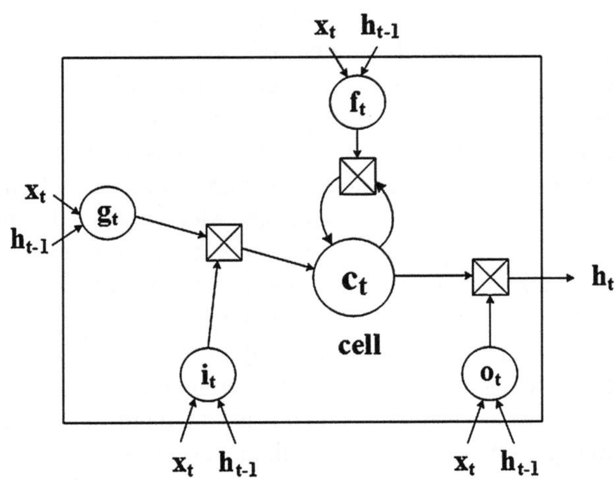

Fig. 16.1 An LSTM memory block

16.3 Deep Learning Framework with EEG Signals: Two Examples

In this chapter, we introduce two examples of DL-based studies to verify the superiority of deep learning methods for multi-channel EEG signals.

Opening and closing the eyes are fundamental behaviors for directing attention to the external versus internal world. However, it remains indistinct whether the resting state of eyes-open relative to eyes-closed are associated with different topological organizations of functional brain networks. Studies on resting-state functional networks from an electrophysiological perspective can take advantage of high temporal resolution. On consideration, EEG signal is the chief source for input, which contains a great deal of physiological information of a working brain. Here, we give two baseline methods to focus on eye state detection, convolutional neural networks, and recurrent neural networks, respectively. The section EEG acquisition and preprocessing is same with the section complex network analysis.

16.3.1 Convolutional Neural Networks

In the experiment, we reach a network with depth of 4 layers. Table 16.1 shows the details of the baseline method CNN. In the framework, a convolutional layer with a size of $[61 \times 1]$ is initially performed as input layer. Another $[1 \times 10]$ convolutional layer is employed with it, and then a dense layer with 100 nodes is appended ended with a softmax classifier.

A unit in the CNN is denoted by $x(l, m, j)$, where l is the layer, m is the feature map, and j is the position of the unit in this feature map. Likewise, $\sigma(l, m, j)$ is denoted as the scalar product between a group of input neurons and the weight connection between these neurons with l, m and j sharing equal meanings:

$$x(l, m, j) = f(\sigma(l, m, j)) \tag{16.6}$$

where f is the rectified linear units function (Nair et al. Nair and Hinton 2010) used for whole network layers.

Notably, each neuron of one feature map in each convolutional layer shares the same set of weights, which aims to decrease the amount of weight parameters. And they are attached to a subset of the neurons of former layer, which depend on the exact position of this neuron. Or rather, the neuron weights are trained independently

Table 16.1 Details of the CNN framework

Layers	Output size	CNN
Input	61×100	–
Convolution	1×100	61×1, map 8
Convolution	1×10	1×10, map 16, stride 10
Dense layer	100	fully connected
Softmax	2	Softmax

to their corresponding receptive fields. Let layer n as Ln, then the information transmission process could be described as:

1. For layer L1

$$\sigma(1,\mathrm{m},\mathrm{j}) = \omega(1,\mathrm{m},0) + \sum_{i=0}^{i<Ne} I_{i,j} w_{(1,m,i)} \tag{16.7}$$

where $w(1,m,0)$ is a threshold and $w(1,m,i)$ denotes a set of weights with $N_e = 61$. Here, m corresponds to the convolutional kernel used in this framework.

2. For layer L2

$$\sigma_{(2,m,j)} = w(2,m,0) + \sum_{\mathrm{i}=0}^{i<10} x(1,m,j*10+i)\cdot w(2,m,i) \tag{16.8}$$

where $w(2,m,0)$ is a threshold. This layer is employed to extract valid temporal features.

3. For layer L3

$$\sigma_{(3,j)} = w(3,0,j) + \sum_{\mathrm{i}=0}^{i<100} x(3,i)\cdot w(3,i) \tag{16.9}$$

where $w(3,0,j)$ is a threshold, and each neuron of L3 is fully connected to each neuron of L2.

We use the classical back propagation as learning algorithm to tune up the thresholds and weights of the network (D. E. Rumelhart et al. 1986), which is reflected by the promotion of model accuracy on validation set. As a loss function, cross-entropy objective function is employed for model performance estimation. Figure 16.2 shows the architecture of the proposed convolution neural network.

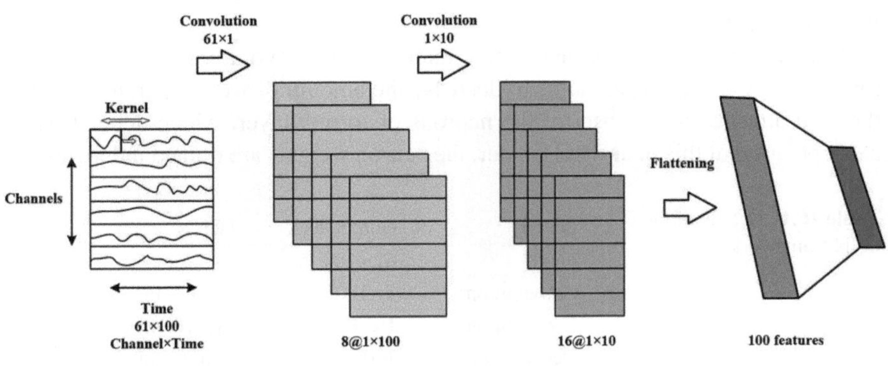

Fig. 16.2 The architecture of the proposed convolution neural network

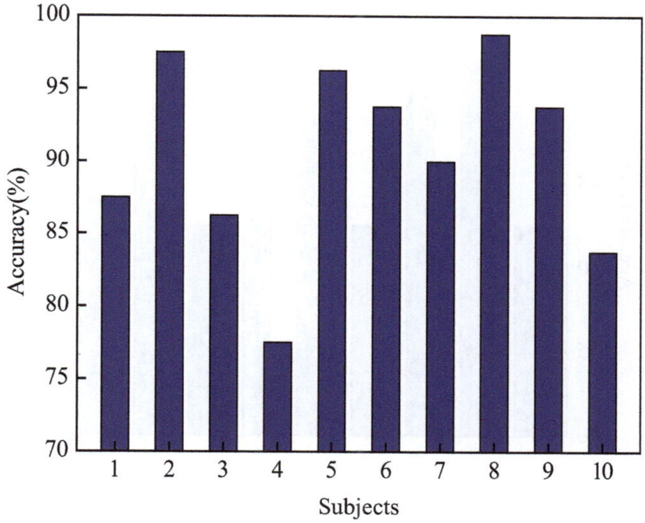

Fig. 16.3 Performances of the CNN framework on each subject

Table 16.2 Details of the LSTM framework

Layers	Output size	LSTM
Input	61×100	–
LSTM	61×100	100 units
LSTM	61×10	50 units
Dense layer	100	Fully connected
Softmax	2	Softmax

The developed CNN is trained to recognize different eye states for each subject. Individual performances are obtained on 10 subjects, and they are shown in Fig. 16.3. We find that the CNN framework is effective on the EEG dataset with an average accuracy of 90.5%. The performances of five subjects are over the average accuracy while the other five are below it. These results reflect that the CNN model provides an effective relationship between eye states and EEG signals.

16.3.2 Recurrent Neural Networks

The deep LSTM architecture is illustrated in Table 16.2. It consists of an input layer, the first sequence-to-sequence LSTM layer, a many-to-one LSTM layer, a 20% dropout layer, and a final sigmoid activation function for binary classification. The first hidden layer contained 50 LSTM units while the second hidden layer used 10 units. Dropout on the input gates to each LSTM layer and between the final LSTM and fully connected sigmoid layer served as a method of regularization and is

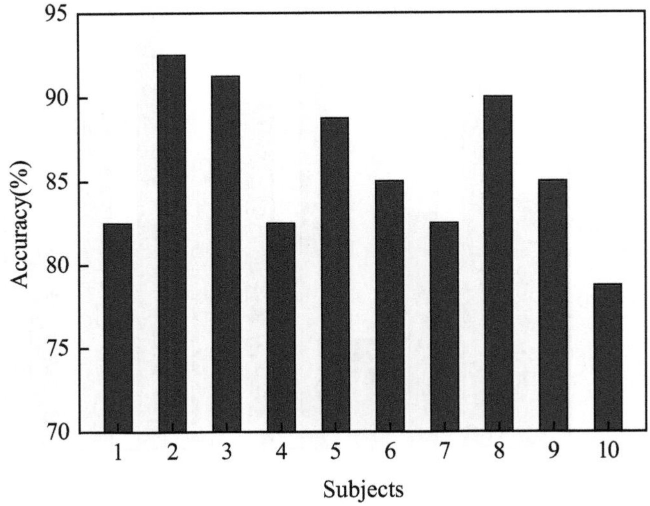

Fig. 16.4 Performances of the LSTM framework on each subject

set to 20%. The purpose of the recurrent connection in a LSTM is to store important long-term dependencies.

The developed LSTM is trained to recognize different eye states for each subject. Individual performances are obtained on 10 subjects, and they are shown in Fig. 16.4. As can be also seen, the LSTM framework receives an average accuracy of 85.88% when evaluated on the EEG dataset. The performances of four subjects are over the average accuracy while the other six are below it. The LSTM framework clearly enables learning effective representations from EEG signals to classify eyes-closed and eyes-open states.

References

Golmohammadi M, et al. Gated recurrent networks for seizure detection. In: Signal Processing in Medicine and Biology Symposium (SPMB). 2017. p. 1–5.

Graves A, Schmidhuber J. Framewise phoneme classification with bidirectional LSTM networks. In: International Joint Conference on Neural Networks. 2005. 2047–2052.

Gulli A, Pal S. Deep Learning with Keras. Birmingham: Packt Publishing Ltd; 2017.

Ioffe S, Szegedy C. Batch normalization: Accelerating deep network training by reducing internal covariate shift. In: International Conference on Machine Learning. 2015. p. 448–456.

Nair V., Hinton GE. Rectified linear units improve restricted boltzmann machines. In: International conference on machine learning. 2010. 807–814.

Rumelhart DE, Hinton GE, Williams RJ. Learning representations by back-propagating errors. Nature. 1986;323(6088):533.

Sakhavi S, Guan C, Yan S. Learning temporal information for brain-computer interface using convolutional neural networks. IEEE Trans Neural Netw Learn Syst. 2018;29(11):5619–29.

Salazar-Gomez AF, et al. Correcting robot mistakes in real time using EEG signals. In: IEEE International Conference on Robotics and Automation (ICRA). 2017. p. 6570–6577.

Schirrmeister RT, et al. Deep learning with convolutional neural networks for EEG decoding and visualization. Hum Brain Mapp. 2017;38(11):5391–420.

Chapter 17
Statistical Analysis

Gan Huang

Abstract Statistics is a collection of methods that you can get answers to important questions from EEG signal analysis. In this chapter, we firstly introduced the descriptive statistical methods for presenting the result from the raw data. Furthermore, analysis techniques comprised of parametric strategies like t-test, ANOVA, regression, and nonparametric procedures, such as permutation test, are introduced with their implementation in MATLAB and SPSS. Selecting the right statistical procedures would be important in EEG signal analysis. Finally, we discussed the problem of multiple comparison, which may increase the probability for the researcher to get to false positive conclusion due to the multiple comparison and remains a challenge in statistical method.

Keywords T-test · ANOVA · Regression · Nonparametric procedures · Cluster-based permutation test

17.1 Statistics Elementary

17.1.1 Introduction

The purpose of this section is to provide the basic idea about statistics. Two important questions are discussed: (1) What is the use of statistics? (2) How to describe and display the EEG results by statistics?

Electronic supplementary material The online version of this chapter (https://doi.org/10.1007/978-981-13-9113-2_17) contains supplementary material, which is available to authorized users.

G. Huang (✉)
School of Biomedical Engineering, Health Science Center, Shenzhen University, Shenzhen, China

L. Hu, Z. Zhang (eds.), *EEG Signal Processing and Feature Extraction*,
https://doi.org/10.1007/978-981-13-9113-2_17

In the previous chapters, all kinds of methods in signal processing and feature extraction are introduced in EEG signal analysis. On this foundation, statistical methods would be used to present data in a suitable tabular and graphical form for easy and clear comprehension of the result, and further draw valid inferences. Hence, according to the different use of statistics, we have (1) *descriptive statistics*, to describe the basic feature of the data in a study, and (2) *inferential statistics*, to make inferences by using a random sample of data from a population (Bickel and Doksum 2015; DeGroot and Schervish 2012; Devore 2001; Johnson et al. 2000; Montgomery et al. 2009).

Take the case of resting state EEG with eyes open and closed for example. We have a dataset with 93 subjects. Comparing the EEG signal on channel Oz from a certain subject in Fig. 17.1(a), it is found that a stronger oscillation of the alpha wave (around 10 Hz) happened in the condition of eyes closed than that in the condition of eyes open. However, showing individual results on a single subject is not representative. We do not know the strong alpha oscillation with eyes closed is a common phenomenon on all subjects, or a special case on the single subject. In this case,

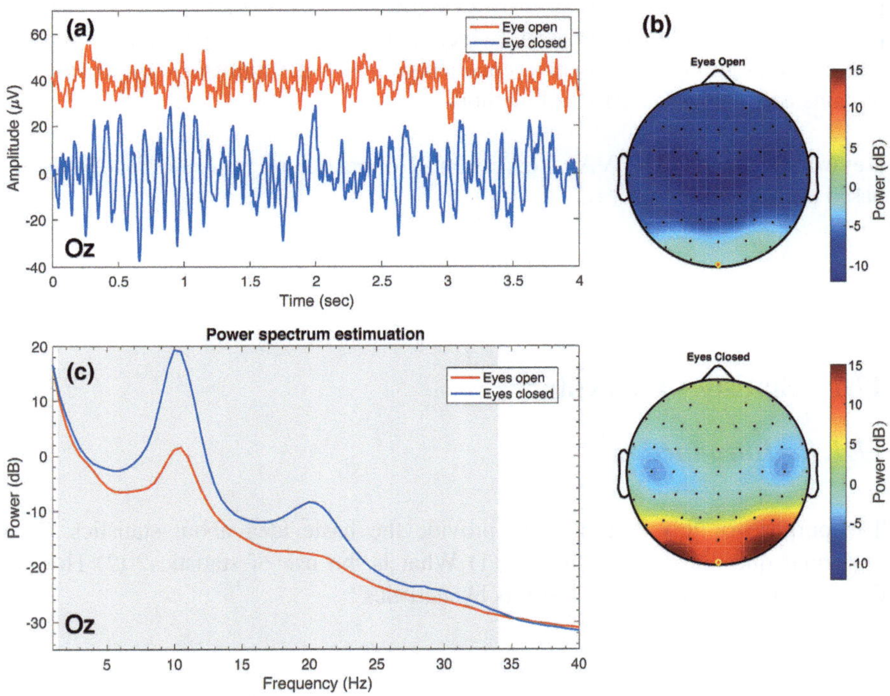

Fig. 17.1 The resting state EEG with eyes open and closed. (**a**) An EEG data fragment from one subject on channel Oz. (**b**) The topography of mean EEG power in alpha band (8–12 Hz). (**c**) The mean EEG power in different frequency bands on channel Oz. The gray-shaded interval (1.5Hz–34 Hz) indicated there is a significant difference between the two conditions

showing the results of all the subjects are too redundant and impractical on a dataset with a large sample size. Hence, we have to find some ways to summarize the results to general characteristics of the dataset. In statistics, displaying the mean value of the samples provides a simple and effective way to do this. Figure 17.1(b) illustrates the topography of the mean alpha power (8–12 Hz) from all subjects, and Fig. 17.1(c) shows the mean EEG power with eyes open and closed in different frequency bands on channel Oz. With the comparison of the mean value of EEG power in both spatial domain and frequency domain, we can have a comprehensive understanding about the difference of EEG power between the conditions of eyes open and closed. Further, the gray-shaded interval indicated there is statistically significant difference for the EEG power between the two conditions from 1.5 to 34 Hz, in which paired sample *t*-test is applied.

In this example, we illustrated the use of the simplest methods in descriptive statistics and inferential statistics in the EEG signal analysis. The mean statistic is used to describe the result of resting state EEG. *T*-test is used to further infer some conclusion based on the results. In the following section, we mainly discuss the descriptive statistics. In the following section, the methods of inferential statistics will be introduced.

17.1.2 Mean and Variance

1. **Measure of the Location**

A statistic is a single measure of some feature of the sample data. The mean is the most commonly used statistic, which can be obtained by the sum of all the sample value and divided by the sample size.

$$\bar{x} = \frac{1}{n}\sum_{i=1}^{n} x_i,$$

in which $\bar{x}$ is the mean, x_i is the value of the sample i, n is the sample size, and the symbol Σ is the sum sign in mathematics, which means add up all the value of x_i from $i = 1$ to $i = n$.

The mean, also called arithmetic mean, represents the central tendency of the samples. In addition to the mean, median and mode provide another measurement of the "average". The median is the middle value in the sorted sample values, which divided the data sample into two halves. The mode is the value that occurs the most often.

Take the dataset

$$\mathrm{A} = [1, 3, 5, 4, 3, 6],$$

for example. To calculate the mean, median, and mode value of dataset A, we can use the following MATLAB codes,

```
clc;clear;close all;
A=[1,3,5,4,3,6];
mean(A)                          % mean
median(A)                        % median
mode(A)                          % mode
```

As a result, the mean value is 3.67, the median is 3.5, and the mode is 3.

2. **Measure of the Variability**

In addition to the central tendency, variance is another commonly used statistic to measure the difference of each sample data from the mean value. The *variance* of a finite dataset with size n can be calculated by the following formula:

$$\sigma^2 = \frac{1}{n}\sum_{i=1}^{n}(x_i - \mu)^2,$$

in which μ is the mean value. Hence, we also have the *standard deviation*

$$\sigma = \sqrt{\frac{1}{n}\sum_{i=1}^{n}(x_i - \mu)^2}.$$

In some definition of standard deviation, we use $n - 1$ instead of n to be the normalization factor. Hence, we have the *sample standard deviation*

$$s = \sqrt{\frac{1}{n-1}\sum_{i=1}^{n}(x_i - \bar{x})^2},$$

and also s^2 is called *sample variance*. To distinguish, σ and σ^2 are called *population standard deviation* and *population variance*, respectively.

The difference between the terms *sample* and *population* is how observations are assigned to the dataset. For a dataset, population means all the elements from the dataset, and sample just includes part of elements in the dataset. Normally, the number of elements in the sample is relatively small with respect to the whole population. To calculate the standard deviation of the dataset, we can use the population standard deviation σ. Once you only have part of elements in the dataset (the sample), the sample standard deviation s is an unbiased estimation for the population standard deviation, since the degree of freedom is $n - 1$. If the sample size n is larger enough, the difference between σ and s is minor. Similarly, in the equation for the definition of σ and s, μ is the population mean and $\bar{x}$ is called sample mean, which is an unbiased estimation of the population mean. In MATLAB, the functions `var` and `std` with the default parameters refer to the sample variance and the sample standard deviation.

To describe the variability of the dataset, we can also use the statistics range and interquartile range. *Range*: the range of a dataset refers to the largest value minus the smallest value. *Interquartile range (IQR)*: Divide a dataset into quartiles to get three points which are Q1, Q2, and Q3. Q2 is the median value of the dataset which divides the dataset into two halves; Q1 and Q3 are the middle point of the first and second half dataset. The interquartile range of the dataset refers to Q3 minus Q1.

Still take the dataset

$$\mathrm{A} = [1, 3, 5, 4, 3, 6],$$

for example. The MATLAB functions to calculate all these statistics are as follows,

```
clc;clear;close all;
A=[1,3,5,4,3,6];
var(A,1)      % population variance
std(A,1)      % population standard deviation
var(A)        % sample variance
std(A)        % sample standard deviation
range(A)      % range
iqr(A)        % interquartile range
```

As a result, with the sample size $n = 6$, we have $\sigma^2 = 2.56$, $\sigma = 1.60$, $s^2 = 3.07$, $s = 1.75$, range $= 5$, and IQR $= 2$.

17.1.3 Probability Distributions

In statistics, a probability distribution is a mathematical function that provides the probabilities of occurrence of different possible outcomes in an experiment. In the following, we will introduce the discrete probability distribution, the normal distribution, and three sampling distributions. Finally, the statistics, skewness and kurtosis, are introduced which describe the shape of the distribution.

1. **Discrete Probability Distribution**

Take the experiment of coin toss for example. If we toss the coin for only one time and we assume the coin is fair, the result with head ($X = 1$) or tail ($X = 0$) follows the *Bernoulli distribution* with $p = 0.5$ that

$$P(X = 1) = 0.5,$$
$$P(X = 0) = 0.5.$$

If we toss the coin for n times, the probability of achieving k times head follows the *binomial distribution* with $p = 0.5$ that

$$P(X = k) = \frac{n!}{k!(n-k)!} \times \frac{1}{2^n}$$

It can be calculated the mean value of the distribution is $n/2$, and the variance is $n/4$. Bernoulli distribution is a special case of binomial distribution with $n = 1$. If n is large enough, the distribution function will approach to the normal distribution.

2. **Normal Distribution**

Normal distribution is the most well-known distribution, which is often referred as $N(\mu, \sigma^2)$ with the mean μ and variance σ^2. The central limit theorem establishes that, in certain situations, when independent random variables are added, their sum tends toward a normal distribution, even if the original variables are not normally distributed. Hence, normal distribution occurs quite frequently in economics, nature and social science. In addition, the statistical methods with the normally distributed assumption, like t-test, ANOVA, correlation and regression analysis, can also be applied to many conditions with other types of distributions. All these make the normal distribution very important in statistics.

The probability density function of the normal distribution with the mean μ and variance σ^2 is a Gaussian function:

$$f(x) = \frac{1}{\sqrt{2\pi\sigma^2}} e^{-\frac{(x-\mu)^2}{2\sigma^2}}.$$

Hence, the normal distribution is also known as Gaussian distribution. Sometimes, the normal distribution is informally called the bell curve, since it looks bell-shaped. The normal distribution with different value of μ and σ is illustrated in Fig. 17.2(a), in which the red curve with $\mu = 0$ and $\sigma = 1$ is called *standard normal distribution*, noted as $N(0, 1)$.

To calculate the probability of a score in the normal distribution, *z-score* is an important statistic. Considering the different parameter settings for μ and σ in the normal distribution, we need to normalize the normal distribution values for comparison under uniform standards. Z-score can do this by converting the value in a normal distribution into the value following the standard normal distribution:

$$z = \frac{x - \mu}{\sigma}.$$

As a measure of standard deviation, z-score represents the number of standard deviations a given value x falls from the mean μ. If z is equal to +1.96, then x is 1.96σ above the mean μ, which means x is in the top 2.5% of the distribution.

To generate n random numbers with normal distribution with the mean μ and the variance σ^2 in MATLAB, we can use the function `randn`. The corresponding script is as follows.

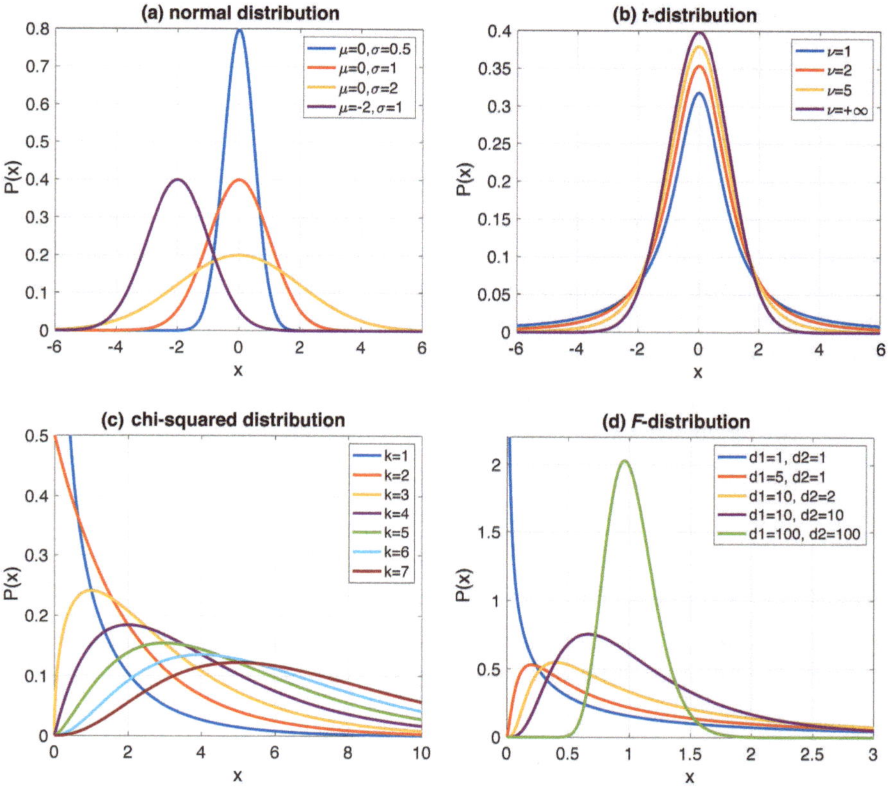

Fig. 17.2 The probability density function of (**a**) normal distribution, (**b**) *t*-distribution, (**c**) chi-squared distribution, and (**d**) *F*-distribution with different parameters

```
clc;clear;close all;
mu=0;          % mean value \mu
sigma=1;       % standard deviation \sigma
n=10;          % sample size n
x=sigma*(randn(1,n)+mu);
```

3. **Sampling Distributions**

In the above section, we have introduced two discrete probability distributions, which are Bernoulli distribution and binomial distribution, and normal distribution, which is the most common continuous distribution. In the following, three other famous distributions will be introduced, which play important roles in the statistical test.

- *T-distribution (Student's t-distribution)*: Given n independent measurement x_i of a normally distributed population. Let t-statistic be

$$t = \frac{\bar{x} - \mu}{s/\sqrt{n}},$$

where μ is the population mean, $\bar{x}$ is the sample mean, s is the sample standard deviation and n is the sample size. The t-distribution is defined as the distribution of the t-statistic, where the population standard deviation σ is unknown. The t-distribution depends on the degrees of freedom $\nu = n - 1$. With $\nu \to +\infty$ ($n \to +\infty$), the t-distribution approaches to the normal distribution.

- *Chi-squared distribution (χ^2distribution)*: Given k independent, standard normal random variables, $x_1, x_2, \ldots, x_k$. The sum of their squares,

$$\chi^2 = \sum\nolimits_{i=1}^{k} x_i^2,$$

is distributed according to the Chi-squared distribution with k degrees of freedom.

- *F-distribution*: Given two independent chi-squared random variables χ_1^2 and χ_2^2 with d_1, d_2 degrees of freedom, respectively, Let F-statistic be

$$F = \frac{\chi_1^2/d_1}{\chi_2^2/d_2}$$

distributed according to the F-distribution with parameters d_1 and d_2.

To generate probability density function for each distribution in MATLAB, we can use the function `pdf` with different parameters setting. The probability density function of normal distribution, t-distribution chi-squared distribution, and F-distribution are illustrated in Fig. 17.2.

4. **Skewness and Kurtosis**

To describe the shape of a distribution or check the non-Gaussianity of a distribution, the statistics *skewness* and *kurtosis* can be used. Given n independent variables $x_1, x_2, \ldots, x_n$, we can have their mean $\bar{x}$, and their sample variance s. Skewness and kurtosis are calculated as follows:

$$Skew = \frac{1}{n}\sum_{i=1}^{n}\left(\frac{x_i - \bar{x}}{s}\right)^3,$$

$$Kurt = \frac{1}{n}\sum_{i=1}^{n}\left(\frac{x_i - \bar{x}}{s}\right)^4.$$

Skewness is a measure of the asymmetry of the probability distribution. A distribution with the skewness close to zeros is a symmetry distribution. A negative skew implies a long right tail, in which the mass of the distribution is concentrated on the right part of the figure, and vice versa. Kurtosis is a measure of whether the data are heavy-tailed or light-tailed relative to a normal distribution. The kurtosis of the normal distribution is equal to 3. Distribution with kurtosis less than 3 is said to be platykurtic, in which the tails are very light compared to the normal distribution, and distribution with kurtosis greater than 3 is said to be leptokurtic with the heavy tails.

17.1.4 Graphical Display of Data

Normally, EEG signal processing involves hundreds of thousands of data. Presenting the raw data is redundant and impractical. Statistical graphs provide a tool to represent the data in a clear and intuitive way. Well-presented graphs can greatly enhance the readability of research results.

In the following, we use the resting state EEG data with eyes open and closed to illustrate the use of typical statistical graphs. One minutes resting state EEG signal with eyes open and closed were collected from 93 subjects. The EEG signal is recorded by a 64-channel BrainAmp system (Brain Products GmbH, Germany) with a sampling rate 1000Hz and common average was applied for reference. To calculate EEG power, the welch method was used with a window size of 2 seconds and the sliding step was set to 1 second. In the excel file "Resting State.xlsx", the alpha band EEG power with eyes open and closed at channel Oz are listed in the first two columns. The third column is the difference between column one and column two. Column four is the recording time, in which "1," "2," and "3" mean the experiment was run in the morning, afternoon, and evening, respectively. The gender information of the subjects is listed in column five, in which "1" stands for male and "0" stands for female.

1. **Pie Chart**

In the dataset, 82.8% (77 out of 93) of the subjects are male. Female subjects accounted for a minority. To visually describe this property, we will use a pie chart. As illustrated in Fig. 17.3(a), a pie chart is a circular statistical graphic which proportionally divides the number of male and female subjects into two slices. To make a pie chart, we can directly use the MATLAB function `pie`.

2. **Bar Graph**

Bar graph is a commonly used statistical graph to compare discrete items. As illustrated in Fig. 17.3(b), we use the bar graph to compare the mean value of the resting EEG power in the condition of eyes open and closed. Furthermore, error bars are added to represent the variability of the data by using the standard deviation in the two conditions. In average, the power of the EEG signal on channel Oz is

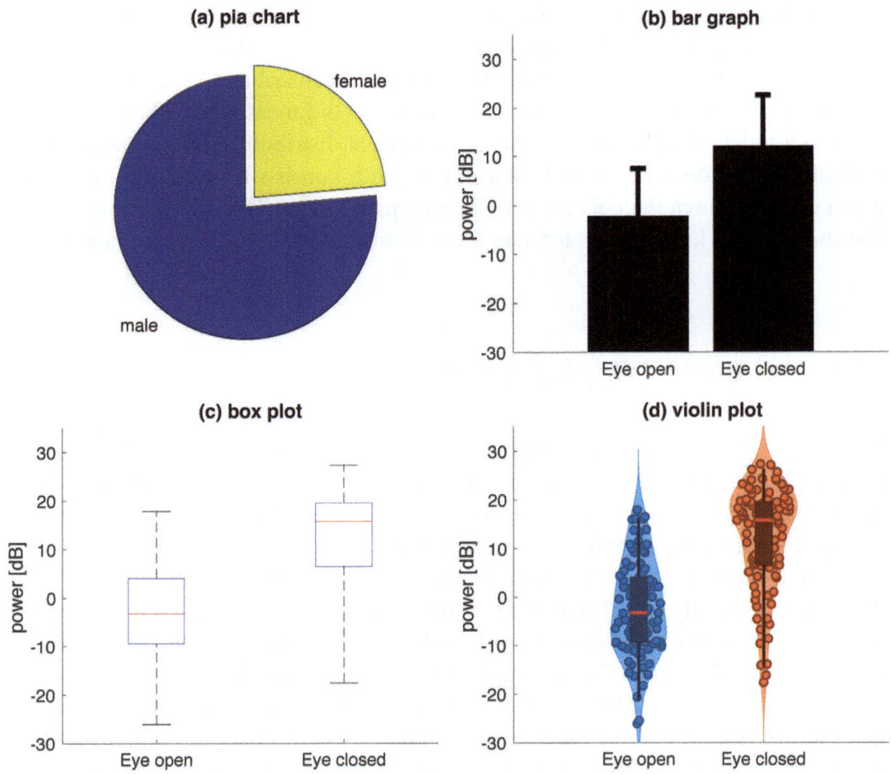

Fig. 17.3 The typical statistical graphs. (**a**) The pie chart to show the male to female ration in the subjects. (**b**) The bar graph for the mean power of the resting EEG with the eyes open and closed. Error bars indicate standard deviation. (**c**) The box plot with the minimum (Q0), first quartile (Q1), median (Q2), third quartile (Q3), and maximum (Q4) value from bottom to up. (**d**) The violin plot shows the box plot with the rotated probability density function

-2.29 ± 9.79 for eyes open and 11.99 ± 10.60 for eyes closed. MATLAB function `bar` and `errorbar` can be used to make the bar graph.

3. **Box Plot**

Bar graph only shows the mean and standard deviation information of a dataset. If we want to display more information about the dataset, box plot can be used. In the box plot, five key values are used to describe the dataset, which are the minimum (Q0), first quartile (Q1), median (Q2), third quartile (Q3), and maximum (Q4). By drawing the box from Q1 to Q3, the interquartile range is represented by the length of the box. The red line in the box is the median value. The line, also called whiskers, from Q0 to Q4, extending vertically from the box indicates the range of the dataset. Hence, the box plot is also termed as box-and-whisker plot. MATLAB function `boxplot` can be used to make a bar graph as Fig. 17.3(c).

4. **Violin Plot**

Violin plot is a combination of the box plot and the probability density. The name violin plot originated from one of the first graphs with appearance of a violin. Typically, violin plot includes the red line as the median and the box as the interquartile range. Instead of the range, the whisker indicates the 95% confidence interval, which is different from box plot. The probability density that is rotated and placed on each side supplements the box plot by graphically showing the distributional characteristics of the dataset. Sometimes, you can also add dots into the violin plot to represent the values for each sample.

A simple line graph and the histogram can be used to illustrate the probability density by MATLAB functions `plot` and `histogram`. Functionality for the violin plot does not exist in MATLAB. But you can still make a short script to do the violin plot by your own codes, in which `ksdensity` is used for probability density estimation. A sample script for violin plot in Fig. 17.3(d) is attached.

17.2 Hypothesis Testing

17.2.1 General Idea

Take the resting state EEG with eyes open and closed, for example. In the bar graph Fig. 17.3(b) in previous Sect. 1.4, we have calculated that the mean power of the EEG signal is -2.29 ± 9.79 for eyes open, which is larger than the mean power of the EEG signal 11.99 ± 10.60 for eyes closed. Naturally, we have a question whether the difference of the mean EEG power is significant between the two conditions or it is just produced by random effects? Answering this question is in the scope of inferential statistics. More specifically, hypothesis testing can be used to make statistical decision based on the experimental data. In the introduction of hypothesis testing, we firstly start with the general idea. Then several types of commonly used hypothesis testing approaches, like t-test, ANOVA, correlation and regression analysis, would be introduced in the later sections.

For hypothesis testing, we normally should have an initial hypothesis. For example, we assume that the mean value of the EEG power between eyes open and closed is the same in the resting state EEG experiment. The hypothesis is called the null hypothesis, H_0. A violation of the null hypothesis, that the mean value of the EEG power between eyes open and closed is different, is often called the alternative hypothesis H_1. A test statistic is calculated and used to check the degree of agreement between a sample of data and the null hypothesis. As a result, the possibility *p*-value is the representation of the degree of the agreement. If the p-value is less than previous given significant level α, like $\alpha = 0.05$, we infer that the null hypothesis H_0 has a small possibility to be true. Hence, we reject the null hypothesis and accept the alternative hypothesis H_1. Otherwise, we fail to reject the null hypothesis, and infer as if the null hypothesis is true (Casella and Berger 2002; Lehmann and Romano 2006).

1. **Steps in Hypothesis Testing**

To illustrate the steps in hypothesis testing, we firstly generate a small dataset of random numbers with their mean not zero. With the following MATLAB script, we have 10 samples

$$x = [1.34, 2.63, -1.46, 1.66, 1.12, -0.51, 0.37, 1.14, 4.38, 3.57];$$

from the normal distribution *N*(0.8,1) with $\mu = 0.8$, $\sigma = 1$. In the following, we will use these samples to illustrate the basic steps in hypothesis.

```
clc;clear;close all;
rng(0);                %make the output of randn repeatable
n=10;                  %sample size n=10
x=randn(1,n)+0.8;      %generate the normal distribution
disp(x);               %display the value of the samples
```

To perform the hypothesis on the dataset *x*, normally we have the following four steps:

Step 1: Make the initial hypothesis. Assume *x* is an independent sample of observations from a normal distribution; we want to check whether the distribution has a zero mean ($\mu = 0$). Hence, we have

- The null hypothesis H_0: $\mu = 0$.
- The alternative hypothesis $H_1 : \mu \neq 0$.

Step 2: Calculate the test statistic. Select a test statistic that can be used to assess the truth of the null hypothesis H_0. Here, we have the *t*-statistic

$$\mathrm{t} = \frac{\bar{x}}{s/\sqrt{n}},$$

in which $\bar{x}$ and *s* are the mean value and standard deviation of the dataset *x* and $n = 10$ is the sample size. Since $\bar{x} = 1.42$ and $s = 1.77$, it can be calculated that $t = 2.54$. Assuming the null hypothesis is true that $\mu = 0$, the *t*-statistic should follow the *t*-distribution with $n - 1$ degrees of freedom.

Step 3: Calculate the *p*-value. The *p*-value is the probability of finding the observed, or more extreme, results when the null hypothesis is true. Under the null hypothesis H_0, the location of *t*-statistic $t = 2.54$ in the *t*-distribution with $n - 1$ degrees of freedom is shown in Fig. 17.4. The probability of $T \leq -|t|$ or $T \geq |t|$ is the very small. Hence, we have the *p*-value

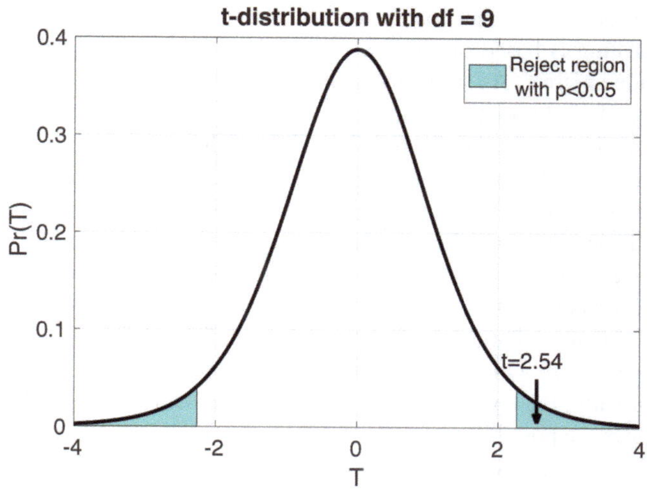

Fig. 17.4 The t-distribution with degree of freedom $df = 9$, the t-statistic $t = 2.54$ is located in the reject region with $p < 0.05$

$$p = \Pr(T \geq |t|) = 0.03.$$

Step 4: Compare the p-value to the significance level α. The significance level α is a pre-chosen probability. Conventionally, $\alpha = 0.05$, 0.01 and 0.001 have been used. If $p \leq \alpha$, then probability that the null hypothesis H_0 holds is less than α. H_0 would be a small probability event. Hence, we reject the null hypothesis H_0 and accept the alternative hypothesis H_1 that statistically significantly $\mu \neq 0$. If $p > \alpha$, then we fail to reject the null hypothesis and infer as if $\mu = 0$.

2. **False Positive and False Negative**

In the example above, we have the p-value $p = 0.03$. Hence, if we set the significance level $\alpha = 0.05$, we will reject the null hypothesis H_0 and accept the alternative hypothesis H_1. However, if we set the significance level $\alpha = 0.01$, we will fail to reject the null hypothesis. Hence, it is found that with different levels of α, the conclusion would be reversed. The question is how to understand the hypothesis with different level of α.

In fact, it should be noticed that we use the phrase "statistically significantly" and "as if" in the inference results, which indicates that we do not have a 100%

Table 17.1 The error types in the hypothesis tests

	Truth	
Decision	H_0 is true	H_0 is false
Do not reject H_0 (Negative result)	True negative ($1 - \alpha$, specificity)	False negative (β, Type II error)
Reject H_0 (Positive result)	False positive (α, Type I error)	True positive ($1 - \beta$, sensitivity, power)

guaranteed proof about the conclusion. In statistics, the conclusions drawn by hypothesis testing is based on the probability. Hence, there is always a possibility what we made an error. As summarized in Table 17.1, there are two types of errors that can be made in hypothesis testing.

Type I error is the error of rejecting the null hypothesis H_0 when it is actually true. $p \leq \alpha$ indicates that the probability that the null hypothesis H_0 is true is less than the possibility of α. In this case, we reject the null hypothesis, but with a possibility of α to make the Type I error. On the other hand, we have the possibility of $1 - \alpha$ to make a true negative decision, which is also called the *specificity* of a test.

Type II error is the error of failing to reject null hypothesis H_0 when alternative hypothesis H_1 is true. The Type II error rate normally is denoted by β and the true positive rate is $1 - \beta$, which is called the *specificity* or the statistical *power* of a test.

Conventionally, the significance level α would be set to be 0.05, 0.01, or 0.001, which indicates the possibility to make the Type I error is 5%, 1%, or 0.1%. There is no necessary connection between the Type I error rate α and the Type II error rate β. But normally lowering the value of α will decrease the possibility of the Type I error and also increase the Type I error rate β, which will make the test more conservative.

In our example, the sample is generated from the normal distribution $N(0.8, 1)$. Hence, we know the truth that the null hypothesis H_0 is false. If we set the significance level $\alpha = 0.05$, reject the null hypothesis H_0, and accept the alternative hypothesis H_1, we have a true positive result. However, if we set the significance level $\alpha = 0.01$, we fail to reject the null hypothesis and the result would be a false negative. Hence, in this case once we decrease the α level, we could have a better Type I error rate control, but the Type II error rate rises.

3. **One- and Two-Tailed Tests**

In statistical significance testing, there are three possible alternative hypotheses to compare the mean value of the sample to zeros, which are $\mu \neq 0$, $\mu > 0$, or $\mu < 0$. In the example above, the alternative hypothesis is $H_1 : \mu \neq 0$, which means we have to test both if the mean value is significantly larger than 0 and if the mean value is significantly smaller than 0. Hence, the test is two-tailed, since the reject region consists of the bottom $\alpha/2$ and the top $\alpha/2$ of the *t*-distribution two parts with the significant level α. With the alternative hypothesis $H_1 : \mu > 0$, the reject region is the bottom α of the *t*-distribution. With the alternative hypothesis $H_1 : \mu < 0$, the reject region is the top α of the *t*-distribution. Hence, these two hypotheses are called one-tailed test. In summary, we have all the conditions in Table 17.2.

Table 17.2 The *p*-value conducted in the three types of tests

Type	Null $\boldsymbol{H_0}$	Alternative $\boldsymbol{H_1}$	*p*-value
Left-tailed	$\mu = 0$	$\mu < 0$	$\Pr(T \leq -\lvert t\rvert) = 1$
Right-tailed	$\mu = 0$	$\mu > 0$	$\Pr(T \geq \lvert t\rvert) = 0.0157$
Two-tailed	$\mu = 0$	$\mu \neq 0$	$2 * \Pr(T \geq \lvert t\rvert) = 0.0315$

17.2.2 T-Test

In sect. 2.1, the example with the simulation data, using t-statistic for the hypothesis testing, is a type of t-test, which is used to compare the mean value of the samples under the normal distribution assumption. Mainly, there are three types of frequently used t-test:

- *One-sample t-test*: check whether the mean value of the samples has a specified value.
- *Paired-samples t-test*: compare the mean value of the samples from the same group in different condition.
- *Independent two-samples t-test*: compare the mean value of the samples from two different groups.

Under the null hypothesis H_0, the t-statistic is calculated based on experimental data. By checking the location of the t-statistic in the corresponding t-distribution, we can have a probability, p-value, associated with that t-statistic. By comparing the p-value to the significant level α, we can decide whether to reject or not reject the null hypothesis H_0.

In the following, we still use the real dataset of the resting state EEG with eyes open and closed as an example to show the process of t-test in both MATLAB and SPSS.

1. **One-Sample *T*-Test**

Question

To check the mean value of the difference in EEG power between eyes open and closed against the constant $m = 0$, in which we have the sample $x = \{x_i\}$ for $i = 1, 2, \ldots, n$ as the difference of EEG power between eyes open and closed for all subjects, the sample size $n = 93$.

Hence, we have the hypothesis that,

- H_0: the mean value of the difference in EEG power between eyes open and closed is equal to zero, i.e.

$$\mu = m,$$

- H_1: the mean value of the difference in EEG power between eyes open and closed is not equal to zero, i.e.

$$\mu \neq m.$$

in which we have the sample $x = \{x_i\}$ for $i = 1, 2, \ldots, n$ is the difference of EEG power between eyes open and closed for all subjects, the sample size $n = 93$.

The t-statistic is

$$t = \frac{\bar{x} - m}{s/\sqrt{n}}$$

where the $\bar{x}$ and s is the mean and sample standard deviation, $m = 0$ is the prechosen constant in the hypothesis, $n = 93$ is the sample size. Under the null hypothesis H_0, the t-statistic has a t-distribution with $n - 1$ degrees of freedom.

MATLAB implementation

```
clc;clear;close all;
[num,txt,raw] = xlsread('Resting State.xlsx');

% one sample test
x=num(:,3);
[h1,p1,ci1,stats1]=ttest(x);
disp(['t = ',num2str(stats1.tstat,'%0.2f')]);
disp(['p = ',num2str(p1,'%0.2e')]);
```

MATLAB function `ttest` can be used for the one-sample *t*-test. With the default parameter setting, two-tailed *t*-test is applied. We can also specify the parameter “Tail” to “left” or “right” for the left- or right tailed *t*-test respectively.

```
[h1,p1,ci1,stats1]=ttest(x,'Tail','left');
```

SPSS implementation

1. Open the data file “**Resting State.sav**”.
2. Click **Analyze -> Compare means -> One Sample T Test**... on the top menu.
3. In the One-Sample T Test dialog, select the variable “**Difference**” into the **Test Variable(s):** listbox. Set the **Test Value** as **0**.
4. Press Button “**OK**”.

Result

With both MATLAB and SPSS, we have the t-value $t = -19.97$. Hence, in the t-distribution with df $= 92$, we have the p-value $p = 3.32 \times 10^{-35}$. If we set the significant level $\alpha = 0.05$, we reject the null hypothesis H_0, accept the alternative hypothesis H_1, and conclude that the difference for the EEG power between eyes open and closed is significant different from zero.

2. **Paired-Samples *T*-Test**

Question

To check whether the mean value of EEG power with eyes open is different from the mean value of EEG power, in which we have the sample x_{EO} and x_{EC} as the EEG power between eyes open and closed respectively, and the sample size $n = 93$.

Hence, we have the hypothesis that,

- H_0: the mean value of EEG power with eyes open is the same with the mean value of EEG power, i.e.

$$\mu_{EO} = \mu_{EC},$$

- H_1: the mean value of EEG power with eyes open is different from the mean value of EEG power, i.e.

$$\mu_{EO} \neq \mu_{EC}.$$

Since, the null hypothesis H_0 can be translated into whether the difference between the two conditions is equal to zeros,

$$\mu = \mu_{EO} - \mu_{EC} = 0.$$

The paired-samples t-test problem can be solved by transforming it to the one-sample t-test problem.

MATLAB implementation

```
clc;clear;close all;
[num,txt,raw] = xlsread('Resting State.xlsx');

% paired-sample test
X_EO=num(:,1);
X_EC=num(:,2);
[h2,p2,ci2,stats2]=ttest(x_EO,x_EC);
disp(['t = ',num2str(stats2.tstat,'%0.2f')]);
disp(['p = ',num2str(p2,'%0.2e')]);
```

SPSS implementation

1. Open the data file "**Resting State.sav**".
2. Click **Analyze** -> **Compare means** -> **Paired-Samples T Test**... on the top menu.
3. In the Paired-Sample T Test dialog, select the variable "**Eyes Open**" and "**Eyes Closed**" into the **Paired Variable(s):** listbox.
4. Press Button "**OK**".

Result

Since the paired-samples t-test question is the same as the one-sample t-test question above, we have the t-value $t = -19.97$, and the p-value $p = 3.32 \times 10^{-35}$. If we set the significant level $\alpha = 0.05$, we reject the null hypothesis H_0, accept the alternative hypothesis H_1, and conclude that the mean value of EEG power with eyes open is significant different from the mean value of EEG power with eyes closed.

3. Independent Two-Samples *T*-Test

In the example of paired-samples *t*-test, the subjects for the two conditions, eyes open and closed, are from the same group. If the subjects are from two different groups, then independent two-samples *t*-test should be used.

Question

To check whether the mean values of the EEG power with eyes open are the same between male subjects and female subjects, in which we have the sample x_M and x_F as the EEG power from male and female subjects respectively, and sample size $n_M = 71$ and $n_F = 22$.

The hypothesis is that,

- H_0: the mean values of EEG power with eyes open between male subjects and female subjects are the same, i.e.

$$\mu_{\mathrm{M}} = \mu_{\mathrm{F}},$$

- H_1: the mean values of EEG power with eyes open between male subjects and female subjects are different, i.e.

$$\mu_{\mathrm{M}} \neq \mu_{\mathrm{F}}.$$

In the independent two-samples t-test, the *t*-statistic is

$$t = \frac{\bar{x}_M - \bar{x}_F}{\sqrt{s_M^2/n_M + s_F^2/n_F}},$$

where the $\bar{x}_M, \bar{x}_F$ are the means, s_M, s_F are the sample standard deviations, and $n_M$and n_F are the sample sizes of the two groups.

- In the case of equal variance for the two groups, the *t*-statistic under the null hypothesis H_0 has a *t*-distribution with $n_M + n_F - 2$ degrees of freedom and the sample standard deviations are replaced by the pooled standard deviation

$$s = \sqrt{\frac{(n_M - 1)s_M^2 + (n_F - 1)s_F^2}{n_M + n_F - 2}}$$

- In the case where the variances of the two groups are not equal, the *t*-statistic under the null hypothesis H_0 has an approximate Student's t distribution with a number of degrees of freedom given by Satterthwaite's approximation.

We should test the homogeneity of variance before the independent two sample t-test. Several methods can be used to test the homogeneity of variance, like Hartley's test, Bartlett's test and Levene's test. For simplicity, only Levene's test is applied in the both MATLAB and SPSS implementation of the independent two sample t-test.

MATLAB implementation

```
clc;clear;close all;
[num,txt,raw] = xlsread('Resting State.xlsx');

%% indenpendent two sample ttest
idx=num(:,5);
x=num(:,1);
x_M=x(idx==1);
x_F=x(idx==0);
[p3,stats3] = vartestn(x,idx,...
    'TestType','LeveneAbsolute','Display','off');
disp('Independent t-test with Eyes open:');
disp(['Levene's test: p = ',num2str(p3,'%0.2f')]);
if p3<0.05
    disp('Equal variances not assumed');
    [h4,p4,ci4,stats4]=ttest2(x_M,x_F,...
        'Vartype','unequal');
else
    disp('Equal variances assumed');
    [h4,p4,ci4,stats4]=ttest2(x_M,x_F);
end
disp(['t = ',num2str(stats4.tstat,'%0.2f')]);
disp(['df = ',num2str(stats4.df,'%0.2f')]);
disp(['p = ',num2str(p4,'%0.2f')]);
disp(' ');
```

SPSS implementation

1. Open the data file "**Resting State.sav**".
2. Click **Analyze** -> **Compare means** -> **Independent-Samples T Test**… on the top menu.
3. In the Independent-Samples T Test dialog, select the variable "**Eyes Open**" in to the **Test Variable(s):** listbox; Select variable "**Gender**" into the **Grouping Variable(s):** box.
4. Click the "**Define Groups**…" Button, in the Define Groups dialog, set Group 1: "**0**" and Group 2: "**1**"; press "**Continue**" to go back.
5. Press Button "**OK**".

Result

Since both p-value of Levene's test in the two case $p = 0.07$ is larger than $\alpha = 0.05$, we perform the independent t-test under the equal variances assumption. As a result, with the t-statistic $t = 1.08$, we have p-value $p = 0.28$, which is larger than $\alpha = 0.05$. Hence, we infer that there is not statistically significant difference between the male and female subjects with eyes open.

17.3 Analysis of Variance (ANOVA)

17.3.1 Introduction of ANOVA

ANOVA is also a form of statistical hypothesis testing to test whether the mean value from two or more groups is the same or not. Similar with *t*-test, ANOVA is also based on several common assumptions, like normal distribution, homogeneity of variance, and independence of observations. *T*-test is used to check the mean value from one group or compare the mean values of two groups. When the mean values of more than two groups are compared, ANOVA is preferred, for example, if we want to check whether the mean value of EEG power would be different with recording times: the recording times would be morning, afternoon, and evening. Pair-wised *t*-test with morning vs. afternoon, morning vs. evening, and afternoon vs. evening would answer this question. But repeatedly applying *t*-test for the pair-wised comparison would lead to severe Type I error problem and is not recommended. ANOVA is able to test the difference among several groups without increasing the Type I error rate.

In an EEG study, several *factors* would possibly influence the experimental response. The experimenter normally manipulates the factors to check the change of the result. Each factor has two or more *levels* (i.e., different values of the factor). Take the resting state EEG experiment for example. Eye, gender, and recording time are all the factors in the study, in which eyes open and closed are the two levels within the factor of eye, male and female are the two levels within the factor of gender, and morning, afternoon, and evening are the three levels within the factor of recording time.

The study with only one factor is called *one-way ANOVA*. With two or more factors in the study, it is called *multi-way ANOVA*. How changing the level of a factor changes the result is called *effect*. The effect of a single factor is also called a *main effect*. The effect that one factor has on the other factor is called *interaction effect*. And the *treatment* is the specific combination of factor levels whose effect will be compared with the other treatments. Similar with paired-samples *t*-test, if each treatment for certain factors from the same subjects in ANOVA, the study is called *repeated measures ANOVA*, and the factor is called *repeated measures factor*, or *within-subjects factor*. Similar to the independent two samples *t*-test, if the subjects at different levels of a factor are from different groups in the ANOVA, the factor is called *between-subjects factor*.

In the following, two examples are used to show the process of ANOVA. The dataset of the resting state EEG is used. One example uses one-way ANOVA to check whether the mean power of the resting state EEG is different in the different recording time (morning, afternoon, and evening). In another example, the repeated measures two-way ANOVA is used to study the effects of two factors in the resting state EEG, which is the eyes (open and closed) and recording time (morning, afternoon, and evening).

17.3.2 One-Way ANOVA

Question

Check whether the mean value of the EEG power with eyes open is the same with different recording times, which are in the morning, afternoon, and evening.

The hypothesis is that

- H_0: the mean values of eyes open EEG power are the same with different recording times, i.e.

$$\mu_1 = \mu_2 = \mu_3$$

- H_1: the mean values of eyes open EEG power are not the same with different recording times, i.e.

  At least one of the means μ_1, μ_2 and μ_3 is not equal to the others.

where μ_1, μ_2, and μ_3 are the EEG power recording in the morning, afternoon, and evening with eyes open.

MATLAB implementation

```
clc;clear;close all;
[num,txt,raw] = xlsread('Resting State.xlsx');

%% one-way ANOVA
x=num(:,1);
group=num(:,4);
[p,tbl,stats] = anova1(x,group)
```

SPSS implementation

1. Open the data file “**Resting State.sav**”.
2. Click **Analyze** -> **Compare means** -> **One-Way ANOVA**... on the top menu.
3. In the One-Way ANOVA dialog, select the variable “**Eyes Open**” into the **Dependent List:** box; select the variable “**Recording Time**” into the **Factor:** box.
4. In the One-Way ANOVA dialog, press button “**OK**”.

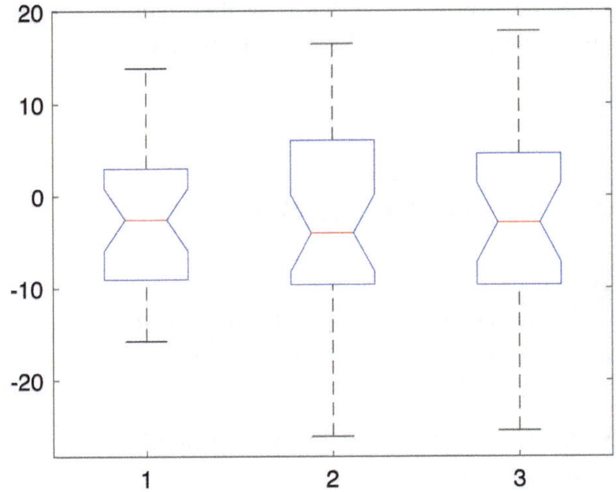

Fig. 17.5 The box plot of the resting state EEG power in different recording time: 1, morning; 2, afternoon; 3, evening. With the p-value $p = 0.981$ in the one-way ANOVA test, there is no significant difference among the three levels

Result

With the between-groups variability $s_b^2 = 1.86$, and the within-groups variability $s_w^2 = 97.89$, we can calculate the F-statistic $F = 0.019$. Using the F-distribution with $d_1 = 2$ and $d_2 = 90$, we have the p-value $p = 0.981$, which indicates there is no significant difference for the mean values of the EEG power with eyes open among the three recording times. Figure 17.5 is the box plot from MATLAB for the three levels in the one-way ANOVA test.

17.3.3 Repeated Measures Two-Way ANOVA

Question

Considering the two factors eyes' condition and recording time together, check whether the mean values of the EEG power are the same or not.

In the repeated measures two-way ANOVA problem, eyes' condition is the within-subjects factor with two levels, eyes open and eyes closed, and recording time is the between-subjects factor with three levels: morning, afternoon, and evening. To apply the repeated measures two-way ANOVA, we are actually testing three null hypotheses:

- There is no difference in the means of factor eyes' condition.
- There is no difference in the means of factor recording time.
- There is no interaction between factor eyes' condition and recording time.

The third null hypothesis is equal to the one-way ANOVA with the null hypothesis:

- H_0:for the difference of the EEG power between eyes open and closed, there is no difference in the means of factor recording time.

MATLAB implementation

```
clc;clear;close all;
[num,txt,raw] = xlsread('Resting State.xlsx');

%% repeated measures two-way ANOVA
t=table(num2str(num(:,4)),num(:,1),num(:,2),...
    'VariableNames',{'time','EO','EC'});
Meas = table([1 2]','VariableNames',{'eye'});
rm = fitrm(t,'EO, EC ~ time','WithinDesign',Meas);
ranovatbl = ranova(rm)
anovatbl = anova(rm)
```

To run the repeated measures two-way ANOVA, we firstly put all the data into a table *t*, in which we should specify all the recording time is a categorical variable, not a numerical variable. Then we fit the repeated model with the MATLAB function `fitrm`. Then `ranova` is used for the analysis of within-subjects factor "eyes" and the interaction "eyes∗ recording time"; and `anova` is used for the analysis of between-subjects factor "recording time."

SPSS implementation

1. Open the data file "**Resting State.sav**".
2. Click **Analyze** -> **General Linear Model** -> **Repeated Measures**... on the top menu.
3. In the Repeated Measures Define Factor(s) dialog, set the Within-Subject Factor Name: as "**eyes**"; set Number of Levels: as **2**, and press button "**Add**", then press button "**Define**".
4. In the Repeated Measures dialog, select the variable "**EyesOpen**" and "**EyesClosed**" into the **Within-Subjects Variables (eyes):** box, select the variable "**Recording Time**" into the **Between-Subjects Factor(s):** box.
5. Press the **Plots**... button, select **"eye"** into the **Horizontal Axis:** box and select **"Recording Time"** into the **Separate Lines:** box, and press button "**Add**", then press button "**Continue**".
6. In the Repeated Measures dialog, press button "**OK**".

Result

With the significant level $\alpha = 0.05$, the main effects for the within-subjects factor "eyes" is significant with *p*-value $p = 1.64 \times 10^{-34}$. But no significant difference for the between-subjects factor "recording time" with *p*-value $p = 0.82$, and also no significant difference for their interaction "eyes∗ recording time" with *p*-value $p = 0.44$. Figure 17.6 shows the interaction of the two factors. The result is consistent with the two-way repeated measures ANOVA.

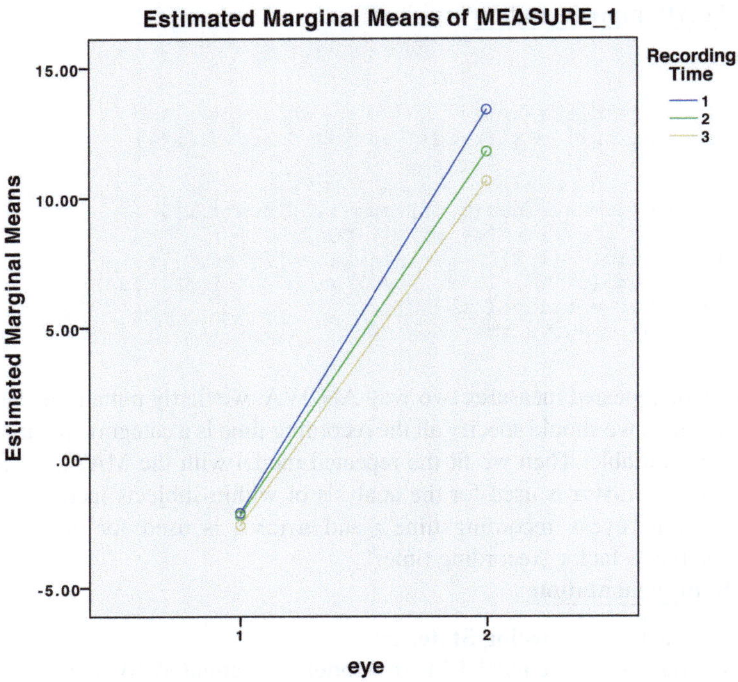

Fig. 17.6 The interaction of the factors of eyes' condition and the recording time. It can be observed that there is no significant difference among the recording time ($p = 0.82$), but the EEG power between eyes open and closed are significantly different ($p = 1.64 \times 10^{-34}$), and the interaction between the two factors is not significant ($p = 0.44$)

17.4 Correlation and Regression Analysis

Both correlation and regression analysis are used to describe the relationship between two variables. But they are not the same. In the correlation analysis, the relationship between variables x and y is symmetrical and also mutual. Linear regression analysis fixes a best line that how y can be predicted from variable x, which can be used in prediction and forecasting. Hence, regression analysis has wider applications than correlation analysis.

17.4.1 Correlation Analysis

Correlation analysis aims to quantify the strength and direction of the relation between two variables. A positive correlation is a relationship between two variables where if one variable x increases, the other variable y also increases, and vice versa. Correlation coefficient is used to describe the relationship between the two variables, which ranges from -1 to 1.

Pearson correlation coefficient is the most commonly used method for correlation analysis. Assume the normal distribution for both the two variables x and y, Pearson correlation coefficient accesses the linear relationship between x and y. Without the normally distributed assumption, Spearman's ρ coefficient and Kendall's τ coefficient can be used for the correlation analysis. In the following, only the Pearson correlation coefficient is introduced.

Pearson correlation coefficient can be calculated as follows,

$$r = \frac{\Sigma_{i=1}^{n}\left(x_i - \bar{x}\right)\left(y_i - \bar{y}\right)}{\sqrt{\Sigma_{i=1}^{n}\left(x_i - \bar{x}\right)^2\left(y_i - \bar{y}\right)^2}},$$

in which n is the sample size, and x_i and y_i are the samples for variables x and y. With the null hypothesis $H_0 : x$ and y are two uncorrelated normal distributed variables, the t-statistic

$$t = r\frac{\sqrt{n-2}}{\sqrt{1-r^2}}$$

should follow the t-distribution with degrees of freedom of $n-2$. Hence, according to the location of the t-statistic in the t-distribution, we can calculate the p-value to determine whether the variables x and y are correlated or not.

It should be noted that the correlation between the two variables x and y does not imply the causality. It is possible that x causes y, or y causes x. It is also possible that both x and y are caused by another common variable z. In the EEG recording, the signals between adjacent channels are always highly correlated. It does not indicate signal from one channel is caused by another channel. With great possibility, both the two channels' signals are caused by the common source.

To check the Pearson correlation coefficient of EEG power between eyes open and closed in resting state, we run the correlation analysis in both MATLAB and SPSS.

MATLAB implementation

```
clc;clear;close all;
[num,txt,raw] = xlsread('Resting State.xlsx');
x=num(:,1);
y=num(:,2);
[r_corr,p_corr]=corr(x,y);
```

SPSS implementation

1. Open the data file "**Resting State.sav**".
2. Click **Analyze -> Correlate -> Bivariate**... on the top menu.
3. In the Bivariate Correlations dialog, select the variable "**Eyes Open**" and "**Eyes Closed**" into the **Variable(s):** listbox.
4. Press Button "OK".

Result

As a result, the correlation coefficient is $r = 0.774^{**}$ with the p-value $p = 9.40 \times 10^{-20}$ from both MATLAB and SPSS. Here "**" indicates the correlation is significant at the 0.01 level. Hence, we can conclude that EEG power between eyes open and closed in resting state is correlated.

17.4.2 Regression Analysis

Unary linear regression is the simplest linear regression model, which involves a single dependent variable **y** and a single independent variable **x**. x and y keep the linear relationship that

$$y_i = \beta_0 + \beta_1 x_i + \epsilon_i, i = 1,2, \ldots, n$$

in which parameters β_0 and β_1 are the intercept and slope of the linear regression model respectively, and ϵ_i is called residuals.

The goal of the regression analysis is to find the best parameters β_0 and β_1 to minimize the residual sum of squares (RSS)

$$\mathrm{RSS} = \sum_{i=1}^{n} \epsilon_i^2.$$

Under the normally distributed assumption of ϵ_i, this least square problem can be solved by setting the gradient to zero. Hence, we have the equations that

$$\frac{\partial RSS}{\partial \beta_0} = -2\sum_{i=1}^{n} (y_i - \beta_0 - \beta_1 x_i) = 0,$$

$$\frac{\partial RSS}{\partial \beta_1} = -2\sum_{i=1}^{n} (y_i - \beta_0 - \beta_1 x_i)x_i = 0,$$

Solving the equations, we have the estimations of β_0 and β_1

$$\hat{\beta}_1 = \frac{\Sigma_{i=1}^{n} (x_i - \bar{x})(y_i - \bar{y})}{\Sigma_{i=1}^{n} (x_i - \bar{x})^2}, \text{and } \hat{\beta}_0 = \bar{y} - \bar{x}\hat{\beta}_1.$$

Hence, the estimation of y_i is

$$\hat{y}_i = \hat{\beta}_0 - \hat{\beta}_1 x_i.$$

The coefficient of determination R^2 is commonly used to assess the goodness-of-fit of the regression, which describes the proportion of the total variance can be explained by the model. Denote the total sum of squares (TSS) and the explained sum of squares (ESS) as

$$\text{TSS} = \Sigma_{i=1}^{n}(y_i - \bar{y})^2$$

$$\text{ESS} = \Sigma_{i=1}^{n}(\hat{y}_i - \bar{y})^2.$$

It can be proved that

$$\text{TSS} = \text{ESS} + \text{RSS}.$$

Hence, we have

$$R^2 = \frac{\text{ESS}}{\text{TSS}} = 1 - \frac{\text{RSS}}{\text{TSS}}.$$

Meanwhile, we can also prove that the coefficient of determination R^2 is equal to the square of the Pearson's correlation coefficient r^2.

Further, with hypothesis in the regression analysis

$$H_0 : \beta_1 = 0, H_1 : \beta_1 \neq 0$$

the F-statistic

$$F = \frac{\Sigma_{i=1}^{n}(\hat{y}_i - \bar{y})^2/1}{\Sigma_{i=1}^{\text{n}}(y_i - \hat{y}_i)^2/(n-2)} = \frac{R^2}{1-R^2}(n-2)\tilde{}F(1, n-2)$$

should follow the F-distribution $F(1, n-2)$. Hence, according to the location of the F-statistic in the F-distribution, we can calculate the p-value to determine whether β_1 equals to zeros or not.

To check whether the EEG power in resting state with eyes open can predict the EEG power in resting state with eyes closed, we use both MATLAB and SPSS to do the same regression analysis.

MATLAB implementation

```
clc;clear;close all;
[num,txt,raw] = xlsread('Resting State.xlsx');
x=num(:,1);
y=num(:,2);
[b_reg,~,~,~,stats_reg] = regress(y,[x,ones(93,1)]);
```

SPSS implementation

1. Open the data file "**Resting State.sav**".
2. Click **Analyze -> Regression -> Linear**... on the top menu.
3. In the Linear Regression dialog, select the variable "**Eyes Closed**" into the **Dependent:** box and select the variable "**Eyes Open**" into the **Independent(s):** listbox, and press Button "OK".

Result

The linear relationship between the resting state EEG power with eyes open and closed is illustrated in Fig. 17.7. Both MATLAB and SPSS get the coefficient $\hat{\beta}_0 = 13.91$ and $\hat{\beta}_1 = 0.84$, coefficient of determination $R^2 = 0.599$, which is equal to the square of the correlation coefficient r^2 obtained above. With the F-statistic $F = 135.93$, we have $p = 9.40 \times 10^{-20}$, which is also the same as the p-value in the correlation analysis. Hence, $\hat{\beta}_1$ is statistical significantly different from zeros, and the EEG power with eyes open can explain 59.9% of the total variance in the EEG power with eyes closed.

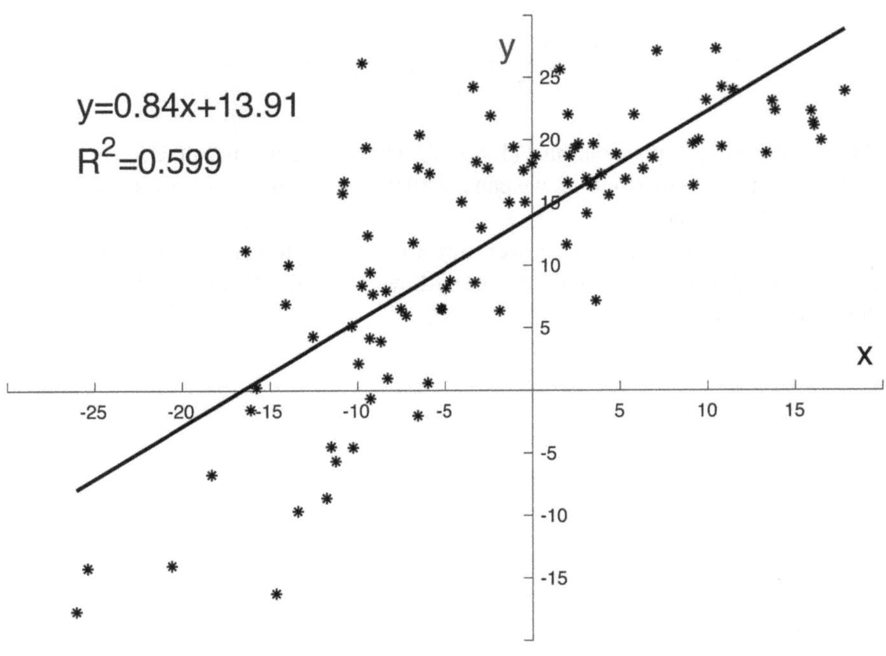

Fig. 17.7 The linear relationship between x and y, where x and y are EEG power of eyes open and closed, respectively, in the resting state from 93 subjects

17.5 Nonparametric Test

17.5.1 Parametric Test vs. Nonparametric Test

In the above sections, we have introduced several types of hypothesis testing methods, like t-test, ANOVA, and correlation analysis. All these are based on the normally distribution summation, hence is called parametric test. Without any specific assumption about the distribution of the data, we can still run the hypothesis testing by nonparametric tests. Normally, there are three ways to run nonparametric tests.

- Compare the distribution. Some methods like Kolmogorov-Smirnov test can be applied to run this kind of test.
- Compare the median instead of the mean value. Paralleling to the parametric test, we have the corresponding nonparametric tests in different conditions. Table 17.3 summarized the methods used in parametric tests and nonparametric tests.
- Generate the distribution with data-driven methods. Without the normal distribution assumption, we can still use the t-statistic and F-statistic, but there is no corresponding t-distribution and F-distribution. In this case, permutation methods can use the sample data to generate the distribution under the null hypothesis. The data-driven distribution can be used instead of the t-distribution and F-distribution under the normal distribution assumption in t-test and ANOVA.

Without the assumption of normal distribution, nonparametric tests seem to be nicer and more straightforward than parametric tests. However, in practice, as a cost of fewer assumptions, nonparametric tests are normally less powerful than their corresponding parametric methods (i.e., with higher false negative rate β). In contrast, parametric tests can even perform well with skewed and non-normal distributions when the sample size is large. That is because the distribution of sample means should be approximately normally distributed according to the Central Limit Theorem. Hence, nonparametric tests are usually applied when you have a small sample size with an unknown distribution. With a large sample size, parametric tests are robust in the presence of violations of the normality assumption (Hollander et al. 2013).

Table 17.3 The comparison of the methods in parametric tests and nonparametric tests

Parametric tests (mean)	Nonparametric tests (medians)
One-sample t-test, paired-samples t-test	Sign test, Wilcoxon signed-rank test
Independent two-sample t-test	Mann-Whitney test
One-way ANOVA	Kruskal-Wallis, Mood’s median test
One-way ANOVA with repeated measures	Friedman test
Spearman’s correlation	Spearman and Kendall correlation

In the following, the methods of Wilcoxon signed-rank test and permutation test will be introduced. Again, the dataset of resting state EEG with eyes open and closed is used as an example.

17.5.2 Wilcoxon Signed-Rank Test

Question

Considering the problem in the one-sample t-test again, checking the difference in EEG power between eyes open and closed against zero, in which we have the sample $x = \{x_i\}$ for $i = 1, 2, \ldots, n$ as the difference of EEG power between eyes open and closed for all subjects, the sample size $n = 93$.

In the Wilcoxon signed-rank test, we have the hypothesis that:

H_0: x follows a symmetric distribution around zero.

H_1: x does not follow a symmetric distribution around zero.

With the null hypothesis, we can run the Wilcoxon signed-rand test in the following steps:

1. Calculate $|x_i|$ and $\text{sgn}(x_i)$ for each $i = 1, \ldots, N$, where sgn is the sign function.
2. Sort $|x_i|$ in ascending order.
3. Rank $|x_i|$ for 1 to n. Let R_i denote the rank.
4. Calculate the w-statistic.

$$W = \sum_{i=1}^{n} [\text{sgn}(x_i)\, R_i],$$

5. Under the null hypothesis, the w-statistic follows a distribution with zero mean and the variance of $\sigma_W = n(n + 1)(2n + 1)/6$. The distribution does not have a simple expression. As n increases, the sampling distribution converges to a normal distribution. Hence,

   - With $n < 15$, compare W with a critical value $W_{critical,\ n}$ from a reference table. Reject H_0 if $|W| > W_{critical,\ n}$.
   - With $n \geq 15$, calculate the z-score $z = W/\sigma_W$ and reject the null hypothesis H_0 if $|z| > z_{ctritcal}$.

MATLAB implementation

```
clc;clear;close all;
[num,txt,raw] = xlsread('Resting State.xlsx');
x=num(:,3);
[p,h,stats] = signrank(x);
```

SPSS implementation

1. Open the data file "**Resting State.sav**".
2. Click **Analyze** -> **Nonparametric Test** -> **One Sample**... on the top menu.
3. In the One-Sample Nonparametric Tests dialog, in the panel **Objective**, select the variable "**Customize analysis**" into the **What is your objective?** box.
4. In the panel **Fields**, select "**Use custom field assignments**", and in the **Fields:** listbox, select "**Difference**" into the **Test Fields:** listbox.
5. In the panel Settings, select "**Choose Tests**" in the **Select an item** listbox, and on the right select "**Customize tests**" and choose "**Compare median to hypothesized (Wilcoxon signed-rank test)**"; set the **Hypothesized median:** as **0**.
6. Press Button "**Run**".

Result

With both MATLAB and SPSS, we have a p-value $p = 5.76 \times 10^{-17}$. Hence, we reject the null hypothesis H_0, and conclude that the median of the difference in EEG power between eyes open and closed is different from 0.

17.5.3 Permutation Test

Question

Considering the problem again, that checking the difference in EEG power between eyes open and closed against zero, in which we have the sample $x = \{x_i\}$ for $i = 1, 2, \ldots, n$ as the difference of EEG power between eyes open and closed for all subjects, the sample size $n = 93$.

Here, we still used the hypothesis in the t-test, that

- H_0: the mean value of the difference in EEG power between eyes open and closed is equal to zero, i.e.

$$\mu = 0,$$

- H_1: the mean value of the difference in EEG power between eyes open and closed is not equal to zero, i.e.

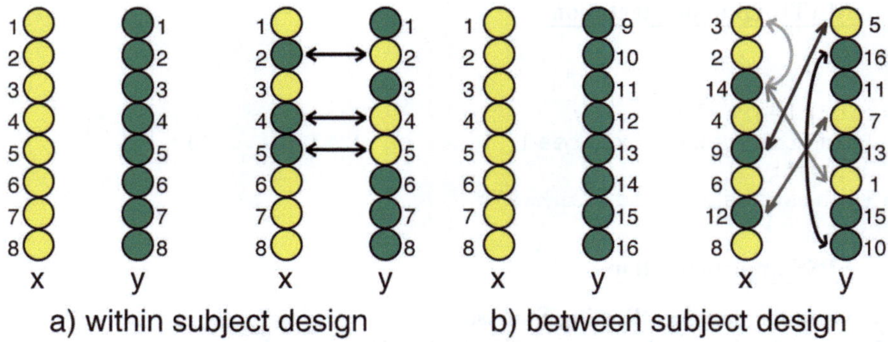

Fig. 17.8 The difference exchange process between within-subjects design and between-subjects design with two conditions

$$\mu \neq 0.$$

With the normal distribution assumption in *t*-test, we can calculate the *t*-statistic, which should follow the *t*-distribution under the null hypothesis H_0. Hence, we can determine the significance of the result by checking *p*-value, the probability that the *t*-statistic would be at least as extreme as we observed in the *t*-distribution. Without the normal distribution assumption, the *t*-statistic may not follow the *t*-distribution under the null hypothesis. Hence, permutation test can be used to generate the sampling distribution.

An important assumption behind the permutation test is that the observations are exchangeable under the null hypothesis. In one-sample test of comparing the mean value to zero, exchanging the observation is equivalent to randomly changing the sign of the samples. If we compare the mean values in two or more conditions, exchanging the observation is swapping experimental conditions, in which the exchanging would be different between within-subjects design and between-subjects design. In a within-subjects design where the observations for different conditions are from the same group of subjects, the observations are exchanged by swapping the condition labels within each subject but not between subjects. In a between-subjects design where the observations for different conditions are from different subjects, the observations are exchanged by swapping the condition labels between the subjects. Figure 17.8 illustrates the difference exchange process between within-subjects design and between-subjects design with two conditions. With randomly permuting the samples for a certain large number of times, we can calculate the *t*-statistic for each permutation and further generate the sampling distribution of the *t*-statistic. The sampling distribution is also called permutation distribution, which is used instead of the *t*-distribution and *F*-distribution under the normal distribution in the hypothesis testing.

The detailed process of the permutation test with the resting state EEG is as follows:

1. Calculate the t-statistic.

$$t = \frac{\bar{x}}{s/\sqrt{n}}$$

2. Permutation. Randomly change the sign of the samples x to x_{perm}, like from $x = \{x_1, x_2, x_3, x_4, \ldots, x_n\}$ to $x_{perm} = \{x_1, -x_2, -x_3, x_4, \ldots, x_n\}$, for example.
3. Repeat step 2 for N times, and calculate the t-value for each permutation.

$$t_{perm} = \frac{\bar{x}_{perm}}{s_{perm}/\sqrt{n}}$$

4. Generate the sampling distribution of the t_{perm}.
5. Calculate the p-value and compare the p-value with the significant level α. The p-value is the probability that

$$p = \Pr\left(t_{perm} \geq |t|\right)$$

MATLAB implementation

```
clc;clear;close all;
[num,txt,raw] = xlsread('Resting State.xlsx');
x=num(:,3);
n=size(x,1);
tval=mean(x)./(std(x)/sqrt(n));

%% permutation
N=1000;
y=x*ones(1,N).*((randn(n,N)>0)*2-1);
t_permute = mean(y,1)./(std(y,1)/sqrt(n));
p=sum(abs(tval)<abs(t_permute))/N;
disp(p);
```

Result

With the permutation times $N = 1000$, the histogram of the t_{perm} is shown in Fig. 17.9. The corresponding kernel distribution is close to the t-distribution with the degrees of freedom $df = 92$. As a result, it is calculated that the t-statistic $t = -19.97$. There is not any permutation with $|t_{perm}| > |t|$. Hence, we have the p-value $p = 0$ and further reject the null hypothesis H_0, and accept the alternative hypothesis H_1 that statistically significantly $\mu \neq 0$.

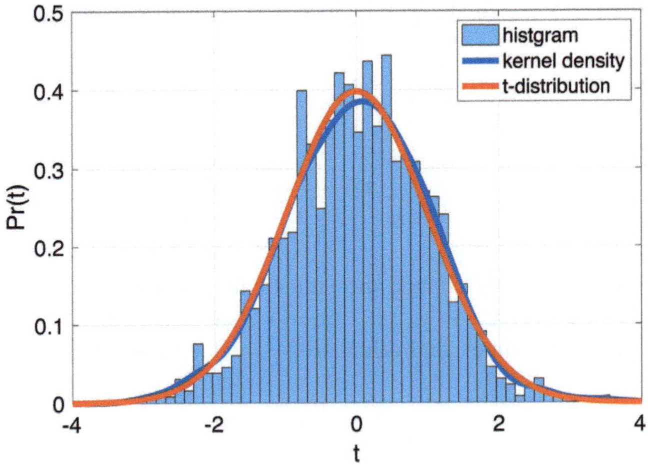

Fig. 17.9 For the permutation times $N = 1000$, the histogram of the t_{perm} is displayed in blue bars. Its kernel density (blue curve) is compared with the t-distribution with degrees of freedom $df = 92$ (orange curve)

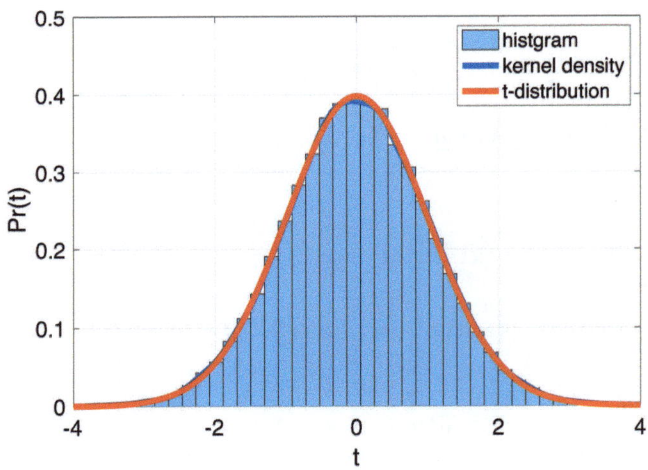

Fig. 17.10 For the permutation times $N = 100000$, the histogram of the t_{perm} is displayed in blue bars. Its kernel density (blue curve) is compared with the t-distribution with degrees of freedom $df = 92$ (orange curve)

Here, we randomly permute the data for $N = 1000$ times, the smallest interval of the p-values is 0.001, and the uncertainty near $p = 0.05$ is about 1%. If we increase the permutation times to $N = 100000$, the kernel distribution will be closer to the t-distribution with degrees of freedom $df = 92$, which is shown in Fig. 17.10. The smallest interval of the p-values is 0.00001 and the uncertainty near $p = 0.05$ is about 0.01%.

17.6 Multiple Comparison Problem

17.6.1 Hypothesis-Driven vs. Data-Driven EEG Analysis

In the EEG signal analysis, usually we have two ways to make statistical inferences, which are *hypothesis-driven* and *data-driven* approaches, respectively. Take the well-studied P300 experiment, for example. Normally, a positive component happened around 300 ms after the stimulus, which is elicited in the process of decision-making. Visual oddball experiment was conducted with the red as the target stimuli and the white as the nontarget stimuli. Each stimulus lasts 80ms with the ISI 200 ms and the target stimuli come with the possibility of 5%. Sixteen subjects took participation in the experiment. The EEG signal is recorded by a 64-channel BrainAmp system (Brain Products GmbH, Germany) with a sampling rate 1000 Hz and reference to TP9 and TP10. As a result, the grand averaged ERPs on channel Pz for both target and nontarget conditions are illustrated in Fig. 17.11(a).

With the hypothesis-driven approach used in the analysis of P300, we empirically propose a hypothesis based on our observation of the data or the previous experience, such as the hypothesis that the mean of ERP from the pre-defined interval 0.2–0.7s would be different between target and nontarget conditions. With the hypothesis, a limited number of t-test or ANOVA would be conducted to test the proposed hypothesis. Hypothesis-driven approach is an effective way to explore neuropsychological mechanisms based on the EEG analysis. However, in the hypothesis-driven approach, the result highly depends on researchers' subjective experience. Firstly, the interval selection from 0.2 to 0.7 s is based on the subjective experience, in which the result would be sensitive to the parameter selection. Secondly, if the interval selection is based on the data observation, then it is easy to make a false positive error since the difference may be caused by random effect. Thirdly, for more complex data, researcher's observation usually is not sufficient, which means some interesting result would be missing, since the data outside the interval has not been analyzed.

Data-driven approach provides a more objective approach for the EEG analysis, by comparing the ERPs at an exhaustive of time, frequency, and spatial points. Such a point-wise comparison would make a more comprehensive understanding of the data. Hence, the data-driven approach can effectively avoid the bias caused by the subjective experience, which also improves the repeatable of the data analysis. Further, the point-wise analysis can tell you exactly where and when an effect occurs. As computers become faster and cheaper, data-driven approach is becoming more widely used. However, with the drastic increase of the number of hypothesis testing in the data-driven approach, we have to face the Family-Wise Error Rate in the multiple comparison problem.

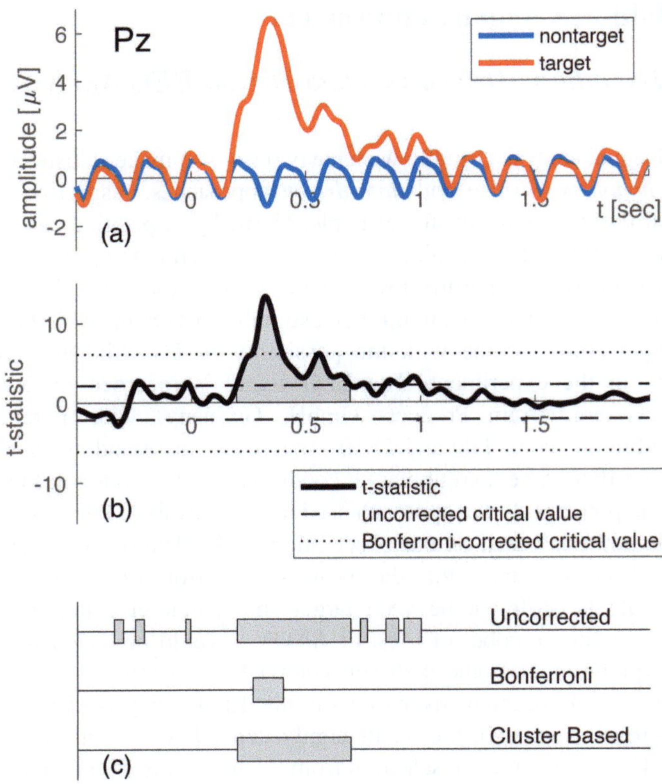

Fig. 17.11 The target and nontarget responses of P300 with the corresponding statistical testing. (**a**) The grand average P300 with the target and nontarget responses from 16 subjects on channel Pz. (**b**) *T*-statistic for the point-wise comparison between the target and nontarget conditions. The interval in gray indicates the cluster-based permutation test result. (**c**) The significant intervals under the three different statistical procedures: (1) point-wise *t*-test with uncorrected significant level 0.05; (2) point-wise *t*-test with Bonferroni-corrected significant level $0.05/2500=2 \times 10^{-5}$; and (3) cluster-based permutation test with cluster-level threshold 0.05

17.6.2 Family-Wise Error Rate

1. **The Problem of FWER**

In Sect. 17.2.1, we have analyzed the false positive (i.e., Type I error) in hypothesis testing, which is controlled by the significant level α. If we consider a set of hypothesis testing problem simultaneously, the probability for making one or more false positive errors is called *Family-Wise Error Rate* (FWER). With one comparison problem, the rate of the false positive error is α. With m independent comparison problems, the FWER is $1 - (1 - \alpha)^m$. Table 17.4 lists the values of FWER with the different value of m. It can be observed that with the test number $m = 1$, the FWER is 5%. With $m = 100$, the FWER is 99.41%. While with $m = 2500$ in our

Table 17.4 The value of FWER with different number of test m, in which the significant level $\alpha = 0.05$

m	FWER
1	$1-(1-\alpha)^1 = 0.0500$
2	$1-(1-\alpha)^2 = 0.0975$
3	$1-(1-\alpha)^3 = 0.1426$
4	$1-(1-\alpha)^4 = 0.1855$
5	$1-(1-\alpha)^5 = 0.2262$
10	$1-(1-\alpha)^{10} = 0.4013$
100	$1-(1-\alpha)^{100} = 0.9941$
2500	$1-(1-\alpha)^{2500} = 1-2\times 10^{-56}$

P300 experiment, the FWER is almost 100%. Hence, if there is no effect in the experiment, it is almost a 100% probability to make a mistake in the hypothesis testing.

The point-wise comparison result on the P300 data is shown in Fig. 17.11(b) and (c). Comparing the target and nontarget result, seven clusters are detected to be significant with the significant level $\alpha = 0.05$. Except the largest cluster at around 300ms, the others are supposed to be false positive, especially the first three clusters before the 0 second.

2. **Bonferroni Correction**

Several methods can be used to control the FWER, in which Bonferroni correction is famous for its simplicity. Considering a family of hypothesis $H_1, H_2, \ldots, H_m$, Bonferroni correction can be used just by decreasing the significant level α to α/m, where m is the number of the hypothesis. Following the Boole's inequality, it can be proved that

$$FWER = 1 - \left(1 - \frac{\alpha}{m}\right)^m \leq m \times \frac{\alpha}{m} = \alpha.$$

Bonferroni correction strictly controls the global false positive rate by setting the α level to α/m for the entire set of m comparisons. Bonferroni correction is a strong FWER control, which means the FWER is guaranteed to be less than α whether the null hypotheses are true or false. However, EEG data are correlated with their neighbors in the time, frequency, or spatial domain, which violates the independent assumption in Bonferroni correction. This correction comes at the cost of greatly increasing the probability of producing false negatives (i.e., Type II error) and consequently reducing statistical power. Hence, Bonferroni correction is not commonly used in the data-driven approach, since it is too conservative. Considering m = 2500 in the point-wise comparison of the P300 data analysis, the corrected significant level is $\alpha/m = 2 \times 10^{-5}$, with $\alpha = 0.05$. Thereby, the critical value for the t-statistic is increased from 2.13 to 6.11 as is shown in Fig. 17.11(b). As a result, all the clusters before 0 are excluded, but only the interval 0.26 to 0.39s in the main cluster has been reserved, which is much narrower than the uncorrected result. Based on the Bonferroni correction, several methods, like Holm's step-down procedure and

Hochberg's step-up procedure, have been developed to increase statistical power under a strong α level FWER control. Alternatively, false discovery rate methods, such as Benjamini-Hochberg procedure and Benjamini-Hochberg-Yekutieli procedure, were developed with a weak control of FWER (i.e., FWER is only guaranteed if all null hypotheses are true).

17.6.3 Cluster-Based Permutation Test

1. **Permutation Test**

Permutation test can also be used in the multiple comparison problem. The procedure of permutation test for single point comparison has been introduced in the previous section. Unlike parametric tests, such as *t*-test and ANOVA, permutation test does not make specific assumptions about the population distribution. For multiple points comparisons, we compute the *t*-statistic for each test and the statistic t_{max} (the most extreme positive or negative value) is used for the permutation test. For a two-tailed test, the statistic t_{max} equals the maximum absolute value of the *t*-statistic for each test due to the symmetric permutation distribution. The distribution of t_{max} adaptively reflects the degree of the correlation among the data points. With the global distribution of t_{max}, calculate the threshold corresponding to a certain α level. For each test, once the *t*- statistic is higher than the threshold, we reject the null hypothesis and conclude that there is significant difference.

With the statistic t_{max}, point-wise permutation test can automatically adjust to the degree of the correlation among the data points. By eliminating the independence assumption in Bonferroni correction, point-wise permutation test increases the statistical power in the multiple comparison problem. Similar to Bonferroni correction, point-wise permutation test provides a strong FWER control. However, when the number of tests is extremely large, the permutation test will also become conservative.

2. **Cluster-Based Permutation Test**

Cluster-based permutation test, proposed by Maris, uses the cluster-level statistic t_{max} instead of the point level statistic t_{max}, so that it can drastically increase the sensitivity of the statistical test while strictly controlling the FWER. As a weak FWER control method, cluster-based permutation test provides a higher sensitivity than false discovery rate methods. The cluster-based permutation test includes two major steps.

1. Calculate the cluster-level statistic: Calculate *t*-statistic, noted as t_{point}, for every point of interest in the temporal, frequency, or spatial domain. All points with t_{point} not exceeding the point-level threshold corresponding to the certain α level (denoted as α_{point}) are ignored. Next, cluster the remaining points in connected sets on the basis of temporal, frequency, or spatial adjacency. Then, calculate

cluster-level statistics, noted as $t_{cluster}$, by taking the sum of t_{point} within a cluster. Let t_{max} be the most extreme value of $t_{cluster}$.

2. Perform the permutation test: Repeat the permutation process and calculate their t_{max} to generate the permutation distribution. With the distribution of t_{max}, the cluster-level threshold is determined by a certain α level (denoted as $\alpha_{cluster}$). If there is any cluster under the true labels with its $t_{cluster}$ larger than the threshold, we reject the null hypothesis and conclude that there is significant difference.

For cluster-based permutation test, no MATLAB function can be used directly. Some toolboxes can be used. Here, we introduce the implementation of the cluster-based permutation test in the toolbox of letswave.

Letswave implementation

1. Open the letswave toolbox in MATLAB by typing "**letswave**" in the command window. Set the **folder** with the P300 files as the **filepath** in the Manage Module of the letswave toolbox.
2. Select the files "**P300 Nontarget**" and "**P300 Target**" in the **Datasets:** listbox in Manage Module. Select **Statistics -> Compare Two datasets (paired sample/ two samples ttest)**.
3. In the Batch module, **enable** the checkbox for Cluster-Based Permutation Test in the **ttest** panel.
4. Click the "**Run**" button in the bottom of the Batch module, to run the cluster-based permutation test. After a few seconds for the computation, the result will appear in the Manage Module with the filename "**ttest P300 Target**".
5. Double-clicking the data file "**ttest P300 Target**", the result is illustrated in the viewer Module. Select the different items in the **Index:** pop-up menu; we can see the result of the raw p-value and t-value without any correction and the corrected p-value and t-value after the cluster-based permutation test.
6. To export the result to MATLAB, select "**ttest P300 Target**" and select the "**send to workspace**" item in the **right-click menu**; then the result appears in the workspace of MATLAB.

Here, the P300 dataset only includes one channel EEG signal from Pz. With multi-channel EEG analysis, we can also run the cluster-based permutation test in the spatial domain, in which we need to enable the "**multiple sensor analysis**" and define the neighbors by setting the **connection threshold** in step 3.

Letswave can also generate the MATLAB script automatically. With the auto-generated scripts, we can run the *m*-file "code6_1.m" in the command windows of MATLAB. The result is the same as it is from the GUI of the Letswave toolbox.

MATLAB implementation

```
%% cluster-based permutation test
LW_init();
option=struct('filename',...
    {{'P300 Nontarget.lw6','P300 Target.lw6'}});
lwdataset= FLW_load.get_lwdataset(option);
option=struct('test_type','paired sample',...
    'tails','both','ref_dataset',1,'alpha',0.05,...
    'permutation',1,'cluster_threshold',0.05,...
    'num_permutations',2000,'show_progress',1,...
    'suffix','ttest','is_save',1);
lwdataset= FLW_ttest.get_lwdataset(lwdataset,option);
```

Result

Select the item "**p-value**" in the **Index:** pop-up menu; the uncorrected *p*-value for each time point of the comparison is shown in the right panel. By setting the **Y-axis** from **0** to **0.05**, we can find seven clusters with their p-value lower than 0.05. Similarly, by setting the **Y-axis** from **0** to **0.00002**, only one cluster with the interval from 0.26 to 0.39s is survived after the Bonferroni correction. The result is the same as it is illustrated in Fig. 17.11(c). Select the item "**t-value**" in the **Index:** pop-up menu, and set the **Y-axis** as **auto**; the *t*-value as illustrated in Fig. 17.11(b) is displayed in the right panel.

Select the item "**cluster p-value**" in the **Index:** pop-up menu; the *p*-value for each cluster based on the value of $t_{cluster}$ in the permutation distribution of t_{max} is shown in the right panel. The cluster with their cluster *p*-value less than 0.05 is excluded with their p-value set to be 1. As a result, only the main cluster with the interval from 0.2 to 0.96s is survived after the cluster-based permutation test.

With the cluster-based permutation test, the result of the P300 experiment is shown in Fig. 17.11(c). As a result, only the largest cluster is reserved, which is not so narrow as the result after Bonferroni correction. As reviewed by Groppe, cluster-based permutation test is possibly the most powerful multiple comparison procedure for providing neat results and detecting the presence of broad effects. But cluster-based permutation test has lower sensitivity for smaller clusters so that some meaningful small EEG effects might be neglected.

References

Bickel PJ, Doksum KA. Mathematical statistics: basic ideas and selected topics, volumes I-II package. New York: Chapman and Hall/CRC; 2015.

Casella G, Berger RL. Statistical inference, vol. 2. Pacific Grove: Duxbury; 2002.

DeGroot MH, Schervish MJ. Probability and statistics. New York: Pearson Education; 2012.

Devore JL. Probability and statistics for engineering and the sciences. New York: Cengage Learning; 2001.

Hollander M, Wolfe DA, Chicken E. Nonparametric statistical methods, vol. 751. New York: Wiley; 2013.
Johnson RA, Miller I, Freund JE. Probability and statistics for engineers, vol. 2000. New York: Pearson Education; 2000. p. 642p.
Lehmann EL, Romano JP. Testing statistical hypotheses. New York: Springer; 2006.
Montgomery DC, Runger GC, Hubele NF. Engineering statistics. New York: Wiley; 2009.

Chapter 18
Simultaneous EEG-fMRI

Xu Lei

Abstract Simultaneous EEG-fMRI combines the advantages of high temporal resolution of EEG with high spatial resolution of fMRI. In addition, it is a noninvasive technique for the study of human brain function. However, it remains many challenges such as the low signal-to-noise ratio, poor individual comfort, and difficulty in data analysis. In this chapter, we first introduce the hardware of simultaneous EEG-fMRI system. Then a review about the advance of this technique is given, including the EEG artifacts correction, the EEG-fMRI data fusion method, and the application of EEG-fMRI. Specifically, we provide a systematic classification for the fMRI-constrained EEG and the EEG-informed fMRI from simple to complex level. Then we provide program practice for the EEG artifacts correction, which may contribute to the widespread application of this new technique. Finally, we discuss the prospects of simultaneous EEG-fMRI for future research.

Keywords EEG-fMRI · Fusion · EEG artifacts correction · fMRI-constrained EEG · EEG-informed fMRI

Both functional magnetic resonance imaging (fMRI) and electroencephalography (EEG) are preeminent techniques for noninvasive brain mapping. Since the first seminal study on data quality and patient safety (Ives et al. 1993), simultaneous EEG/fMRI has matured in technology and reached commercialization level (Laufs et al. 2008). Simultaneous EEG-fMRI is a powerful approach not only to study the endogenous brain oscillations during various mental states but also to study the neuronal changes in cognitive neuroscience (Debener et al. 2006; Abreu et al. 2018).

Most studies were based on the perspective that EEG and fMRI recordings may obtain the same underlining neural activity, and the simultaneous measurement may reveal more convergence evidence for experimental investigation. However, the

X. Lei (✉)
Sleep and Neuroimaging Centre, Faculty of Psychology, Southwest University, Chongqing, China
e-mail: xlei@swu.edu.cn

L. Hu, Z. Zhang (eds.), *EEG Signal Processing and Feature Extraction*,
https://doi.org/10.1007/978-981-13-9113-2_18

mechanism for the generation of EEG and fMRI signals is still largely unknown. Currently, a widely excepted concept that EEG and fMRI can be fused is that both modalities have the same neurophysiologic substrate. Animal study based on monkeys revealed that BOLD responses correlate stronger with the local field potentials (LFPs), when compared with multiunit activities (Logothetis et al. 2001). Because LFP is link to pyramidal neurons that generate scalp EEG potentials (Nunez 1995), the neurovascular coupling may be still subsist for scalp EEG and fMRI.

In this chapter, we first introduce the hardware of simultaneous EEG-fMRI system. Then a systematically review is given in the advance in signal processing of this technique, including the EEG artifacts correction, the EEG-fMRI data fusion, and the application of EEG-fMRI. Especially, we provide a systematic classification for the fMRI-constrained EEG and the EEG-informed fMRI. This framework is based on the complexity level and may enlighten a more flexible application of EEG-fMRI. Then we provide computer practice for the EEG artifacts correction, which may contribute to the popularization of this new techniques. Finally, we discuss the prospects for future research of simultaneous EEG-fMRI.

18.1 EEG-fMRI System

EEG instrumentation comprises electrodes, an acquisition system to amplify and digitize the EEG signals, and displays facilities for the observation and analysis of the recorded signals (Fig. 18.1). The design of EEG instrumentation has considered a number of safety factors of the MR condition, which is greatly distinct from a conventional EEG system. These factors include the high field static condition, the time-varying magnetic gradient fields, the radio frequency (RF), and the head movement. These considerations result in the unsuitability of traditional EEG monitoring equipment for EEG-fMRI monitoring.

Simultaneous EEG-fMRI system is usually wired as the flow-process diagram that illustrated in Fig. 18.1. The EEG electrode cap connector box is connected to a MR-compatible amplifier by a strip cable. The electric amplifier is powered by a MR-compatible battery or a rechargeable power device. The amplified EEG signal is transmitted through the optical fiber to the optoelectronic converter, which is in the outside of the scanner room. Simultaneously, the photoelectric converter receives the gradient conversion clock information from a synchronous box. The synchronization box is a device which matches the scan time of MR and the analog to digital converter time of EEG. This is very important because it provides a precise time for MR artifact correction. Comparing with no synchronization box, the gradient artifacts are more consistent in each slice duration with the synchronization box. This will facilitate the correction of the gradient noise during the off-line EEG data processing. The synchronization box is connected to the MRI master control unit through a device called synchronous box scanner interface (see Fig. 18.1). Finally, the digitized EEG signal is transmitted to the EEG-recording computer through USB data cable, and the scanned fMRI image outputs independently through the MR system.

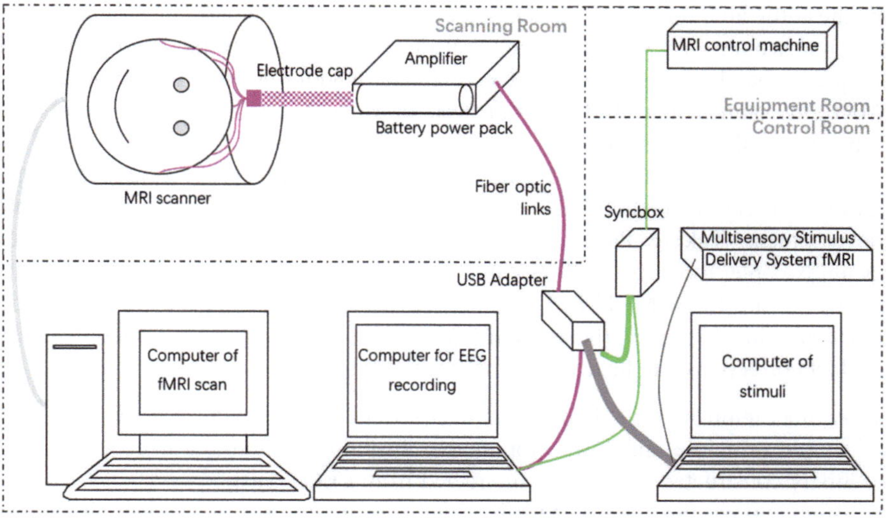

Fig. 18.1 The schematic of the simultaneous EEG-fMRI system. A commercial EEG-fMRI instrumentation has a battery power pack, a connector box containing current-limiting resistors, an electrode cap, and a 32-channel EEG amplifier/digitizer. This instrumentation is sited adjacent to the scanner bore and transmits data to a receiver computer outside with fiber optic links

Base on Faraday law of electromagnetic induction, it is important to minimize the area of any loop formed by the electrode leads to reduce the artifacts induced by the changing magnetic fields. Usually, electrode leads are bunched together closely around the head to minimize the loop circuit. For the distance from the subject's head to the amplifier inputs, the wires are typically twisted together as far as possible. This procedure, on the one hand, keeps the leads in proximity to each other. On another hand, it cancels the induced electromotive forces in adjacent leads. However, there are some loop areas that are inevitable, as scalp EEG is generated by separate points on the head surface. Using the electrode cap, multichannel referential recordings are introduced to further reduce the loop areas (Fig. 18.1).

18.2 Artifact Correction

EEG artifact correction is the largest challenge for simultaneous EEG-fMRI, because this is the first step for the following data fusion. There are three main artifacts in the simultaneously recorded EEG in the MR scanner: (1) electric signal caused by radio frequency pulse. The frequency of this kind of artifact differs greatly from the frequency of EEG signal in 1 MHz range and can be eliminated by low-pass filter. (2) The gradient artifact is caused by the fast switching of the gradient magnetic field. This artifact has a fixed pattern, and its amplitude is typically 100–1000 times of the amplitude of physiology signal. (3) ECG artifacts, which is the artifacts caused by the aortic arch flow impact and heartbeat under the MRI environment. In contrast, the first two are easy to be corrected, but ECG artifact may coincide with the EEG frequency bands and is difficult to be corrected. In addition, the artifacts induced by MRI environment may further from head movement, helium cooling pump, and scanner ventilation system.

In recent years, thanks to the rapid development of signal processing technology, artifact correction methods of simultaneous EEG-fMRI have made considerable development. Taking PubMed database (https://www.ncbi.nlm.nih.gov/pubmed) as an example, more than 170 research papers can be found by searching with the keywords of "EEG-fMRI" and "artifact." Especially after 2008, there were more than ten artifact removal articles published on average in every year. Generally, these artifact corrections can be divided into three categories:

1. Artifact template time-domain subtraction. The technique has the advantages of simple principle and low computation. In these methods, artifact template is firstly constructed and then is subtracted from EEG signal.
2. Blind source separation. Generally, independent component analysis (ICA) and principal component analysis (PCA) are used to separate the EEG signal into different components. Then the artifact components are identified, and finally they are removed.
3. Hardware method. The artifact is measured directly with some professional hardware such as carbon fiber loop and then is subtracted from the contaminated EEG signal.

Since gradient artifacts and ECG artifacts are the main sources of noise for simultaneous EEG signals, we mainly introduce the removal of these two artifacts, as shown in Table 18.1. In present, there are more than ten different methods for artifact correction.

18.2.1 MRI Scanner Artifact

Artifact template subtraction is a widely applied processing method to correct MRI scanner artifact, and this method was proposed by Allen et al. (2000). In this method, the shape of the gradient artifact was assumed to be static in a certain period of time and to be uncorrelated with the physiological signal. By averaging the EEG over a pre-specified number of TR-related epochs, channel-specific artifact templates are estimated and then are subtracted from the EEG signal in the current epoch. There are two type of epochs, one marks the slice acquisition and another marks the volume acquisition. And both are identified by a recording channel from the MRI scanner. Currently, artifact template subtraction is able to implement in real time (Allen et al. 2000). Image acquisition artifact template subtraction has been successfully adopted for the reconstruction of spontaneous EEG signatures such as alpha rhythm (Laufs et al. 2003; Moosmann et al. 2003), high-frequency bursts (Ritter et al. 2008), epileptic activity (Bénar et al. 2003), and evoked potentials in the visual (Becker et al. 2005) and somatosensory system (Schubert et al. 2008).

Table 18.1 EEG artifact correction methods. These methods are designed to correct the MRI scanner artifact and heart pulse artifact. And they are divided into three categories: software, hardware, and other methods

Artifact type	Hardware/ software	Methods	Related toolbox or product	Reference and link
MRI scanner artifact	Software	Artifact template subtraction	BrainVision Analyzer	Allen et al. (2000)
		Optimal basis sets	FMRIB	Niazy et al. (2005)
		Independent component analysis	EEGLAB toolbox	Grouiller et al. (2007)
		Independent vector analysis	—	Acharjee et al. (2015)
	Hardware	Artifact in reference level	New type of electrode cap	Chowdhury et al. (2014)
	Other	Head movement	Down movement to the foot direction with 4 cm	Mullinger et al. (2011)
Heart pulse artifact	Software	Artifact template subtraction	BrainVision Analyzer	Allen et al. (1998)
		Independent component analysis	EEGLAB toolbox	Bénar et al. (2003)
	Hardware	Adaptive filter	Piezoelectric motion sensor	Bonmassar et al. (2002)
		Recursion least square	Carbon-wire loop	Masterton et al. (2007)
		Orthogonal Matching Pursuit	High-density electrode cap	Xia et al. (2014)
	Other	Maximum likelihood algorithm	Simulation pulse artifact based on harmonic theory	Krishnaswamy et al. (2016)

The averaging procedures are a little different in these algorithms, especially the number of TR-related epochs, the weight of each epoch, and the calculation procedure. In the original implementation, the template consisted of a sliding average of artifact epochs to account for the possible changes of the artifact waveform over time. Considering the timing error, adaptive noise cancellation (ANC) is further employed to reduce residual image acquisition artifacts (Allen et al. 2000). A least mean square algorithm is used to adjust the weights of the ANC filter. This approach needed an extremely high sampling frequency; however, some unsatisfactory results were obtained even at sampling rates of 10 kHz (Niazy et al. 2005). Another modified approach for dynamic template estimation was proposed according to a spectrum-based similarity measure (Freyer et al. 2009). In this method, artifact epochs in the template are weighted according to the similarity. This approach allowed the recovery of ultrahigh-frequency EEG signatures with amplitudes in the nanovolt range even during image acquisition periods.

The Vision Analyzer algorithm (V.2.1.2 Brain Products, Munich, Germany) offers multiple methods for artifacts template estimation: (1) all epochs during entire session, (2) a sliding average during a certain period of time, or (3) a predefined number of initial scan epochs combining with some subsequent epochs exceeding a predefined cross-correlation with the initial template. The mark of epochs is identified from a specific scanner-generated signal or from searching for steep gradients in the EEG signal. Notice that the quality of artifact removal by template subtraction depends on the assumption of a stationary artifact, which is best satisfied when using synchronization system (including synchronization box and the scanner interface, see Fig. 18.1). Setting the repetition time (TR) as a multiple of the sampling rate of the EEG recording is a good choice, because the template is consistent sampled in this condition. In contrast, jitter between MR acquisition and EEG sampling leads large residual artifacts, and the dominate frequency is approximately above 50 Hz. However, this effect can be reduced with low-pass filtering (with cutoff frequency around 50 and 80 Hz). Another method, adaptive finite impulse response (FIR) filter, is proposed to identify the artifact frame-by-frame (Wan et al. 2006). This method is an alternative to the artifact template subtraction method, and the core assumption is also that the image acquisition artifacts are temporally stationary.

18.2.2 Pulse Artifact

Pulse artifact correction utilized very similar method as imaging-artifact correction: the average artifact subtraction (AAS) (Allen et al. 1998). The assumption of this algorithm is that the artifact is stable across a small set of successive heartbeats, and the ECG are not correlated with the EEG signals of neuronal activity. The template of pulse artifact is averaged over a set of preceding heartbeats and then subtracted from the ongoing EEG.

The main procedure of the AAS method includes the following steps. First, the AAS approach requires knowledge of the precise onset of each cardiac cycle, which is usually obtained by the simultaneously recorded ECG channel. An R peak detection method is used to initialize the onset of each cardiac cycle. Second, an artifact template was estimated for each EEG channel separately. The template is achieved by averaging the EEG over a predefined epoch preceding and time-locked to the R peak. The resulting artifact template represents the evoked pulse artifact while averaging out the ongoing EEG activity. Third, the artifact template is then subtracted from each EEG epoch, and the remaining signal is the corrected signal. AAS procedure can be implemented in real time and can provide satisfactory EEG data quality. This algorithm has been implemented in commercially available and open-source software packages (e.g., Hamandi et al. 2008).

The limitations of the AAS algorithm result from the deviations of above assumptions: correlation between event potential and cardiac activity, instability of the artifact, and precise detection of each cardiac cycle. The problems inherent in the AAS approach have been investigated further by Niazy and his colleagues (Niazy et al. 2005). In order to relax the stability requirement, they recommended estimating artifact template based on principal components analysis (PCA). They proposed the optimal basis set (OBS) to represent several distinct pulse artifacts. A recent comparison among different methods revealed that OBS and K-means tended to outperform AAS, because of the inability of the latter in modeling short-timescale variability (Jorge et al. 2019).

Other channel-by-channel correction approaches considered the template duration of pulse artifact. Kalman filter approach was utilized in some methods, with some extra motion or electrooculogram (EOG) sensor to be employed as a reference signal (Bonmassar et al. 2002). Wavelet-based methods were another choice; however, its nonlinear version of reduction is computationally demanding and is therefore unlikely to supplant AAS in the near future (Wan et al. 2006). Recently, Vincent and colleagues proposed a moving general linear model (mGLM) approach, along with evidence of improved performance compared to the AAS (Vincent et al. 2007).

18.3 fMRI-Constrained EEG Imaging

At present, the fusion of simultaneous EEG-fMRI can be divided into two categories: symmetric fusion and asymmetric fusion. Symmetric fusion is based on the establishment of a common generative model or the use of interactive information to explain two modalities (Trujillo-Barreto et al. 2001; Valdes-Sosa et al. 2009). Asymmetric fusion mainly uses one of the modality information to guide the analysis of another modality. The most influential asymmetric fusion methods include (1) time prediction, that is, fMRI analysis based on the temporal information of EEG: a specific EEG feature is convoluted with the hemodynamic response function (HRF) to model the fMRI waveform (Debener et al. 2005); (2) spatial constraints, that is, the EEG imaging based on the spatial prior of fMRI: the EEG source reconstruction is constrained by the spatial activity information obtained from the fMRI (Lei et al. 2011b, 2012; Liu et al. 1998).

FMRI-constrained EEG imaging has evolved from being mostly empirical to Bayesian framework. A large number of studies use statistic activity map to construct the EEG source space (Liu et al. 1998) or to initially equivalent current dipoles for further source fittings. Perhaps the latter one is the earliest and simplest method to reconstruct the source of event-related potentials. The estimated time series of each dipole reports the response dynamics at the corresponding fMRI activation region. Although overly simplified, this method is valuable in revealing the temporal sequence of task-evoked neural responses underlying perception or cognition (Auranen et al. 2009; Stancak et al. 2005).

Another way is to use the empirical Bayesian (EB) framework to employ statistic activity map as priors to relax the direct correspondence between EEG and fMRI sources (Phillips et al. 2002). The PEB framework has been proved to be a promising tool for reliable estimation of EEG sources, because various priors can be used for source reconstruction. Based on the EB framework, multiple spatial patterns, derived from functional connectivity analysis of fMRI data, were utilized as priors in EEG source model, and this method is named network-based source imaging (NESOI) (Lei et al. 2011b). Different from the statistic activity map, the patterns generated by ICA are free from any assumption about the task response. In another word, transiently task-related and non-task-related components can facilitate the source imaging.

Here, we emphasize that there are many different types of match between EEG source and fMRI activation region. Hence the choice of the hyperparameter that controls the fMRI bias is critical. In fMRI, a voxel being activated means that the voxel's signal is significantly different from noise and is predictable by the stimuli and tasks of interest. Note that this statistical meaning does not inform any physical characteristic about neuroelectric activity. In respond to this, fMRI spatial information can be utilized as property of source with different strategies (Lei et al. 2012): (*i*) fMRI priors only provide spatial information or quantitative information about magnitude of neural activity; (*ii*) within each fMRI network, the corresponding EEG sources have similar time courses. The resulting covariance components (CC) are different in the EB framework: continuous versus binary and covariance versus variance CC. The influence of different structures of CC is systematically investigated with synthetic data and real data to reveal the effect of structures of CC in each condition. Our results suggest that Bayesian model comparison may be more informative for source imaging.

An interesting topic for fMRI-constrained EEG imaging is the coupling relationship between modalities during resting state (see Fig. 18.2). The term "resting state" mainly refers to the coherent fluctuations of brain oscillations in different brain regions, while the subject is at rest without any stimulus or task. It is feasible to model the size of the BOLD effect that is approximately proportional to the integral of the power of the stimulus-evoked synaptic activity for event-related or block-design paradigms. However, this model fails to account for BOLD signals in spontaneous activity (Makeig et al. 2002). The limitation is worth noting, given that spontaneous activity consumes most energy (Raichle and Mintun 2006), drives BOLD fluctuations (Leopold and Maier 2012), and interacts with tasks or stimuli (He 2013).

18.4 EEG-Informed fMRI Analysis

Compared with fMRI-constrained EEG imaging, using EEG features to inform fMRI analysis is more straightforward. The central idea is to extract features from EEG and then relate them to the voxel-wise fMRI signal. Based on the linear hypothesis of neurovascular coupling, the characteristics of the neural activity extracted in EEG form the predicted functional MRI signals after convoluting the HRF and then find the related active region in the fMRI signal of the whole brain (Lange and Zeger 1997). With this idea, one may generate high-resolution activity maps presumably underlying the EEG features of interest (Murta et al. 2015) or

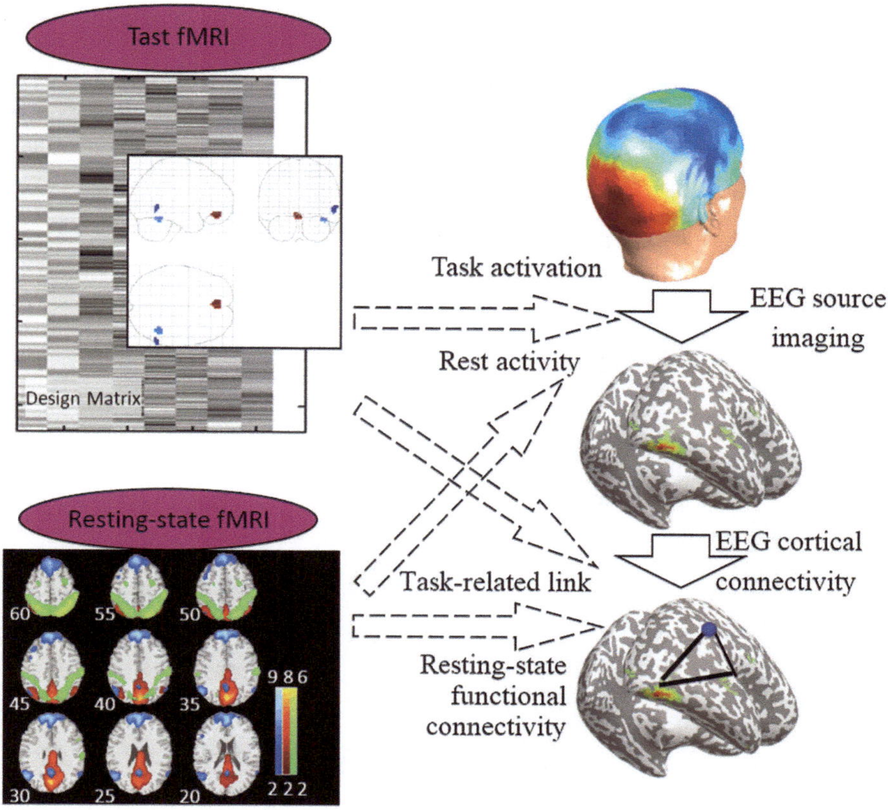

Fig. 18.2 fMRI-constrained EEG imaging employs the spatial priors derived from task-evoked or resting-state fMRI. Effective and functional connectivity may benefit the derivation of EEG cortical connectivity

address the EEG correlates to fMRI activity at specific voxels, regions, or networks (Mantini et al. 2007). A summary of previous studies, as shown in Table 18.2, is based on the purpose of fusion of fMRI data based on the different indicators of EEG signals. This classification is based on the number of electrodes, head table coverage, time precision, difficulty, and fMRI analysis based on EEG information.

18.4.1 Classify Brain State

EEG signals present distinct pattern from wake to deep sleep. These EEG features can be used to separate the different brain states, and then the corresponding fMRI was analyzed to reveal the high spatial brain functional dynamics. Using EEG or its derived polysomnography (PSG), sleep can be further divided into non-rapid eye movement (NREM) and rapid eye movement sleep (REM). Each sleep period is

Table 18.2 EEG-informed fMRI analysis from simple section classification to complex ERPs-fMRI

Fusion level	Number of electrode	Coverage of scalp surface	Time resolution	Cole scientific problem	Complexity	Purpose	References
Classify brain state	<10	Partial	~10 s	fMRI analysis in special EEG-defined brain state	Very simple	Classification of the wake, sleep, or vigilant state	Kaufmann et al. (2006)
							Horovitz et al. (2009)
							Lei et al. (2015)
							Uehara et al. (2014)
Identify electrophysiological event	<30	Partial/ whole brain	~1 s	Identify the onset time of spontaneous electrophysiology event; construct the design matrix	Simple	Interictal epileptiform discharges, spindle, K-complex, EEG microstate	Gotman et al. (2004)
							Schabus et al. (2007)
							Tamminen et al. (2010)
							Jahnke et al. (2012)

Obtain the power of rhythm	30–60	Partial/ whole brain	~1 s	Obtain the power wave of specific rhythm; construct regression term	Complex	α-rhythm; γ-rhythm	Goldman et al. (2002)
							Mantini et al. (2007)
							Jann et al. (2009)
							Lei et al. 2014
ERPs-fMRI	>64	Whole brain	~1 ms	Extract the amplitude or power of specific ERPs component	Very complex	P300 localization N170 localization	Eichele et al. (2005)
							Nguyen et al. (2014)
							Bénar et al. (2007)
							Goldman et al. (2009)

mainly determined according to its characteristic waveform, ocular electricity, and jaw electromyographic activity. EEG-based sleep staging can be used to label fMRI section and then used to compare the fMRI activation or functional connectivity at different stages of sleep.

Kaufmann and his colleagues explored the changes of local activation of BOLD signals from awake to NREM sleep (Kaufmann et al. 2006). The results showed that the overall activity during NREM sleep stages was declined in many brain regions, including the cerebral cortex, the thalamus, the limbic lobe and the caudate nuclei, compared to the wake period. The decreased brain regions during N1 stage include the frontal lobe, limbic lobe, occipital lobe, and insula, and the N2 stage was further reduced in the frontal, marginal, temporal, and right insula. The activity of N3 phase was reduced to the lowest level, involving the frontal lobe, limbic lobe, temporal lobe, occipital lobe, and insula. This indicates that with the gradual deepening of sleep, brain areas with decreased activity gradually increase, which is conducive to the formation and maintenance of deep sleep (Kaufmann et al. 2006). Horovitz et al. further explored the spatial distribution of the default mode network (DMN) in the slow-wave sleep stage and found the phenomenon of the loss of the prefrontal region in DMN during the slow-wave sleep (Horovitz et al. 2009). Their study indicates that DMN plays an important role in the awareness maintenance. In addition to the above analysis of fMRI data from the perspective of activation and large-scale brain networks, Lei et al. analyzed the brain scale-free characteristics in the wake and NREM stages (Lei et al. 2015). Using "graph theory," Uehara et al. explored the "small-world network" attribute (Uehara et al. 2014) of the wake and the N1 stages. Because the frequency range of MRI's RF pulse and gradient artifacts and the frequency range of electromyography and eye power overlap greatly, it is very difficult to distinguish the REM phase at present study. It is still a challenge to explore the activity characteristic of fMRI related to REM.

18.4.2 Identify Electrophysiological Event

In this application, EEG is used to characterize the changes in endogenous alertness and identify spontaneous electrical activities or even, such as epileptic discharge, spindles, and K-complex waves. The onset time of these events is further inputted as regressor to identify fMRI correlative. Simultaneous EEG-fMRI was first used to explore the changes in functional MRI signals related to interictal epileptiform discharges (IEDs) and further to determine the origin and conduction process of epileptic waves (Gotman et al. 2004). The occurrence of IEDs is random; hence the onset time of spontaneous EEG event is crucial for general linear model of fMRI analysis. In sleep neuroimaging, similar idea is applied to the localization spindle waves. Sleep spindles are mainly involved in sleep-dependent memory consolidation and the integration of new and old knowledge (Tamminen et al. 2010). Using the simultaneous EEG-fMRI to obtain the sleep data of the subjects and to denoise the EEG data described above, the complete EEG data need to be staged from the N1 phase of the NREM to the N3 phase and the REM phase, and then the spindle wave is identified from the N2 period, and finally the fMRI related to the spindle wave activity was found (Schabus et al. 2007).

18.4.3 Obtain the Power Wave of Rhythm

The EEG signal contains rich time and spectrum features. The time-frequency decomposition of EEG data can separate typical bands or rhythms, including delta (2–4 Hz), theta (4–8 Hz), alpha (8–13 Hz), beta (13–30 Hz), and gamma (30–40 Hz). These band powers can be used to explore the fMRI activity of the frequency spectrum energy covariance. First, we need to calculate the average power time series of each frequency band on each electrode of EEG, and then convolute these time series with the HRF, and downsample the time series after convolution and standardize the time series. Finally, the average power time series of each frequency band after convolution is analyzed. Regression analysis was performed with each voxel fMRI signal to explore the corresponding brain activity in each frequency band. The researchers firstly focused on alpha rhythm. Goldman et al. identified the core brain regions of alpha rhythm by simultaneous EEG-fMRI (Goldman et al. 2002). The results showed that the enhancement of alpha rhythm was associated with the negative activation of the occipital lobe, the dorsal frontal gyrus, the ventral temporal and cingulate gyrus, and the positive activation in the insula and thalamus. By taking the EEG power of multiple frequency bands as the regression term in the generalized linear model of fMRI data, the contribution of these bands to the functional MRI signals and the interaction between the signals are also discussed (de Munck et al. 2009). In addition to the EEG power of a specific frequency band, the other features of the spectrum can also be used to explain the fMRI activity, such as the total power (Wan et al. 2006), the linear combination of specific frequencies (Goense and Logothetis 2008), the average frequency (Rosa et al. 2010), the root mean square frequency (Jann et al. 2009), and so on.

Exploring the relationship between EEG rhythm and fMRI large-scale brain network is currently one of the hot topics. Jann et al. measured the phase-locking activity of all the electrodes in the specific band of the whole brain based on global field synchronization (GFS) and then discussed the relationship between the alpha rhythm GFS and the resting-state brain network. The results show that the alpha rhythmic phase-locking component is related to the activity of the DMN (Jann et al. 2009). Mantini further extended five typical rhythms (delta, theta, alpha, beta, and gamma) and large-scale resting networks (Mantini et al. 2007). They found that a resting-state brain network was related to multiple rhythms. For example, DMN is related to alpha and beta rhythms, and the visual network is related to delta, theta, alpha, and beta rhythms. These results indicated the common effects of nerve oscillations in different frequency bands, in the same functional system. In recent years, a large number of studies have begun to pay attention to the statistical correlation between EEG rhythm and functional connectivity in long-distance resting-state networks. The effects of alcohol intake on EEG rhythm and resting-state brain network function connection were studied by simultaneous EEG-fMRI (Lei et al. 2014). The results revealed that drinking would increase the energy of theta rhythms and be associated with the correlation of the default mode network with dorsal attention network, suggesting that the slow-wave rhythm of the brain may reflect the interaction of resting-state networks. The underlining electrophysiological mechanisms for functional connectivity of fMRI need further be investigated (Scholvinck et al. 2013).

18.4.4 ERPs-fMRI

Event-related potentials (ERPs) is the evoked potential of the human brain in processing cognitive stimuli. The low SNR is the main challenge for single-trial analysis. For some weak brain waves, especially the early ERPs, it is very difficult to extract stable amplitude information in single trials. At present, with the help of feature extraction, pattern recognition, and machine learning technology, there has been considerable progress in weak signal extraction. For example, the ICA method can be used to extract the weak event-related potential measured on multiple electrodes into a clear independent component. Based on the sparse representation and the external input autoregressive model algorithm, the latent period and the amplitude can be extracted under the condition of low signal-to-noise ratio. After further integration of trial-and-error information, a single-trial amplitude diagram can be formed.

By observing the time delay information of amplitude map potential intensity, ERPs related to cognitive processing, such as N170, P300, were found. Some features extracted from the single test EEG data, such as the amplitude, incubation period, and spatial topological distribution of ERP components, can be used to further explore the relationship between electrophysiological characteristics and functional magnetic resonance signal activity (Eichele et al. 2005; Nguyen et al. 2014; Bénar et al. 2007). Furthermore, the feature identified by machine learning with task resolution can be used as a quantitative index for a single trial (Goldman et al. 2009; Lei et al. 2009). In order to align the time of occurrence of stimulation from EEG and sequence scanning time of fMRI, the "single experiment quantization curve" can be obtained according to the amplitude of the single experiment, and the design matrix is constructed, and the statistical parameters are generated.

18.5 Multimodal Brain Network

EEG and fMRI measure the functional activity of multiple networks distributed in the whole cerebral cortex. Estimating network communication from both the neuroelectric and hemodynamic signals may help to explain the complex

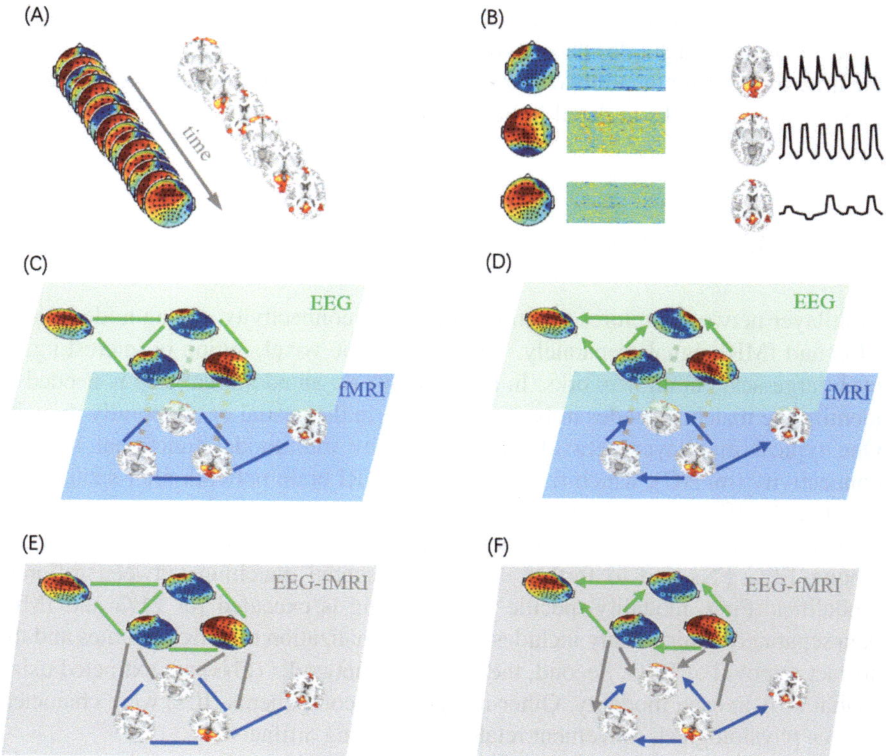

Fig. 18.3 Illustration of the multimodal brain network. (**a**) EEG and fMRI original signals are preprocessed and separated to (**b**) independent components. These components represent nodes in our following network construction, and their time courses are utilized to inference the links (edges) between brain networks. (**c**) Functional and (**d**) causality connectivity with the framework of multilayer network or (**e**) functional and (**f**) causality network with the framework of hybrid network. The green and blue links represent the edges estimated within the modality of EEG and fMRI, respectively. The gray solid lines (or arrows) and dotted lines represent the identified links and matching nodes between modalities, respectively

relationships between brain regions. Currently, multimodal brain network can be mainly subdivided into functional and effective connectivity (Friston and Price 2001). Functional connectivity demonstrates how different areas of the brain with similar patterns of activation enable brain functions at rest and in response to external stimuli. On the other hand, effective connectivity is to infer a causal relationship between functionally linked activated areas and how they can be related through structural connections depicted independently (see Fig. 18.3).

When considering the fusion of the networks reconstructed from EEG and fMRI modalities, two frameworks were considered in the network space. First is the multilayer network. In this framework, EEG and fMRI construct modal-specific network separately, and then both networks were matched to share some common node in the spatial or temporally domain (Lei et al. 2011a). Second is an alternative framework which is the hybrid network analysis. In this framework, EEG and fMRI were collapsed to a single space, and the hybrid network was constructed based on the time course with the unified temporal scale (Yu et al. 2016). Here we introduced those two frameworks in detail in the following sections.

18.5.1 Multilayer Network

Multilayer network explores functional or causal connectivity among nodes within EEG and fMRI signals separately. The node can be voxel, region of interesting, or even large-scale brain networks. In this framework, an additional step is needed to identify the matching nodes in both modalities, in the spatial or temporally domain. One typical multilayer network is estimated by multimodal functional network connectivity (mFNC), which fused EEG and fMRI brain network after spatial ICA (see Fig. 18.3D) (Lei et al. 2011a).

Multimodal FNC is a natural extension of the fMRI FNC to cover the interaction among EEG FNs and to further explore the spatial matching between different modalities. First, modality-specific preprocessing is executed for EEG and fMRI data separately. This mainly includes spatial normalization of fMRI volumes and the artifact removal of EEG. Second, the functional networks (FNs) are extracted using spatial ICA in each modality. Other independent components (ICs) were characterized as physiological, movement related, or imaging artifacts.

Third, the interactions among FNs in each modality are explored by Granger causality analysis (GCA). The time courses of EEG (or fMRI) components are employed to explore the networks' interactions. Finally, fMRI FNs are matched to EEG FNs in the spatial domain using network-based source imaging (NESOI). This step stitches EEG-FNC and fMRI-FNC to construct multimodal FNC, which is a node with multilayer information. Each node of mFNC has a "common substrate" of neuronal activity, though the link in each modality may be district different.

The simulation demonstrated the potential of mFNC to reveal the correct networks when EEG and fMRI have the same spatial location of neural sources. Further analysis of visual experiment shown that mFNC has the potential to reveal the information flow during simple visual processing. With the utilization of mFNC, comprehensive relationships among FNs might be unveiled for the deep exploration of neural activities and metabolic responses in a specific task or neurological state.

18.5.2 Hybrid Network

Hybrid network analysis collapse both EEG and fMRI modalities to a single space, and the hybrid network was constructed based on the time course with the unified temporal scale (Yu et al. 2016). EEG signal is usually down sampled to the same time resolution of fMRI. For example, EEG time series are segmented into epochs that correspond to one concurrently recorded fMRI volume. Then each epoch was converted to the frequency domain by the fast Fourier transform. The spectral power was averaged within typical frequency bands, which include delta, theta, alpha, beta, and gamma.

Yu and his colleague decomposed fMRI data into FNs with associated time courses by group ICA (Yu et al. 2016). Both EEG electrodes and fMRI FNs served as nodes, and EEG-fMRI brain graphs are built by computing Pearson correlation coefficients within and between fMRI FNs time courses and EEG spectral power time courses. Similar to mFNC, EEG power time courses were convolved with a canonical HRF to account for the delayed hemodynamic response.

This method can be further extended to dynamic EEG-fMRI graphs, which are built using a sliding window method. A window width of 40 s was used because a previous study indicates that cognitive states may be correctly identified with as little as 30–60 s of data (Shirer et al. 2012). Shorter time windows may result in a lower number of statistically significant correlations in brain connectivity and greater variability of correlation values (Allen et al. 2014; Yu et al. 2015), whereas a window size of about 40 s provides a good trade-off between the ability to resolve dynamics and the quality of connectivity estimation (Allen et al. 2014; Yu et al. 2015).

This method provides a new approach to examine EEG-fMRI associations within a graph theoretic framework. Previous investigation found this method incorporate fMRI spatial localization and EEG frequency information which could not be obtained by examining only one modality (Yu et al. 2016).

18.6 Example Application

The FMRIB (FMRI Artifact Slice Template Removal) is an EEGLAB plug-in toolbox (http://users.fmrib.ox.ac.uk/~rami/fmribplugin). It is a toolbox developed by the University of Oxford Centre for Functional MRI of the Brain, and this toolbox allow the removal of MRI artifacts from EEG signal. This toolbox is implemented within the EEGLAB environment, providing an interface to remove gradient artifacts, detect R peak, and remove pulse artifacts from the EEG. All of the algorithms can also be utilized as MATLAB command line. The following section provides a very short introduction to use GUI to remove artifacts from EEG data. I assumed the reader know the basics of EEGLAB, e.g., loading a dataset, viewing, events, data structures, etc. If not, refer to the EEGLAB website for more basics about EEGLAB.

The methods used in FMRIB are based on the experimentations of simultaneous EEG-fMRI (Niazy et al. 2005). It requires that a volume trigger event (or alternatively a slice timing event) is present in the data, i.e., an event for each fMRI volume acquired. The sampling rate of the data should be sufficient as to not have any aliasing from high-frequency gradient noise. Our experiment data were collected at 5000 Hz, which is much larger than the maximum gradient artifact frequency in a typical EPI sequence, about 700–800 Hz.

18.6.1 Experiment

One healthy subject (female, age 22) participated in the simultaneous EEG-fMRI study, which was approved by the Ethics Committee of the Southwest University. She was without any history of psychiatric or neurological illness as confirmed by psychiatric clinical assessment. Written informed consent was obtained after detailed explanation of the study protocol. The study contained an eyes-open (EO) and an eyes-closed (EC) one-minute resting state. In the EO condition, the subject was required to concentrate on the "+" in the middle of the screen and then the sign term to "•" in the middle of the screen; the subject needed to close her eyes. While in the EC condition, she was informed that they needed to calm their mind into a relaxed state and avoid consciously thinking about something. All procedures were in accordance with the sixth revision of the Declaration of Helsinki.

The sample rate is 5 kHz, and the FCz is the reference in a nonmagnetic MRI-compatible EEG system (BrainAmp MR plus, Brain Products, Munich, Germany). All 9 Ag/AgCl electrodes were ring-type sintered nonmagnetic electrodes, and they were placed on the scalp according to the international 10/20 system. An additional electrode was dedicated to the electrocardiogram (ECG). Eighty fMRI functional volumes were acquired using a 3T Siemens Trio scanner. These data used an EPI sequence with the following parameters: TR/TE of 1500/29 ms, FOV of 192×192 mm^2, flip angle of 90°, acquisition matrix of 64×64, thickness/gap of 5/0.5 mm, in-plane resolution of 3.0×3.0 mm^2, and 25 axial slices.

Download the compressed FMRIB plug-in file into the "plugins" directory of EEGLAB distribution (https://fsl.fmrib.ox.ac.uk/eeglab/fmribplugin/fmrib1.21.zip). Then download the compressed example dataset (12MB, http,//www.leixulab.net/data.asp) and uncompressed it. Here, we loaded the example data set (EOEC.set). As illustrated in Fig. 18.4, there is an event "R128" indicating fMRI volume timings for the removal of gradient artifacts, and this event occurs every time a volume is acquired. Total fMRI experiment time was 2 minutes. There were 80 fMRI volumes collected, i.e., 80 "R128" markers. An ECG channel is also included to remove the ballistocardiographic artifacts. View the data and scroll to the beginning of the gradient artifacts. You can see that the data is contaminated with gradient artifacts and other noise.

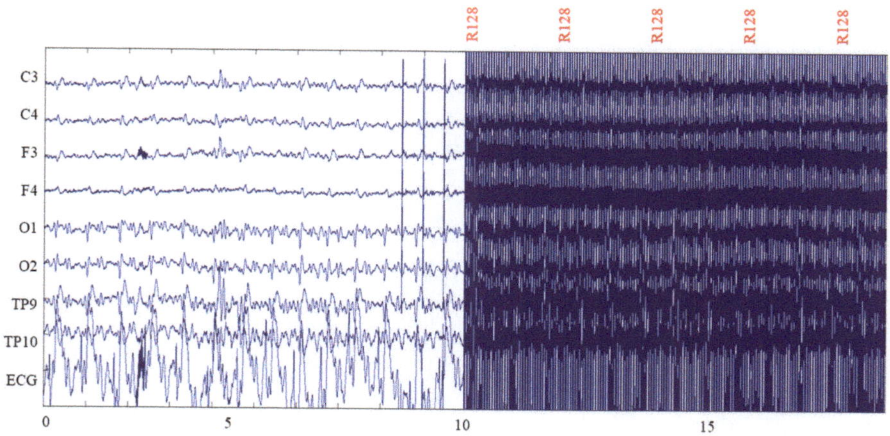

Fig. 18.4 The loaded data contains an eyes-open and an eyes-closed one-minute resting state. Obviously, the signal is contaminated with gradient artifacts and other noise. "R128" is fMRI volume marker which is important for gradient artifacts correction

18.6.2 Remove fMRI Gradient Artifacts

In the EEGLAB GUI, go to Tools > FMRIB Tools > FASTR: Remove FMRI gradient artifacts. The interface illustrated in Fig. 18.5 will now show up. Enter the values as in the figure below, and press OK. Here, because our data was sampled with high resolution of 5000 Hz, we do not upsample the data, and set "1" in the second text field. The MATLAB window will now show the selected settings and show the progress of the process. When FASTR is finished, you will be asked if you want to save the data; keep the default setting here. The data is downsample to 250 Hz to save memory. Go to Tools > Change sampling rate, and then enter 250 as our new sampling rate. Then click OK. Store the date in the current file with the name "ECEOcMRI250.set." Now you can view the cleaned data. As illustrated in Fig. 18.6, the red lines are the volume triggers indicating the start of the fMRI acquisition. You can also notice the high amplitude ECG signal, which contains the pulse artifacts.

The process contains several steps. FASTR can automatically align the volume artifacts to correct for any jitter in the exact time of the onset of volume. This operation is done on the first EEG channel, and the adjustments are then applied to all the channels. This may useful if there is not synchronization system between EEG and fMRI. Notice our data skipped this step. The main step is computing an average template for the artifact then subtracted. FASTR does this by taking a moving-

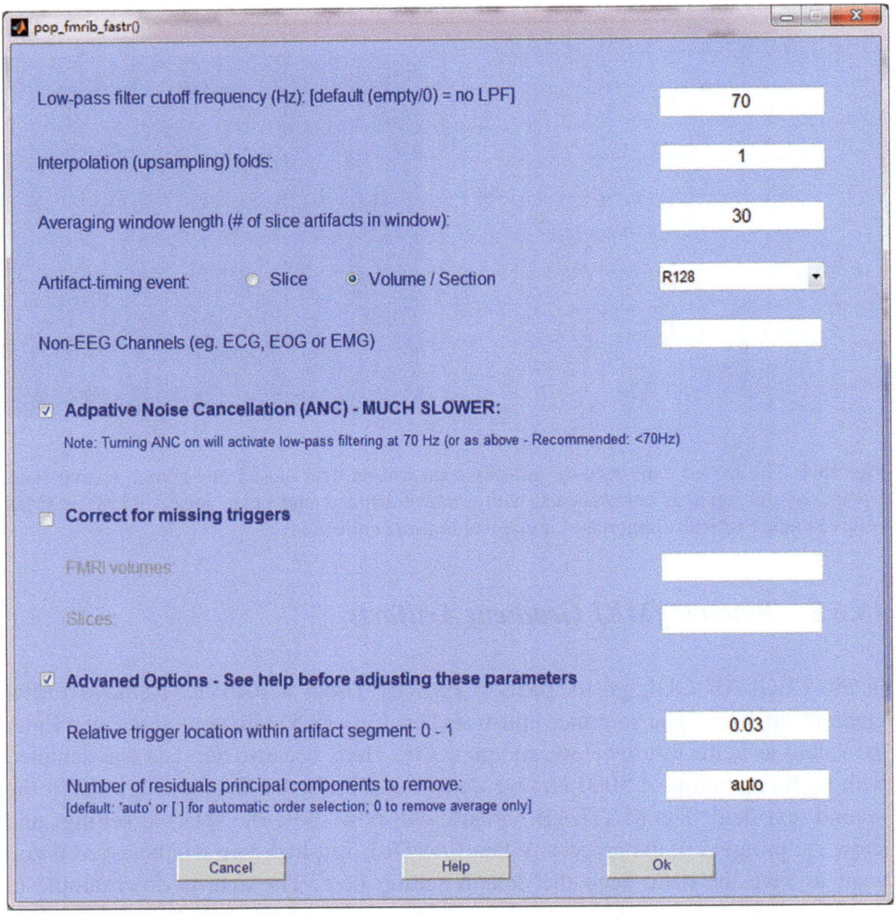

Fig. 18.5 Parameter in the main GUI of FASTR

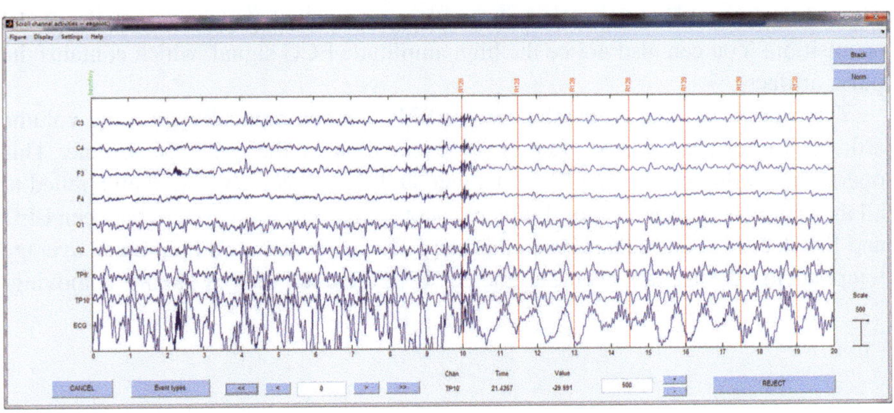

Fig. 18.6 EEG signal after fMRI gradient artifacts correction

window average of slice artifacts and then subtracting the average template from the center contaminated data. This is implemented with OBS algorithm. Then the data are low-pass filtered. Finally, adaptive noise cancellation (ANC) is used to remove any remaining residuals.

18.6.3 Detect QRS Events

FMRIB plug-in can detect heartbeats robustly. Its algorithm aligns all events and corrects for false positives and negatives. The QRS/heartbeat detection tool has an average sensitivity and specificity of 99%. Now we detected the QRS complexes. Go to Tools > FMRIB Tools > Detect QRS events. Then an interface was shown and we enter the following parameters. The ECG channel number is 9, and the default value of "qrs" was used to name the QRS events, and select "Yes" to delete the ECG channel when finished. The QRS events are now stored as type "qrs"; see Fig. 18.7 top panel for more details.

18.6.4 Remove Pulse Artifacts

We can now proceed with removing the pulse artifacts. Click on Tools > FMRIB Tools > Remove pulse artifacts. In the interface, enter the parameters with Heart Beat Event = qrt, Optimal Basis Set with Number of PCs =4. When the process is done, store the data as a new dataset, and call it "EOEC – CLEAN." Now we can scroll through the data to see the cleaned data (Fig. 18.7 bottom panel).

18.6.5 Batch Code for Remove Artifacts

As illustrated in Code 18.1, there are at least four steps to remove artifacts.

The functions pop_fmrib_fastr, pop_fmrib_qrsdetect and pop_fmrib_pas were used to remove fMRI gradient artifacts, detect QRS events, and remove pulse artifacts, respectively.

The batch analysis in Code 18.1 has the same output as our prior setting in GUI. Here, you can find that removement fMRI gradient artifacts and pulse artifacts have several options.

(A)

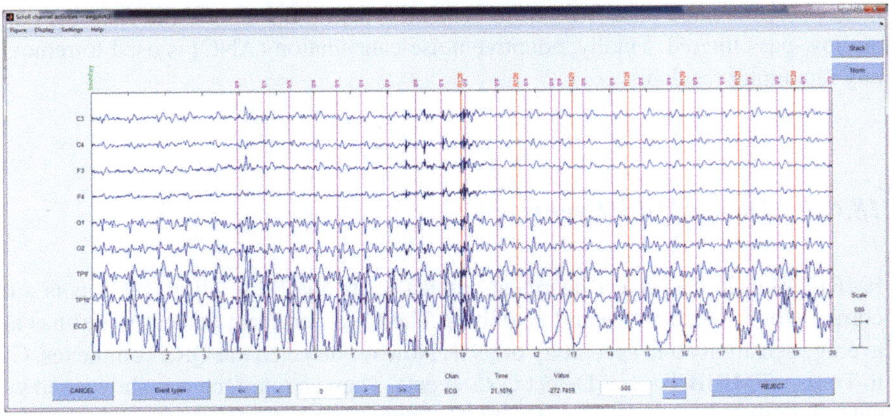

(B)

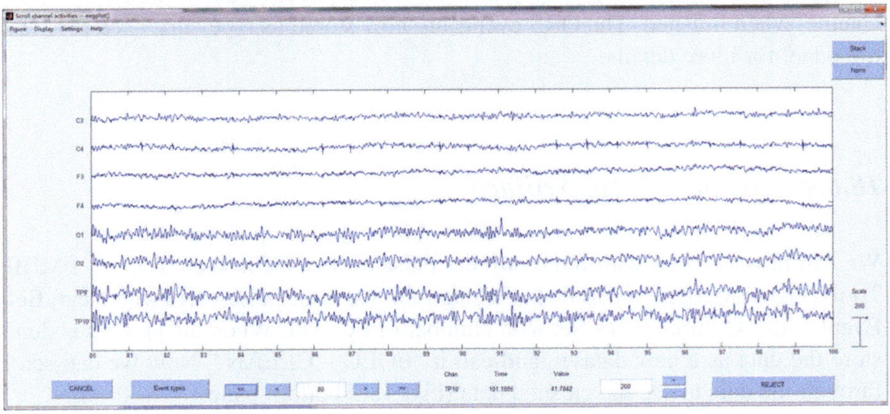

Fig. 18.7 The main results of FASTR during pulse artifacts correlation

```
% % FMRIB to remove fMRI gradient and pulse artifacts
EEG = pop_loadset('filename','EOEC.set','filepath','');% Load data
% % Remove fMRI gradient artefacts
EEG = pop_fmrib_fastr(EEG,70,1,10,'R128',0,0,0,0,0,0,[4],3);
EEG = pop_resample(EEG, 250);% Down sample to 250 Hz
EEG = pop_fmrib_qrsdetect(EEG,4,'R','no');% Detect QRS events
EEG = pop_fmrib_pas(EEG,'R','obs',3);% Remove pulse artifacts
% % Save data as EOEC - CLEAN.set
EEG = pop_saveset( EEG, 'filename','EOEC - CLEAN.set','filepath','');
```

Code 18.1 Batch analysis code for remove artifacts

18.7 Conclusion

Both spatial and temporal information are very important for understanding the basic principle of brain activity and to explore the neural and pathological mechanism of mental diseases. Simultaneous EEG-fMRI has broad prospects for application and has developed rapidly in the past decade. The number of papers continues to grow, and the scope of application continues to expand.

18.7.1 Future Direction

Although the advantages of simultaneous EEG-fMRI are obvious, there are many technical difficulties in the combination of them. Because the data is multimodal and collected at the same time, the signal will interfere with each other, and some artifacts can hardly be removed. In the algorithm of artifact removal, the average artifact subtraction has achieved some success (Green et al. 2017; Steyrl et al. 2017), and the ICA based on blind source analysis has a wide application (Wang et al. 2018). However, how to further combine the pattern recognition, depth learning, and some other brand-new methods, to develop a better artifact removal method? It is still a difficult problem and need to be solved. In terms of hardware, there is no mature artifact removal hardware system, and many parameters, such as the number and distribution of carbon-wire loops, are still uncertain. Future research needs to integrate both software and hardware and their interaction to further improve the artifact removal effect.

With the improvement of hardware facilities and the innovation of software methods, simultaneous EEG-fMRI is becoming a powerful tool and is enriching the researchers' tool kits (Table 18.3). Some representative research directions in the future are:

1. Simultaneous EEG-fMRI combined with brain stimulation technology. Transcranial magnetic stimulation (TMS) and transcranial direct current stimulation (TDCS) are the most widely used noninvasive brain stimulation techniques. It can stimulate the excitability of the cerebral cortex by stimulating the specific parts of the brain and affect the metabolism of the brain and the electrical activity of the brain. The combination of synchronous EEG-fMRI and transcranial electrical/magnetic stimulation can provide electrophysiological and metabolic evidence for the intervention effect. Bharath and so on used the synchronous EEG-fMRI technique to study the changes in the brain function connection of patients with writing spasm by repetitive TMS. The results showed that after the stimulation, the two modes of EEG and fMRI showed the clustering coefficient and the enhanced (Bharath et al. 2017) of the small-world network properties (Ladenbauer et al.). Transcranial electrical stimulation was used to stimulate 0.75Hz in mild cognitive impairment patients during N2 sleep. They found that electrical stimulation promoted the production of slow waves and spindle waves, thereby enhancing and consolidating memory (Ladenbauer et al. 2017).

Table 18.3 The MATLAB software for EEG-fMRI fusion

Toolbox	Application in EEG-fMRI fusion	Other functions	References and Links
SPM	Preprocess of EEG and fMRI data; EEG-informed fMRI analysis (general linear model); fMRI-constraints EEG imaging	Group inference of EEG and fMRI data; general linear model; dynamic causal model	Friston et al. (1994) www.fil.ion.ucl.ac.uk/spm
GIFT EEGIFT FIT	Fusion of EEG-fMRI in feature level; group ICA of EEG and fMRI; joint ICA and parallel ICA	Fusion of fMRI and DTI, fMRI and gene data; visualization, sorting, and average of ICs	Calhoun et al. (2001) http://icatb.sourceforge.net
EEGLAB FMRILAB	Artifact correction; independent component analysis	EEG time-frequency decomposition; single trail analysis	Delorme and Makeig (2004) http://sccn.ucsd.edu/eeglab
fMRIB	Artifact correction		http://users.fmrib.ox.ac.uk/~rami/fmribplugin
BERGEN			http://fmri.uib.no/tools/bergen_plugin.htm
STEFF	Simulation fusion in EEG and fMRI		Lei et al. (2010) http://www.leixulab.net/software.asp
NIT	Artifact correction; EEG-informed fMRI analysis; fMRI-constraints EEG imaging	The fMRI module has fMRI specific preprocessing; the EEG module has EEG specific preprocessing	Dong et al. (2018) http://www.neuro.uestc.edu.cn/NIT.html

2. Real-time artifact removal. At present, most simultaneous EEG-fMRI studies use offline processing technology to explore the related neural mechanisms, but many studies need to remove the nuclear magnetic artifacts online. For example, the synchronous EEG-fMRI technique is used to explore the memory reactivation process in the sleep stage, which requires the activation of memory during the slow-wave sleep through sound, odor, and other stimuli. This requires a clear slow wave can be seen in the synchronous acquisition to read the sleep phase. Real-time artifact removal techniques such as reference layer adaptive filtering (Steyrl et al. 2017) and carbon fiber cable sheath artifact correction (van der Meer et al. 2016) lay the foundation for these studies.

3. Multimodal neural feedback. Neural feedback based on EEG signals has a history of several decades and has accumulated a lot of experience, but the brain mechanism related to the effect of intervention is still unclear. On the basis of real-time noise removal, Zich used motion imagination to train stroke patients through EEG neural feedback, and collect fMRI synchronously, and explore the relationship between neural feedback based on motion imagination and activation of sensory motor cortex (Zich et al. 2015). A closed loop based on two modes of man-machine interaction and the evaluation of the effect of intervention with fMRI can help to promote the rehabilitation training and effect tracking of the related diseases.

18.7.2 Scientific Questions Answered by EEG/fMRI Fusion

FMRI is sufficient for questions of where, and EEG is sufficient for questions of when in some cognitive neuroscience researches (Friston 2009). Then, what is the sort of scientific question which really requires EEG/fMRI fusion? This question is particularly important for the extensive application of simultaneous EEG-fMRI.

At present, simultaneous EEG-fMRI is widely used in the fields of epilepsy and sleep. This is because in the epileptic study, simultaneous EEG-fMRI can be a noninvasive, precise location of the epileptic focus. In the sleep study, sleep stages must rely on the synchronous acquisition of EEG signals, which makes the simultaneous EEG-fMRI technology irreplaceable in these two fields. In the future research, researchers should give full play to the advantages of simultaneous EEG-fMRI, select the appropriate analytical methods, and solve the scientific problems of concern.

In initial, people used simultaneous EEG-fMRI technique to reveal the areas of the brain with changed BOLD signal in response to epileptic spikes detected in the EEG signal. Established diagnostic application in the last 10 years improves our understanding of the spatiotemporal characteristics of epileptic networks (Vulliemoz et al. 2010). Another promising application is the study of rest-state network. Researches both in EEG (Chen et al. 2008; Laufs et al. 2003) and fMRI (Mantini et al. 2007) have illustrated that these networks are pervasive in resting state and during task performance and hence provide robust measures of the interacted brain activity.

References

Abreu R, Leal A, Figueiredo P. EEG-informed fMRI: a review of data analysis methods. Front Hum Neurosci. 2018;12:29.

Acharjee PP, Phlypo R, Wu L, Calhoun VD, Adali T. Independent vector analysis for gradient artifact removal in concurrent EEG-fMRI data. IEEE Trans Biomed Eng. 2015;62:1750–8.

Allen PJ, Polizzi G, Krakow K, Fish DR, Lemieux L. Identification of EEG events in the MR scanner: the problem of pulse artifact and a method for its subtraction. Neuroimage. 1998;8:229–39.

Allen PJ, Josephs O, Turner R. A method for removing imaging artifact from continuous EEG recorded during functional MRI. Neuroimage. 2000;12:230–9.

Allen EA, Damaraju E, Plis SM, Erhardt EB, Eichele T, Calhoun VD. Tracking whole-brain connectivity dynamics in the resting state. Cerebral Cortex. 2014;24:663–76.

Auranen T, Nummenmaa A, Vanni S, Vehtari A, Hamalainen MS, Lampinen J, Jaaskelainen IP. Automatic fMRI-guided MEG multidipole localization for visual responses. Hum Brain Mapp. 2009;30:1087–99.

Becker R, Ritter P, Moosmann M, Villringer A. Visual evoked potentials recovered from fMRI scan periods. Hum Brain Mapp. 2005;26:221–30.

Bénar C, Aghakhani Y, Wang Y, Izenberg A, Al-Asmi A, Dubeau F, Gotman J. Quality of EEG in simultaneous EEG-fMRI for epilepsy. Clin Neurophysiol. 2003;114:569–80.

Bénar CG, Schon D, Grimault S, Nazarian B, Burle B, Roth M, Badier JM, Marquis P, Liegeois-Chauvel C, Anton JL. Single-trial analysis of oddball event-related potentials in simultaneous EEG-fMRI. Hum Brain Mapp. 2007;28:602–13.

Bharath RD, Panda R, Reddam VR, Bhaskar MV, Gohel S, Bhardwaj S, Prajapati A, Pal PK. A single session of rTMS enhances small-worldness in writer's cramp: evidence from simultaneous EEG-fMRI multi-modal brain graph. Front Hum Neurosci. 2017;11:443.

Bonmassar G, Purdon PL, Jaaskelainen IP, Chiappa K, Solo V, Brown EN, Belliveau JW. Motion and ballistocardiogram artifact removal for interleaved recording of EEG and EPs during MRI. Neuroimage. 2002;16:1127–41.

Calhoun VD, Adali T, Pearlson GD, Pekar JJ. A method for making group inferences from functional MRI data using independent component analysis. Hum Brain Mapp. 2001;14:140–51.

Chen AC, Feng W, Zhao H, Yin Y, Wang P. EEG default mode network in the human brain: spectral regional field powers. Neuroimage. 2008;41:561–74.

Chowdhury ME, Mullinger KJ, Glover P, Bowtell R. Reference layer artefact subtraction (RLAS): a novel method of minimizing EEG artefacts during simultaneous fMRI. Neuroimage. 2014;84:307–19.

de Munck JC, Goncalves SI, Mammoliti R, Heethaar RM, Lopes da Silva FH. Interactions between different EEG frequency bands and their effect on alpha-fMRI correlations. Neuroimage. 2009;47:69–76.

Debener S, Ullsperger M, Siegel M, Fiehler K, von Cramon DY, Engel AK. Trial-by-trial coupling of concurrent electroencephalogram and functional magnetic resonance imaging identifies the dynamics of performance monitoring. J Neurosci. 2005;25:11730–7.

Debener S, Ullsperger M, Siegel M, Engel AK. Single-trial EEG-fMRI reveals the dynamics of cognitive function. Trends Cognit Sci. 2006;10:558–63.

Delorme A, Makeig S. EEGLAB: an open source toolbox for analysis of single-trial EEG dynamics including independent component analysis. J Neurosci Methods. 2004;134:9–21.

Dong L, Luo C, Xiaobo L, Sisi J, Fali L, Hongshuo F, Jianfu L, Diankun G, Dezhong Y. Neuroscience information toolbox: an open source toolbox for EEG–fMRI multimodal fusion analysis. Front Neuroinf. 2018;12:56.

Eichele T, Specht K, Moosmann M, Jongsma ML, Quiroga RQ, Nordby H, Hugdahl K. Assessing the spatiotemporal evolution of neuronal activation with single-trial event-related potentials and functional MRI. Proc Natl Acad Sci U S A. 2005;102:17798–803.

Freyer F, Becker R, Anami K, Curio G, Villringer A, Ritter P. Ultrahigh-frequency EEG during fMRI: pushing the limits of imaging-artifact correction. Neuroimage. 2009;48:94–108.

Friston KJ, Price CJ. Dynamic representations and generative models of brain function. Brain Res Bull. 2001;54:275–285.

Friston KJ. Modalities, modes, and models in functional neuroimaging. Science. 2009;326:399–403.

Friston KJ, Holmes AP, Worsley KJ, Poline JP, Frith CD, Frackowiak RSJ. Statistical parametric maps in functional imaging: a general linear approach. Hum Brain Mapp. 1994;2:189–210.

Goense JB, Logothetis NK. Neurophysiology of the BOLD fMRI signal in awake monkeys. Curr Biol. 2008;18:631–40.

Goldman RI, Stern JM, Engel J Jr, Cohen MS. Simultaneous EEG and fMRI of the alpha rhythm. Neuroreport. 2002;13:2487–92.

Goldman RI, Wei CY, Philiastides MG, Gerson AD, Friedman D, Brown TR, Sajda P. Single-trial discrimination for integrating simultaneous EEG and fMRI: identifying cortical areas contributing to trial-to-trial variability in the auditory oddball task. Neuroimage. 2009;47:136–47.

Gotman J, Bénar CG, Dubeau F. Combining EEG and FMRI in epilepsy: methodological challenges and clinical results. J Clin Neurophysiol. 2004;21:229–40.

Green JJ, Boehler CN, Roberts KC, Chen L-C, Krebs RM, Song AW, Woldorff MG. Cortical and subcortical coordination of visual spatial attention revealed by simultaneous EEG-fMRI recording. J Neurosci Off J Soc Neurosci. 2017;37:7803–10.

Grouiller F, Vercueil L, Krainik A, Segebarth C, Kahane P, David O. A comparative study of different artefact removal algorithms for EEG signals acquired during functional MRI. Neuroimage. 2007;38:124–37.

Hamandi K, Laufs H, Noth U, Carmichael DW, Duncan JS, Lemieux L. BOLD and perfusion changes during epileptic generalised spike wave activity. Neuroimage. 2008;39:608–18.

He BJ. Spontaneous and task-evoked brain activity negatively interact. J Neurosci. 2013;33:4672–82.

Horovitz SG, Braun AR, Carr WS, Picchioni D, Balkin TJ, Fukunaga M, Duyn JH. Decoupling of the brain's default mode network during deep sleep. Proc Natl Acad Sci U S A. 2009;106:11376–81.

Ives JR, Warach S, Schmitt F, Edelman RR, Schomer DL. Monitoring the patient's EEG during echo planar MRI. Electroencephalogr Clin Neurophysiol. 1993;87:417–20.

Jahnke K, von Wegner F, Morzelewski A, Borisov S, Maischein M, Steinmetz H, Laufs H. To wake or not to wake? the two-sided nature of the human K-complex. Neuroimage. 2012;59:1631–8.

Jann K, Dierks T, Boesch C, Kottlow M, Strik W, Koenig T. BOLD correlates of EEG alpha phase-locking and the fMRI default mode network. Neuroimage. 2009;45:903–16.

Jorge J, Bouloc C, Bréchet L, Michel CM, Gruetter R. Investigating the variability of cardiac pulse artifacts across heartbeats in simultaneous EEG-fMRI recordings: a 7T study. Neuroimage. 2019;191:21–35.

Kaufmann C, Wehrle R, Wetter TC, Holsboer F, Auer DP, Pollmächer T, Czisch M. Brain activation and hypothalamic functional connectivity during human non-rapid eye movement sleep: an EEG/fMRI study. Brain. 2006;129:655–67.

Krishnaswamy P, Bonmassar G, Poulsen C, Pierce ET, Purdon PL, Brown EN. Reference-free removal of EEG-fMRI ballistocardiogram artifacts with harmonic regression. Neuroimage. 2016;128:398–412.

Ladenbauer J, Ladenbauer J, Kulzow N, de Boor R, Avramova E, Grittner U, Floel A. Promoting sleep oscillations and their functional coupling by transcranial stimulation enhances memory consolidation in mild cognitive impairment. J Neurosci. 2017;37:7111–24.

Lange N, Zeger SL. Non-linear Fourier time series analysis for human brain mapping by functional magnetic resonance imaging. J R Stat Soc Ser C (Appl Stat). 1997;46:1–29.

Laufs H, Krakow K, Sterzer P, Eger E, Beyerle A, Salek-Haddadi A, Kleinschmidt A. Electroencephalographic signatures of attentional and cognitive default modes in spontaneous brain activity fluctuations at rest. Proc Natl Acad Sci U S A. 2003;100:11053–8.

Laufs H, Daunizeau J, Carmichael DW, Kleinschmidt A. Recent advances in recording electrophysiological data simultaneously with magnetic resonance imaging. Neuroimage. 2008;40:515–28.

Lei X, Yang P, Yao D. An empirical Bayesian framework for brain-computer interfaces. IEEE Trans Neural Syst Rehabil Eng. 2009;17:521–9.

Lei X, Qiu C, Xu P, Yao D. A parallel framework for simultaneous EEG/fMRI analysis: methodology and simulation. Neuroimage. 2010;52:1123–34.

Lei X, Ostwald D, Hu J, Qiu C, Porcaro C, Bagshaw AP, Yao D. Multimodal functional network connectivity: an EEG-fMRI fusion in network space. Plos One. 2011a;6:e24642.

Lei X, Xu P, Luo C, Zhao J, Zhou D, Yao D. fMRI functional networks for EEG source imaging. Hum Brain Mapp. 2011b;32:1141–60.
Lei X, Hu J, Yao D. Incorporating fMRI functional networks in EEG source imaging: a Bayesian model comparison approach. Brain Topogr. 2012;25:27–38.
Lei X, Wang Y, Yuan H, Mantini D. Neuronal oscillations and functional interactions between resting state networks: effects of alcohol intoxication. Hum Brain Mapp. 2014;35:3517–28.
Lei X, Wang Y, Yuan H, Chen A. Brain scale-free properties in awake rest and NREM sleep: a simultaneous EEG/fMRI study. Brain Topogr. 2015;28:292–304.
Leopold DA, Maier A. Ongoing physiological processes in the cerebral cortex. Neuroimage. 2012;62:2190–200.
Liu AK, Belliveau JW, Dale AM. Spatiotemporal imaging of human brain activity using functional MRI constrained magnetoencephalography data: Monte Carlo simulations. Proc Natl Acad Sci U S A. 1998;95:8945–50.
Logothetis NK, Pauls J, Augath M, Trinath T, Oeltermann A. Neurophysiological investigation of the basis of the fMRI signal. Nature. 2001;412:150–7.
Makeig S, Westerfield M, Jung TP, Enghoff S, Townsend J, Courchesne E, Sejnowski TJ. Dynamic brain sources of visual evoked responses. Science. 2002;295:690–4.
Mantini D, Perrucci MG, Del Gratta C, Romani GL, Corbetta M. Electrophysiological signatures of resting state networks in the human brain. Proc Natl Acad Sci U S A. 2007;104:13170–5.
Masterton RA, Abbott DF, Fleming SW, Jackson GD. Measurement and reduction of motion and ballistocardiogram artefacts from simultaneous EEG and fMRI recordings. Neuroimage. 2007;37:202–11.
Moosmann M, Ritter P, Krastel I, Brink A, Thees S, Blankenburg F, Taskin B, Obrig H, Villringer A. Correlates of alpha rhythm in functional magnetic resonance imaging and near infrared spectroscopy. Neuroimage. 2003;20:145–58.
Mullinger KJ, Yan WX, Bowtell R. Reducing the gradient artefact in simultaneous EEG-fMRI by adjusting the subject's axial position. Neuroimage. 2011;54:1942–50.
Murta T, Leite M, Carmichael DW, Figueiredo P, Lemieux L. Electrophysiological correlates of the BOLD signal for EEG-informed fMRI. Hum Brain Mapp. 2015;36:391–414.
Nguyen VT, Breakspear M, Cunnington R. Fusing concurrent EEG-fMRI with dynamic causal modeling: application to effective connectivity during face perception. Neuroimage. 2014; https://doi.org/10.1016/j.neuroimage.2013.06.083.
Niazy RK, Beckmann CF, Iannetti GD, Brady JM, Smith SM. Removal of FMRI environment artifacts from EEG data using optimal basis sets. Neuroimage. 2005;28:720–37.
Nunez P. Neocortical dynamics and human EEG rhythms. Oxford: Oxford University Press; 1995.
Phillips C, Rugg MD, Fristont KJ. Systematic regularization of linear inverse solutions of the EEG source localization problem. Neuroimage. 2002;17:287–301.
Raichle ME, Mintun MA. Brain work and brain imaging. Annu Rev Neurosci. 2006;29:449–76.
Ritter P, Freyer F, Curio G, Villringer A. High-frequency (600 Hz) population spikes in human EEG delineate thalamic and cortical fMRI activation sites. Neuroimage. 2008;42:483–90.
Rosa MJ, Kilner J, Blankenburg F, Josephs O, Penny W. Estimating the transfer function from neuronal activity to BOLD using simultaneous EEG-fMRI. Neuroimage. 2010;49:1496–509.
Schabus M, Dang-Vu TT, Albouy G, Balteau E, Boly M, Carrier J, Darsaud A, Degueldre C, Desseilles M, Gais S, Phillips C, Rauchs G, Schnakers C, Sterpenich V, Vandewalle G, Luxen A, Maquet P. Hemodynamic cerebral correlates of sleep spindles during human non-rapid eye movement sleep. Proc Natl Acad Sci. 2007;104:13164–9.
Scholvinck ML, Leopold DA, Brookes MJ, Khader PH. The contribution of electrophysiology to functional connectivity mapping. Neuroimage. 2013;80:297–306.
Schubert R, Ritter P, Wustenberg T, Preuschhof C, Curio G, Sommer W, Villringer A. Spatial attention related SEP amplitude modulations covary with BOLD signal in S1--a simultaneous EEG--fMRI study. Cerebral Cortex. 2008;18:2686–700.
Shirer WR, Ryali S, Rykhlevskaia E, Menon V, Greicius MD. Decoding subject-driven cognitive states with whole-brain connectivity patterns. Cerebral Cortex. 2012;22:158–65.

Stancak A, Polacek H, Vrana J, Rachmanova R, Hoechstetter K, Tintra J, Scherg M. EEG source analysis and fMRI reveal two electrical sources in the fronto-parietal operculum during subepidermal finger stimulation. Neuroimage. 2005;25:8–20.

Steyrl D, Krausz G, Koschutnig K, Edlinger G, Müller-Putz GR. Online reduction of artifacts in EEG of simultaneous EEG-fMRI using reference layer adaptive filtering (RLAF). Brain Topogr. 2017;31(1):129–49.

Tamminen J, Payne JD, Stickgold R, Wamsley EJ, Gaskell MG. Sleep spindle activity is associated with the integration of new memories and existing knowledge. J Neurosci. 2010;30:14356–60.

Trujillo-Barreto N, Martinez-Montes E, Melie-Garcia L, Valdes-Sosa P. A symmetrical Bayesian model for fMRI and EEG/MEG neuroimage fusion. Int J Bioelectromag. 2001;3:1.

Uehara T, Yamasaki T, Okamoto T, Koike T, Kan S, Miyauchi S, Kira J-I, Tobimatsu S. Efficiency of a "Small-World" brain network depends on consciousness level: a resting-state fMRI study. Cerebral Cortex. 2014;24:1529–39.

Valdes-Sosa PA, Sanchez-Bornot JM, Sotero RC, Iturria-Medina Y, Aleman-Gomez Y, Bosch-Bayard J, Carbonell F, Ozaki T. Model driven EEG/fMRI fusion of brain oscillations. Hum Brain Mapp. 2009;30:2701–21.

van der Meer JN, Pampel A, Van Someren EJW, Ramautar JR, van der Werf YD, Gomez-Herrero G, Lepsien J, Hellrung L, Hinrichs H, Moller HE, Walter M. Carbon-wire loop based artifact correction outperforms post-processing EEG/fMRI corrections--a validation of a real-time simultaneous EEG/fMRI correction method. Neuroimage. 2016;125:880–94.

Vincent JL, Larson-Prior LJ, Zempel JM, Snyder AZ. Moving GLM ballistocardiogram artifact reduction for EEG acquired simultaneously with fMRI. Clin Neurophysiol. 2007;118:981–98.

Vulliemoz S, Rodionov R, Carmichael DW, Thornton R, Guye M, Lhatoo SD, Michel CM, Duncan JS, Lemieux L. Continuous EEG source imaging enhances analysis of EEG-fMRI in focal epilepsy. Neuroimage. 2010;49:3219–29.

Wan X, Riera J, Iwata K, Takahashi M, Wakabayashi T, Kawashima R. The neural basis of the hemodynamic response nonlinearity in human primary visual cortex: Implications for neurovascular coupling mechanism. Neuroimage. 2006;32:616–25.

Wang K, Li W, Dong L, Zou L, Wang C. Clustering-constrained ICA for ballistocardiogram artifacts removal in simultaneous EEG-fMRI. Front Neurosci. 2018;12:59.

Xia H, Ruan D, Cohen MS. Removing ballistocardiogram (BCG) artifact from full-scalp EEG acquired inside the MR scanner with Orthogonal Matching Pursuit (OMP). Front Neurosci. 2014;8:218.

Yu Q, Erhardt EB, Sui J, Du Y, He H, Hjelm D, Cetin MS, Rachakonda S, Miller RL, Pearlson G, Calhoun VD. Assessing dynamic brain graphs of time-varying connectivity in fMRI data: application to healthy controls and patients with schizophrenia. Neuroimage. 2015;107:345–55.

Yu Q, Wu L, Bridwell DA, Erhardt EB, Du Y, He H, Chen J, Liu P, Sui J, Pearlson G, Calhoun VD. Building an EEG-fMRI multi-modal brain graph: a concurrent EEG-fMRI study. Front Hum Neurosci. 2016;10:476.

Zich C, Debener S, Kranczioch C, Bleichner MG, Gutberlet I, De Vos M. Real-time EEG feedback during simultaneous EEG-fMRI identifies the cortical signature of motor imagery. Neuroimage. 2015;114:438–47.

Chapter 19
EEG/ERP Data Analysis Toolboxes

Gan Huang

Abstract The development and application of new methods allow us to perform more advanced analyses on EEG signals. However, the understanding of the mathematical and methodological details about these new methods would be no easy for the researchers without engineering and mathematics background. As a collection of tools for EEG signal processing and data visualization, EEG/ERP analysis toolboxes make the researchers be able to perform the complex analysis by simply clicking buttons or running some lines of MATLAB script. In this chapter, we firstly make a brief overview about the currently popular toolboxes in EEG/ERP analysis, such as EEGLab, FieldTrip, BrainVision Analyzer, etc., and then focus on the introduction of Letswave, which is an intuitive and streamlined tool to process and visualize EEG data, with a shallow learning curve. Examples are provided for a better understanding of Letswave7 in EEG/ERP data analysis.

Keywords Toolbox · Letswave · Signal processing · Data visualization

19.1 Brief Overview

In the recent decades, a variety of new methods have been developed and applied in the area of EEG analysis. By integrating these methods, the development of EEG/ERP toolboxes provides the researcher with powerful and convenient tools for EEG signal visualization and processing. In the following, we make a brief overview of these most frequently used toolboxes in EEG/ERP analysis.

EEGLAB (Delorme and Makeig 2004): Currently the most widely used open-source toolbox for EEG/ERP analysis, which provides the basic operations from data importing, segmentation, ICA, bandpass filtering, baseline correction to time-frequency analysis, etc. EEGLAB has been developed by Arnaud Delorme and

G. Huang (✉)
School of Biomedical Engineering, Health Science Center, Shenzhen University, Shenzhen, China

L. Hu, Z. Zhang (eds.), *EEG Signal Processing and Feature Extraction*,
https://doi.org/10.1007/978-981-13-9113-2_19

Scott Makeig at the Salk Institute (La Jolla CA, USA) since 1997. Based on different levels of programming sophistication, EEGLAB provides both menu-based graphic user interface (GUI) and script-based command-line interface (CLI) on the platform of MATLAB. Rich plugins make it easier for users to complete more diverse tasks with EEGLAB.

FieldTrip (Oostenveld et al. 2011): Another famous MATLAB-based open-source EEG analysis toolbox. The development team is from the Donders Institute for Brain, Cognition and Behaviour (Radboud University, the Netherlands), led by Robert Oostenveld and Jan-Mathijs Schoffelen. FieldTrip provides a set of functions with some advanced operations for EEG analysis, such as source analysis and cluster-based permutation test. However, no graphic interface makes it difficult for beginners to use, especially those without any programming foundation.

Letswave: As another comprehensive EEG/ERP analysis toolbox, Letswave was developed by the team of André Mouraux (Institute of Neuroscience, Université catholique de Louvain, Belgium). Compared to other EEG signal processing toolboxes, Letswave emphasizes on intuitive and streamlined tools to process and visualize EEG data, with a shallow learning curve. More detailed introduction about Letswave is arranged in the following section.

Some other toolboxes may be not developed just for EEG signal processing but also provide excellent tools on EEG/ERP analysis. For example, BrainStorm (Tadel et al. 2011) shares a comprehensive set of user-friendly tools with the scientific community using MEG/EEG as an experimental technique. Statistical parametric mapping (SPM) (Litvak et al. 2011), designed for the analysis of brain imaging data sequences, is compatible for the analysis of fMRI, PET, SPECT, EEG, and MEG. Different from most of the open-source toolboxes, MNE-Python (Gramfort et al. 2013) provides a collection of functions for MEG/EEG processing and visualization on the Python platform.

In addition to these free academic toolboxes, there are some excellent EEG analysis commercial programs that are often used with some specialized amplifiers, such as BrainVision Analyzer for Brain Product and Scan for Neuroscan. In most cases, the commercial tools for EEG analysis are developed with programming languages, which should be called program since they normally exactly generate an executable program. While the free tools should be called as toolboxes, because it is a set of scripts on certain platform, not a program in the strict sense. Compared with free toolboxes, commercial program normally provides a more efficient and friendlier user interface, but the higher prices limit their use in academia community. To the best of the authors' knowledge, we have seen any research group packaged their newly developed methods into a plugin for these programs for the others to use.

Some free toolboxes and commercial programs may not provide full functionality for EEG analysis but only provide analysis functions in some specialized areas, such as SIFT (Delorme et al. 2011), MVGC (Barnett and Seth 2014), Hermes (Niso et al. 2013), TRENTOOL (Lindner et al. 2011), EEGNET (Lawhern et al. 2016), and Chronux (Bokil et al. 2010) for connectivity estimation and LORETA, BESA, and Curry for source analysis.

The web links of all toolboxes and programs are listed as follows:

- EEGLAB: https://sccn.ucsd.edu/eeglab/
- FieldTrip: http://www.fieldtriptoolbox.org/
- Letswave: https://letswave.cn/
- BrainStorm: https://neuroimage.usc.edu/brainstorm
- SPM: https://www.fil.ion.ucl.ac.uk/spm/
- MNE-Python: https://mne-tools.github.io/
- Analyzer: https://www.brainproducts.com/promo_analyzer2.php
- Scan: https://compumedicsneuroscan.com/tag/scan/
- SIFT: https://sccn.ucsd.edu/wiki/SIFT
- MVGC: http://www.sussex.ac.uk/sackler/mvgc/
- Hermes: http://hermes.ctb.upm.es/
- TRENTOOL: http://www.trentool.de/
- EEGNET: https://sites.google.com/site/eegnetworks
- Chronux: http://chronux.org/
- LORETA: http://www.uzh.ch/keyinst/loretaOldy.htm
- BESA: http://www.besa.de/
- Curry: https://compumedicsneuroscan.com/curry-8-released/

19.2 The Introduction of Letswave

19.2.1 General Overview

Letswave is a free, open-source MATLAB toolbox to analyze EEG and other neurophysiological signals, covered under the terms of the GNU General Public License. The project is managed by André Mouraux (Institute of Neuroscience, Université catholique de Louvain, Belgium), in collaboration with Gan Huang (Shenzhen University, China), Bruno Rossion (Institute of Neuroscience, Université catholique de Louvain), Li Hu (Institute of Psychology, Chinese Academy of Sciences, China), and Giandomenico Iannetti (University College London, UK). The development of Letswave started in 2006. The first four versions are developed in Borland Delphi. It is switched to MATLAB platform since Letswave5. Up to today, the newest version of the Letswave is Letswave7, and it continues to be developed.

Letswave can be used on Windows, Mac, or Linux system with MATLAB 2010 or later version. Considering the major updating of graphics system in MATLAB 2014b, MATLAB with the version later than 2014b is highly recommended for running Letswave. The basic processing requirement is 4GB RAM. For larger dataset analysis, 8GB RAM is recommended, and 16GB or more of RAM would be helpful in time-frequency analysis.

Letswave includes the following four modules, which is shown in Fig. 19.1:

- **Manage module** is the main interface of Letswave. It can be launched by typing "letswave" in the command window of MATLAB. The role of the manage

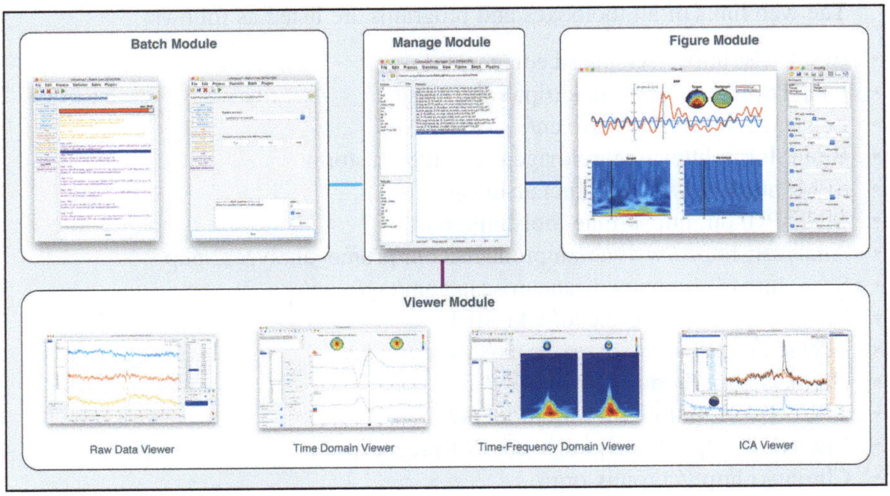

Fig. 19.1 The framework of Letswave. There are four modules in the GUI design, which are manage module, batch module, viewer module, and figure module

module is mainly data management, such as dataset import/export/delete/rename, epochs/channels/events management, and also multiple datasets selection for further operation. In the manage module, all the datasets are managed by the affixes in their filenames, which are separated by space. Multiple datasets can be easily and quickly selected from tens of thousands of datasets by the affixes filtering in the left side. After the selection, the interface to the other modules is provided in the menu of the manage module and right-clicking menu.

- **Batch module** contains multiple functions for EEG signal analysis, including data preprocessing, segmentation, time/frequency domain analysis, and statistical analysis. Batch module allows the GUI-based users to do the same processing on a single dataset or multiple datasets at the same time. It is also easier for the users to do multiple processing steps on the same dataset(s). This processing history can also be viewed in the batch module. Furthermore, users can define their own processing flow, which may involve multiple datasets and multiple processing steps. The frequently used processing flow can also be saved in the menu of the manage module. For another similar dataset(s), the users can call out the processing flow and just modify the input datasets to run the analysis. More interestingly, once the processing flow has been made, the script has also been generated automatically. Just by clicking the button of "script", users can get the corresponding script without writing any code.
- **Viewer module** is used for the observation of one or multiple dataset(s). Viewer module includes a set of viewers, which are used for the observation of the continuous data and the result in time domain and time-frequency domain, respectively. For example, there is a batch of time domain datasets with multiple channels and several epochs. The viewer allows the users to display the result in

their own way, that is, show the datasets in separated windows in different rows or columns for different datasets/channels/epochs or superimpose them in the same windows. Spatial topography information can be jointly displayed for a single time/frequency point or the mean value from certain interval. Simple statistic of the mean, maximum, and minimum value with their location on the selected interval would make the work of peak detection with their latency information become simple and intuitive.

- **Figure module** is used for figure generation, that is, to customize the figure to show the waveform, time-frequency response, and the topography which could be used in the publication or presentation or importing to the other graphics software for further editing. Different from the other existence toolboxes, the figure module is very open. The GUI-based user can make their graphics from certain templates or directly from a blank canvas. It allows user to flexibly customize the layout, free to the style of font, line, and map. Further, the graphics can also be saved as a template. The user can generate a new graphic in the same style with different data sources by using this template.

19.2.2 To GUI-Based Users

With the graphic interface, the users can easily to run the whole progress for EEG analysis, from data importing, preprocessing, time-frequency analysis to statistical analysis. Compared with the existence toolboxes, there are two major features in the design of Letswave7.

Firstly, the function of data management is strengthened in Letswave. With number of subjects increasing, the lack of data management and batch processing functions makes the existence toolbox be difficult to handle the large number of datasets by GUI. Take an experiment with a two-way ANOVA design for example. If there are 3 levels for each factor and 120 subjects were included in the experiment and 1 dataset for 1 condition from 1 subject, then there would be more than 1000 raw datasets involved in the study. Including intermediate files and the final results, the number of the datasets involved in this study would be three or four times larger than 1000. Even selecting the target datasets for the corresponding processing would be difficult for GUI-based user. Hence, the datasets management and data batch processing are necessary in the design of the EEG signal processing toolbox. Hence, in Letswave, the datasets are managed by affixes; two list boxes in the left side of the manage module make it easy to dataset selection.

Secondly, the function for data visualization is separated into two parts, which are data observation during data analysis in viewer module and figure generation for publication in figure module. Data observation does not require beauty of the image, but is highly demanding in terms of convenience and observability during the whole process of the data processing, from data import, artifact rejection in the preprocessing to the joint result observation in time-frequency-spatial domain.

19.2.3 To Script-Based Users

In addition to the GUI, script may be the most commonly used feature in Letswave. The following unique designs make the script writing super easy in Letswave:

19.3 Full Transparency Between Script and GUI

In batch module, Letswave runs a processing flow by running the corresponding MATLAB script. Once the processing flow has been made, the corresponding MATLAB script has already been generated automatically. Users can click the "script" button in the batch module to get the full script. Hence, the users, who have learned the use of GUI, almost already understand the writing of script.

The script for the single-step operation can also be gotten with the "script" button in each step, which can tangibly reduce the users' dependence on the tutorial and the help document. This feature is essential, since it can effectively help the script-based user to speed up the script writing and improve the script quality. More importantly, it can really reduce the script-based user's entry threshold. Even the users without a solid background for MATLAB programming are able to write high-quality script for batch EEG signal processing.

Not only batch module, the operations in other modules can also become scripts automatically. Hence, from data import, to preprocessing, time/frequency analysis, and statistical analysis, to the final figure generation, the whole process can be easily and quickly written as a script for the EEG signal analysis.

19.4 Easy Access to the Data

Each dataset can be easily accessed by clicking "send to workspace" in the right menu in the manage module, which would be appeared as a variable with the name of "lwdata" in the workspace of MATLAB. Then the script-based user can make their own operation on the dataset in MATLAB. After the operation, the user can also save the dataset in the format of Letswave by clicking "read from workspace" in the right menu in the manage module.

Similarly, the other information is also easily to be accessed in the other modules, such as the event information, the channel name, and statistical result about the mean, maximum, and minimum value.

19.5 The Simplest Grammar

In Letswave, we provide a rich library of functions for the script-based users, which are starting with `FLW_`. The grammar of all the functions in this library is uniform. For one step of operation, we just need to define the option and then call the corresponding FLW function.

```
option=struct('XXX',xxx,'XXX',xxx);
lwdata= FLW_XXX.get_lwdata(lwdata,option);
```

The parameters setting in the option is identical to the parameter's selection in the GUI. It just needs to be noticed whether the input and output of the function is a single dataset or multiple datasets.

19.5.1 To Developers

Developers can also make their own methods integrated into Letswave by developing the FLW file. For these advanced users, the requirements on the programming will be much higher.

Before the developing, the understanding of the data structure of Letswave is important (Fig. 19.2). In Letswave, each dataset is saved in the hard driver by two files, which are "lw6" and "mat" files. "Lw6" file is a basic description about the dataset, including some simple information such as filename, data size, the starting point, and step size for each dimension. Some information, like the channel, event, and history, are kept in the "lw6" file as structure variables. "Mat" file is using a six-dimension matrix to save the data with respect to epoch, channel, and index, z, y, x correspondingly. In MATLAB, the dataset "lwdata" is composed of header and data, which are form lw6 and mat file, respectively.

The developing of FLW function mainly includes four parts of work. Firstly, design the UI which would be integrated into batch module. Secondly, define the parameter setting for Letswave generating the script automatically. Thirdly, implement their methods. Finally, put the FLW file into the corresponding place for Letswave to call it. A guide about how to write a FLW file is also included in the toolbox.

```
>> lwdata

lwdata =

  struct with fields:

    header: [1×1 struct]
      data: [6-D double]

>> lwdata.header

ans =

  struct with fields:

        filetype: 'time_frequency_amplitude'
            name: 'avg cwt ep_s1 oc_rm butt continuous_EEG_LEPs_S06'
            tags: {}
        datasize: [1 32 1 1 100 2000]
          xstart: -1
          ystart: 1
          zstart: 0
           xstep: 0.0013
           ystep: 0.2929
           zstep: 1
        chanlocs: [1×32 struct]
          events: [0×0 struct]
         history: [1×4 struct]
    index_labels: {'index 1'}
       epochdata: [0×0 struct]

>> size(lwdata.data)

ans =

           1          32           1           1         100        2000
```

Fig. 19.2 The data structure for a dataset in Letswave

19.6 Download and Setup

19.6.1 Download

The latest version can be downloaded as an archive from the NOCIONS Github repository (https://github.com/NOCIONS/letswave7/archive/master.zip). The users can also download the zip file by visiting the NOCIONS Github repository (https://github.com/NOCIONS/letswave7) and manually download it.

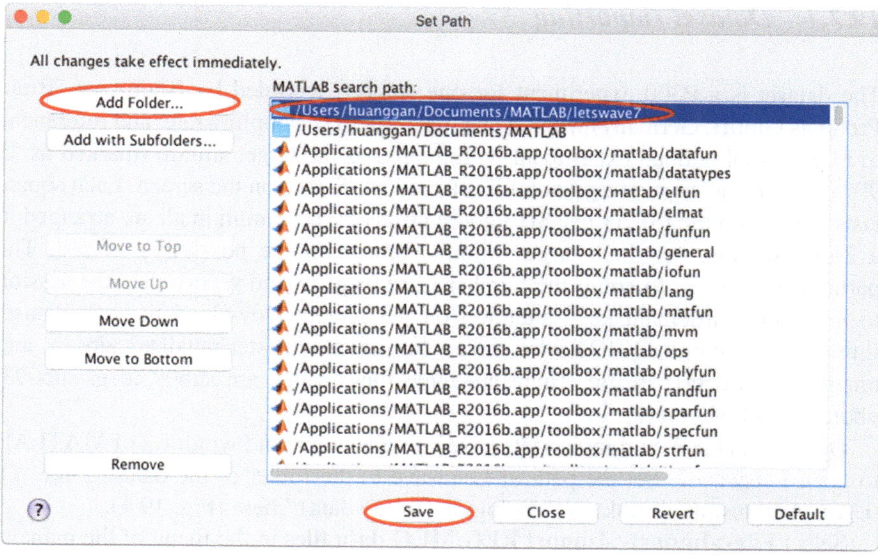

Fig. 19.3 In the Set Path window, click the "Add Folder" button

19.6.2 Installation

- Unzip the zip archive of Letswave7 in the folder of your choice.
- Launch MATLAB.
- In the MATLAB menu ribbon, click on "Set Path".
- In the Set Path window, click button "Add Folder" (Fig. 19.3).
- Select the location of the Letswave7 folder.
- Save the modified path and close.

To check whether the installation is successful or not, input "letswave7" in the command windows of MATLAB. The pop-up of manage module, which is the main interface in Letswave7, indicates that the installation is successful.

19.7 Example for Single-Subject Analysis

In this part, a dataset with P300 experiment will be used to illustrate the data processing for single subject. After the preprocessing with frequency filtering, bad electrodes interpolation, and ICA decomposing, both time domain analysis and time-frequency domain analysis will be performed on the single-subject level.

19.7.1 Dataset Importing

The dataset is a P300 experiment for one subject, recorded by BrainAmp (Brain Products GmbH, Germany) with 64 channels, 1000Hz sampling rate, and referenced to FCz. Visual oddball experiment with the red as the target stimuli (marked as "S 9") and white as the nontarget stimuli (marked as "S 10") on the screen. Each square lasts 80ms with the ISI 200ms. Six hundred trials of the stimuli in all are arranged in a 2-min session, in which target stimuli come with the possibility of 5%. The participant is asked to count the number of red square and report after the session to make the participant keep attention on the screen. Download the raw dataset (https://github.com/NOCIONS/letswave7_tutorial/raw/master/rawdata1.zip), and unzip the rawdata1.zip file. There are three files, which are sub093.eeg, sub093.vhdr, and sub093.vmrk.

Open MATLAB, and input "letswave7" in the command windows of MATLAB to open Letswave7. Set the path of Letswave to the folder of the dataset, like "C:\Users\Adminstrator\Desktop\SynologyDrive\rawdata1" here (Fig. 19.4).

Select **File->Import->Import EEG/MEG data files** in the menu of the manage module, and then the dialog of import data will pop up. Press the button **add files** to add the file *sub093.eeg*. Press the button **import files** to import the dataset. During the importing, it will display "processing." Once the importing is finished, the corresponding dataset will turn red and display "done." Close the dialog of import data; the dataset *sub093* will be appeared in the manage module.

Select dataset *sub093*, and click **View->Continuous Data Viewer** in the menu to check the data quality of the imported dataset. In the viewer of the continuous data, it can be seen the channel P1 is obviously abnormal (Fig. 19.5). A strong 50Hz power frequency interference covers the signal. Even after a notch filter, there is still a strong noise. Hence in the following processing, channel **P1** is treated as a bad electrode, which would be interpolated by the surround channels.

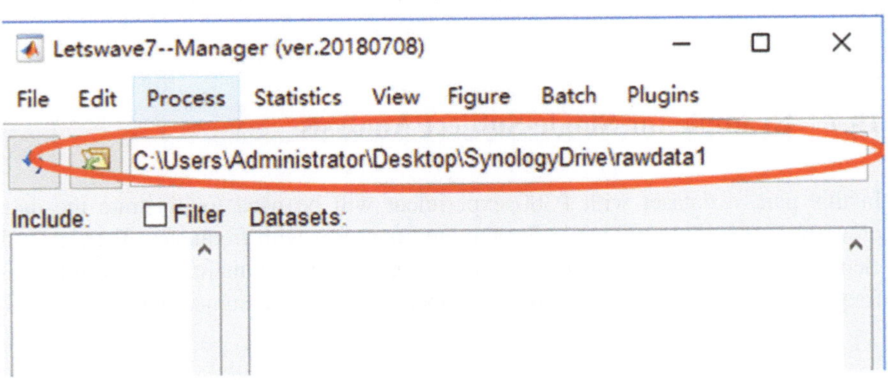

Fig. 19.4 Set the path of Letswave to the folder of the dataset

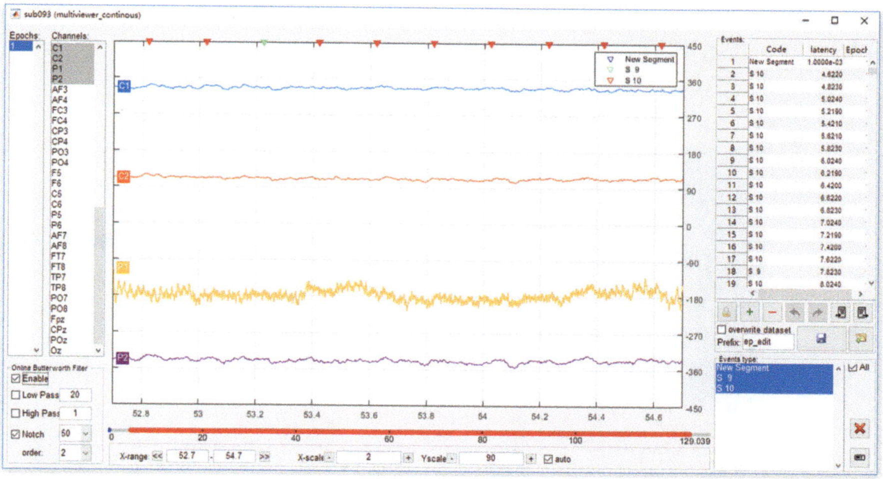

Fig. 19.5 Check the quality of the imported dataset

19.7.2 Useless Channels Removing

Before removing useless channels, usually it is necessary to assign the coordinate information for each channel in the dataset firstly. But Letswave7 can automatically assign the channel location with commonly used 10–20 system. If the other EEG coordinate system is used, the users can still manually assign the correct channel location by clicking **Edit->Electrodes->Edit electrode coordinates** in the menu.

To make an efficient analysis and save the storage space, removing the useless channels is necessary. In this example, channel **IO** will be removed, which records the electrooculogram (EOG) signals. To remove channel IO, select the dataset *sub093* in the manage module, and click **Edit->Arrange signals->Rearrange or delete epochs, channels, indexes** in the menu. In the batch module (Fig. 19.6), press the button **Add all** to add all the channels into the left list box. Then select channel IO, and press button **Remove** to delete channel **IO.** Click the button **Run** in the bottom of batch module, and then a new dataset with the name *sel_chan sub093* will be appeared in the data list of the manage module.

19.7.3 Frequency Filtering

Frequency filtering is an effective way to filter out the high-frequency artifact, low-frequency drift, and the 50/60Hz power-line interference. In this case, the bandpass filter is set as 0.05–30Hz. Select the dataset *sel_chan sub093* in the manage module, and click **Process->Frequency analysis and filters->Butterworth filters** in the menu. In the batch module, set the low cutoff frequency (Hz) as **0.05Hz**, and

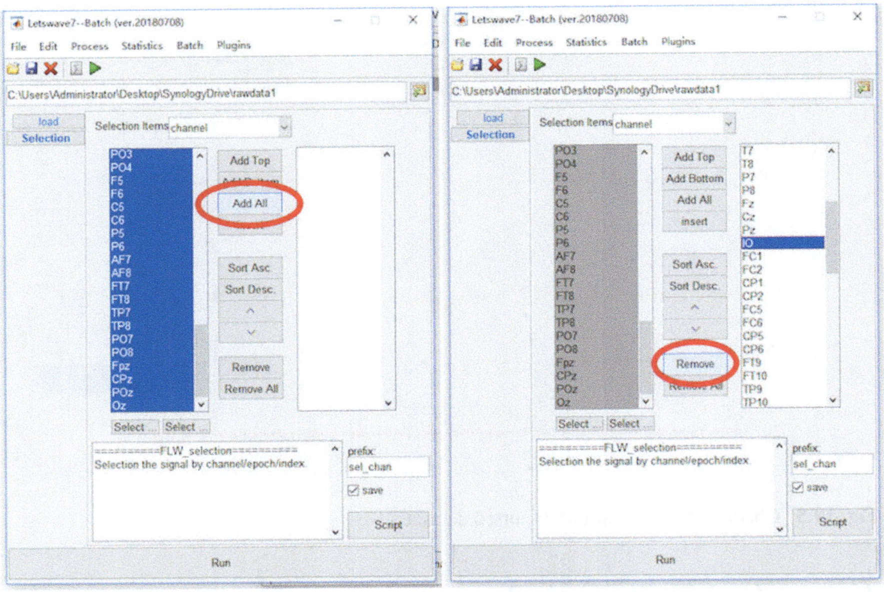

Fig. 19.6 Useless channels removing

click the button **Run** to run the bandpass filtering. A new dataset with the name *butt sel_chan sub093* will be appeared in the data list of the manage module.

19.7.4 Bad Electrodes Interpolation

In this step, channel P1 would be interpolated with the surround channels. Select the dataset *butt sel_chan sub093* in the manage module, and click **Edit->Electrodes->Interpolate channel using neighboring electrodes** in the menu. In the batch module, select channel **P1** in the **Channel to Interpolate** list box, and then click the button **Find closest electrodes**. Since the **number of channels used for interpolation** is **3** for the default setting, Letswave7 will find the closest electrodes **P3**, **Pz**, and **CP1** for interpolation automatically according to the location of the channels. Click the button **Run** for the bad electrodes interpolation. A new dataset with the name *chan_interp butt sel_chan sub093* will be appeared in the data list of the manage module.

19.7.5 ICA Decomposing

Independent component analysis (ICA) is a kind of blind signal separation (BSS) method. Linearly, ICA can be modeled as in Fig. 19.7. Consider **X** is the recorded

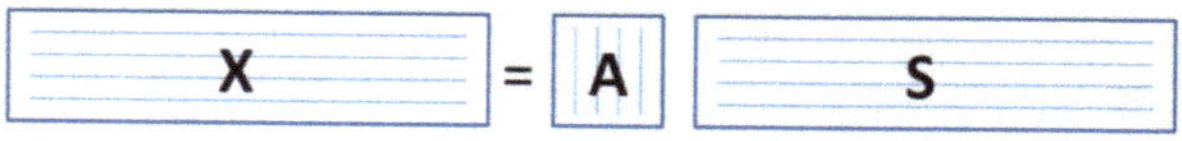

Fig. 19.7 Linear model for ICA

EEG signal with the dimension *channel* × *time*, **S** is the source signal with the dimension *component* × *time*, and **A** is the mixing matrix with the dimension *channel* × *component*. The aim of ICA is to find the mixing matrix **A** to make each component (each row) to be independent of each other. According to the linear model, ICA can be used for artifact removing in EEG signal processing by the following steps:

1. Run ICA algorithm to get the mixing matrix **A**.
2. Automatically, we can have the source signal **S** = **pinv(A)** × **X**.
3. Identify the artifact component manually in **S**. By setting the corresponding row as **0**, we have **S_bar**.
4. By **S_bar** with the artifact removed, we can automatically get **X_bar** = **A** × **S_bar**.

The signal **X_bar** is the result by ICA artifact removal. In Letswave7, steps 2 and 4 can be completed automatically, in which **pinv(A)**, also called unmixing matrix, is the pseudo-inverse matrix for matrix **A**. Step 1 (computer ICA matrix) and step 3 (identify artifact component) need to be performed manually. Hence, in Letswave7, we need two steps to finish the work of artifact removal by ICA.

Select the dataset *chan_interp butt sel_chan sub093* in the manage module, and click **Process->Spatial filters (ICA/PCA)->Compute ICA matrix** in the menu. In the batch module, select number of components as **decide by user**, and set the components numbers as **40**. Click the button **Run** for compute ICA matrix. A new dataset with the name *ica chan_interp butt sel_chan sub093* will be appeared in the data list of the manage module.

After computing the ICA matrix, we need to manually identify the components with artifact. Select the dataset *ica chan_interp butt sel_chan sub093* in the manage module, and click **Process->Spatial filters(ICA/PCA)->Apply ICA/PCA spatial filter** in the menu. The interface is popped up for manually removing component for spatial filter (Fig. 19.8). Different types of information is marked with different colors. The black color represents the original signal **X**, color blue for source **S**, and color orange for the filtered signal **X_bar**. In the left panel (black), we can select the dataset, epoch, and channel to check the original signal **X** as the black curve in the middle panel. Next, we can select the component in the left panel (blue) to check the time, frequency, and spatial feature of each component of source **S** in the bottom panel (blue). After identifying the component as the artifact in the right panel in orange, the orange curve in the middle panel would show the corresponding filtered signal **X_bar**. We can check the result of ICA filtering immediately. In this study, **component 1** is identified as the eye blink artifact. The scalp topography suggests the "equivalent current dipole" (ECD)s close to the eyes. The waveform in the time domain looks like a spike. The power in the frequency domain concentrates at low-frequency band (<5Hz). All these suggest **comp 1** as the eye blink artifact. After removing this component by selecting **comp 1** in the right panel, the blink

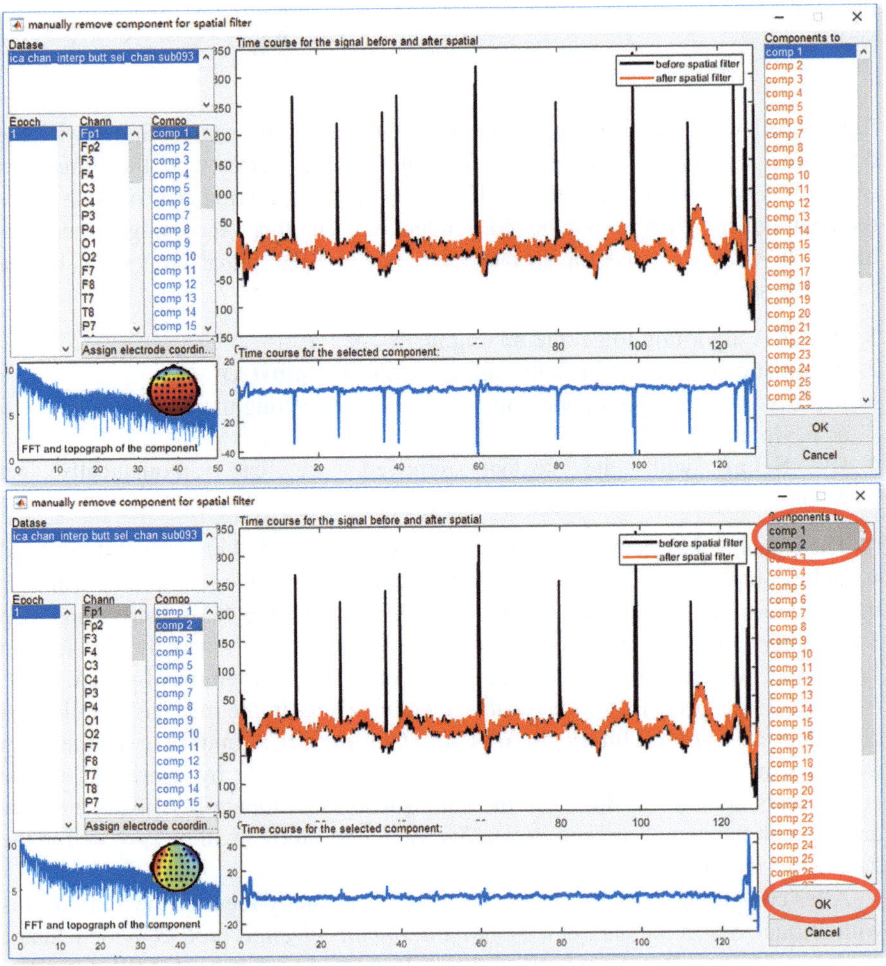

Fig. 19.8 Artifact component identification

artifact has been effectively removed as compared the orange curve **X_bar** with the black curve **X** in the middle panel. Similarly, **component 2** as the artifact with lateral eye movement. The clear evidence can be observed from the scalp topography. Select **comp 1** and **comp 2** in the right panel, and click the button **OK**. A new dataset with the name *sp_filter ica chan_interp butt sel_chan sub093* will be appeared in the data list of the manage module, which is a result of artifact removing by ICA.

The order for ICA and segmentation should also be discussed in preprocessing. Since ICA is a data-driven method for artifact removal, we need enough data to run ICA. While, excessive amount of data can greatly increase computing time, but the improvement of the accuracy in the result is limited. Normally, we run the ICA after segmentation, since segmentation could shorten the data length and remove irrelevant noises. However, in this case of the P300 study, the overlap between trials is very serious; running ICA after segmentation would unnecessarily increase calculation time. Hence, ICA is applied before segmentation.

In this case, we run the ICA by selecting the number of the components as **40**. Normally, the maximum number of independent components that can be separated is equal to the number of channels in the original signal **X**. If **X** has been referenced, the maximum number should minus 1. For each channel has been interpolated, the maximum number should minus 1 again. In addition, decreasing the number of independent components to separate is a way to reduce the computing time for the ICA matrix. In this case, since the channel number 64 is large enough, we set the number of independent components as 40 is OK.

19.7.6 Segmentation

Select the dataset *sp_filter ica chan_interp butt sel_chan sub093* in the manage module, and click **Process->Epoch segmentation->Segment relative to events (one file per event code)** in the menu. In the batch module, select event codes **S 9** and **S 10**, and set the epoch starting time and duration as **-1** and **3**. Click the button **Run** in the bottom of batch module, and then two new datasets with the name *ep_S 9 sp_filter ica chan_interp butt sel_chan sub093* and *ep_S 10 sp_filter ica chan_interp butt sel_chan sub093* will be appeared in the manage module.

In the menu, there are two items for segmentation, which are **Segment relative to events** and **Segment relative to events (one file per event code)**. They have similar function, but the output would be different. If multiple event codes have been selected, **Segment relative to events** will segment all the epochs with different event codes into one dataset. However, with **Segment relative to events (one file per event code)**, separated dataset will be generated according to different event codes. For example, in this case of P300 dataset, the target and nontarget events are marked as **S 9** and **S 10**. Hence with **Segment relative to events (one file per event code)**, two datasets have been generated.

Since no obviously artifact has been observed for both the two datasets, it is not necessary to perform artifact rejection on these datasets. Hence, this step is omitted in the preprocessing of this P300 dataset. If necessary, the operation of artifact rejection can be done by clicking **Edit->Arrange signals->Rearrange or delete epochs, channels, indexes** in the menu.

19.7.7 Rereference

In the study of P300, the average of bilateral mastoids regions normally is selected as reference. Hence, we will rereference to the mean value of TP9 and TP10. Select the datasets *ep_S 9 sp_filter ica chan_interp butt sel_chan sub093* and *ep_S 10 sp_filter ica chan_interp butt sel_chan sub093*, and click **Process->Rereference signals-**

>Rereference in the menu. In the batch module, select the **TP9** and **TP10** as the **new reference** in the left list box, and select all channels in the right list box for **apply reference to**. Click the button **Run** in the bottom of batch module to finish the work of artifact rejection. Two new datasets with the name *reref ep_S 9 sp_filter ica chan_interp butt sel_chan sub093* and *reref ep_S 10 sp_filter ica chan_interp butt sel_chan sub093* will be appeared in the manage module.

Here, TP9 and TP10 are selected for using the mean value of the two channels as the new reference. For the common average reference, we can select all channels in the left list box, and then the average of all channels will be used as the new reference.

19.7.8 Baseline Correction

In the previous processing, the epochs are segmented from **-1s** to **2s**. Hence, the baseline is set from **-1s** to **0s** for baseline correction. Select the datasets *reref ep_S 9 sp_filter ica chan_interp butt sel_chan sub093* and *reref ep_S 10 sp_filter ica chan_interp butt sel_chan sub093*, and click **Process->Baseline operation-> Baseline correction** in the menu. Keep the default setting in the batch module, and click the button **Run** in the bottom of batch module to finish the work of artifact rejection. Two new datasets with the name *bl reref ep_S 9 sp_filter ica chan_interp butt sel_chan sub093* and *bl reref ep_S 10 sp_filter ica chan_interp butt sel_chan sub093* will be appeared in the manage module.

19.7.9 Averaging

After the ten steps of preprocessing, we can simply average the epochs to do time domain analysis. Select the datasets *bl reref ep_S 9 sp_filter ica chan_interp butt sel_chan sub093* and *bl reref ep_S 10 sp_filter ica chan_interp butt sel_chan sub093*, and click **Process->Average->Compute averag, std, median across epochs** in the menu. Keep the default setting in the batch module, and click the button **Run** in the bottom of batch module for averaging. Two new datasets with the name *avg bl reref ep_S 9 sp_filter ica chan_interp butt sel_chan sub093* and *avg bl reref ep_S 10 sp_filter ica chan_interp butt sel_chan sub093* will be appeared in the data list of the manage module.

To observe the waveform of time domain analysis result in the multiviewer, select datasets *avg bl reref ep_S 9 sp_filter ica chan_interp butt sel_chan sub093* and *avg bl reref ep_S 10 sp_filter ica chan_interp butt sel_chan sub093*, and click the **view** in the right-click menu. Selecting both the two datasets, selecting channel **Pz**, opening the topography in the toolbar, and setting the cursor to **0.35**, the P300 result from single subject is shown as it is in Fig. 19.9.

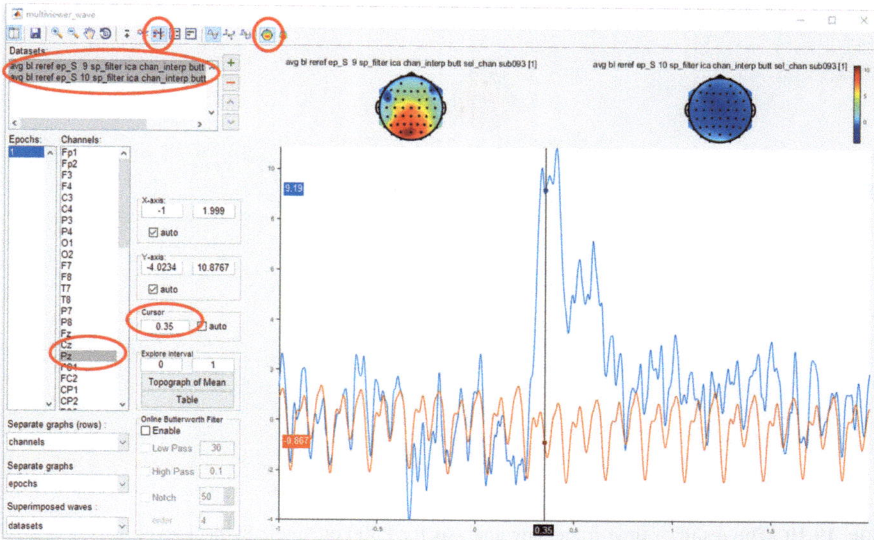

Fig. 19.9 The result of time domain analysis for P300 on single-subject level

19.7.10 Continuous Wavelet Transform

To perform CWT for time-frequency analysis before the time domain averaging, select the datasets *bl reref ep_S 9 sp_filter ica chan_interp butt sel_chan sub093* and *bl reref ep_S 10 sp_filter ica chan_interp butt sel_chan sub093*, and click **Plugins->my_tfa->Averaged CWT** in the menu. Keep the default parameter setting in the batch module, and click the button **Run** in the bottom of batch module. Two new datasets with the name *avg cwt bl reref ep_S 9 sp_filter ica chan_interp butt sel_chan sub093* and *avg cwt bl reref ep_S 10 sp_filter ica chan_interp butt sel_chan sub093* will be appeared in the manage module.

The operation of **Plugins->my_tfa->Averaged CWT** combines the steps of CWT and averaging together, which can be performed by **Process->Frequency analysis and filters->CWT (Continuous Wavelet Transform)** and **Process->Average->Compute average, std, median across epochs**, respectively, in the menu. Since the time-frequency analysis is time-consuming and also need a larger space for storage, the computer with small memory would easily go to the error of "out of memory". Hence, we combine the steps of time-frequency analysis and averaging together as a plugin for saving the storage space.

After the time-frequency analysis, baseline correction in time-frequency domain is necessary as it is in time domain. Select the datasets *avg cwt bl reref ep_S 9 sp_filter ica chan_interp butt sel_chan sub093* and *avg cwt bl reref ep_S 10 sp_filter ica chan_interp butt sel_chan sub093*, and click **Process->Baseline operation-> Baseline correction** in the menu. It is suggested to shrink the interval for baseline correction in time-frequency domain. Hence, we set the **-0.75** to **-0.25**

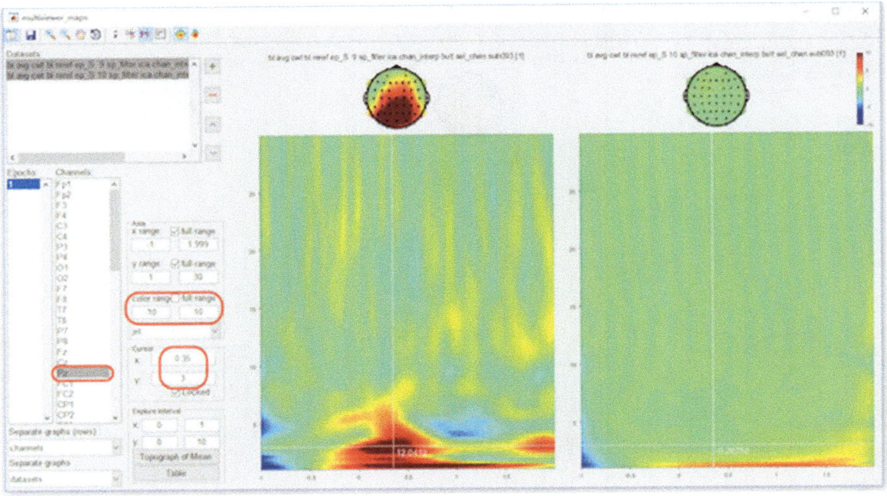

Fig. 19.10 The result of time-frequency analysis for P300 on single-subject level

seconds as the baseline in the batch module. Click the button **Run** in the bottom of batch module for baseline correction. Two new datasets with the name *bl avg cwt bl reref ep_S 9 sp_filter ica chan_interp butt sel_chan sub093* and *bl avg cwt bl reref ep_S 10 sp_filter ica chan_interp butt sel_chan sub093* will be appeared in the manage module.

Select datasets *bl avg cwt bl reref ep_S 9 sp_filter ica chan_interp butt sel_chan sub093* and *bl avg cwt bl reref ep_S 10 sp_filter ica chan_interp butt sel_chan sub093*. Click the **view** in the right-click menu, set the separate graphs (columns) as **datasets**, select both the two datasets, select channel **Pz**, and set color range from **-10** to **10**; the result of time-frequency analysis can be observed in the multiviewer (Fig. 19.10). Open the topography, and set the cursor to **x=0.35** and **y=3**; we can observe the phase-locked P300 activity on the topography.

19.8 Example for Multiple-Subject Analysis

19.8.1 Datasets Merge

Based on the single-subject analysis, we will focus on the multiple-subject analysis in this part. Before this, we need to arrange the datasets. For ease of the group analysis, we need firstly to downsample the datasets from 1000Hz to 250Hz by clicking **Edit->Resample Signals->Downsample signals (integer ratio)** and setting the down sampling ratio to 4. And then we need to rename these datasets, since the filename is too long for displaying. Select these dataset *ds avg bl reref ep_S 9 sp_filter ica chan_interp butt sel_chan sub093*, and press the **rename** in the right-

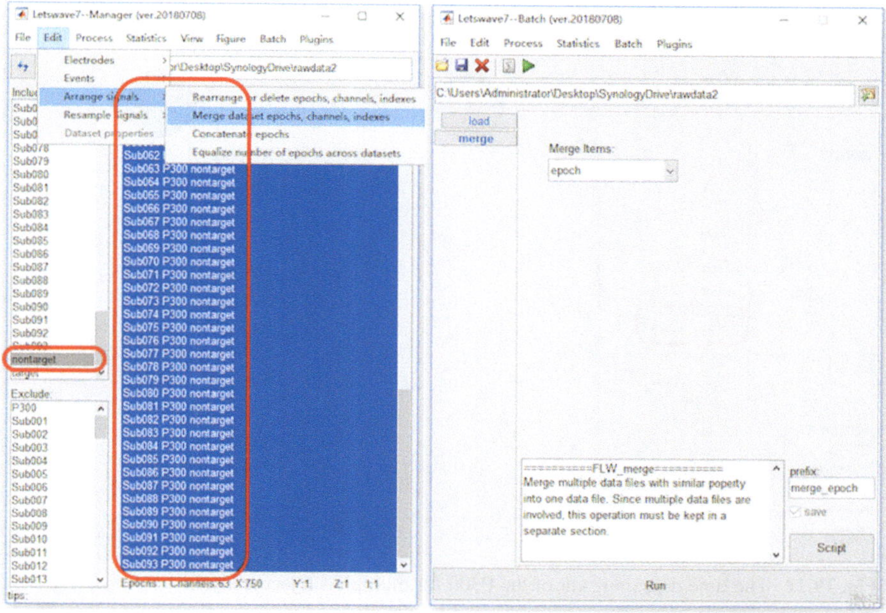

Fig. 19.11 Datasets merge

clicked menu, and rename the selected dataset to *Sub093 P300 target*. Similarly, rename the dataset *ds avg bl reref ep_S 10 sp_filter ica chan_interp butt sel_chan sub093* to *Sub093 P300 nontarget*.

Copy all the results on single-subject level to a new folder such as **rawdata2** here. In this example, all 93∗2 files, which are the average of all the epochs in the two conditions of target and nontarget from 93 subjects, can be downloaded here (https://github.com/NOCIONS/letswave7_tutorial/raw/master/rawdata2.zip). For multiple-subject analysis, we need to merge the averaged P300s from 93 subjects into one dataset, in which each subject in the new dataset is treated as an epoch. Select the tag **nontarget** in the **include** list box from the left part of the manage module (Fig. 19.11), and select all the datasets in the dataset list box in the right part of the manage module. Select **Edit->Arrange signals->Merge dataset epochs, channels, indexes**, keep the default setting in the batch module, and click the button **Run** to merge all these selected datasets in the condition of nontarget into one dataset *merge_epoch Sub001 P300 nontarget*. In the same way, we can merge all the datasets in the condition of target by selecting the tag **target** in the **include** list box and get the dataset *merge_epoch Sub001 P300 target* with 93 epochs.

Select the tag **merge_epoch** in the **include** list box, and select all the datasets in the manage module. Similar with the averaging operation on the single-subject level, select **Process->Average->Compute average, std, median across epochs**. Keep the default setting in the batch module, and click the button **Run** to average all the epochs in the two datasets, which is actually run the grand average for all the subjects

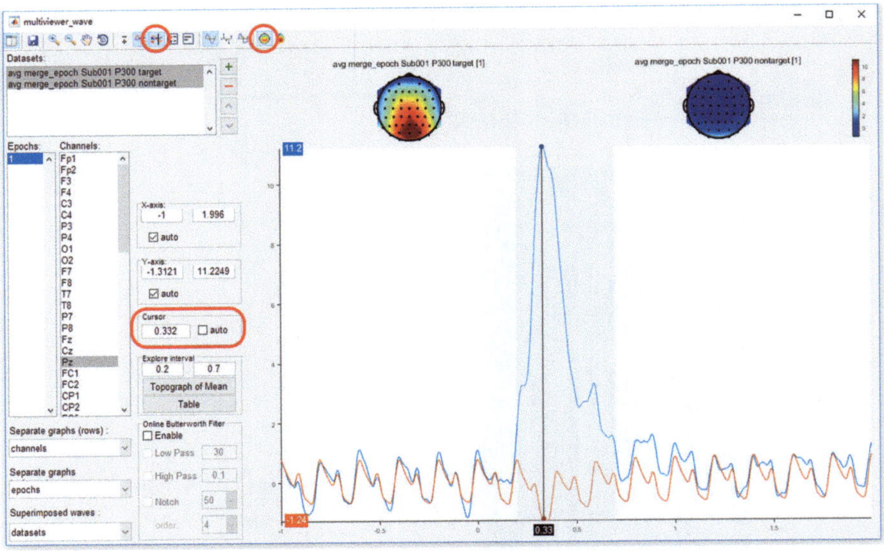

Fig. 19.12 The time domain result of the P300 for multiple-subject analysis

in the conditions of target and nontarget. Two new datasets with the name *avg merge_epoch Sub001 P300 nontarget* and *avg merge_epoch Sub001 P300 target* will be appeared in the manage module.

To view the result of grand average, select datasets *avg merge_epoch Sub001 P300 nontarget* and *avg merge_epoch Sub001 P300 target*, click the **view** in the right-click menu, and select channel **Pz**. To observe the results with the same order as it is in the single-subject level, select the dataset **avg merge_epoch Sub001 P300 target**, and click the button **dataset up**. Enabling the **cursor** and the **2D topography** in the toolbar and setting the location of the cursor to be **0.332**, the grand average topography of the P300 experiment can be observed in both the conditions of target and nontarget at the peak time point of the P300 component (Fig. 19.12).

19.8.2 Statistical Analysis on Predefined Interval

With the hypothesis that **the maximum or mean value of ERP from the predefined interval 0.2 to 0.7s at channel Pz would be different between target and nontarget conditions**, paired-sample t-test will be performed on the 93 subjects.

Firstly, we need to pick up the maximum and mean value from the interval 0.2 to 0.7s at channel Pz for both target and nontarget conditions. Select the datasets *merge_epoch Sub001 P300 nontarget* and *merge_epoch Sub001 P300 target*, and click the **view** in the right-click menu. Select dataset **merge_epoch Sub001 P300 target** and channel **Pz**, set the **Superimposed waves** as **epochs**, enable the **interval selection** in the toolbar, and set the explore interval as **0.2–0.7**s. Press the button

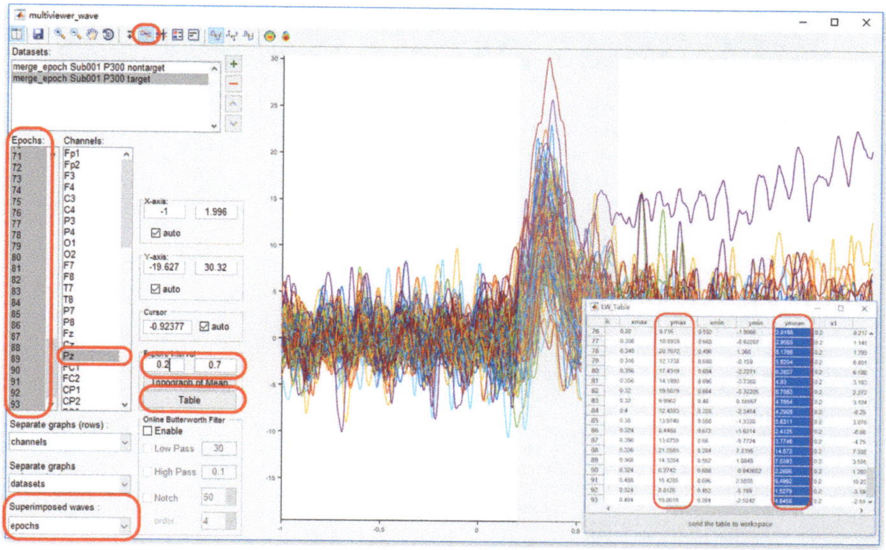

Fig. 19.13 Detect the peak and mean value of the P300 for all subjects in the interval of 0.2–0.7s at channel Pz

Table; a table is popped up with the statistic of the maximum and mean value; copy these data to the external software, such as Excel or SPSS, for the further statistical analysis. Similarly, select the dataset *merge_epoch Sub001 P300 nontarget*, and press the button **Table**; pick up the same value on another dataset, and copy it to the external software (Fig. 19.13).

Here the statistical inferences are done in Excel. Previously, we copy the maximum and mean value in the condition of **target** to columns A and B and copy the maximum and mean value in the condition of **nontarget** to columns C and D. In item **E1**, input "**=T.TEST(A:A,C:C,2,1)**" for the paired-sample t-test on the maximum value. The result **p = 6.8 ∗ 10^-40** indicates that for the maximum value of the EEG signal in the interval 0.2–0.7s on channel Pz, there is a significant difference between the conditions of target and nontarget. Similarly, input "**=T.TEST(A:A,C:C,2,1)**" in item **E2** for the paired-sample t-test on the mean value. The result **p = 3.2 ∗ 10^-28** indicates that for the mean value of the EEG signal in the interval 0.2–0.7s on channel Pz, there is a significant difference between the conditions of target and nontarget (Fig. 19.14).

19.8.3 Point-Wised Statistical Analysis

Different from the hypothesis-driven approach as is shown above, we can also do the point-wised statistical analysis, which is a data-driven method without any previous experience for interval selection. Select the datasets *merge_epoch Sub001 P300*

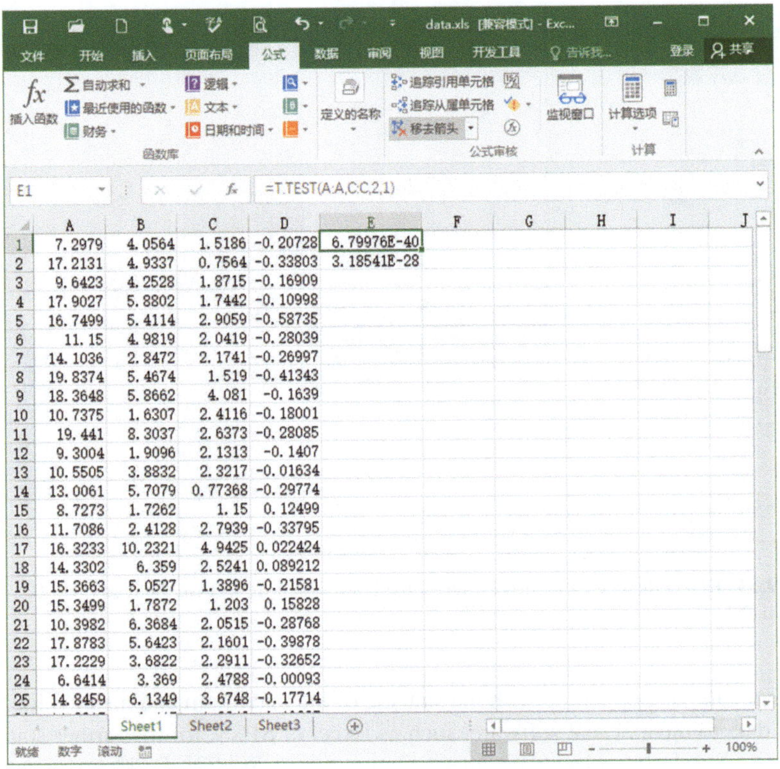

Fig. 19.14 Paired-sample t-test in Excel

nontarget and *merge_epoch Sub001 P300 target*, and click **Statistics->Compare two datasets (paired sample/two sample t-test)**. Keep the default setting in the batch module, and click the button **Run** to get the point-wise t-test result *ttest merge_epoch Sub001 P300 target* in the manage module.

Select dataset *ttest merge_epoch Sub001 P300 target*, and click the **view** in the right-click menu; select the channel as **Pz**, and keep the index as **p-value**. Since we set the significant level $\alpha = 0.05$, we can set the Y-axis as **0–0.05** to watch the interval with the p-value lower than 0.05. It could be found except the main cluster around 0.2–0.7 seconds; there are still several clusters in the other time intervals, even before the stimulus, which indicates a high family-wise error rate (Fig. 19.15). Bonferroni correction can be done by simply setting the y-axis as **0–0.000001**. Since there are 750 time points and 63 channels in the test, the corrected α value by Bonferroni method is **0.05/750/63 = 10^-6.**

Cluster-based permutation test is an efficient way to reduce family-wise error rate. In the same way in the point-wise t-test, select the datasets *merge_epoch Sub001 P300 nontarget* and *merge_epoch Sub001 P300 target*, and click **Statistics->Compare two datasets (paired sample/two sample t-test)**. Enable the bottom panel for the **cluster-based permutation test**. To get a more precise result, set the number of

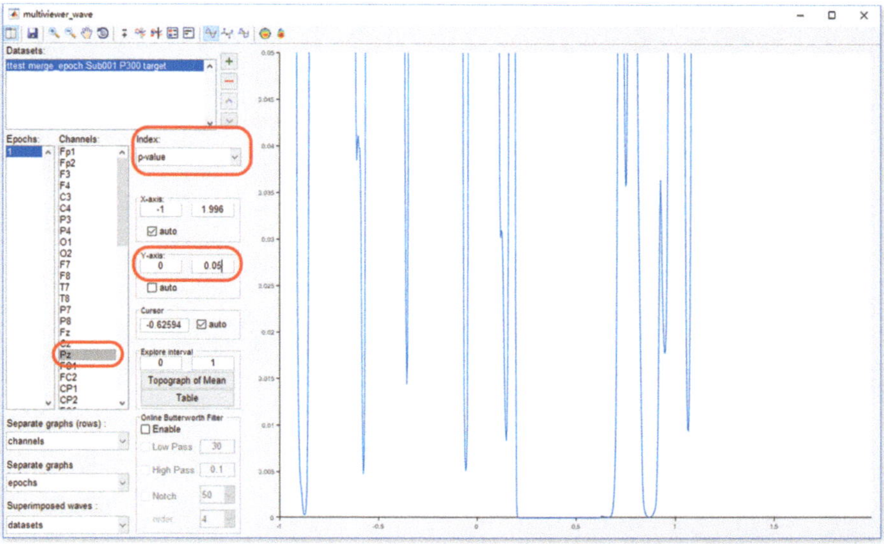

Fig. 19.15 Point-wised paired-sample t-test result on channel Pz

permutations as **20000**, which would take more time for computing. Click the button **Run**; the result would be kept in the dataset with the same name *ttest merge_epoch Sub001 P300 target.*

Open the dataset *ttest merge_epoch Sub001 P300 target* by clicking the **view** in the right-click menu. The uncorrected result is exactly the same as it is in the point-wise t-test. Set the index as "**cluster p-value**", and set the cursor as **0.332**, which is the peak of the P300 component. It can be found that only the main cluster on channel Pz is reserved after cluster-based permutation test; the other clusters are excluded as false positive (Fig. 19.16).

19.9 Figure Generation and Batch Processing

19.9.1 Figure Generation

Figure generation is an important feature in Letswave7. Based on the result of multiple-subject analysis and statistical analysis, we can generate Fig. 19.17 by the following steps:

Step1: *Open the figure module*

- Select the datasets *avg Sub001 P300 nontarget* and *avg Sub001 P300 target* in the folder of rawdata1.
- Click **Figure->General Figure creator** to open the figure module with a blank canvas.

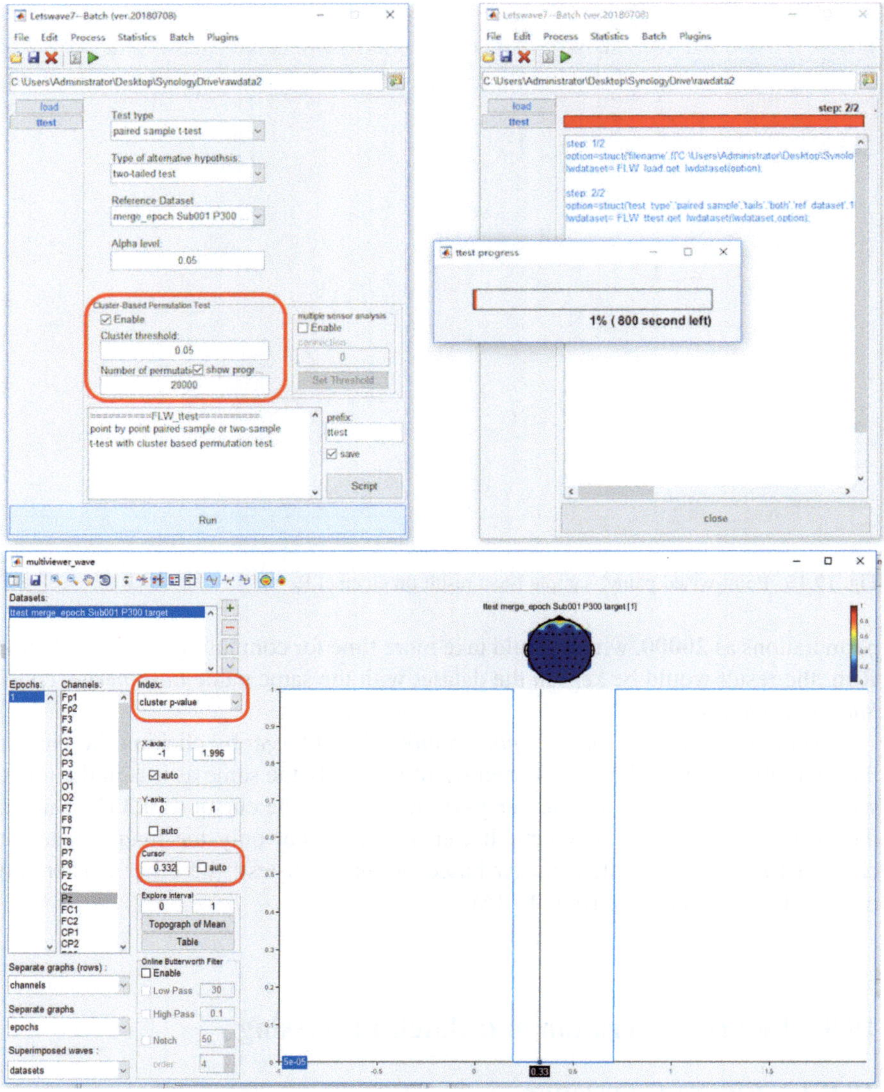

Fig. 19.16 The result from cluster-based permutation test

Step2: *Create the subfigures*

- Set the width and height of the figure as **1000** and **400**.
- Add a **curve** with the title **P300**, font size **12**, and position **x=80**, **y= 70**, **w=870**, and **h=300**.
- Add a **topography** with the title **0 ms**, font size **12**, and position **x= 100**, **y=180**, **w=120**, and **h=120**.

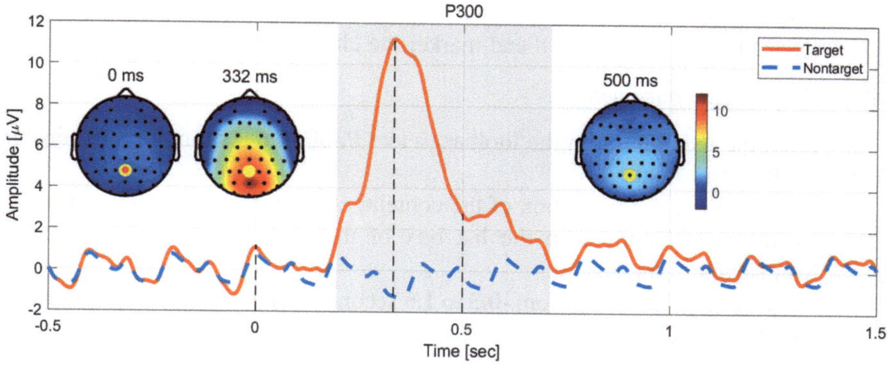

Fig. 19.17 The result from cluster-based permutation test

- Add a **topography** with the title **332 ms**, font size **12**, and position **x= 230**, **y=180**, **w=120**, and **h=120**.
- Add a **topography** with the title **500 ms**, font size **12**; enable the color bar in the **content** panel and position **x=100**, **y=180**, **w=120**, and **h=120**.

Step3: *Add the content*

- Click the **content** button in the toolbar, and select **P300** in the subfigure. Add a **curve** with width as **3**, the data source as **avg merge_epoch Sub001 P300 target**, and channel as **Pz**.
- Add a **curve** with the color as blue **[0,0.45,0.74]**, width as **3**, line style as **dashed**, data source as **Sub093 P300 nontarget**, and channel as **Pz**.
- Add a **rect** with face opacity **0.25**, edge opacity **0**, position **x = 0.196**, **y= -2**, **w = 0.516**, and **h = 14**. And sort the rect above the **curve1** and **curve2**.
- Add a **line** with width as **1**, line style as **dashed**, and the position as **x1=0**, **y1=-2**, **x2=0**, and **y2=1.2**.
- Add a **line** with width as **1**, line style as **dashed**, and the position as **x1=0.332**, **y1=-2**, **x2=0.332**, and **y2=11.3**.
- Add a **line** with width as **1**, line style as **dashed**, and the position as **x1=0.5**, **y1=-2**, **x2=0.5**, and **y2=2.8**.
- Select **0 ms** in the subfigure. Set the data source as "**avg merge_epoch Sub001 P300 Target**", x from **0** to **0**, head radius as **0.5**, shrink as **0.95**, range from **-2** to **12** to keep it the same as the range of P300 curve. For the electrodes, set the size as **8** and marker the channel **Pz**.
- Select **332 ms** in the subfigure. Set the data source as "**avg merge_epoch Sub001 P300 Target**", x from **0.332** to **0.332**, head radius as **0.5**, shrink as **0.95**, range from **-2** to **12** to keep it the same as the range of P300 curve. For the electrodes, set the size as **8** and marker the channel **Pz**.
- Select **500 ms** in the subfigure. Set the data source as "**avg merge_epoch Sub001 P300 Target**", x from **0.5** to **0.5**, head radius as **0.5**, shrink as **0.95**,

range from **-2** to **12** to keep it the same as the range of P300 curve. For the electrodes, set the size as **8** and marker the channel **Pz**.

Step4: *Set the axis parameters*

- Click the **axis** button in the toolbar, select **P300** in the subfigure, and enable the **Box** and **Legend**.
- Select **curve1** in the list box of the content; change its name to **Target** in the legend. Select **curve2** in the list box of the content; change its name to **NonTarget** in the legend.
- For x-axis, set the x-lim from **-0.5** to **1.5** second. Enable the **grid** and the **label**; set label as "**Time [sec]**".
- For y-axis, enable the **grid** and the **label**; set the label as "**Amp [\muV]**".

19.9.2 Batch Processing

In this part, all the time domain analysis on single-subject level will be done again but using batch processing. After data importing, remove channel IO, which is exactly the same operation as what we do in Sect. 19.3. After that, click the tab **selection** in the left, and select **Process -> Frequency analysis and filters -> Butterworth filters** in the menu of the **batch module**, and set the low cutoff frequency (Hz) as **0.05Hz** (Fig. 19.18). From this step, the operation in the batch processing would be different. We call out the Butterworth filter from the batch module, **NOT** the manage module.

Step by step, add all the steps in the batch module to get the final processing flow as it is shown in Fig. 19.19. By pressing the button "**Run**", we can run the whole batch processing. In the step of **Identify Artifact Component**, we still manually select **comp 1** and **comp 2** in the right panel in orange color and click the button **OK**. It needs more than 2 minutes for all the processing on the testing, in which the step of computing ICA matrix is time-consuming.

To reuse the batch, we can save the entire process. Click the **save** button in the toolbar of the batch module, and save the process as "P300.lw_script", for example. For the next time, we can click the **open** button to load the process. It should be noticed that since the processing object has been changed, the users need to change the input datasets in all the steps of **load**.

If the batch processing is frequently used, we can put the saved **.lw_script** file, like "P300.lw_script" in this case, into the path of the Letswave installed "**../letswave7/plugins/**". After that, restart Letswave7, and the batch operation will come out in the menu of the batch in both the manage module and batch module.

Once the processing flow has been made, the corresponding MATLAB script has already been generated automatically. Users can click the "script" button in the batch module to get the full script. Once the users have learned the use of GUI, they almost already understand the writing of script.

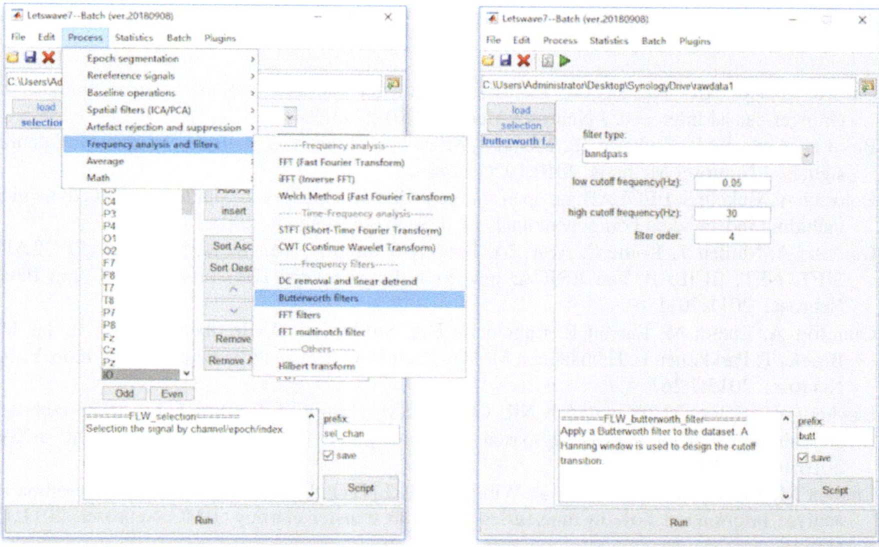

Fig. 19.18 Add the step of frequency filtering in batch module

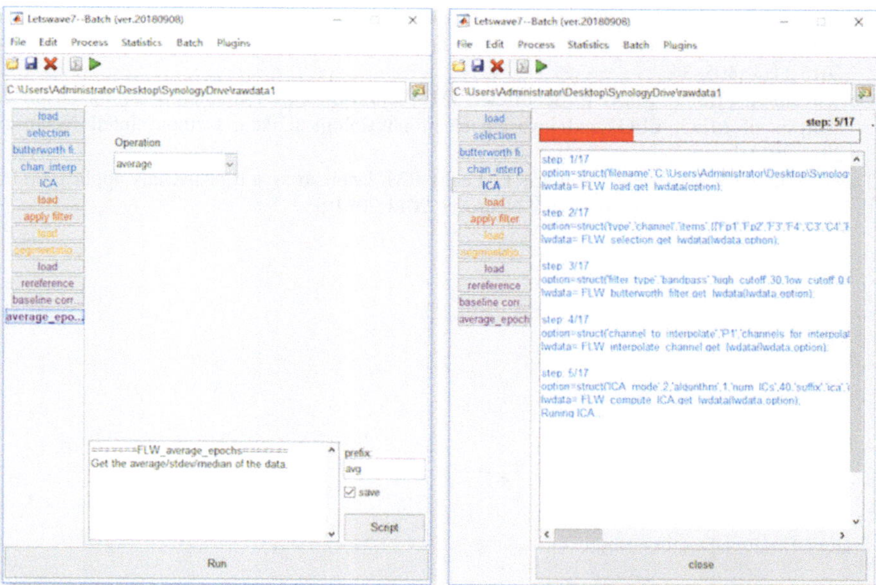

Fig. 19.19 Add the step of frequency filtering in batch module

References

Barnett L, Seth AK. The MVGC multivariate Granger causality toolbox: a new approach to Granger-causal inference. J Neurosci Methods. 2014;223:50–68.

Bokil H, Andrews P, Kulkarni JE, Mehta S, Mitra PP. Chronux: a platform for analyzing neural signals. J Neurosci Methods. 2010;192(1):146–51.

Delorme A, Makeig S. EEGLAB: an open source toolbox for analysis of single-trial EEG dynamics including independent component analysis. J Neurosci Methods. 2004;134(1):9–21.

Delorme A, Mullen T, Kothe C, Acar ZA, Bigdely-Shamlo N, Vankov A, Makeig S. EEGLAB, SIFT, NFT, BCILAB, and ERICA: new tools for advanced EEG processing. Comp Intell Neurosci. 2011;2011:10.

Gramfort A, Luessi M, Larson E, Engemann DA, Strohmeier D, Brodbeck C, Goj R, Jas M, Brooks T, Parkkonen L, Hämäläinen M. MEG and EEG data analysis with MNE-Python. Front Neurosci. 2013;7:267.

Lawhern VJ, Solon AJ, Waytowich NR, Gordon SM, Hung CP, Lance BJ. Eegnet: a compact convolutional network for eeg-based brain-computer interfaces. arXiv preprint arXiv. 2016:1611.08024.

Lindner M, Vicente R, Priesemann V, Wibral M. TRENTOOL: a Matlab open source toolbox to analyse information flow in time series data with transfer entropy. BMC Neurosci. 2011;12 (1):119.

Litvak V, Mattout J, Kiebel S, Phillips C, Henson R, Kilner J, Barnes G, Oostenveld R, Daunizeau J, Flandin G, Penny W. EEG and MEG data analysis in SPM8. Comp Intell Neurosci. 2011;2011

Niso G, Bruña R, Pereda E, Gutiérrez R, Bajo R, Maestú F, del-Pozo F. HERMES: towards an integrated toolbox to characterize functional and effective brain connectivity. Neuroinformatics. 2013;11(4):405–34.

Oostenveld R, Fries P, Maris E, Schoffelen JM. FieldTrip: open source software for advanced analysis of MEG, EEG, and invasive electrophysiological data. Comput Intell Neurosci. 2011;2011:1.

Tadel F, Baillet S, Mosher JC, Pantazis D, Leahy RM. Brainstorm: a user-friendly application for MEG/EEG analysis. Comput Intell Neurosci. 2011;2011:8.

Chapter 20
Summary and Conclusions

Zhiguo Zhang and Li Hu

Abstract EEG remains useful and irreplaceable in multiple clinical applications and scientific researches, with regard to its massive advantages. As EEG continues to spread widely over time, EEG signal processing is still a highly promising and evolving field. Recent developments of EEG processing techniques, such as advanced machine learning for EEG and EEG-related multimodality brain imaging, are expected to make EEG a more powerful and versatile tool in the future.

Keywords EEG signal processing · Machine learning · Multimodality imaging

20.1 EEG: Still Useful and Irreplaceable

Among all existing techniques for assessing the brain functions, EEG could still be the most widely used one (Biasiucci et al. 2019). Some people may have thought that EEG is sort of outdated and limited of utility, as many advanced functional brain imaging techniques that can provide higher temporal and/or spatial resolutions have been well developed and applied in recent decades. However, the role of EEG in clinical applications and scientific researches can hardly be replaced by any other functional brain imaging techniques, considering that EEG's advantages (e.g., noninvasiveness, easiness-to-operate, and cheapness) have not been challenged by other techniques (Cavanagh 2019). These advantages of EEG make it particularly suitable for the acquisition of "big brain data," which should be collected from a

Z. Zhang (✉)
School of Biomedical Engineering, Health Science Center, Shenzhen University, Shenzhen, Guangdong, China
e-mail: zgzhang@szu.edu.cn

L. Hu
CAS Key Laboratory of Mental Health, Institute of Psychology, Chinese Academy of Sciences, Beijing, China

Department of Psychology, University of Chinese Academy of Sciences, Beijing, China
e-mail: huli@psych.ac.cn

L. Hu, Z. Zhang (eds.), *EEG Signal Processing and Feature Extraction*,
https://doi.org/10.1007/978-981-13-9113-2_20

large number of participants with many times (Sejnowski et al. 2014). For example, if a researcher intends to collect functional brain data from a large group of normal children or the elderly in a longitudinal study, EEG will definitely be on the top of his/her selection of neuroimaging techniques, because other techniques such as functional near-infrared spectroscopy (fNIRS) and fMRI are much more expensive and more difficult to operate. In the foreseeable future, EEG will still be a powerful and widespread technique and will be used by more and more researchers with different backgrounds and purposes (Biasiucci et al. 2019; Cavanagh 2019).

Another advantage of EEG, which is often overlooked, is that EEG is equipped with well-established data analysis methods and pipelines. However, students and researchers in the fields of psychology, clinical neuroscience, and psychiatry often have difficulty in the analysis of EEG data. Although many powerful and user-friendly EEG processing toolboxes exist, most of them are focused on one or several aspects of EEG signal processing. So, it is difficult for people without an engineering background to learn and to use many toolboxes. On the other hand, it is also difficult for students and researchers in engineering areas, like electrical engineering, computer sciences, and artificial intelligence, to understand the nature and characteristics of EEG, which hinders them from applying their knowledge and skills of signal processing on EEG signals. Therefore, there is a huge gap between the knowledge about EEG (its origin, acquisition, properties, experiments, etc.) and the skills of signal processing and feature extraction. A manual providing both essential knowledge about EEG and basic skills of EEG signal processing in a comprehensive, simple, and easy-to-understand way is urgent and helpful to bridge the gap.

20.2 Current Trends and Future Directions

EEG signal processing is still a highly dynamic and evolving field. The development of EEG signal processing methods is driven by two major factors. First, there are increasing and new expectations and requirements from EEG users. Second, the rapid developments of data analytics techniques (such as machine learning, data fusion, complex network analysis, etc.) enable more and more potential applications of EEG. Although our goal in writing this book has been to present a comprehensive introduction of EEG signal processing and feature extraction methods, it is impossible to include all related researches in this book because of the fast emergence of new EEG analysis methods and new applications. In the following, we discuss two trends of EEG signal processing, machine learning and multimodality, and also mention some topics that are not covered by this book.

- **Machine learning:** Undoubtedly, machine learning is the most popular and fast-developed data analysis technique in recent years. Particularly, deep learning, as an important and useful branch of machine learning techniques, has been widely applied to predict behavioral variables and psychophysiological states from EEG data in a variety of fields, especially in the field of brain-computer interface. In this book, Chap. 15 is devoted to elaborate basic concepts and classical algorithms of machine learning, whereas Chap. 16 briefly introduces the applications of deep learning on EEG. Admittedly, only two chapters cannot cover all topics related to the applications of machine learning on EEG. Emerging methods of machine learning, such as transfer learning, reinforcement learning, and ensemble learning, have been gradually used on EEG data. For example, some new deep neural networks, such as generative adversarial networks and spiking neural networks, have already been applied as powerful tools for EEG decoding, and transfer learning is often adopted by researchers in the area of brain-computer interface to increase the accuracy of cross-individual prediction. Readers who are interested in these new techniques should keep a close eye on related journals and preprint repositories (e.g., arXiv and bioRxiv).

- **Multimodality:** Simultaneous EEG-fMRI is an important multimodal neuroimaging technique, which is introduced in Chap. 18. Multimodal neuroimaging can normally provide a more complete and more complementary picture of the brain and the interaction between the brain and other organs, thus making it become more favorable to researchers. EEG can be acquired (1) with some other brain imaging techniques, such as MRI and fNIRS; (2) with some types of physiological signals, such as ECG and EMG; and (3) with some brain stimulation techniques, such as transcranial magnetic stimulation (TMS) and transcranial direct current stimulation (tDCS). EEG-related multimodality imaging has two specific difficulties in EEG signal processing. First, EEG may introduce artifacts into other signals (such as EMG) or be contaminated by other signals (such as MRI) or stimulation (such as TMS). So, noise removal is of particular importance in EEG-related multimodal imaging. Second, how to fuse EEG with other modalities to provide complimentary information also poses a great challenge for data analyses. Data-driven multivariate methods and machine learning methods can play a role in the analyses of multimodal brain imaging data.

Besides the above two major topics that are not completely covered by this book, there are still some other related contents, such as real-time implementation (which is highly desired in practical applications such as brain-computer interface and neurofeedback) and clinical uses, which have not been fully discussed in this book. Anyhow, we believe this book has already covered almost all mainstream EEG signal processing and feature extraction methods. We hope what readers learn from the book can make them better prepared for learning new knowledge and skills that are not included in this book.

References

Biasiucci A, Franceschiello B, Murray MM. Electroencephalography. Curr Biol. 2019;29(3): R80–5.

Cavanagh JF. Electrophysiology as a theoretical and methodological hub for the neural sciences. Psychophysiology. 2019;56(2):e13314. https://doi.org/10.1111/psyp.13314.

Sejnowski TJ, Churchland PS, Movshon JA. Putting big data to good use in neuroscience. Nat Neurosci. 2014;17(11):1440–1.

MIX
Papier aus verantwortungsvollen Quellen
Paper from responsible sources
FSC® C105338

If you have any concerns about our products,
you can contact us on
ProductSafety@springernature.com

In case Publisher is established outside the EU,
the EU authorized representative is:
Springer Nature Customer Service Center GmbH
Europaplatz 3, 69115 Heidelberg, Germany

Printed by Libri Plureos GmbH
in Hamburg, Germany